Methods for faecal sludge analysis

Methods for faecal sludge analysis

Konstantina Velkushanova
Linda Strande
Mariska Ronteltap
Thammarat Koottatep
Damir Brdjanovic
Chris Buckley

IWA PUBLISHING

Published by: **IWA Publishing**
Alliance House
12 Caxton Street
London SW1H 0QS, UK
T: +44 (0) 20 7654 5500
F: +44 (0) 20 7654 5555
E: publications@iwap.co.uk
I: www.iwapublishing.com

First published 2021

British Library Cataloguing in Publication Data
A CIP catalogue record for this book is available from the British Library

Library of Congress Cataloguing in Publication Data
A catalogue record for this book is available from the Library of Congress

Cover design: Peter Stroo
Cover image: WASH R&D Centre
Graphic design: Synopsis d.o.o.
English language editor: Claire Taylor

ISBN: 9781780409115 (Hardback)
ISBN: 9781780409122 (eBook)

Editors:
Konstantina Velkushanova
Linda Strande
Mariska Ronteltap
Thammarat Koottatep
Damir Brdjanovic
Chris Buckley

Authors:
Amédé Ferré
Andreas Scheidegger
Barbara J. Ward
Bhekumuzi Gumbi
Carlos M. Lopez Vazquez
Colleen Archer
Damir Brdjanovic
Francisco J. Rubio Rincon
Juan Pablo Carbajal
Konstantina Velkushanova
Krailuck Fakkaew
Linda Strande
Mariska Ronteltap
Max Maurer
Merlien Reddy
Miriam Englund
Nienke Andriessen
Roni Penn
Santiago Septien Stringel
Saroj Chapagain
Stanley Sam
Thabiso Zikalala
Thammarat Koottatep

Contributors:
Araya Wicheansan
Barbara J. Ward
Christopher Friedrich
Claire Furlong
Damir Brdjanovic
Dayna Hamilton
Hector A. Garcia
Jamie Radford
Jonathan Wilcox
Kerry Lee Philp
Konstantina Velkushanova
Lungi Zuma
Naomi Korir
Nienke Andriessen
Principal Mdolo
Rebecca Sindall
Samuel Renggli
Samuel Tenaw Getahun
Santiago Septien Stringel
Sudhir Pillay
Suparat Jampathong
Thabiso Zikalala
Thammarat Koottatep
Tracy Ratidzaishe Mupinga

Reviewers:
Aissatou Ndoye
Andy Peal
Berend Lolkema
Caetano Dorea
Chris J. Bouckaert
Damir Brdjanovic
Daniela A. Peguero
Eberhard Morgenroth
Isabel Blackett
James Madalitso Tembo
Kapanda Kapanda
Kathelyn Sellgren
Peter Hawkins
Sudhir Pillay
Susan Mercer

Chris, thank you for everything that you have done for the underserved.

Authors | Contributors | Reviewers

1. **Setting the stage**
   Linda Strande (*Aut.*)
   Konstantina Velkushanova (*Aut.*)
   Damir Brdjanovic (*Aut.*)

2. **Faecal sludge properties and considerations for characterisation**
   Konstantina Velkushanova (*Aut.*)
   Linda Strande (*Aut.*)
   Principal Mdolo (*Con.*)
   Rebecca Sindall (*Con.*)
   Thabiso Zikalala (*Con.*)
   Barbara J. Ward (*Con.*)
   Samuel Renggli (*Con.*)
   Christopher Friedrich (*Con.*)
   Damir Brdjanovic (*Con.*)
   Claire Furlong (*Con.*)
   Thammarat Koottatep (*Con.*)
   Sudhir Pillay (*Rev.*)

3. **Faecal sludge sample collection and handling**
   Thammarat Koottatep (*Aut.*)
   Amédé Ferré (*Aut.*)
   Saroj Chapagain (*Aut.*)
   Krailuck Fakkaew (*Aut.*)
   Linda Strande (*Aut.*)
   Sudhir Pillay (*Con.*)
   Konstantina Velkushanova (*Con.*)
   Jamie Radford (*Con.*)
   Nienke Andriessen (*Con.*)
   Barbara J. Ward (*Con.*)
   Caetano Dorea (*Rev.*)

4. **Experimental design for the development, transfer, scaling-up, and optimisation of treatment technologies: case studies of dewatering and drying**
   Barbara J. Ward (*Aut.*)
   Santiago Septien (*Aut.*)
   Mariska Ronteltap (*Aut.*)
   Linda Strande (*Aut.*)
   Naomi Korir (*Con.*)
   Jonathan Wilcox (*Con.*)
   Hector A. Garcia (*Con.*)
   Damir Brdjanovic (*Con.*)
   Eberhard Morgenroth (*Rev.*)

5. **Estimating quantities and qualities (Q&Q) of faecal sludge at community to city-wide scales**
   Linda Strande (*Aut.*)
   Miriam Englund (*Aut.*)
   Nienke Andriessen (*Aut.*)
   Juan Pablo Carbajal (*Aut.*)
   Andreas Scheidegger (*Aut.*)
   Barbara J. Ward (*Con.*)
   Samuel Renggli (*Con.*)
   Isabel Blackett (*Rev.*)
   Andy Peal (*Rev.*)
   Peter Hawkins (*Rev.*)

6. **Towards city-wide inclusive sanitation (CWIS) modelling: modelling of faecal sludge containment/treatment processes**
   Carlos M. Lopez Vazquez (*Aut.*)
   Francisco J. Rubio Rincon (*Aut.*)
   Damir Brdjanovic (*Aut.*)
   Chris J. Bouckaert (*Rev.*)
   Susan Mercer (*Rev.*)

7. **Faecal sludge simulants: review of synthetic human faeces and faecal sludge for sanitation and wastewater research**
   Roni Penn (*Aut.*)
   Barbara J. Ward (*Aut.*)
   Linda Strande (*Aut.*)
   Max Maurer (*Aut.*)
   Konstantina Velkushanova (*Con.*)
   Damir Brdjanovic (*Rev.*)
   Susan Mercer (*Rev.*)

8. **Laboratory procedures and methods for characterisation of faecal sludge**
   Konstantina Velkushanova (*Aut.*)
   Merlien Reddy (*Aut.*)
   Thabiso Zikalala (*Aut.*)
   Bhekumuzi Gumbi (*Aut.*)
   Colleen Archer (*Aut.*)
   Barbara J. Ward (*Aut.*)
   Nienke Andriessen (*Aut.*)
   Stanley Sam (*Aut.*)
   Linda Strande (*Aut.*)
   Suparat Jampathong (*Con.*)
   Araya Wicheansan (*Con.*)
   Thammarat Koottatep (*Con.*)
   Santiago Septien Stringel (*Con.*)
   Tracy Ratidzaishe Mupinga (*Con.*)
   Samuel Tenaw Getahun (*Con.*)
   Dayna Hamilton (*Con.*)
   Kerry Lee Philp (*Con.*)
   Lungi Zuma (*Con.*)
   Berend Lolkema (*Rev.*)
   Kathelyn Sellgren (*Rev.*)
   Aissatou Ndoye (*Rev.*)
   Daniela A. Peguero (*Rev.*)
   James Madalitso Tembo (*Rev.*)
   Kapanda Kapanda (*Rev.*)

Editors

Konstantina Velkushanova, PhD

Dr Konstantina Velkushanova is a senior lecturer in non-sewered sanitation at IHE-Delft Institute for Water Education. Previously she was a senior research associate at the Pollution Research Group of the University of KwaZulu-Natal (UKZN PRG), South Africa. Her areas of expertise include faecal sludge management, sanitary engineering, centralized and decentralized waste water treatment and solid waste management. Her work has been focused on a number of research projects in the field of water and sanitation in developing countries, particularly on non-sewered sanitation, improved faecal sludge management solutions and the development and evaluation of innovative sanitation technologies. She has been actively involved in different programs, activities and initiatives in that support the global development agenda of the faecal sludge management and non-sewered sanitation, such as: the formation of the FSM Alliance, acting as a board member and FSM6 conference chair; the formation of the Non-sewered Sanitation Specialist Group of IWA, acting as a former deputy chair; capacity building through the delivery of FSM online courses and masters programs under the Global Sanitation Graduate School and the Global Partnership for Faecal Sludge Analysis; acting as a technical committee expert of ISO standards for non-sewered sanitation, water and sanitation (ISO31800; ISO 30500). In 2019, she received a UKZN Wonder Woman in Science in 2019 award.

Linda Strande, PhD

Dr. Linda Strande leads a research group at Eawag (Swiss Federal Institute of Aquatic Science and Technology), in Sandec (Department Sanitation, Water and Solid Waste for Development). Her focus is globally relevant solutions for sustainable city-wide sanitation, through developing fundamental scientific knowledge, translating it to innovative technology solutions, and ensuring uptake through implementation strategies. Current research includes governing mechanisms controlling dewaterability, affordable methods for characterisation and quantification, optimization of treatment technologies, and innovation in resource recovery. Dr. Strande has been working in the WASH sector for over 20 years, has had research collaborations with local and international stakeholders from over 20 countries, and published over 60 international, refereed publications. She is passionate about mentoring upcoming generations of engineers and practitioners, and has had over 10,000 learners in her online course in faecal sludge management. She is a founding member of the SFD Promotion Initiative and FSM conferences, a contributor to the WHO Guidelines on Sanitation and Health, World Bank FSM Tools, and lead editor of the IWA Publishing book Faecal Sludge Management: Systems Approach for Implementation and Operation. Her wide-ranging international experiences have given her a global perspective, and an ability to research and apply engineering fundamentals in complex, interdisciplinary situations.

Mariska Ronteltap, PhD

Dr. Mariska Ronteltap is an environmental engineer from Wageningen University. After a few consulting years she pursued a PhD at Eawag (the Swiss Federal Institute of Aquatic Science and Technology) and ETH (Swiss Federal Institute of Technology), specialising in resource recovery from sanitation. In 2006 she joined IHE Delft Institute for Water Education (formerly UNESCO-IHE), the Netherlands. She developed a large quantity of new teaching material and initiated research in the field of resource recovery and non-sewered sanitation, and inspired many students, practitioners and decision makers to make sanitation a priority. Dr. Ronteltap mentored in excess of 40 MSc students and 3 PhD fellows. With more than 1,500 citations and publications of high impact, her work is globally recognized; she has been invited as speaker to several conferences and was one of the editors of the book Faecal Sludge Management: Systems Approach for Implementation and Operation published by IWA Publishing. She is a member of the management team of the IWA Specialist Groups on Resource Oriented Sanitation and Non-sewered Sanitation, of the Circular Water Technology of the Royal Dutch Water Network, and of the Program Committee on Wastewater of the Foundation of Applied Water Research (STOWA). Dr. Ronteltap currently works for Delfland Water Authority in The Netherlands as a senior wastewater expert and technology innovator and she continues to teach, publish, present and supervise students.

Prof. Thammarat Koottatep, PhD

Dr. Thammarat Koottatep is a Professor of the Environmental Engineering Management of the Asian Institute of Technology (AIT), Thailand. He is an internationally recognized professional on faecal sludge management, sanitation systems, and wastewater treatment technology. His major scholarly contributions include publications of more than 60 refereed international journal papers, 3 books, and 9 book chapters. He has invented sanitation technologies, one of which is patented to his credit and several are filing. He has jointly developed a professional master degree program in Regenerative Sanitation and mentored 18 doctoral students. He has secured significant funded projects including, research and training grants, and most notably, the Bill & Belinda Gates Foundation grants on "Decentralized Wastewater Management in Developing Countries: Design, Operation and Monitoring". He has contributed significantly to capacity building in faecal sludge management and decentralized wastewater treatment systems in Thailand and abroad, including capacity strengthening for policy makers.

Prof. Damir Brdjanovic, PhD

Dr. Damir Brdjanovic is Professor of Sanitary Engineering at IHE Delft Institute for Water Education and Endowed Professor at Delft University of Technology. Areas of his expertise include pro-poor and emergency sanitation, faecal sludge management, urban drainage, and wastewater treatment. He is a pioneer in the practical application of models in wastewater treatment practice in developing countries. He is co-inventor of DEMOS©: Digital Epidemic Observatory and Management System, and of the Shit Killer® device for excreta management in emergencies, the award-winning eSOS® Smart Toilet and the medical toilet MEDiLOO®, with funding by the Bill & Melinda Gates Foundation (BMGF). He has initiated the development and implementation of innovative didactic approaches and novel educational products (including e-learning) at IHE Delft. Brdjanovic is co-founder and director of the Global Sanitation Graduate School and co-founder of Global Partnership of Laboratories for Faecal Sludge Analysis. In addition to dozen of PhD students, in excess of 150 MSc students have graduated under his supervision so far. Prof. Brdjanovic has a sound publication record, is co-initiator of the IWA Journal of Water, Sanitation and Hygiene for Development, and is the initiator, author and editor of eight books in the wastewater treatment and sanitation field. In 2015 he became an IWA Fellow, in 2018 received the IWA Publishing Award, and in 2019 IWA Water and Development Research Award 2019 Runner Up.

Prof. Chris Buckley

Chris Buckley is a Research Professor and Head of the Pollution Research Group at the University of KwaZulu-Natal (UKZN PRG). He is a chemical engineer, a Professional Engineer and a Fellow of IWA. He has been a member of the ISO/ANSI/SABS panel for the development of ISO30500 standards, Chair of the Water SA journal editorial board and a member of numerous Water Research Commission (WRC) project reference groups. Prof. Buckley's public sector activities have been directed for decades to providing water and sanitation services to the 'unserved'. Since 2009 he and his team have participated in a number of Bill & Melinda Gates Foundation (BMGF) projects related to sanitation for the poor (including the flagship Reinvent the Toilet Challenge program). The current grant has provided a fully staffed, dedicated faecal sludge laboratory and an engineering field testing platform for reinvented toilets and other advances in real-world conditions with the aim to facilitate the roll-out of innovative pro-poor WASH solutions. He has supervised more than 100 Master and PhD students and his current research projects are funded by the WRC, BMGF, Emory University, Swiss Development Cooperation, Swedish Research Council, National Science Foundation with San Diego State University, and IHE Delft. Because of his dedicated work in sanitation, Prof. Buckley describes himself as a 'Shit Manager'; he continues to inspire generation of young professionals to join the sanitation field.

Foreword

Since 2015, with the adoption of the Sustainable Development Goals (SDGs) by the United Nations (UN), the importance of non-sewered sanitation service provision and the major inequalities of sanitation service delivery have been highlighted. In 2017 the Joint Monitoring Program (JMP) of the UN and the World Health Organisation (WHO) estimated that around 4.2 billion people, or 55% of the global population, did not have access to safely managed sanitation. This is the equivalent of 6 people out of 10, with the vast majority living in developing countries, in communities where sewer-based approaches are not feasible, practical, or too expensive.

For several decades, attempts to develop sanitation solutions in developing countries focused on adapting treatment solutions from centralised, sewer-based management solutions. These approaches assumed that faecal sludge characteristics are similar to those of sewage or urban wastewater. As a result, several treatment plants and technologies resulted in failure, and did not meet communities' need for the protection of public health, as they were designed based on the wrong assumptions.

However, since 2011 we have seen a growing number of academic institutions, industries, and sector professionals engaging in the development of sanitation solutions that meet the needs of communities relying on non-sewered sanitation solutions. This growing community is among the target audience for this book.

These global efforts have inspired some strategic organisations and platforms that are serving the non-sewered sanitation community. In 2011 several leading organisations in non-sewered sanitation supported the creation and the establishment of the Faecal Sludge Management (FSM) conference series to help consolidate and disseminate best practices. These conferences continue to be supported by the Faecal Sludge Management Alliance (FSMA). The Toilet Board Coalition launched a business-led partnership and platform in 2015 that has the ambition to address the global sanitation crisis by accelerating the Sanitation Economy through enabling private sector engagement and collaboration between private, public and non-profit sectors. In 2016, 24 organisations launched the Indian National Faecal Sludge and Sewerage Alliance with the goal of sharing best practices on non-sewered sanitation. The Pan-African Association of Sanitation Actors (PASA) was launched in 2019 by pit latrine emptiers in Africa (non-sewered sanitation service providers), in partnership with the African Water Association (AfWA, the water and sanitation utility operators), to better organise their industry and business. In 2018 the Global Sanitation Graduate School (GSGS) initiative managed by IHE Delft Institute for Water Education was launched with the aim to establish reference training centres in national universities in developing countries; this program supports the next generation of academic leaders in the development of sustainable solutions for communities on city-wide inclusive sanitation (CWIS) systems. Also in 2018 the Global Partnership of Laboratories for Faecal Sludge Analysis (GPLFSA) was established to improve the communication between different faecal sludge laboratories and to address together challenges related to analysis of faecal sludge and the generation of a uniform database for faecal sludge characteristics. Most recently, in 2019 the International Water Association (IWA) launched a new specialist group on non-sewered sanitation, with the focus to consolidate the generation of technical and scientific evidence on faecal sludge management solutions.

This book consolidates three decades of evidence gathering on methods for sampling and analysing faecal sludge collected from non-sewered sanitation facilities. It addresses the needs of inventors who are working on innovative sanitation technologies such as systems that meet ISO 30500 and ISO 31800 specifications, new pit emptying technologies or decentralised faecal sludge processing technologies for resource recovery. The chapter on faeces simulants is particularly designed to help inventors and laboratories that do not have access to testing facilities with raw human excreta and also for replication in scientific studies. For national and university analytical laboratories and researchers, the book provides a step-by-step approach to upgrade faecal sludge analytical laboratories. For conformity assessment laboratories and standardisation bodies, this book will serve as a reference for methods to test technology performance and compliance with standards. National programs focusing on accelerating access to non-sewered sanitation will find references in the book to develop analytical capabilities and reference to training materials.

The authors have compiled the latest data to fill an important gap for the sanitation sector, as the chapters highlight strong scientific evidence on why and how faecal sludge differs to sewage, and provide clear recommendations for sampling and analytical methods. These recommendations are derived from a consensus of global leading academic centres: the University of KwaZulu-Natal (South Africa), the Asian Institute of Technology (Thailand), the Swiss Federal Institute of Aquatic Science and Technology - Eawag (Switzerland) and the IHE Delft Institute for Water Education (Netherlands).

My special thanks go to Dr. Konstantina Velkushanova and the team at the University of KwaZulu-Natal for coordinating the edition and publication of this reference book. As the community of sanitation professionals grows, it is my expectation that this book will be regularly updated to capture new evidence and stand as a reference for the community, and for the growing number of sanitation industry players.

Dr. Doulaye Koné

Bill & Melinda Gates Foundation

Preface

Until recently, publishing on the topic of faecal sludge from onsite sanitation systems (known as the *unserved*) has been a neglected area. On the other hand, the topic of water supply for the previously *unserved* has been well covered by many publications by different authors and organisations. Some of the early publications in the field were related to aid agencies and humanitarian organisations, and these publications focused on practical planning, construction, implementation and maintenance. (*e.g.* Unicef, WaterAid, Oxfam). Publications on urine separation were motivated by environmental improvement and sustainability issues and generally focused on the health, social acceptance and agricultural aspects (Stockholm Environment Institute - SEI, German Corporation for International Cooperation GmbH - GIZ, IHE Delft Institute for Water Education - IHE Delft). The *Compendium of Sanitation Systems and Technology* from Eawag (the Swiss Federal Institute of Aquatic Science and Technology) marked the start of publications in faecal sludge, followed by the book *Faecal Sludge Management: Systems Approach for Implementation and Operation*, published collaboratively between Eawag and IHE Delft.

The Bill & Melinda Gates Foundation (BMGF) research programme into sanitation started by funding sanitation programmes for the unserved at Eawag, London School of Hygiene and Tropical Medicine, Asian Institute of Technology (AIT) and IHE Delft. The faecal sludge management (FSM) conference series initiated by the South African Water Research Commission (WRC) was the start of regular conferences on the topics which brought researchers and practitioners together. Up to this point, the main reference source for sampling, analytical methods and data interpretation was the well-known and frequently updated publication: Standard Methods for the Examination of Water and Wastewater[1]. The Reinvent the Toilet Challenge in 2011 brought a whole range of new researchers into the wonderful world of faecal sludge management. During the different events and convenings which were organised for the sharing of the results from these inventions, the question of different synthetic simulants, characterisation techniques and appropriate methods of analysis became a heated discussion point. The need for a *common language* became evident.

The conception and framing for this book was the end result of such discussions which were being held in parallel whenever FSM researchers got together and data was being exchanged. The editorial team came together on many occasions in different parts of the world and finally this publication came out as the first step towards the development of mutually agreed methods for faecal sludge analysis.

The finalisation of the two ISO standards related to non-sewered sanitation (ISO 30800 and ISO 31800) will be the spur for future updates in order to ensure the analytical methods keep up with the new and emerging technologies employed in the novel systems. The first intention of editors was to focus on a joint publication of methods for laboratory-based faecal sludge analysis but during the course of content development, it was identified the necessity to expand the scope and share important practical developments in the field, divided into eight chapters of the book. These include setting the scene, innovations and current trends in the field (Chapter 1), considerations for the measurement of properties and characterisation of faecal sludge (Chapter 2), different methods and techniques for faecal sludge sampling and handing (Chapter 3) and practical tips of how to set up a laboratory experimental design in a meaningful way to be able to support the design improvement of large-scale treatment technologies (Chapter 4). It is also discussed the importance and knowledge required to estimate the qualities and quantities at scale - from community to city-wide (Chapter 5), and modelling

[1] Rice E.W., Baird R.B. and Eaton A.D. (eds.) (2017). Standard Methods for the Examination of Water and Wastewater, 23rd edition. *American Public Health Association, American Water Works Association, Water Environment Federation.* ISBN: 9780875532875.

frames and approaches of faecal sludge processes taking place in the containments (Chapter 6). The importance of faecal sludge simulants and their use for technology development testing has also been provided along with examples of faecal sludge simulants for different purposes (Chapter 7). A guideline and considerations of how to set up a faecal sludge laboratory are provided along with case studies of already established faecal sludge laboratories (Chapter 2) and other partnering organisations under the Global Partnership of Laboratories for Faecal Sludge Analysis (GPLFSA)[2]. This consortium consists of laboratories to address the current challenges in the field and continuously work towards standardisation of methods for faecal sludge characterisation and quantification (Annex 1). The GPLFSA falls under the umbrella of the Global Sanitation Graduate School[3] (GSGS) platform committed to capacity building and knowledge dissemination in the field of faecal sludge management through postgraduate programs, online and campus-based courses and training programmes, aiming to improve the communication between sanitation practitioners, provide a comparative faecal sludge database, and greater confidence in the methods and obtained results. The GSGS and GPLFSA play a central role as dissemination platforms of this book.

The book is not necessarily intended to be read from cover to cover, but consulted as the need arises. The editors aim at a wide audience, represented by researchers and students, laboratory personnel and practitioners.

Students new to faecal sludge will find chapters 1, 2, 3 and 4 particularly useful for orientating themselves into the quantitative issues of the topic. City planners, designers and consultants will find chapters 5 and 6 essential reading. Researchers will be continually consulting all chapters of the book depending on the phase of the research being undertaken at the time. Laboratory personnel will be particularly interested in Chapter 8 augmented with information from chapters 2 and 3. Regulators will be guided by chapters 2, 5 and 6 to set standards and guidelines based on techniques in Chapter 8. Practitioners and developers of new technologies will find chapters 1, 2, 3, 4 and 8 particularly useful.

The editorial team have learnt a lot during the journey of collating this information and sincerely hopes it is of value to all involved in the important task of faecal sludge management. During this concerning period of the COVID-19 pandemic and the detection of SARS-CoV-2 RNA it is realised that this publication is not the final word in this important field and edited versions will be issued when the need arises.

Prof. Chris Buckley

University of KwaZulu-Natal

[2] https://sanitationeducation.org/laboratories/

[3] https://sanitationeducation.org/

Acknowledgements

The inspiration for this book originated a few years ago following conversations between different practitioners in the faecal sludge management field, and the realisation of the importance of having a standardised approach for the analysis, sampling and handling practices of faecal sludge. Back then, different groups of professionals and academics were working individually on the development of internal and standard operating procedures, so bringing this knowledge and experience together into a single publication was a major step forward to set the foundation needed for the development of standard methods for faecal sludge analysis.

The Methods for Faecal Sludge Analysis book is truly a team achievement. Therefore, I would like to extend my gratitude towards the editorial team for their never-ending enthusiasm, creativity, dedication, and inspirational approaches in dealing with the issues we encountered. Particularly, I would like to thank Dr. Linda Strande and Prof. Damir Brdjanovic for their unstinting support and mentorship throughout the entire period of the book development. Dr. Strande and the MEWS team from Eawag/Sandec made an excellent and invaluable contribution to the writing, editing and production of all the chapters in the book. Prof. Brdjanovic played a crucial role in the editing and coordinating the technical production of the book.

I would like to extend my thanks to the entire team of the Pollution Research Group (WASH R&D Centre[1]) of the University of KwaZulu-Natal for their endless support in the delivery of this book, including editing, authorship, managerial and moral support. Particularly I would like to recognize the importance of the mentorship of Prof. Chris Buckley who was the real inspiration to start my work in this field. I wish to express my gratitude to the most important people - the international group of authors, contributors and reviewers who produced, contributed and carefully reviewed one or more chapters of the book. A vote of thanks goes to the English editor, Ms. Claire Taylor, for her high level of professionalism and excellence, to the graphic designers of Synopsis d.o.o., and to IWA Publishing for their support in the publication of the book, particularly Mr. Rod Cookson, Mr. Mark Hammond, and Mr. Niall Cunniffe. Last but not least, I would like to thank the funders of this project - the Bill & Melinda Gates Foundation. Particularly I would like to thank the project's programme officer Dr. Carl Hensman and Dr. Doulaye Koné for trust and continuous support. Without their support this book would not have been possible.

I hope that you will enjoy reading this book as much as we enjoyed writing it.

Dr. Konstantina Velkushanova

IHE Delft Institute for Water Education

---

[1] The Pollution Research Group at the University of KwaZulu-Natal (UKZN PRG) has been rebranded to Water, Sanitation and Hygiene Research and Development Centre (WASH R&D Centre) since December 2020. The rebranding occurred during the advanced stage of production and as such has not been reflected further in this book.

Table of contents

1

Setting the stage

Linda Strande
Konstantina Velkushanova
Damir Brdjanovic

OBJECTIVES

The objectives of this chapter are to:

- Introduce city-wide inclusive sanitation (CWIS)
- Define faecal sludge
- Explain the need for standard methods
- Provide an overview of the book chapters
- Present additional resources

 Methods for Faecal Sludge Analysis. Edited by Velkushanova K., Strande L., Ronteltap M., Koottatep T., Brdjanovic D. and Buckley C. ISBN: 9781780409115. Published by IWA Publishing, London, UK.

1.1 CITY-WIDE INCLUSIVE SANITATION

City-wide inclusive sanitation (CWIS) is becoming the new paradigm in thinking about globally appropriate solutions for urban sanitation. The goal of CWIS is equitable, safe, and sustainable sanitation for everyone. Access to safely managed sanitation can be achieved through implementation of a range of appropriate technologies tailored to the realities of rapidly growing cities, with integrated combinations of sewered and non-sewered, and onsite, decentralised, and centralised technologies (Gambril *et al.*, 2020; Schrecongost *et al.*, 2020). Faecal sludge management (FSM) refers to the safe containment and treatment of non-sewered sanitation, and as illustrated in Figure 1.1, integrated faecal sludge management plays a vital role in CWIS.

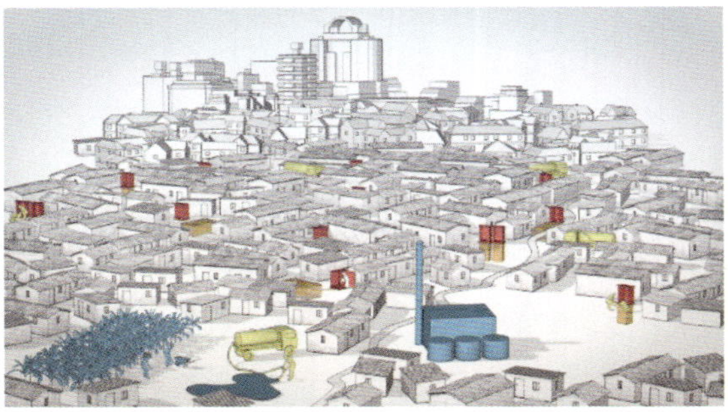

Figure 1.1 An example of the role of integrated faecal sludge management in city-wide inclusive sanitation (CWIS), with networked solutions including sewer-based and non-sewer-based sanitation solutions. Colours illustrate the faecal sludge management service chain. Red: user interface; orange (underground): onsite containment; yellow: collection and transport; green: treatment; blue: end use as resource recovery (source: Eawag).

Centralised, sewer-based technologies are well established, with a long record of research, knowledge, and implementation (Jenkins and Wanner, 2014), and guidelines for onsite containment of excreta in rural areas are well accepted (Wagner and Lanoix, 1958). In comparison, the concept of integrated faecal sludge management in urban and peri-urban areas is relatively new. Some technology solutions exist at the level of 'established' (*e.g.* existing guidelines for implementation and operation), whereas others are at the level of 'transferring' (*e.g.* not yet established, being adapted from other applications), or 'innovative' (*e.g.* still in the development phase) (WHO, 2018). There is a need for greater scientific knowledge to move solutions for faecal sludge management forward, a need that this book is designed to address through methods of data collection, analysis, and interpretation.

The importance and need for faecal sludge management has been recognised worldwide, and with rapidly occurring developments, it is an exciting time in the sanitation sector. The incorporation of the entire faecal sludge management service chain in the Sustainable Development Goals (SDGs), launched in 2015, has further established acknowledgement of its importance. There has also been an increase in the incorporation of faecal sludge management in national regulations and development agency agendas, increased funding from foundations and governments, and implementation of infrastructure and service provision. Curricula in CWIS are being developed and implemented, there is an increase in evidence-based research and journal publications, and an emerging new generation of students, practitioners, and scientists, the future champions in developing and implementing sanitation solutions.

It is important that the professional sanitation community maintains the momentum of these positive developments, and continues to focus efforts on these drivers of change, as there is still much work ahead. Cities in low- and middle-income countries are rapidly growing, with only a fraction of faecal sludge safely managed. For example, based on shit-flow diagrams (SFDs) in 39 cities, over 50% of excreta in urban areas remains untreated, with discharge of faecal sludge into open drains, onsite containments that are not emptied and are overflowing, and dumping of faecal sludge directly into urban environments (Peal *et al.*, 2020). To achieve CWIS, there is a need for further development of more policies and institutional frameworks with clear responsibilities; integrated planning methodologies for drinking water, wastewater, grey water, rainwater, and solid waste (Narayan *et al.*, 2021); sustainable business models and revenue fee structures (Otoo and Drechsel, 2018); and increasing knowledge dissemination and capacity development. There is a need for the development of improved and sustainable solutions for the future, while managing in parallel the existing faecal sludge crisis.

One of the goals of this book is that increased scientific knowledge will lead to an increased understanding of faecal sludge characteristics, its quantification, and correlation to source populations. Through increased scientific knowledge, and by remaining open and inquisitive, optimal new solutions can be developed. Based on this, research can lead to an understanding of treatment mechanisms in order to advance technologies from the level of innovative and transferring to established, and to reduce required footprints for treatment in urban areas (Gold *et al.*, 2016). Reliable data can improve projections and modelling, which are needed for the design of treatment plants and transfer stations (Englund *et al.*, 2020). Laboratory experience can lead to the development of methods that are lower cost and easier to implement, which could be used for the dynamic operation of faecal sludge treatment plants to adapt to highly variable loadings (Klinger *et al.*, 2019). Monitoring of treatment performance can lead to treatment plants designed for appropriate levels of treatment based on type of resource recovery (Andriessen *et al.*, 2019). Established methods for scaling up laboratory- and pilot-scale solutions can facilitate increased uptake and advancement of knowledge and experience. The closed-loop solutions being investigated with the Reinvent the Toilet Challenge (RTTC) can be based on onsite treatment technologies such as hydrothermal carbonisation, microwave technology, supercritical oxidation, pyrolysis, and electrochemical processes (Hiolski, 2019). Established methods of data collection and analysis can be used to establish guidelines and monitoring for the protection of public and environmental health, and advancements in scientific knowledge will elevate the perception of onsite sanitation as a sustainable component of CWIS.

1.2 WHAT IS FAECAL SLUDGE?

Faecal sludge management refers to the storage, collection, transport, treatment, and safe end use or disposal of faecal sludge (Strande *et al.*, 2014). Faecal sludge is defined very broadly as what accumulates in onsite sanitation technologies and specifically is not transported through a sewer. It is composed of excreta, but also anything else that goes into an onsite containment technology, such as flushwater, cleansing materials and menstrual hygiene products, grey water (*i.e.* bathing or kitchen water, including fats, oils and grease), and solid waste. Hence, faecal sludge is highly variable, with a very wide range of quantities (*i.e.* produced and accumulated volumes) and qualities (*i.e.* characteristics). In this book, faecal sludge is grouped by consistency as 'liquid' (TS <5%), 'slurry' (TS 5-15%), 'semi-solid' (TS 15-25%), and 'solid' (TS >25%).

To better understand what faecal sludge is, it is helpful to look at the different definitions for excreta, faecal sludge, and wastewater (or sewage), and their service chains, as shown in Figure 1.2.

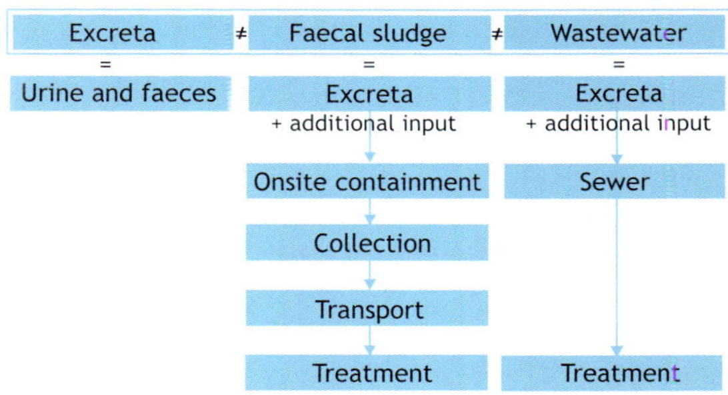

Figure 1.2 Comparison of excreta, and the sanitation service chain for management of faecal sludge and wastewater (source: Eawag).

Excreta is urine and faeces. Faecal sludge and wastewater are composed of excreta, together with additional inputs, and are both designed for the safe management of the resulting waste streams. The main difference between faecal sludge and wastewater is the respective sanitation service chains, which has very significant ramifications for management, cost, appropriate treatment, and quantities and qualities (Q&Q) (Dodane *et al.*, 2012). Faecal sludge is stored onsite, and is periodically collected and transported to a faecal sludge treatment plant, followed by safe disposal or end use. In contrast, wastewater also contains excreta, but it is transported via a sewer and in general contains larger volumes of flush water, grey water, and rainwater, which conveys it to a wastewater treatment plant. Hence, the service chains of wastewater and faecal sludge are entirely different, with faecal sludge management relying on a complex service chain that depends on interactions between people at every step (Englund and Strande, 2019).

In addition to the service chain, Q&Q of faecal sludge and wastewater are very different, with the range of faecal sludge characteristics being 1-2 orders of magnitude higher than wastewater (Figure 1.3, A and B). Wastewater is mixed during transport in the sewer, meaning that what is delivered to treatment is relatively homogenised. In contrast, the heterogeneity of faecal sludge observed at the level of containment is directly transferred to the treatment plant (Figure 1.3, C). It is important to note that wastewater influent and faecal sludge delivered to treatment plants also follow different statistical distributions and deviations, and that faecal sludge typically does not follow a normal distribution, with standard deviations that can be as high as mean values (Figure 1.3, A and B). Hence, when reporting the results of data analysis, more comprehensive summary statistics should be used to convey the variability, such as average, standard deviation, median, and quartile values, and the sharing of complete raw data sets is strongly encouraged[1].

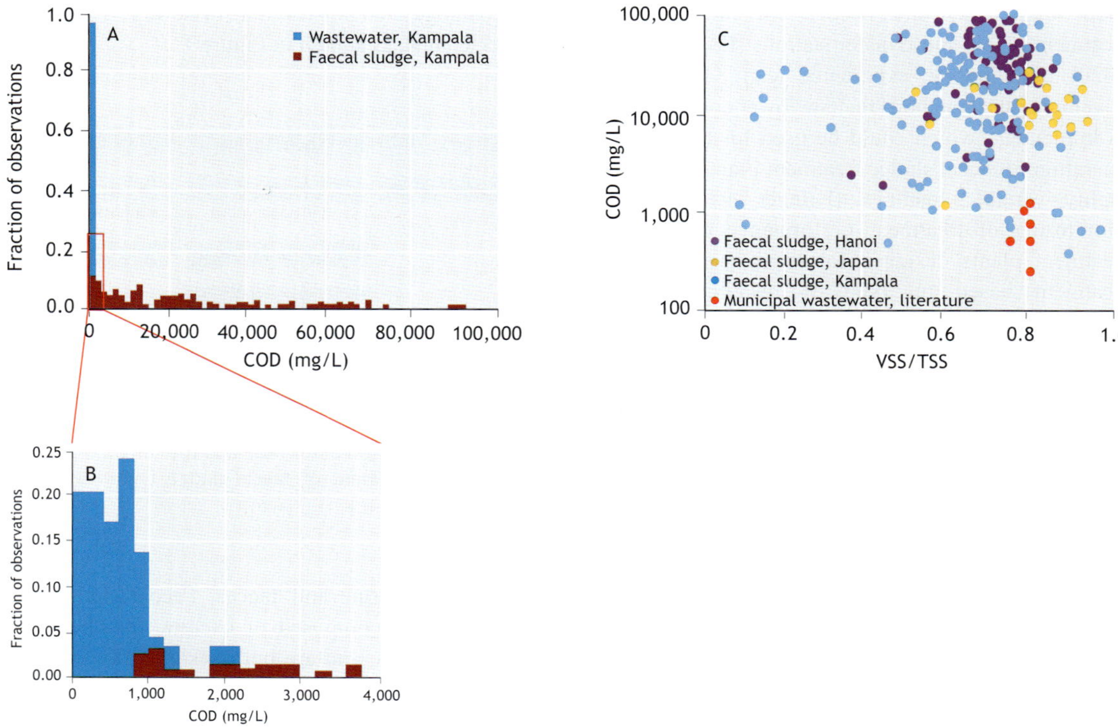

Figure 1.3 A) Histogram showing distribution of influent chemical oxygen demand (COD) values for the Lubigi wastewater and faecal sludge treatment plant in Kampala, Uganda. 32 wastewater and 143 faecal sludge samples, with an average COD for faecal sludge of 23,550 mg/L and standard deviation of 23,433. B) Enlargement of the X axis for comparison to wastewater (Englund and Strande, 2019). C) COD concentration and volatile suspended solids to total suspended solids (VSS/TSS) ratio for untreated faecal sludge collected at faecal sludge treatment plants or from collection trucks (Gold *et al.*, 2018; Englund *et al.*, 2020), and influent of untreated domestic wastewater (Tchobanoglous *et al.*, 2014; Henze *et al.*, 2008; Von Sperling *et al.*, 2020).

[1] For example: https://doi.org/10.25678/0000tt

Usage of the word 'sludge' varies and can be confusing, as there are several different types of sludge, and terminology is used differently by different practitioners. For example, types of wastewater sludges include sludge that settles out in the sewer, or sludges that are separated from liquid flows within wastewater treatment plants (Figure 1.4). In contrast, 'faecal sludge' is conventionally referred to as 'faecal sludge' throughout the service chain, from the time it accumulates in containments, and passes through collection, transport, and delivery to treatment facilities. The terminology of faecal sludge also varies depending on geographic location, professional background, or preference, but different terminology does not change the actual definition or characteristics. For this reason, as the sector continues to develop, agreed terminology for the faecal sludge management service chain will be an important development. In the meantime, during analysis and reporting of results, it is important to clearly describe where and how samples were taken to ensure transferability of results. Different types of sludge have widely varying characteristics and are not comparable, due to different storage conditions and treatment processes (*e.g.* redox conditions, level of stabilisation, biomass, nutrients, particle size, undigested plant fibres, salts and ions, and extracellular polymeric substances (Ward *et al.*, 2019). For example, faecal sludge from septic tanks is also commonly called 'septage', and might or might not include sludge, scum or supernatant layers. In addition, septic tanks commonly do not operate as designed, and/or what many people frequently refer to as 'septic tanks' are in reality more like cess pits. Hence, it is important to develop standard methods of sampling, analysis, and reporting of data, and to report exactly what is referred to when analysing and reporting Q&Q of 'faecal sludge'.

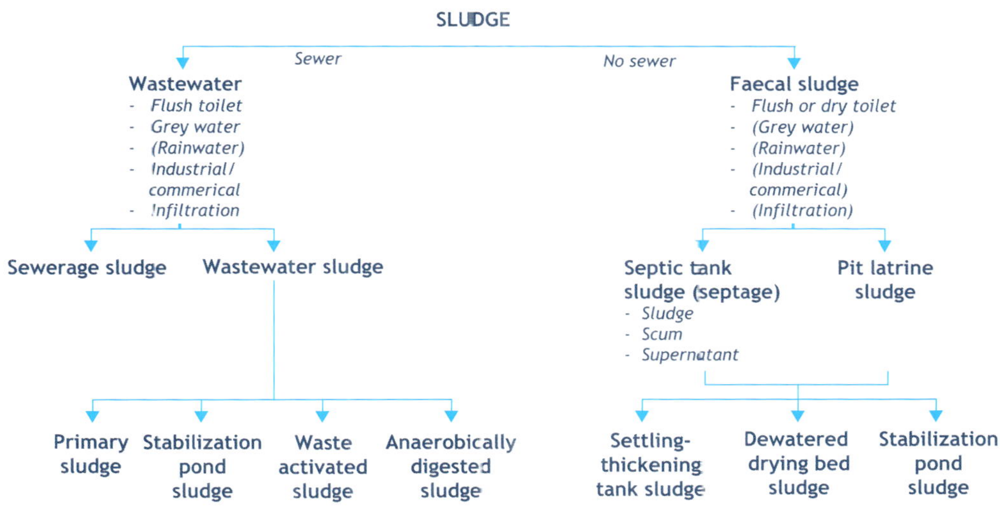

Figure 1.4 Examples of terminology used for types of sludge resulting from 'sewer-based' (wastewater) and 'non-sewer-based' (faecal sludge) sanitation components in an integrated approach to city-wide inclusive sanitation (Englund and Strande, 2019).

1.3 TOWARDS STANDARDISATION OF METHODS FOR FAECAL SLUDGE ANALYSIS

This book, *Methods for Faecal Sludge Analysis*, compiles methods of data collection, analysis, and interpretation specifically for faecal sludge, which until now have been lacking. In contrast, *Standard Methods for the Examination of Water and Wastewater* was first published in 1905, and *Methods of Soil Analysis* in 1965, both with multiple subsequent editions (Rice *et al.*, 2017; Klute, 1986; Page, 1983; Sparks *et al.*, 1996). Due to this lack of standard methods for faecal sludge, methods from water, wastewater, and soil and food science have been applied to faecal sludge. The problem is that they are not necessarily directly transferable to faecal sludge, which has very different characteristics. Methods for faecal sludge sampling are also greatly complicated by the wide range of technologies in each local context, and the heterogeneity within sanitation systems. Due to the lack of standard methods for sampling, laboratory approaches, analytical methods, and projections and modelling, the sector lacks a definitive source of respected guidelines to follow, which has translated into results that are not comparable. In addition, without established methods for data validation, results are not meaningful. This has greatly complicated the transfer of knowledge and data between different regions and institutions, and limited the ability to learn from each other and advance scientific understanding.

One of the goals of this book is to start developing 'standard' methods for faecal sludge analysis to improve communication among practitioners, designers, researchers, students, and teachers, to build comparative faecal sludge databases of information, and to increase confidence in obtained results. With this aim, the electronic version of this book has been made available free of charge for everyone. Although there are not yet 'standard' methods, this book addresses these challenges and provides a basis to start to establish them. The methods presented in this book have been peer-reviewed, and have wide acceptance in the sector. Now that this first round of methods has been compiled, they can start to undergo the required steps to become official 'standard' methods. This will require international, collaborative validation, where blind samples are independently analysed in parallel in different laboratories, and methods are further evaluated for precision and ruggedness (Rice *et al.*, 2017, method 1040). In moving towards standardisation, a committee of members of the Global Partnership of Laboratories for Faecal Sludge Analysis (GPLFSA)[2] is coordinating these processes. The GPLFSA was established in 2018 to address the need for increased communication between the growing number of laboratories equipped for faecal sludge analysis.

A further goal of this book is to be dynamic and keep pace with rapid developments. The list of methods presented in this book is meant as a starting place that will continue to grow and develop. As the need for new methods arises, existing methods will need to be adapted and new methods developed to fill the gaps. Tips for adapting existing methods for application to faecal sludge are included in Chapter 8, and guidelines for developing standard methods are covered in Rice *et al.* (2017), method 1040. As these changes are taking place, publications are needed that can adapt more quickly than the time required for new books and book editions to be published. The GPLFSA webpage provides a platform for exchanging information on method development that is regularly updated and publicly available, together with an ongoing conversation through a dedicated web application channel. The webpage includes video explanations and examples of methods, and online tutorials and courses. The GPLFSA will continue to disseminate this knowledge and experience through different platforms such as the IWA Specialist Groups, the SuSanA forum, the FSM Alliance, and the Global Sanitation Graduate School (GSGS). For information on how to stay updated, or become involved in the process, visit the GPLFSA webpage.

[2] https://sanitationeducation.org/laboratories/

1.4 INTEGRATED APPROACH TO DATA COLLECTION

This book provides an integrated approach to data collection and analysis of faecal sludge in order to generate meaningful data. As illustrated in Figure 1.5, the entire process must be coordinated, and aspects of each chapter are interrelated. The objectives of each of the book chapters are summarised in Table 1.1. The first step in using the book is to define the overall objectives that the methods can fulfil, and what information is required to achieve it. For example, this could be setting up a monitoring program at a treatment plant, defining a research question, or collecting data to design an integrated faecal sludge management plan. Further examples are provided throughout the book chapters. The next step is to make a plan for data collection to carry out the defined objectives. This will include a sampling plan (*e.g.* for monitoring), and could also include a research plan. Information on how to set up a sampling plan is covered in Chapter 3, examples of laboratory strategies for the upscaling of technologies from laboratory or pilot-scale are provided in Chapter 4, macro- and micro-scale projections of Q&Q of faecal sludge in chapters 5 and 6, and the use of simulants (recipes for synthetic faecal sludge) that can be used in research are the focus of Chapter 7.

An integral part of making a data collection strategy, is to determine which specific properties of faecal sludge need to be quantified to achieve the defined objective. Then, in an iterative process, while further refining the plan for data collection, the specific analytical methods that will be used to characterise the selected properties are selected based on the available resources and desired level of accuracy. Information on how properties of faecal sludge are influenced along the service chain are described in Chapter 2, and the analytical methods and procedures related to safe handling, storage, sample preparation, and disposal are provided in Chapter 8. The final steps are carrying out the developed sampling plan, conducting the laboratory analysis, and analysing and interpreting the obtained data. Guidelines of how to set up a faecal sludge laboratory are provided in Chapter 2, further examples of data collection are applications for scaling-up technologies in Chapter 4, and projections and modelling of Q&Q of faecal sludge in chapters 5 and 6.

With the completion of these steps, it is important to transparently share and disseminate the results widely, in order to advance the scientific knowledge of faecal sludge and management solutions.

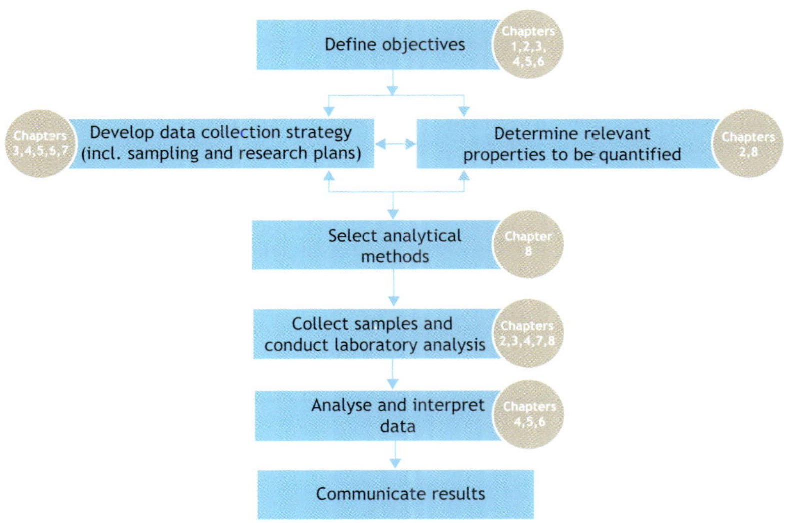

Figure 1.5 The integrated approach to data collection and analysis presented in this book, and relevant chapters at each step of the integrated process.

Table 1.1 Objectives of book chapters.

Chapter 1: Setting the stage • Introduce city-wide inclusive sanitation (CWIS) • Define faecal sludge • Explain the need for standard methods • Provide an overview of the book chapters • Present additional resources	**Chapter 2: Faecal sludge properties and considerations for characterisation** • Present four types of faecal sludge depending on total solids content • Provide a brief overview of factors that can influence characteristics of faecal sludge along the service chain. • Explain the relevance of selecting different faecal sludge properties based on the objectives of characterisation • Explain factors for consideration when selecting characterisation methods • Provide guidelines for setting up faecal sludge laboratories, along with case studies of existing implementations
Chapter 3: Faecal sludge sample collection and handling • Select different sampling techniques depending on objectives • Select sampling devices and locations • Develop appropriate and reliable faecal sludge sampling schemes and plans • Ensure sample representativeness and integrity • Protect health and safety of employees and users of onsite sanitation	**Chapter 4: Experimental design for the development, transfer, scaling-up, and optimisation of treatment technologies: case studies of dewatering and drying** • Introduce scales of experimentation and experimental design for the development, transfer, scaling-up, and optimisation of faecal sludge treatment technologies • Provide examples of experimental approaches for scaling-up conditioners for dewatering and drying for resource recovery • Present case studies that address research questions at different scales of faecal sludge treatment processes and technology development and adaptation
Chapter 5: Estimating quantities and qualities (Q&Q) of faecal sludge at community to city-wide scales • Explain the importance of being able to reasonably estimate Q&Q of faecal sludge • Define the six stages in the faecal sludge service chain where Q&Q of faecal sludge can be estimated • Summarise the existing state of knowledge and future prospects for making projections of Q&Q of faecal sludge • Provide an overview of a methodology to estimate Q&Q of faecal sludge on a scale relevant for the planning of management and treatment solutions, from community scale to city-wide planning	**Chapter 6: Towards city-wide inclusive sanitation (CWIS) modelling: modelling of faecal sludge containment/treatment processes** • Promote modelling of onsite sanitation • Familiarise readers with the basic principles of established modelling approaches applied in sewered sanitation • Introduce ideas on how faecal sludge containment/treatment processes can be modelled using the analogy with modelling practices in sewered sanitation • Bring sewered and onsite sanitation closer together through the integrated approach of community city-wide inclusive sanitation modelling
Chapter 7: Faecal sludge simulants: review of synthetic human faeces and faecal sludge for sanitation and wastewater research • Introduce the concept of simulants and their applications • Present current state-of-the-art knowledge in simulants for faecal sludge, faeces and urine • Compare properties between simulants and typical values observed in the field • Introduce customisation of simulants, including advantages and constraints	**Chapter 8: Laboratory procedures and methods for characterisation of faecal sludge** • Provide methods for protecting health and safety during collection, handling, transportation, storage, and disposal of faecal sludge • Provide information required to adapt and develop standard methods for faecal sludge characterisation, including quality control and quality assurance strategies and selection of appropriate methods. • Provide an overview of existing methods for faecal sludge analysis being used in partner laboratories.

1.5 ADDITIONAL RESOURCES

The editors of this book came together for this project because their research, field, and teaching experience identified the need for standard methods for faecal sludge analysis. This open access book addresses that gap, and it can now be used by practitioners, designers, researchers, students, and teachers as an integrated resource. However, *Methods for Faecal Sludge Analysis* is not intended to be used as a stand-alone reference, but to be used as a companion guide to existing publications, face-to-face teaching, and online courses. Fortunately, over the last decade available tools and references for faecal sludge management have become increasingly available. The editors have assembled a list of additional sanitation resources that are available, with particular focus on faecal sludge management. This includes textbooks and manuals, massive online open courses (MOOCs), online and short courses, professional certificate/diploma programs, undergraduate and postgraduate university curricula, tailor-made training courses, conferences and events, networks, partnerships, toolboxes, toolkits, software, and some key global sanitation/FSM initiatives. The list is not by any means exhaustive or final, and will be regularly updated and expanded in the online version of this book on the GPLFSA website.

Textbooks and manuals

- Faecal sludge management: Systems approach for implementation and operation (Strande *et al.,* 2014)[3,4] available in several languages[5]
- Faecal sludge management: Highlights and exercises (Englund and Strande, 2019)[6]
- Compendium on sanitation systems and technologies (Tilley *et al.,* 2014)[7] available in several languages[8]
- Guidelines for community-led urban environmental sanitation planning (CLUES) (Lüthi *et al.,* 2011)[9]
- Sanitation 21: A planning framework for improving city-wide sanitation services (Parkinson *et al.,* 2014)[10]
- How to design wastewater systems for local conditions in developing countries (Robbins and Ligon, 2014)[11]
- Hygiene and sanitation software - An overview of approaches (Peal *et al.,* 2010)[12]
- Regenerative sanitation (Koottatep *et al.,* 2019)[13]
- Compendium of global good practices – Urban Sanitation (NIUA, 2015)[14]
- Faecal sludge and septage treatment (Tayler, 2018)[15]
- Guidelines on sanitation and health (WHO, 2018)[16]
- Resource recovery from waste (Otto and Drechsel, 2018)[17]
- Innovations for urban sanitation (Myers *et al.,* 2018)[18]
- Sustainable sanitation for all (Bongartz *et al.,* 2016)[19]
- Organic waste recycling: technology, management and sustainability (Polprasert and Koottatep, 2017)[20]

---

[3] https://www.eawag.ch/en/department/sandec/publications/fsm-book/
[4] https://www.un-ihe.org/sites/default/files/fsm_book_lr.pdf
[5] English, French, Spanish, Marathi, Tamil and Hindi
[6] https://www.eawag.ch/fileadmin/Domain1/Abteilungen/sandec/publikationen/EWM/FSM_Book_Highlights_and_Exercises/FSM_Highlights_and_Exercises_Final-compressed.pdf
[7] https://www.eawag.ch/en/department/sandec/publications/compendium/
[8] English, French, Spanish, Arabic, Nepali, Russian, Vietnamese, Korean and Romanian
[9] https://www.eawag.ch/en/department/sandec/projects/sesp/clues/
[10] https://www.eawag.ch/en/department/sandec/
[11] https://iwaponline.com/ebooks/book/385/How-to-Design-Wastewater-Systems-for-Local
[12] https://www.wsscc.org/resources-feed/hygiene-sanitation-software-overview-approaches/
[13] https://www.iwapublishing.com/books/9781780409672/regenerative-sanitation-new-paradigm-sanitation-40
[14] https://pearl.niua.org/sites/default/files/books/GP-GL2_SANITATION.pdf
[15] https://www.susana.org/_resources/documents/default/3-3439-7-1540380071.pdf
[16] https://www.who.int/water_sanitation_health/publications/guidelines-on-sanitation-and-health/en/
[17] http://www.iwmi.cgiar.org/Publications/Books/PDF/resource-recovery-from-waste.pdf
[18] https://practicalactionpublishing.com/book/1139/innovations-for-urban-sanitation
[19] https://practicalactionpublishing.com/book/2130/sustainable-sanitation-for-all
[20] https://practicalactionpublishing.com/book/693/faecal-sludge-and-septage-treatment

- Compendium of sanitation technologies in emergencies (Gensch *et al.*, 2018)[21]
- Resource recovery and reuse series (IWMI series)[22]
- Co-treatment of septage and faecal sludge in sewage treatment facilities (Narayana, 2020)[23]
- Experimental methods in wastewater treatment (Van Loosdrecht *et al.*, 2016)[24] available in several languages[25]

MOOCs / Online courses

- Water supply and sanitation policy in developing countries[26]
- Planning and design of sanitation systems and technologies[27]
- Introduction to faecal sludge management[28]
- Foundations of public health practice: Health protection[29]
- Water supply and sanitation policy in developing countries[30]
- Public health in humanitarian crises[31]
- Introduction to public health engineering in humanitarian contexts[32]
- Municipal solid waste management in developing countries[33]
- Planning and design of sanitation systems and technologies[34]
- Water - Addressing the global crisis[35]
- Faecal sludge management[20,36,37,38,39,40,41]
- Faecal sludge management online course[42]
- Experimental methods in wastewater treatment series[43]
- Sanitation systems and services[44]
- Sanitation and public health[44]
- Sanitation technology[44]
- Sanitation governance[44]
- Sanitation financing[44]
- Behaviour change and advocacy[44]
- Emergency sanitation[44]
- Analysis of sanitation flows[44]
- Various online courses on non-sewered sanitation[44]

---

[21] http://washcluster.net/emersan-compendium
[22] https://www.iwmi.cgiar.org/publications/resource-recovery-reuse/
[23] https://www.iwapublishing.com/books/9781789061260/co-treatment-septage-and-faecal-sludge-sewage-treatment-facilities
[24] https://www.iwapublishing.com/books/9781780404745/experimental-methods-wastewater-treatment
[25] English, Spanish, Croatian, Russian, Marathi and Hindi
[26] http://www.cltsfoundation.org/massive-online-open-courses-moocs-water-supply-sanitation-policy-developing-countries/
[27] https://www.mooc-list.com/course/planning-design-sanitation-systems-and-technologies-coursera
[28] https://www.eawag.ch/en/department/sandec/e-learning/moocs/, in English, Portuguese, Spanish, French, Hindi and Bengali
[29] https://www.mooc-list.com/course/foundations-public-health-practice-health-protection-coursera
[30] https://www.mooc-list.com/course/water-supply-and-sanitation-policy-developing-countries-part-1-understanding-complex-problems
[31] https://www.mooc-list.com/course/public-health-humanitarian-crises-coursera, in English, Portuguese, French, Spanish, Hindi, Indonesian and Arabic
[32] https://www.mooc-list.com/course/introduction-public-health-engineering-humanitarian-contexts-coursera
[33] https://www.eawag.ch/en/department/sandec/e-learning/moocs/
[34] https://www.eawag.ch/en/department/sandec/e-learning/moocs/, in English, Portuguese, French, Spanish, Hindi, Bengali and Russian
[35] https://www.siwi.org/watercourse/
[36] https://www.un-ihe.org/online-course-faecal-sludge-management
[37] https://sanitationeducation.org/courses/faecal-sludge-management/
[38] https://www.fsmonlinecourse-ait.com/
[39] http://prg.ukzn.ac.za/fsm-on-line-course
[40] http://cseindia.org/fsm/onlinecourse.html
[41] http://www.2ie-edu.org/index.php/fr/?page_id=1989
[42] https://www.lboro.ac.uk/research/wedc/news/news/fsm-online-course.html
[43] https://experimentalmethods.org/
[44] https://sanitationeducation.org/online-courses/

Short courses

- Sanitation systems and services[45]
- Sanitation and public health[45]
- Sanitation technology[45]
- Sanitation governance[45]
- Sanitation financing[45]
- Behaviour change and advocacy[45]
- Emergency sanitation[45]
- Analysis of sanitation flows[45]
- Non-sewered sanitation[45,46,47]
- Faecal sludge management in India[48]
- Various short courses on non-sewered sanitation[49]

Professional diploma programs

- Graduate Professional Diploma Program (GPDP) in Sanitation and Sanitary Engineering[50,51]
- Various GPDP programs on non-sewered sanitation[52]

Undergraduate / postgraduate programs

- MSc in Sanitation[53]
- ME Sanitation Science Technology and Management[54]
- MSc in Sanitation[55]
- MS/ME in Sanitation Technology[56]
- MTech in Non-Sewered Sanitation[57]
- MSc in Non-Sewered Sanitation[58]
- MSc in Water Supply and Sanitation Management[59]
- MSc in Sanitary Engineering[60]
- Various postgraduate programs on sanitation[61]

Tailor-made training courses

- Consultant capacity development (ConCaD) for city-wide inclusive urban sanitation[62]
- Non-sewered sanitation[63]
- Training the trainers in MOOC development[64]
- FSM toolbox training[65]

---

[45] http://enpho.org/
[46] https://www.ku.edu.np/
[47] https://www.ukgllp.com/
[48] https://cddindia.org/network
[49] https://sanitationeducation.org/on-campus-courses/
[50] https://sanitationeducation.org/gpdp/
[51] https://www.un-ihe.org/urban-water-and-sanitation
[52] https://sanitationeducation.org/gpdp/
[53] https://sanitationeducation.org/master-program/master-science-program-sanitation/
[54] https://www.bits-pilani.ac.in/goa/biologicalScience/GSGS
[55] http://gsgs-kuet.education/
[56] http://gsgs.ku.edu.np/
[57] https://jaipur.manipal.edu/foe/programs/program-list/MTech-Non-Sewered-Sanitation.html
[58] http://www.2ie-edu.org/index.php/fr/presentation/projets-et-dons/projet-de-master-specialise-en-assainissement-non-collectif
[59] https://gsgs-ina.itb.ac.id/
[60] https://www.un-ihe.org/msc-programmes/specialization/sanitary-engineering-0
[61] https://sanitationeducation.org/msc-in-sanitation/
[62] https://www.eawag.ch/en/department/sandec/projects/sesp/consultant-capacity-development-concad-for-city-wide-inclusive-urban-sanitation/
[63] http://www.2ie-edu.org/index.php/fr/?page_id=1989
[64] http://www.2ie-edu.org/index.php/fr/?page_id=1989
[65] https://www.ait.ac.th/

- Training on non-sewered sanitation in India[66]
- Novel sanitation in India[67]

Conferences / events

- WEDC international conference[68] (since 1973)
- IWA Water and Development Congress and Exhibition[69] (since 2009)
- FSM conference[70] (since 2011)
- IWA Non-sewered Sanitation Conference[71] (starting in 2021)
- AfricaSan conference[72] (since 2008)
- AfWA International Conference and Exhibition[73] (since 2002)

Networks

- Sustainable Sanitation Alliance (SuSanA)[74]
- Faecal Sludge Management Alliance (FSMA)[75]
- IWA specialist groups, *e.g.* Non-Sewered Sanitation, Sanitation and Water Management in Developing Countries, Resources Oriented Sanitation[76]
- National Faecal Sludge and Septage Management Alliance (NFSSMA)[77]
- African Water Association (AfWA)[78]
- Toilet Board Coalition[79]
- Container-based Sanitation Alliance[80]
- Women in Water and Sanitation Network[81]
- Water Supply and Sanitation Collaborative Council[82]
- Asia-Pacific Water Forum[83]

Partnerships

- Global Sanitation Graduate School (GSGS)[84]
- Global Partnership of Laboratories of Faecal Sludge Analysis (GPLFSA)[85]
- Global WASH Cluster[86]

Toolboxes / toolkits / software

- FSM Toolbox[87]
- Sustainable sanitation and water management toolbox[88]

---

[66] https://asci.org.in/
[67] https://www.bits-pilani.ac.in/
[68] https://www.lboro.ac.uk/research/wedc/conferences/
[69] https://iwa-network.org/events/iwa-water-and-development-congress-exhibition-2019/
[70] https://fsm-alliance.org/fsm6/
[71] https://iwa-network.org/events/1st-iwa-non-sewered-sanitation-conference-21-to-24-july-2020-future-africa-campus-university-of-pretoria-pretoria-south-africa/
[72] https://www.africasan.com/
[73] https://www.africasan.com/africasan
[74] https://www.susana.org/en/
[75] https://fsm-alliance.org/
[76] https://iwa-network.org/wp-content/uploads/2015/12/IWA-Specialist-Groups-2018_A4.pdf
[77] https://www.washinstitute.org/nfssm.php
[78] https://www.afwa-hq.org/index.php/fr/
[79] https://www.toiletboard.org/
[80] http://www.cbsa.global/#/
[81] http://wwsn.org
[82] https://www.wsscc.org
[83] http://apwf.org/
[84] https://sanitationeducation.org/
[85] https://sanitationeducation.org/laboratories/
[86] http://washcluster.net/
[87] https://www.fsmtoolbox.com/
[88] https://www.sswm.info/

- Shit (excreta) flow diagram (SFD)[89]
- SaniPath[90]
- SanI-Kit[91]
- SaniPlan[92], SaniTab[92], FSM assessment and planning toolkit[92], Performance assessment toolkit[92], Open defecation free cities model[92], Target-setting model[92], SBM support tool for infrastructure and financial requirement calculations[92], PSP toolkit for IFSM[92]
- World Bank FSM tools[93]
- The Sphere emergency training toolkit[94]

Other initiatives

- Reinvent the Toilet Challenge[95]
- Health, safety and dignity of sanitation workers - WaterAid, World Bank, the World Health Organisation and the International Labour Organisation
- Sanitation Technology Platform (STeP)[96]
- Engineering field testing platform of innovative toilet technologies[97]
- South African Sanitation Technology Demonstration Programme (SASTEP)[98]
- ISO Standards for non-sewered sanitation: ISO 30500, ISO 24521, and ISO PC 318[99]
- Pit emptying challenge under World Skills
- The World Bank[100]
- Asian Development Bank[101]
- Inter-American Development Bank[102]
- African Development Bank[103]
- Islamic Development Bank[104]
- European Bank for Reconstruction and Development[105]
- Bill & Melinda Gates Foundation[106]

---

89 https://sfd.susana.org/
90 http://sanipath.org/
91 https://www.cseindia.org/sanikit/index.html
92 https://www.pas.org.in/web/ceptpas/urbansanitation
93 https://www.worldbank.org/en/topic/sanitation/brief/fecal-sludge-management-tools
94 https://www.spherestandards.org/resources/the-sphere-emergency-training-toolkit/
95 https://www.gatesfoundation.org/Media-Center/Press-Releases/2012/08/Bill-Gates-Names-Winners-of-the-Reinvent-the-Toilet-Challenge
96 https://stepsforsanitation.org/
97 https://www.susana.org/en/knowledge-hub/projects/database/details/474
98 https://www.susana.org/en/knowledge-hub/resources-and-publications/library/details/2201
99 https://sanitation.ansi.org/
100 https://www.worldbank.org/en/topic/sanitation
101 https://www.adb.org/sectors/water/financing-program/sanitation-wastewater-management
102 https://www.iadb.org/en/sector/water-and-sanitation/overview
103 https://www.afdb.org/en/topics-and-sectors/sectors/water-supply-sanitation
104 https://www.isdb.org/sector/sanitation
105 https://www.ebrd.com/what-we-do.html
106 https://www.gatesfoundation.org/what-we-do/global-growth-and-opportunity/water-sanitation-and-hygiene

REFERENCES

Andriessen N., Ward B.J. and Strande L. (2019). To char or not to char? Review of technologies to produce solid fuels for resource recovery from faecal sludge. *Journal of Water, Sanitation and Hygiene for Development*, 9(2), 210-224.

Dodane P.H., Mbéguéré M., Ousmane S. and Strande L. (2012). Capital and Operating Costs of Full-Scale Faecal Sludge Management and Wastewater Treatment Systems in Dakar, Senegal. *Environmental Science & Technology*. 46(7), 3705-3711.

Englund M., Carbajal J.P., Ferré A., Bassan M., Vu T.H.A., Nguyen V.A. and Strande L. (2020. Modelling quantities and qualities (Q&Q) of faecal sludge in Hanoi, Vietnam and Kampala, Uganda for improved management solutions. *Journal of Environmental Management*, 261, 110202.

Englund M. and Strande L. (eds.) (2019). Faecal Sludge Management: Highlights and Exercises. ISBN 978-3-906484-70-9, Eawag: Swiss Federal Institute of Aquatic Science and Technology. Dübendorf, Switzerland. www.sandec.ch/fsm-hebook

Gambrill M., Gilsdorf R.J. and Kotwal N. (2020). Citywide Inclusive Sanitation - Business as Unusual: Shifting the Paradigm by Shifting Minds. *Front. Environ. Sci.,* 11, doi.org/10.3389/fenvs.2019.00201.

Gold M., Dayer P., Faye C., Clair G., Seck A., Niang S., Morgenroth E. and Strande L. (2016). Locally produced natural conditioners for dewatering of faecal sludge. *Environmental Technology*, 1-13.

Gold M., Harada H., Therrien J.-D., Nishida T., Cunningham M., Semiyaga S., Fujii S., Niwagaba C., Dorea C., Nguyen V.-A. and Strande L. (2018). Cross-country analysis of faecal sludge dewatering. *Environmental Technology*, 1-11.

Henze M., Van Loosdrecht M.C.M., Ekama G.A. and Brdjanovic D. (2008). Biological Wastewater Treatment - Principles, Modelling and Design, 1st ed. IWA Publishing, London, UK.

Hiolski E. (2019). The Toilet Gets a Makeover, How chemistry can help solve the sanitation crisis in low-resource areas. *ACS Central Science,* 5(8), 1303–1306.

Jenkins D. and Wanner J. (eds.) (2014). Activated Sludge - 100 Years and Counting. IWA Publishing. London, UK.

Klinger M., Gueye A., Manandhar Sherpa A. and Strande L. (2019). Scoping Study: Faecal Sludge Treatment Plants in South-Asia and Sub-Saharan Africa. eFSTP Project Report. *Gates Open Research* 2019, 3, 1716 (document) (https://doi.org/10.21955/gatesopenres.1116557.1)

Klute A. (ed.) (1986). Methods of soil analysis. Part 1 Physical and mineralogical properties. Madison, WI: American Society of Agronomy; Crop Science Society of America; Soil Science Society of America.

Narayan A.S., Marks S.J., Meierhofer R., Strande L., Tilley E., Zurbrügg C. and Lüthi C. (2021). Advancements in and Integration of Water, Sanitation and Solid Waste in Low and Middle Income Countries. Annual Review of Environment and Resources.

Otoo M. and Drechsel P. (eds.) (2018). Resource recovery from waste: business models for energy, nutrient and water reuse in low- and middle-income countries. Oxon, UK: Routledge - Earthscan.

Page A.L. (ed.) (1983). Methods of Soil Analysis Part 2 Microbiological and Biochemical Properties. Madison, WI: American Society of Agronomy; Crop Science Society of America; Soil Science Society of America.

Peal A., Evans B., Sangaralingam A., Ban R., Blackett I., Hawkins P., Schoebitz L., Scott R., Sleigh A., Strande L. and Roda O. (2020). Estimating safely managed sanitation in urban areas, lessons learned from a global implementation of excreta-flow diagrams. *Front. Environ. Sci.,* https://doi.org/10.3389/fenvs.2020.00001.

Rice E.W., Baird R.B. and Eaton A.D. (eds.) (2017). Standard Methods for the Examination of Water and Wastewater, 23rd edition. American Public Health Association; American Water Works Association; Water Environment Federation. ISBN: 9780875532875.

Schrecongost A., Pedi D., Rosenboom J.W., Shrestha R. and Ban R. (2020). Citywide Inclusive Sanitation: A Public Service Approach for Reaching the Urban Sanitation SDGs. *Front. Environ. Sci.* 8, 1-8. doi:10.3389/fenvs.2020.00019.

Sparks D.L., Page A.L., Helmke P.A., Loeppert R.H., Soltanpour P.N., Tabatabai M.A., Johnston C.T. and Sumner M.E. (eds.) (1996). Methods of Soil Analysis Part 3 Chemical methods. Madison, WI: American Society of Agronomy; Crop Science Society of America; Soil Science Society of America.

Strande L., Ronteltap M. and Brdjanovic D. (eds.) (2014). Faecal Sludge Management - Systems Approach for Implementation and Operation. IWA Publishing, London, UK, doi:10.1007/s13398-014-0173-7.2.

Tchobanoglous G., Stensel H.D., Tsuchihashi R. and Burton F. (2014). Wastewater Engineering: Treatment and Resources Recovery, 5th edn. New York, NY: McGraw-Hill Education.

Tilley E., Ulrich L., Lüthi C., Reymond P. and Zurbrügg C. (2014). Compendium of Sanitation Systems and Technologies. 2nd edn. Eawag: Swiss Federal Institute of Aquatic Science and Technology.

Von Sperling M., Verbyla M.E. and Oliveira S.M.A.C. (2020). Water Quality Data: A Guide for Students, Researchers and Practitioners. IWA Publishing, London, U.K.

Wagner E.G. and Lanoix J.N. (1958). Excreta disposal for rural areas and small communities. World Health Organisation Publication, Geneva, Switzerland.

Ward B.J., Gueye A., Diop B., Traber J., Morgenroth M. and Strande L. (2019). Evaluation of conceptual model and predictors of faecal sludge dewatering performance in Senegal and Tanzania. *Water Research,* 167, 115101.

World Health Organisation (2018). Guidelines on sanitation and health. ISBN 978-92-4-151470-5. World Health Organisation, Geneva, Switzerland.

2

Faecal sludge properties and considerations for characterisation

Konstantina Velkushanova
Linda Strande

OBJECTIVES

The objectives of this chapter are to:

- Present four types of faecal sludge depending on total solids content
- Provide a brief overview of factors that can influence characteristics of faecal sludge along the service chain.
- Explain the relevance of selecting different faecal sludge properties based on the objectives of characterisation
- Explain factors for consideration when selecting characterisation methods
- Provide guidelines for setting up faecal sludge laboratories, along with case studies of existing implementations

 Methods for Faecal Sludge Analysis. Edited by Velkushanova K., Strande L., Ronteltap M., Koottatep T., Brdjanovic D. and Buckley C. ISBN: 9781780409115. Published by IWA Publishing, London, UK.

2.1 INTRODUCTION

Faecal sludge characterisation is the process of measuring and evaluating faecal sludge properties. The characterisation of faecal sludge as a material, and understanding the nature of the physical, biological, and chemical characteristics, is necessary for the research, design, implementation, and operation of faecal sludge management solutions. Common reasons for characterising faecal sludge include understanding biochemical processes of degradation and nutrient cycling, monitoring of treatment efficiency and pathogen removal, determining loadings for the design and operation of a treatment plant, selecting the best technology for emptying of sludge from onsite containments, and evaluating the potential for resource recovery. Based on the defined purpose and objectives of the faecal sludge characterisation, the appropriate methods for measurement of properties need to be selected.

However, defining standardised methods for the characterisation of faecal sludge is challenging due to the high variability of faecal sludge, from the micro- to the macro-scale. In addition to the variability, different methods for sample preparation and analysis are appropriate depending on the 'type' of faecal sludge. For example, samples with higher total solids (TS) content and lower moisture content from 'dry' systems, such as pit latrines, urine-diverting dehydration toilets (UDDTs) and composting toilets, will likely require a different preparation than samples collected from 'wet' systems, such as septic tanks, 'wet' pit latrines, and cesspits that have much lower TS content. The solids content will also affect whether it is relevant to conduct volumetric analysis (*e.g.* milligram of constituent per litre of the sample) or gravimetric (*e.g.* gram constituent per gram TS of the sample). Other complicating factors for standardisation include a wide range of available resources, equipment, and capacity of laboratories. This chapter presents background information that is necessary to understand prior to the use of the methods presented in Chapter 8. It defines types of faecal sludge based on TS concentration, which is necessary for implementing the correct steps in the methods. It introduces factors that affect the variability of characteristics along the entire service chain in order to understand what analyses are relevant. It then provides guidance on how to select appropriate methods for characterisation, based on several criteria such as characterisation objectives, relevant characteristics, desired level of accuracy, laboratory capacity, and available resources. The chapter then presents considerations specific to the characterisation of faecal sludge for setting up a faecal sludge analytical laboratory, and includes four case study examples of how operational laboratories can look when implementing all of the steps presented in this chapter.

2.2 TYPES OF FAECAL SLUDGE

Faecal sludge is highly variable based on its broad definition and decentralised nature, as faecal sludge is anything and everything that is collected and accumulated within containment technologies of onsite sanitation systems (Chapter 1). Qualitative observations of different moisture or TS content of faecal sludge range from dilute and watery, to slurries that are still pumpable, to dewatered sludge that is 'shovelable' or 'spadable'. Although these differences do not have clear boundaries that can be precisely defined, it is useful to define approximate ranges of types of faecal sludge based on TS. The different ranges can have an impact on which methods of analysis and sample preparation are applicable, and also if concentrations are analysed and reported by volumetric or gravimetric concentration. Below are the four types of faecal sludge based on TS concentration.

- *Liquid faecal sludge*
  TS <5%, runny liquid, relatively dilute with the consistency of water or domestic wastewater, readily pumpable. Usually collected from 'wet' containments such as leach pits and septic tanks, or 'wet' pit latrines.
- *Slurry faecal sludge*
  TS 5-15%, thicker than liquid, but still runny, from watery to wet mud consistency, pumpable in the lower range, too runny to shovel, and not spadable. Common in pit latrines (improved or unimproved) with a frequent input of greywater or due to infiltration. Can also be collected from the bottom of septic tanks and leach pits.

- *Semi-solid faecal sludge*
  TS 15-25%, soft paste-like, not pumpable, at the higher range can be spadable, is collected from onsite containments such as pit latrines, composting toilets, and leach pits, or from dewatering treatment technologies.

- *Solid faecal sludge*
  TS >25%: The majority of free water has been removed, can come from dry toilet systems or dewatering treatment technologies. For more details on free water, bound water, and dewatering, refer to Chapter 4.

If TS measurements are taken volumetrically (*e.g.* g/L), then they need to be converted to % as TS. This can easily be done using the density of samples. For example, if a sample with 10 g TS/L faecal sludge has the density of water, then it is equivalent to 1% TS. In this way gravimetric measurements can also be converted to volumetric. When doing such conversions, it is always recommended to measure the actual density of the specific samples, and this becomes even more important with samples at the higher range of % TS.

$$\underbrace{\left(\frac{10\ \text{gTS}}{\text{L}_{\text{sample}}}\right)}_{\text{Volumetric measurement}} \cdot \underbrace{\left(\frac{\text{m}^3}{1{,}000\ \text{kg}}\right)}_{\text{Water density}} \cdot \underbrace{\left(\frac{1{,}000\ \text{L}}{\text{m}^3}\right) \cdot \left(\frac{\text{kg}}{1{,}000\ \text{g}}\right)}_{\text{Conversion units}}$$

$$= \frac{10\ \text{gTS}}{1{,}000\ \text{g}_{\text{sample}}} = \underbrace{\frac{1\ \text{gTS}}{100\ \text{g}_{\text{sample}}}}_{\text{Gravimetric equivalent}} = 1\%\ \text{TS}$$

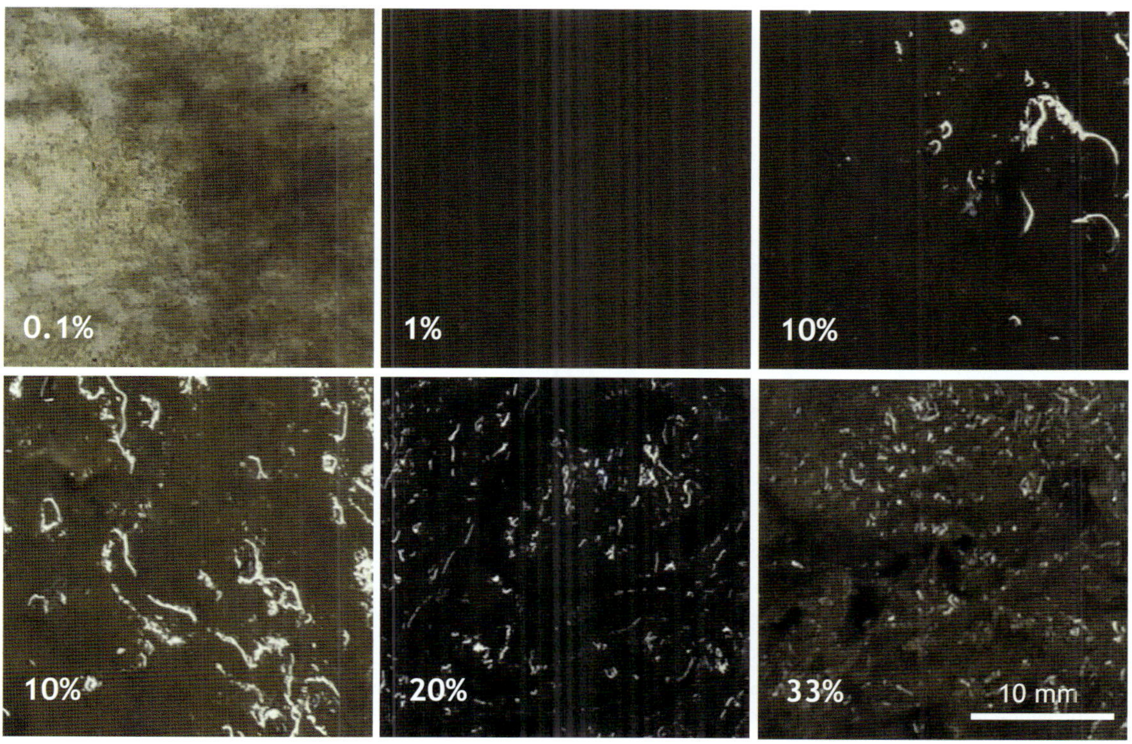

Figure 2.1 Example photos of 10 mL samples of faecal sludge in Petri dishes, in the four different types of faecal sludge by TS concentration: liquid, slurry, semi-solid, and solid. The scale bar is included for reference (Ward *et al.*, 2021).

Presented in Figure 2.1 are examples of what faecal sludge can look like at different TS content. The pictures were taken of 10 mL samples of faecal sludge in a Petri dish, during a study in Lusaka, Zambia (Ward *et al.*, 2021). The samples were collected *in situ* from onsite containments, and have not been treated. In the picture of the 0.1% TS faecal sludge, it is apparent that the sample is fairly dilute. In the 1% TS sample, the colour is more consistent but the texture is still watery. With the 10-33% TS samples there is increasing texture as the solids become more concentrated. In contrast, the appearance is quite different with solid to semi-solid sludge following dewatering (see Figure 2.10 for comparison). It is important to note that although the liquid, slurry, semi-solid, and solid types of faecal sludge are defined by their TS concentration, all the other characteristics do not follow the same trend, and need to be grouped independently. For example, level of stabilisation, or ammonia (NH_3) nitrogen concentration, could be relatively high or relatively low in any of the pictured samples. This is illustrated by the similar texture but difference in colour between the two samples with 10% TS, which is an indication of their differing levels of stabilisation.

2.3 FACTORS INFLUENCING THE FAECAL SLUDGE CHARACTERISTICS ALONG THE SANITATION SERVICE CHAIN

Faecal sludge in general consists of excreta, anal cleansing material, flushwater, greywater, chemicals, and solid waste, in addition to anything else that can end up in the containment, all of which are referred to as 'inputs' to faecal sludge. The diverse practices of individual households, communities, and the commercial sector contribute to the variability of characteristics and volumes of produced, accumulated, and collected faecal sludge. In addition, a wide range of factors along the entire service chain influence the faecal sludge characteristics in a multitude of different ways (Figure 2.2).

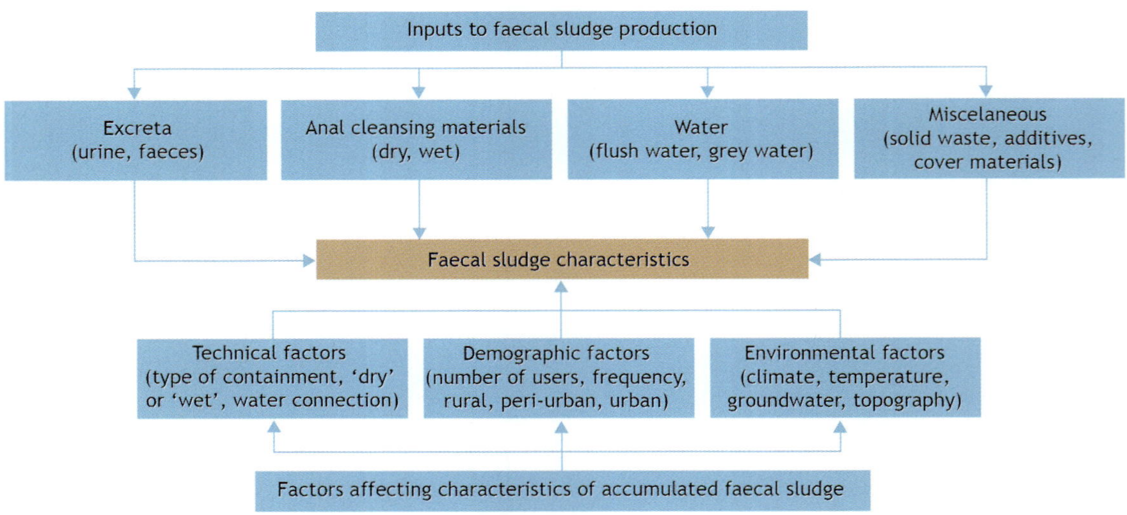

Figure 2.2 Illustration of the inputs to faecal sludge described in Section 2.3.1, and different technical, demographic, and environmental factors affecting and modifying the characteristics of faecal sludge in onsite containment, as described in Section 2.3.2.

What happens during onsite storage of faecal sludge in containment is a complex system. With the current state of knowledge, it cannot be said exactly what the role of each factor is since they are all interrelated. How to start developing this level of knowledge is the topic of Chapter 6, which is focused on developing models of what is occurring at the micro-scale within containment. The important distinction between what is produced versus what actually accumulates in containment is discussed in more detail in Chapter 5. Presented in this section is a brief overview of the overall influence that different factors can have on the diversity of faecal sludge characteristics along the service chain that should be taken into account when determining relevant properties to characterise, together with sampling plans. For specific examples from the literature of the range of reported values of characteristics, the reader is referred to the link of a database provided in Annex 2, open source values for 240 samples in Hanoi and Kampala[1], and the following textbooks: Strande *et al.* (2014), Robbins and Legon (2014), Tayler (2018), and Englund and Strande (2019).

2.3.1 Inputs to faecal sludge production

The first step in the sanitation service chain is the user interface (*e.g.* toilet of any design), which is connected to the onsite containment. In addition to excreta, anal cleansing materials and potentially flushwater, other inputs to the containment can include greywater, solid waste, cover material, and chemicals, as explained in the following sections.

2.3.1.1 Excreta

Excreta consists of urine and faeces that have not been mixed with flushwater, and together are considered to be highly concentrated in both nutrients and pathogens (Tilley *et al.*, 2014, Figure 1.2). Excreta are either collected as mixed urine and faeces, or separately using urine diversion (UD) toilets, with or without the use of flushwater. The amount of urine per person per year ranges from 300 to 550 L/cap.yr, depending on factors such as liquid intake and sweat production (Rose *et al.*, 2015). Yearly production of urine contains 2 to 4 kg of nitrogen depending on the local diet (Tilley *et al.*, 2014). The proportional contribution of urine to faecal sludge will affect the total nutrients and salts, which continue to have an effect on characteristics throughout the service chain. For example, total NH_3 concentration in faecal sludge greater than 3,000 mg/L inhibits anaerobic digestion processes (Colón *et al.*, 2015). The median daily wet mass of faeces produced per person is 128 g, but the reported range is 35-796 g (Rose *et al.*, 2015; Zakaria *et al.*, 2018). Factors affecting the characteristics of faeces include pathogens that can cause diarrhoea, and dietary intake, such as fibres (*i.e.* fruits, grains, vegetables, beans), polysaccharide (*i.e.* starch), and lipid (*i.e.* fats and oils) intake. The type and amount of fibre content can reduce the time that faeces spend in the colon, and increase the size of faeces production and water-holding capacity (Stephen and Cummings, 1979; Stasse-Wolthuis *et al.*, 1980). Although diet has an effect on faeces composition, the overall effect of diet and health on the characteristics of the resulting faecal sludge that accumulates over time in containment has not yet been studied. Detailed information on the chemical and physical properties of faeces and urine are presented in Chapter 7.

2.3.1.2 Water inputs

In some toilet systems, flushwater is used to transport excreta to the containment. The volume of flush depends on the type, there are no standard volumes, but in general the volume increases in the order pour-flush (0.5 L), low-flush (1-2 L), and cistern-flush (6-9 L), with modern versions of cistern-flush as low as 3 L, and older versions of cistern flush going all the way up to 20 L. The mix of excreta, anal-cleansing materials and flushwater that is transported to the containment is called blackwater. If the urine and faeces are collected and/or flushed separately in urine diversion toilets for example, then they are referred to as yellow water and brown water, respectively (Tilley *et al.*, 2014). Additional inputs of water into containment include greywater from food preparation, cleansing, and bathing. Greywater can also contain pathogens from washing diapers, dirty clothes, or food (Gross *et al.*, 2015).

[1] https://doi.org/10.25678/0000tt

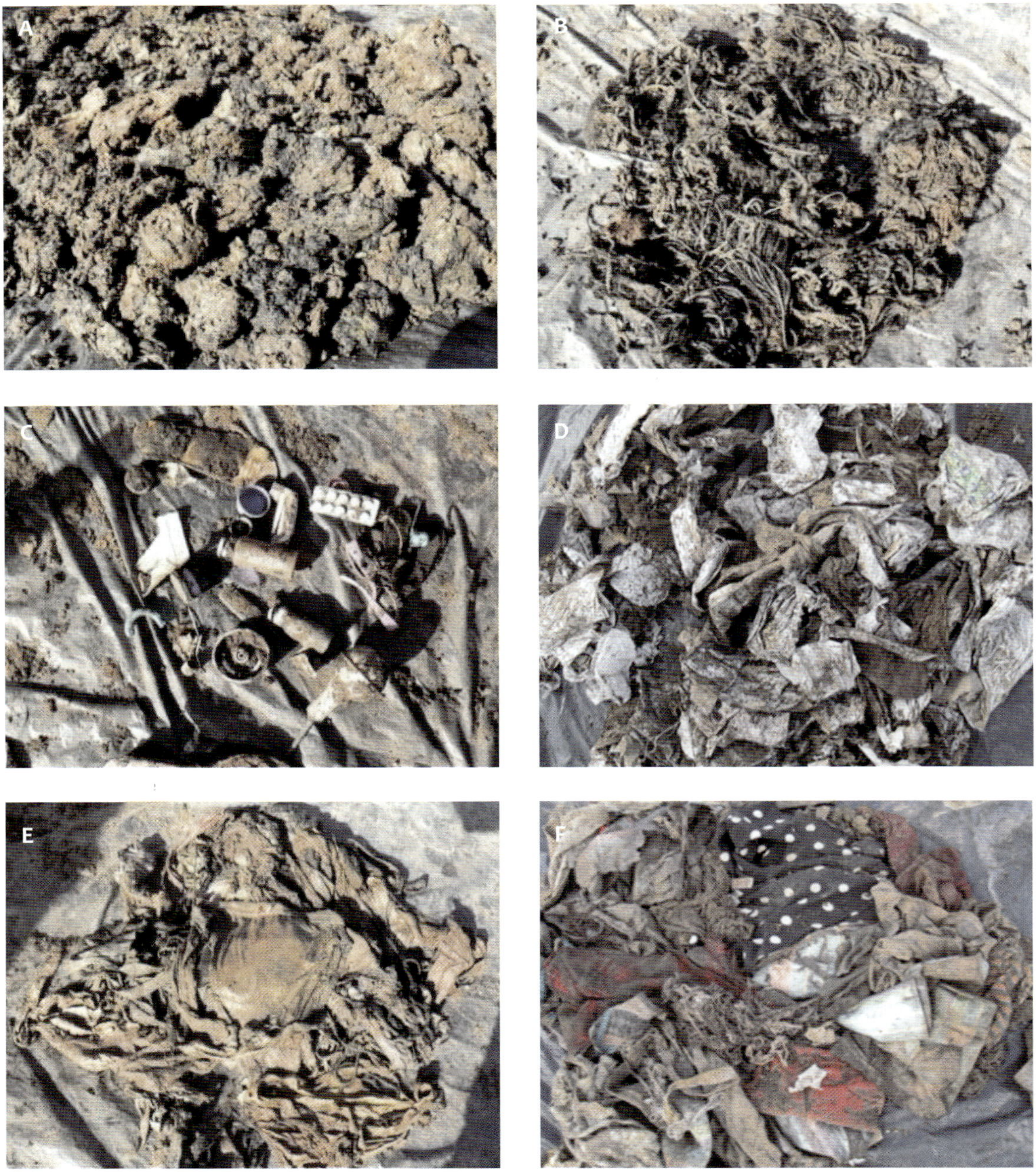

Figure 2.3 Solid waste materials removed from onsite containments in Durban, South Africa: A) paper, B) artificial hair, C) rigid plastics, D) menstrual products and nappies, E and F) textiles (A, B, C and E are from ventilated improved pit latrines; D and F are from a standing UDDT vault, not currently in use), (source: UKZN PRG).

In general, water inputs to containment are much larger with increased availability of water. If, for example, households have to collect water at a standpipe they will tend to use much less water than if they have a direct connection to a water supply pipe. This additional influx of water into containments results in a greater volume of liquid faecal sludge being produced. The resulting increased volumes of liquid faecal sludge are more difficult to safely contain and manage, and can result in increased environmental contamination, whether from outflow of tanks, overflowing containments, or leaching. For an example the reader is referred to the published data set associated with Strande *et al.* (2018) and Englund *et al.* (2020).

2.3.1.3 Anal cleansing materials

Liquid or solid anal-cleansing materials are used by individuals to cleanse themselves after defecating and/or urinating. Liquid materials are water or water mixed with cleansing detergents (Zakaria *et al.*, 2018), usually between 0.5 L and 3 L per use (Tilley *et al.*, 2014). Solid or dry materials can include toilet paper, newspapers, magazines, leaves, and rags, which can be collected and disposed of in the containment or separately from the toilet system. Depending on the culture of anal cleansing, users are in general categorised as 'washers' using liquid, and 'wipers' using solid materials. The accumulation of anal-cleansing materials can affect the characteristics of the faecal sludge, depending on the additional inputs. For example, wet cleansing can result in a higher water content, and dry cleansing a greater concentration of fibres from paper.

2.3.1.4 Additional inputs

The disposal of materials in containments, such as non-biodegradable solid waste (*e.g.* textiles, rags, plastic bags, paper, broken glass, bottles) and food waste is common practice in many low- and middle-income countries (Ahmed *et al.*, 2018). Municipal solid waste management practices also play a role in the amount of solid waste that accumulates in containments. Where affordable solid waste collection exists, there tends to be less waste ending up in the faecal sludge. However, it is difficult to know what is in a containment, without physical removal of the sludge (Bakare *et al.*, 2012). The disposal of solid waste into containments (see the pit latrine example in Figure 2.3) can increase the filling rate, reduce the sludge biodegradation rate, and affect the pit emptying process (Zuma *et al.*, 2015; Radford *et al.*, 2015). Technical factors also play a role, for example there will in general be less solid waste in containment associated with flush toilets such as septic tanks and leach pits, as it is difficult to pass through the water seal syphon (Byrne *et al.*, 2017).

Chemical products also find their way into containments in the form of cleaning materials, or additives that are purposely put into the containment in the belief that they can reduce odours or increase degradation (Anderson *et al.*, 2015). However, there is no evidence that additives are effective. On the contrary, evidence shows that it can have negative results such as impeding the biodegradation process, and the accumulation of undesired gases and odours (Buckley *et al.*, 2008; Grolle *et al.*, 2018; Kemboi *et al.*, 2018).

Cover materials such as soil, ash, sawdust, and garden or agricultural waste are often added to dry systems such as composting and urine diversion and dehydration toilets (UDDT) after each use to combat odour and facilitate the composting process (Stenström, 2004).

2.3.2 Factors affecting characteristics of accumulated faecal sludge

What actually accumulates over time in containment is quite different to the inputs into containment. The difference is the result of a number of demographic, environmental, and technical factors, as depicted in Figure 2.2. Reported examples from the literature include: environmental factors such as oxygen, moisture, climate, inflow and infiltration, soil characteristics; technical factors such as the presence of an overflow pipe, the containment design, sludge age, influent organic matter content, hydraulic retention time, non-biodegradable fraction; and demographic factors such as the number of users, and user behaviour (Brouckaert *et al.*, 2013; Elmitwalli, 2013; Franceys *et al.*, 1992; Gray, 1995; Howard, 2003; Koottatep *et al.*, 2012; Lugali *et al.*, 2016; Nakagiri *et al.*, 2015, Strande *et al.*, 2018). Further factors that affect the resulting quantities and qualities (Q&Q) of accumulated faecal sludge are discussed in Section 2.3.2 on emptying and transport, and chapters 5 and 6.

2.3.2.1 Technical factors

Technical factors such as the type and quality of construction, and whether or not systems are dry or wet (Section 2.3.1.2) will play an interrelated role in contributing to the characteristics of accumulated faecal sludge. Since onsite containments are typically located underground, with little to no manufacturing or construction standards or records, it is difficult to figure out exactly how they were constructed. Care has to be taken, as what is commonly referred to in many countries as a 'septic tank' can actually mean something quite different in the local vernacular, and similarly what is meant by a pit latrine or cesspit is also not standardised. This is discussed in more detail in Example 5.1, and types of onsite containment in Tilley *et al.*, 2014.

Although no clear definitions can be made, major influences on the characteristics of faecal sludge resulting from different types of containment will have to do with whether they are fully-lined, partially-lined, or unlined, and whether or not there is an overflow. If a containment is fully lined with no outlet, it will likely need to be emptied frequently so the sludge will be more 'fresh' or less stabilised and the accumulated faecal sludge will have a lower TS concentration. If a containment is unlined or partially lined, it will be more influenced by soil and groundwater conditions. In more 'wet' systems that include overflows, depending on emptying frequency, layers will form with higher concentrations of TS in a sludge layer at the bottom, and a scum layer at the top, consisting of fats, oil, and grease.

Dry systems are most commonly a type of pit latrine, whereas wet systems can include pit latrines, septic tanks, or cesspits (Nakagiri *et al.*, 2015; Semiyaga *et al.*, 2015; Chiposa *et al.*, 2017). Logically, faecal sludge from dry toilets tends to have higher TS and chemical oxygen demand (COD) content (*i.e.* slurry to solid) than wet systems, and can develop a thick layer at the bottom that is difficult to empty (Brandberg, 2012; Radford and Fenner, 2013). In some regions, composting toilets and UDDT are also common, with accumulated faecal sludge >20% TS (*i.e.* semi-solid to solid). Since the urine is collected separately, UDDT sludge will also have lower concentrations of nitrogen and salts.

The amount of water going into wet systems will depend on the type of flush (Section 2.3.1.2), if greywater goes to the containment, and access to water. The additional water input to the containment means that faecal sludge from wet systems is more dilute (*i.e.* liquid to slurry) than dry systems. In comparison to sludge from pit latrines, septic tank sludge commonly has lower concentrations of TS and COD (Strande *et al.*, 2018; Bassan *et al.*, 2013; Nzouebet *et al.*, 2015; Englund *et al.*, 2020). Faecal sludge with lower TS concentration is more pumpable, which can determine whether or not manual emptying is required (Radford and Fenner, 2013). The level of stabilisation will depend on the emptying frequency, and moisture content will also have an effect on the rates of microbial activity (Byrne *et al.*, 2017; Bakare, 2014).

2.3.2.2 Demographic factors

Studies have found significant differences in faecal sludge and wastewater characteristics based on demographic factors such as number of users and income level (Campos and Von Sperling, 1996; Strande *et al.*, 2018; Englund *et al.*, 2020). Demographic factors may or may not play a direct role in the characteristics of faecal sludge, but can have an indirect effect due to cultural differences, types of dwellings, and land use, for example, septic tanks being located in higher-income areas with more access to household water, and pit latrines in poorer areas with less dilution from greywater (Semiyaga *et al.*, 2015; Strande *et al.*, 2018). In urban areas, pit latrines typically have more users and more frequent emptying than pit latrines in rural areas (Wagner and Lanoix, 1958). This is due to higher population density, increased number of users per household, and increased use frequency. For example, in Kampala there is an average of 30 users per household level latrine, and 82 people per public toilet latrine (Günther *et al.*, 2011). The effect on characteristics can be quite variable, and will also depend on environmental and technical factors.

In addition to faecal sludge that is produced at a household level, it is important to consider sources such as public toilets, restaurants, hotels, schools, hospitals, offices, stores, shopping centres, places of worship, and industrial areas, which will have comparatively different usage patterns. The faecal sludge from restaurants, for example, has a comparatively higher content of fat, oil and grease. Sometimes in establishments with high levels of generated sludge such as commercial areas, hospitals, or industrial areas, the faecal sludge produced is collected in watertight tanks with a very high emptying frequency (Strande *et al.*, 2018), but in contrast, in other locations, industrial and commercial areas have been observed to have lower rates of accumulation (Prasad *et al.*, 2021). Regardless, the non-household contribution represents a significant fraction of generated faecal sludge, and in urban areas the population can double during the day with people commuting in to the city for work. At the Lubigi faecal sludge treatment plant (FSTP) in Kampala, Uganda, 50% of the faecal sludge was found to originate from non-household sources (Strande *et al.*, 2018).

2.3.2.3 Environmental factors

Environmental factors such as climate, geology, groundwater table and topography, and combinations of these factors, can have a direct impact on the characteristics of faecal sludge. The extent of the impact will vary depending on the local conditions and the type of containment. For example, biological degradation of faecal sludge will depend on anaerobic conditions, temperature, total moisture, and inhibitory compounds (Bourgault, 2019; Bourgault *et al.*, 2019; Byrne *et al.,* 2017; Van Eekert *et al.*, 2019; Bakare, 2014). Moisture content is also dependent on the net inflow and outflow (or infiltration) of moisture, which depends on soil type, type of lining used, local topography, and groundwater level. Infiltration into containment from groundwater with a high water table can lead to the 'floating' of faecal sludge fractions in pit latrines and increase the water content of the sludge (Chirwa *et al.*, 2017). Groundwater tables also fluctuate by season, which can result in different groundwater hydraulic conditions that can influence sludge characteristics throughout the year. Sandy soils are more permeable and allow for a higher exchange of water and gases, whereas clay-dominated soils are much less permeable and limit the exchange. Rainfall directly affects the groundwater table, and runoff from steep slopes can enter the containment through toilet openings or access ports. These factors are accounted for in the modelling approaches described in Chapter 6.

2.3.2.4 Variability of accumulated faecal sludge

The result of the demographic, environmental, and technical factors that influence characteristics of faecal sludge is a high level of heterogeneity that complicates characterisation. As shown in Figures 2.4 and 1.3, there is often no 'standard range of variation' for particular properties, and findings from one study cannot necessarily be used as a base of comparison to another. This is shown in Figure 2.4 with the level of variation of COD, ash content, moisture content, and calorific value in Durban, South Africa (Velkushanova *et al.*, 2019; Zuma *et al.*, 2015). Each data point represents the results of analysis from one faecal sludge sample, collected from the following containments: dry ventilated improved pit latrines (red); wet ventilated improved pit latrines (green); community ablution blocks (blue); urine-diverting dehydration toilets (UDDT, yellow); ventilated improved pit latrines in schools (purple); and unimproved pit latrines (turquoise). The mean value for each type of faecal sludge sample is presented as a dotted line in the respective colour. The level of variation is even higher within samples collected from the same type of onsite sanitation system than in comparison to other containments, which raises the question whether it is even possible to find statistical relations or predictors in this data. More details are presented in Chapter 5 on approaches and techniques for collecting and processing community to city-wide data sets of faecal sludge characteristics.

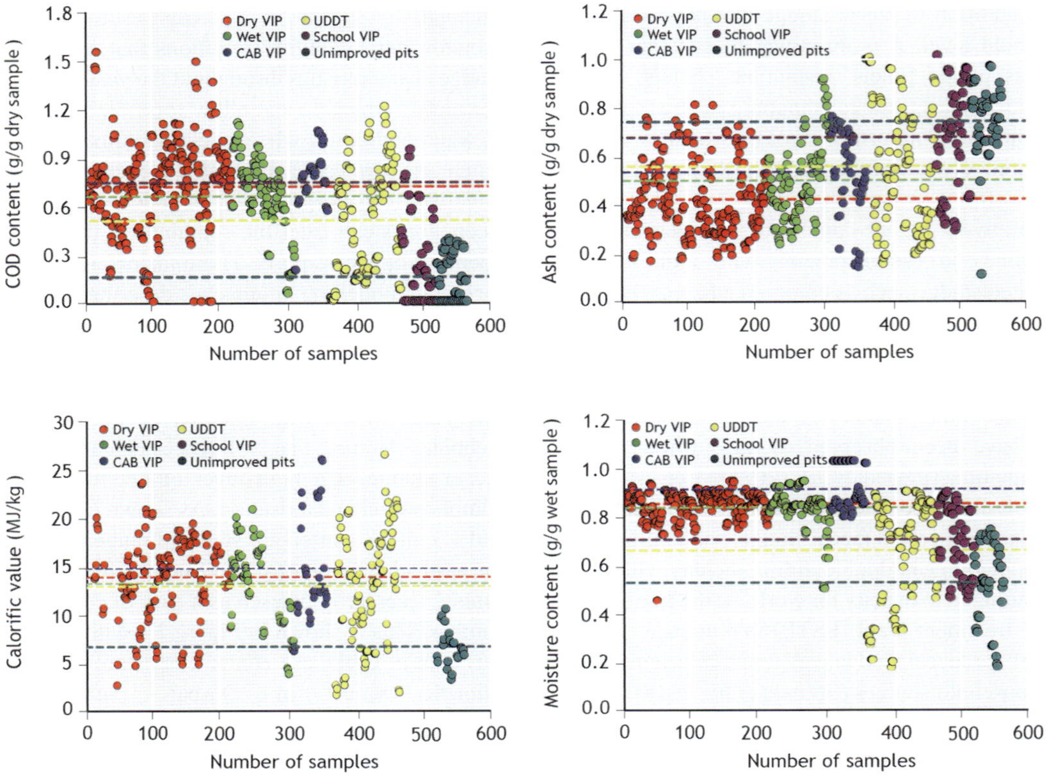

Figure 2.4 Variation of COD, ash content, moisture content, and calorific value properties of faecal sludge from different types of containment, collected in Durban, South Africa. Collectively, the total number of analysed replicates was 564, with a total of 188 samples all together for all the containment types, collected from different sections and depths within the containments. Each of these samples were analysed for properties such as moisture content, TS/VS, organic content such as COD, TKN, pH and electrical conductivity, thermal conductivity, calorific value, nutrient content, rheological properties and viscosity, and helminths (source: UKZN PRG[2]).

2.3.2.5 Developments and innovations in onsite containment

Some emerging innovative sanitation technologies combine the user interface ('front-end') with containment ('back-end'), to simultaneously contain and treat excreta onsite. For example, systems that are based on flush-type toilets can include membrane and other treatment processes to re-use the flushwater. One technology example is the nano-membrane toilet by Cranfield University (Figure 2.5, Parker, 2014). The user interface is a pedestal toilet with a waterless swiping flush mechanism, with waste-processing components housed within the pedestal. The solids are extracted by an auger, and then dried and combusted with only a small amount of ash remaining. The liquids are preheated and purified with a hydrophobic membrane, which is reusable. This system has been tested in communities in Durban, South Africa (Hennigs *et al.*, 2019; Mercer *et al.*, 2018), along with other innovative toilet systems, such as the Blue Diversion Autarky (Reynaert *et al.*, 2020), and a household-scale onsite blackwater treatment system (Sahondo *et al.*, 2020; Welling *et al.*, 2020). If implemented at scale, these types of technologies could have a dramatic impact on the Q&Q of faecal sludge that accumulate, with the goal to eliminate accumulation as much as possible.

[2] https://osf.io/uy7t2/

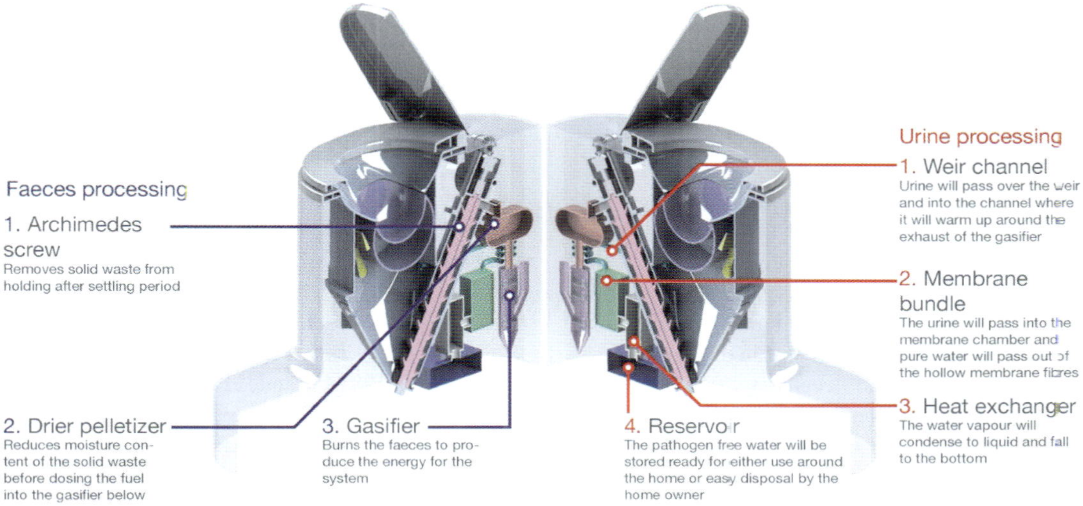

Figure 2.5 Nano Membrane toilet: an example of a waterless self-contained toilet (source: Cranfield University[3]).

Another example is the Solar Septic Tank - a technology aiming to enhance the degradation of solids and increase the quality of effluent by passive solar heating to 50-60 °C (Connelly *et al.*, 2019, Figure 2.6). The heating promotes enhanced microbial degradation of both soluble compounds and retained solids, as well as partial pasteurisation of the liquid effluent prior to discharge. This technology has been installed and tested in Bangkok, Thailand and reported average removal efficiencies of total COD, soluble COD, and total biochemical oxygen demand (BOD) are between 90-99% over one year period (Koottatep *et al.*, 2020).

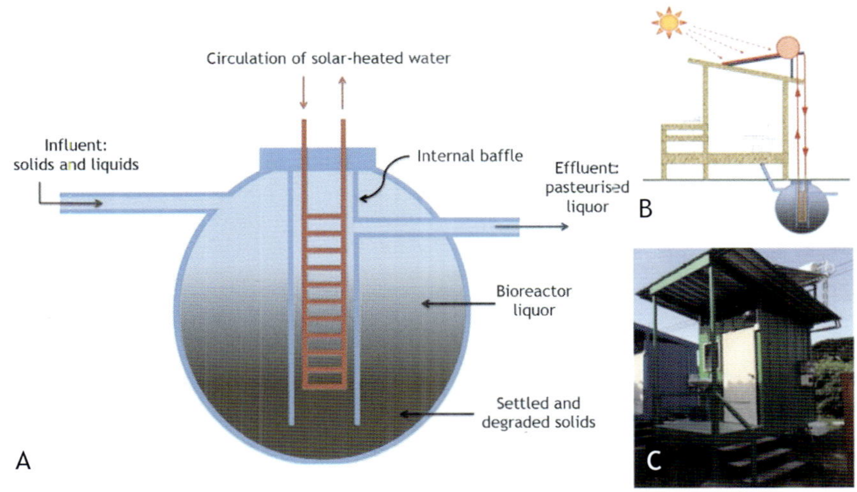

Figure 2.6 Principles of the Solar Septic Tank: (A) principles of solar heating applied to SST; (B) illustration of the buried septic tank and solar collection unit on the toilet roof; (C) installation of the SST in the field test site (source: AIT).

---

[3] http://www.nanomembranetoilet.org

2.3.3 Emptying and transport

The emptying of faecal sludge from the onsite containment, followed by transportation to treatment, is the next step in the sanitation service chain.

2.3.3.1 Storage time or emptying frequency

The emptying frequency of sludge in onsite containments defines the sludge storage time, residence time, or 'age' of accumulated sludge. Depending on the type of containment, accessibility and usage patterns, sludge remains in the containment anywhere from days to weeks, to years or even decades (Taweesan *et al.*, 2015; Strande *et al.*, 2014; Tayler, 2018). With increased residence time in the containment, the sludge will be more stabilised, with rates of stabilisation depending on environmental factors. Rates of biodegradation impact nutrient cycling and stabilisation, which affect the dewaterability properties of faecal sludge and its suitability for treatment with different technologies. Fresher sludge is frequently observed to have poor dewatering performance due to the level of stabilisation (Ward *et al.*, 2019, Chapter 4). Systems with a high number of users such as public toilets or commercial enterprises will also be more frequently emptied, meaning that the sludge will be 'fresher' and not as digested as older sludge. However, the faecal sludge accumulated in public toilets does not fit into one type of faecal sludge and will vary in characteristics depending on the type of containment technology, local context, and other environmental factors (Appiah-Effah *et al.*, 2014; Heinss *et al.*, 1998; Strauss *et al.*, 1997; Strande *et al.*, 2018). An example of public toilets are community ablution blocks (CAB) in Durban, South Africa (Figure 2.7). The CAB is a system that uses old shipping containers as a superstructure equipped with toilets, wash basins and showers (Starkl *et al.*, 2010). Since high volumes of greywater from bathing and laundry are inputs to the containment, the faecal sludge is classified as liquid with low TS.

Figure 2.7 Community ablution blocks in Durban, South Africa (source eThekwini Municipality, photo: UKZN PRG).

2.3.3.2 Manual or mechanical emptying

The method of emptying can influence faecal sludge characteristics, and *vice versa* the characteristics of faecal sludge in the containment can dictate possible methods of emptying (Zziwa *et al.*, 2016; Balasubramanya *et al.*, 2016; Chipeta *et al.*, 2017). If faecal sludge is too thick it is not pumpable and will require manual emptying (e.g. measured as moisture content, viscosity, or rheological properties) (Bosch and Schertenleib, 1985; Radford and Fenner, 2013). Excessive amounts of solid waste can also prevent pumping due to blockage or breakage of the sludge emptying equipment (Ahmed *et al.*, 2018). In addition, if the site is not accessible by larger vehicles (*e.g.* trucks), it will also require manual emptying. Mechanical collection with vacuum trucks is also not possible if the solid content of faecal sludge is too high (Mikhael *et al.*, 2014). Due to these limitations on which type of emptying technologies can be used, faecal sludge that is collected mechanically can have different properties to faecal sludge that is collected manually. In adition, faecal sludge demonstrates shear thinning characteristics (meaning that it can become more liquid with an increasing shear rate), which can result in changes in viscosity of faecal sludge after mechanical collection (Septien *et al.*, 2018a). Another example is the addition of water into containments before emptying to dilute the sludge and make it easier to remove. This results in modified characteristics of faecal sludge, such as higher moisture content and reduced viscosity. Based on factors such as thickness, depth of containment, and affordability of service, the sludge is also not always entirely removed (Nakagiri *et al.*, 2015; Semiyaga *et al.*, 2015; Chiposa *et al.*, 2017). For example, in Durban, it was observed that sludge in the bottom of pit latrines was the oldest and most stabilised, compared to the upper layers of the pit latrine containment (Buckley *et al.*, 2008; Bakare *et al.*, 2012).

2.3.3.3 Transportation

Transportation can be done manually with carts, or motorised with trucks (Mikhael *et al.*, 2014). The effect of transportation on faecal sludge characteristics is not clear, but samples taken from transport trucks have different concentrations of TS and COD than those taken directly from containment (see Case Study 3.3). Solids also separate out in the bottom of vacuum trucks during transport. Another possibility for increasing the efficiency of transport is transfer stations. Possibilities include a tank installed for delivery of sludge by manual emptiers who cannot transport sludge long distances, which could then be transferred to treatment by trucks, and/or as a dewatering step with supernatant going to a sewer and dewatered sludge being transported to treatment (Strande, 2017). There are not yet many examples of successful implementations. However, one that is currently being field-tested in Nairobi, Kenya appears promising (Junglen *et al.*, 2020).

2.3.3.4 Innovations in faecal sludge emptying and transportation

Emptying of faecal sludge, particularly mechanical emptying, is challenging due to inaccessibility, high TS and solid waste content and the high heterogeneity of faecal sludge characteristics, which makes it difficult to have sludge emptying technologies that are uniform for all types of faecal sludge and onsite sanitation technologies.

A number of innovative technologies for faecal sludge emptying are trying to address these challenges, in order to empty sludge with a higher TS and level of stabilisation from pit latrines. For example, the Flexcrevator is a technology developed by North Carolina State University (Sisco *et al.*, 2017; Rogers *et al.*, 2014; Portiolli, 2019; Figure 2.8). It consists of a vacuum tank, external extruder and a flexible screw, operating simultaneously to extract the faecal sludge while pushing away the solid waste materials. In this way the sludge is emptied while the solid wastes remain in the containment. The Flexcrevator is relatively small in size to enable access to containments that normally cannot be reached by vacuum tankers.

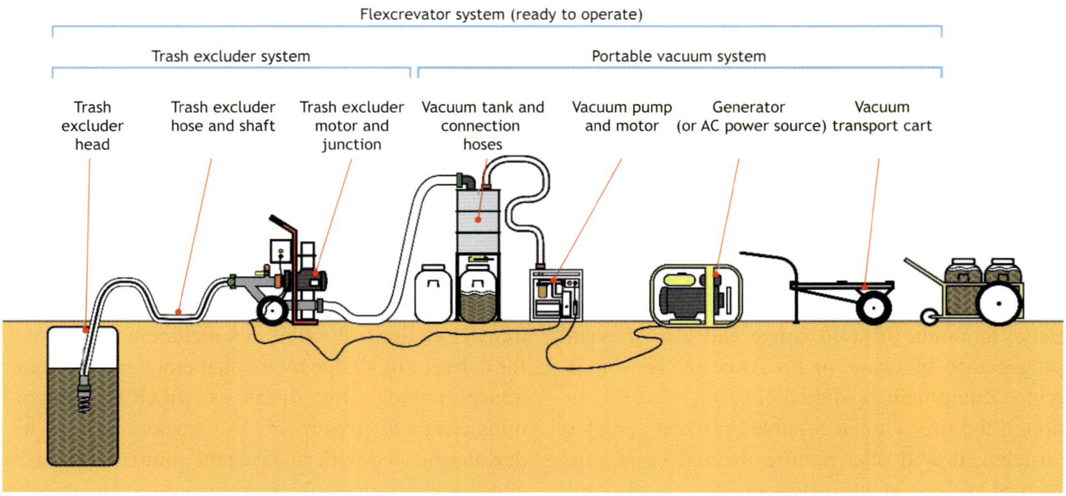

Figure 2.8 Innovative technology for faecal sludge emptying - the Flexcrevator (source: North Caroline State University[4]).

2.3.4 Treatment and end use

2.3.4.1 Faecal sludge treatment plants

There are several technology options for the treatment of faecal sludge. Faecal sludge treatment plants that are currently in operation are commonly decentralised or semi-centralised, with the faecal sludge being delivered by trucks following collection. The four main treatment objectives are stabilisation, nutrient management, pathogen inactivation, and dewatering/drying (Niwagaba *et al.*, 2014; Strande, 2017). The characteristics of faecal sludge during treatment will be significantly different, and depend on the treatment objectives and location in the treatment chain (see example 3.5.3). A typical treatment chain includes preliminary separation, settling-thickening tanks, drying beds, with the leachate going to treatment in stabilisation ponds and/or co-treatment with wastewater, and resource recovery or disposal of the dewatered sludge (Klinger *et al.*, 2019). For concerns related specifically to characteristics of faecal sludge regarding treatment potential, the reader is referred to the following freely available reference books: Tayler (2018); Strande *et al.* (2014); Englund and Strande (2019); Robbins and Legon (2014); Polprasert and Koottatep (2017); and Narayana (2020).

Preliminary separation processes usually include screening to remove large objects and waste from the sludge. Solids that are removed in settling-thickening tanks varies depending on the specific characteristics of faecal sludge (Dodane and Bassan, 2014; Gold *et al.*, 2018; Ward *et al.*, 2019). This is an important distinction, as different types of sludge have widely varying characteristics and are not comparable. How properties such as different redox conditions, level of stabilisation, biomass, nutrients, particle size, undigested plant fibres, salts and ions, and extracellular polymeric substances (EPS) affect dewaterability is not yet fully understood (Bourgault *et al.* 2019; Ward *et al.*, 2019). An example of ranges of dewaterability is provided in Figure 2.9. The turbidity of the supernatant of faecal sludge samples following centrifugation in the laboratory appeared quite different. In addition, the more stabilised sludge dewatered more quickly, and the less stabilised sludge had more clogging of filters, possibly due to higher concentrations of EPS (Ward *et al.*, 2019). As scientific knowledge is advanced, the use of conditioners will be possible to reduce total suspended solids (TSS) in the effluent of settling tanks, and reduce drying time on drying beds (Gold *et al.*, 2016; Ward *et al.*, 2021).

[4] https://www.globalinnovationexchange.org/innovation/flexcrevator-a-pit-emptying-device

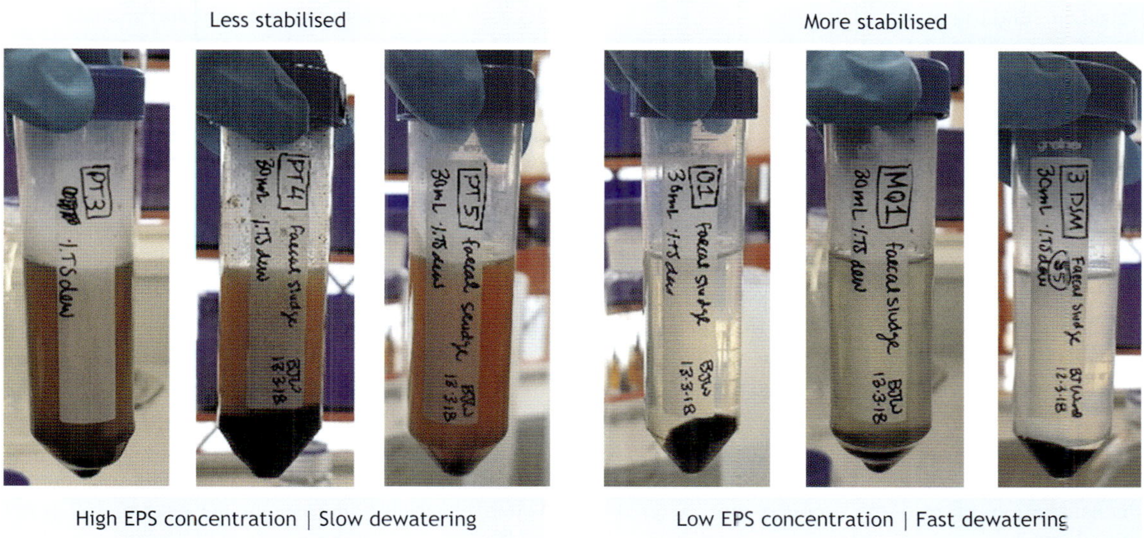

Figure 2.9 Comparison of supernatant turbidity following centrifugation of faecal sludge samples in Dakar, Senegal (source: Ward *et al.*, 2019).

The TSS that separate out in settling-thickening tanks are loaded batch-wise onto unplanted drying beds, or continuously onto planted drying beds (Englund and Strande, 2019). The leachate that percolates through the drying beds requires further treatment, as it is high in salts, organic content, nutrients, and pathogens, with loadings similar to influent concentrations of wastewater treatment plants (Kengne *et al.*, 2014; Seck *et al.*, 2015; Sonko *et al.*, 2014; Thomas *et al.*, 2019). The leachate is usually treated together with the supernatant from the settling-thickening tanks. Following successful dewatering, sludge on drying beds is semi-solid to solid, and can be removed by hand or with a shovel, as shown in Figure 2.10. Other established treatment technologies include co-composting with organic solid waste (Nikiema *et al.*, 2014).

Figure 2.10 Removal of semi-solid to solid dewatered sludge from drying beds at Camberene treatment plant in Dakar, Senegal (photo: Eawag).

2.3.4.2 End use or disposal

As shown in Table 2.1, there are many possibilities for resource recovery from faecal sludge, and research is actively taking place on improving recovery as energy (Andriessen *et al.*, 2019; Krueger *et al.*, 2020; Onabanjo *et al.*, 2016;), nutrients and organic matter (Nikiema *et al.*, 2014; Orner and Mihelcic, 2018; Hashemi and Han, 2019; Roy *et al.*, 2019; Simha *et al.*, 2017), and animal fodder as black soldier fly and plants (Lalander *et al.*, 2013 and Gueye *et al.*, 2016). Characteristics of concern will be dependent on the final end use; for example, for use as a fuel, the water content and calorific value are important to evaluate (Murray Muspratt *et al.*, 2014), whereas for use as a soil amendment, pathogens and heavy metals are important. Further examples of characteristics for the consideration of resource recovery are covered in Section 2.4.

Table 2.1 Potential faecal sludge treatment products and type of resource recovery (source: Schoebitz *et al.*, 2016).

Resource	Treatment product	Product type
Energy	Solid fuel	Pellets, briquettes, powder
Energy	Liquid fuel	Biogas
Energy	Electricity	Conversion of biogas or gasification of solid fuel
Food	Protein	Black soldiers flies, fish meal
Food	Animal fodder	Plants from drying beds, dried aquaculture plants
Food	Fish	Fish grown on effluent from faecal sludge treatment
Material	Building materials	Additive to bricks, road construction materials
Nutrients	Soil conditioner	Compost, pellets, digestate, black soldier fly residual
Nutrients	Fertiliser	Pellets, powder
Nutrients	Soil conditioner	Untreated sludge, dewatered sludge from drying beds
Water, nutrients	Reclaimed water	Effluent from faecal sludge treatment

2.3.4.3 Innovations in treatment and end use

Several innovative and emerging faecal sludge processing technologies have been developed to treat faecal sludge at scale. Some of them are based on unconventional faecal sludge processes such as hydrothermal oxidation, pyrolysis, gasification, combustion thermal drying, infrared irradiation, microwave irradiation, black soldier fly larvae and vermicomposting, to reduce the sludge volume and pollutants, inactivate pathogens and convert the sludge components into valuable resources (Hiolski, 2019; Mawioo *et al.*, 2017; Fakkaew *et al.*, 2018; Septien *et al.*, 2018b; Yadav *et al.*, 2012). For example, the omniprocessor is a faecal sludge treatment technology using combustion that treats human waste and produces drinking-water quality water, electricity and ash. In the case of full water reclamation, it is important to evaluate characteristics for the protection of public health, including pathogens, heavy metals, and pharmaceuticals. The pilot of this technology is installed in Dakar, Senegal (Figure 2.11). Other examples of innovations are included in chapters 4, 5 and 6.

Figure 2.11 The omniprocessor faecal sludge treatment system in operation in Dakar, Senegal (photos: UKZN PRG and Sedron Technologies).

2.3.4.4 Container-based sanitation (CBS)

The business model and technology implementations for container-based sanitation (CBS) have rapidly progressed over the last decade, and are now classified as a type of improved sanitation facility by the Joint Monitoring Programme (Figure 2.12, Russel *et al.*, 2020; World Bank, 2019, Brdjanovic *et al.*, 2015).

Faecal sludge from CBS tends to have a much higher TS content than other faecal sludge, as most CBS toilets do not collect flushwater and grey water, and many are also urine-diverting with a dry desiccant as cover material. For example, the average TS content observed in Sanivation toilets, a Kenyan-based CBS service provider, is 60% (personal communication, Woods E.). Other differences include much less solid waste mixed in with the faecal sludge, and a higher C/N ratio due to carbon-rich cover material (*e.g.* ash, saw dust, bagasse). Faecal sludge is transported manually with trolleys, pickups, or tuk-tuks in containers to treatment plants (Figure 2.12). Therefore, it tends to arrive at the faecal sludge treatment plants in relatively small batches throughout the day.

If off-grid, self-contained solutions are successfully scaled up, it could significantly impact the faecal sludge management service chain. The considerations for characterisation are specific to the technology and operation, and the design of such systems will also potentially be context-specific based on regional characteristics, as described in Section 2.1.1.

2.3.4.5 Summary of technologies along the sanitation service chain

A wide range of technologies that correspond to management of faecal sludge at each step in the sanitation service chain are summarised in Table 2.2. There are varying levels of knowledge as to the effects of different technologies on the characteristics of faecal sludge. As also presented in chapters 1 and 4, based on the current operational experience and practical knowledge, they can be grouped into established, transferring, and innovative technologies (WHO, 2018).

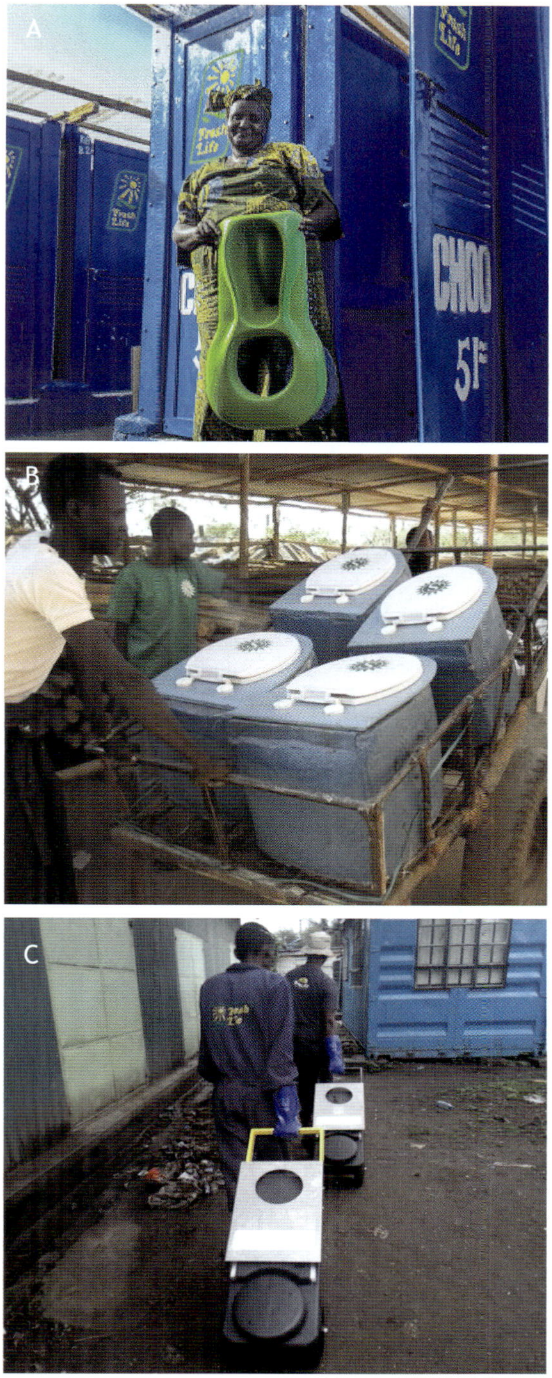

Figure 2.12 Examples for CBS from Sanergy (A) and Sar ivation (B), both based in Kenya, and (C) eSOS Smart Toilet field-testing in Nairob (photos: World Bank and IHE Delft).

Table 2.2 (Part 1 of 2) Examples of established, transferring, innovative and container-based sanitation technologies along the sanitation service chain.

Technology application	Toilet (user interface)	Collection & storage (containment)	Emptying & transport (conveyance)	Treatment	End use and/or disposal
Established and transferring	*Dry* • Open hole pedestal • Open hole squatting • Urine-diversion dry • Urinal *Water-based* • Pour flush toilet • Low flush toilet • Urine-diversion flush toilet • Cistern flush	*Waterless* • Ventilated improved pit latrine (VIP) • Composting toilet • Urine storage tank • Urine diversion and dehydration vault *Water-based* • Pit latrine • Tank • Septic tank • Leach pit • Soak pit • Aqua privy	*Manual* • Shovel • Bucket • Cart for transportation • Sludge gulper • Diaphragm pump • Nibbler • MAPET • Hook and claw *Mechanised* • Vacuum tanker • Vacutug • Micravac • Motorised diaphragm • Trash pump • Gobber • Motorised screw auger	*Established* • Settling-thickening tank • Stabilisation pond • Unplanted drying bed • Planted drying bed • Co-composting *Transferring* • Mechanical dewatering • Conditioners • Alkaline treatment • Lime stabilisation • Incineration • Anaerobic digestion • Pelletising • Thermal drying	*End use products* • Biogas • Compost • Treated leachate • Ash • Fodder/animal feed • Effluent *Disposal* • Landfill • Burial
Emerging and innovative	*Front-end component* • EOOS • Urine-diversion toilets • Waterless flush toilets • Nano Membrane Toilet • eSOS Smart Toilet • MEDiLOO	*Back-end treatment* • Gasification • Biogas reactor • Anaerobic baffled reactor • Peepoo • Compost filter • Black soldier fly larvae • Hydrothermal carbonisation • Microwave treatment • Microbial fuel cells • Nanomembrane • Membrane bioreactor • Bioelectrical processing • Dry combustion • Drying			*End use products* • Hydrochar • Biochar • Biogas • Biodisel • Liquid fertiliser • Protein • Animal feed • Oil • Electricity • Heat • Purified water

Table 2.2 (Part 2 of 2) Examples of established, transferring, innovative and container-based sanitation technologies along the sanitation service chain.

Technology application	Toilet (user interface)	Collection & storage (containment)	Emptying & transport (conveyance)	Treatment	End use and/or disposal
Emerging and innovative			*Emptying of faecal sludge from established onsite containments* • Modified sludge gulper • Extraction auger • Flexcevator	*Treatment at scale of faecal sludge from established onsite containments* • Omni Processor • Supercritical water oxidation • Black soldier fly larvae • Vermicomposting • Infrared radiation (LaDePa) • Urine treatment (struvite reactor) • Hydrothermal carbonisation • Combustion • Pyrolysis • Microwave radiation (Shit Killer, Tehno Sanitizer)	*End use products* • Hydrochar • Biochar • Biogas • Biodisel • Liquid fertiliser • Protein • Animal feed • Oil • Electricity • Heat • Purified water
Emerging and innovative Container-based sanitation (CBS)	*CBS toilets* • Sealed container, often urine-diverting • Waterless • Usually portable	*Emptying* • Regular collection via service provider • Replace full containers with empty clean containers	*Transport* • Push carts • Collection depot • Large transport vehicle • Full containers are sealed and transported to treatment or disposal site	*Treatment* • Various treatment processes from pathogen reduction to full resource recovery (*e.g.* thermophilic composting, urine nitrification) • Containers are emptied, cleaned and disinfected before reuse	*End use and products* • Compost • Biogas • Biomass fuel • Animal feed • Phosphorus and nitrogen from urine

2.4 PROPERTIES OF FAECAL SLUDGE AND SELECTING METHODS OF CHARACTERISATION

The characterisation and understanding of the properties of faecal sludge as a material is crucial for the provision of integrated faecal sludge planning, management and treatment solutions through the entire sanitation service chain. The first step in the characterisation process of faecal sludge is to determine the purpose and the objectives of the characterisation (Figure 2.13). The purpose is the reason, for example, selecting and designing a faecal sludge treatment technology, with the objective to maximise valorisation potential. Common reasons for characterising faecal sludge could involve setting up a monitoring program at a treatment plant, defining a research question, designing and developing new processes or technologies, or collecting data to design an integrated faecal sludge management plan. Specific examples of characterisation objectives include:

- Understanding biochemical processes of degradation and nutrient cycling
- Evaluating faecal sludge stabilisation with location and time in onsite containment technologies
- Planning of emptying services for a community
- Selecting the best technology for emptying of sludge from onsite containments
- Designing an innovative toilet and containment solution
- Designing a new technology for emptying or treatment
- Designing a new faecal sludge treatment plant
- Determining loadings for the operation of a treatment plant
- Evaluating operational parameters during the start-up phase of a faecal sludge treatment plant
- Monitoring a treatment plant for overall treatment efficiency and pathogen removal
- Evaluating potential for resource recovery
- Assessing compliance with requirements for end use
- Quantifying resource recovery value (*e.g.* energy, food, nutrients, water).

Once the purpose and the objectives are defined, then the type of properties to measure in the characterisation process can be determined. For example, if the purpose is the design of a thermal treatment technology for resource recovery as a fuel, important parameters to measure include moisture content, TS, VS, thermal conductivity, heat diffusivity and calorific value. In this particular case, the measurement of COD will be of secondary importance. On the other hand, if the purpose is to design an anaerobic digester, the total bio-degradable organic matter will be important to determine, and can be evaluated with analytical methods such as BOD, COD, and volatile solids. It would also be important to measure moisture content, TS, TSS, NH_3, and other macro- and micro-nutrients. In this case, there is no need to measure thermal conductivity and calorific value of the faecal sludge, because these properties are not directly related to the design parameters of anaerobic digestion.

In this book, faecal sludge properties are grouped into three main groups: (*i*) chemical and physico-chemical, (*ii*) physical, and (*iii*) biological, details of which are provided in Section 2.4.1.

The next step in the characterisation process is the selection of suitable methods for analysis, based on factors such as type of faecal sludge (based on TS), level of accuracy of the required results, costs of analysis, and laboratory capacity (see Section 2.4.2). The selection of methods is an essential part of the planning process before undertaking the sample collection, as it involves considerations such as budget and time restrictions, and the availability of instruments and trained personnel to undertake the analysis. Figure 2.14 provides an overview of this decision-making process.

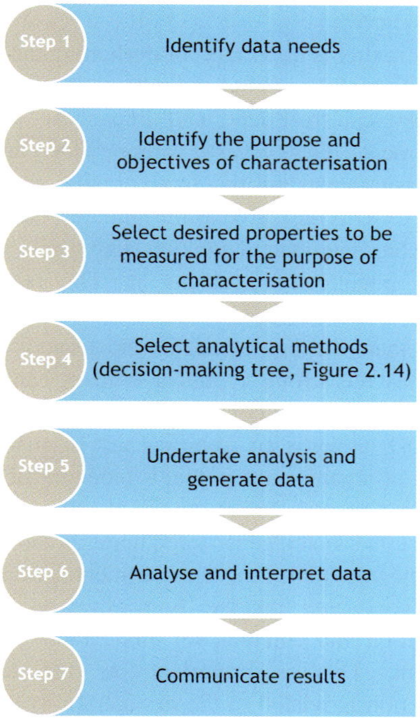

Figure 2.13 Steps in the faecal sludge characterisation process. Further information on how to select analytical methods (Step 4) is provided in Section 2.4.2. Further information on the integrated approach for data collection, analysis and interpretation within the entire book is provided in Chapter 1, and further information on integrating characterisation into a sampling plan is included in Chapter 3.

After selecting suitable methods for the purpose of characterisation, the next steps are undertaking the analysis, followed by data analysis and interpretation to fulfil the purpose of characterisation. The laboratory methods for the analysis of faecal sludge presented in this book are summarised in Section 8.4, Table 8.3, with cross references to where they are located in Chapter 8. Many methods have been adapted from methods for water and wastewater, in addition to soil and food science. The methods presented here are the first step towards standardisation of methods and procedures for faecal sludge analysis. As the need for additional methods arises, they will also need to be developed or adapted from standard methods. One of the challenges of adapting methods is the high heterogeneity of faecal sludge characteristics, which requires special care. Examples are steps for sample homogenisation, filter size due to clogging, and sample volume for representativeness. For more information on developing methods, refer to the tips for adapting methods specific to faecal sludge included in Chapter 8, and standard method 1040 on development and evaluation in Standard Methods for the Examination of Water and Wastewater (Rice *et al.*, 2017). As more methods become established, they will be included in future editions of this book. It is important to keep in mind, even when following established methods for faecal sludge, that they need to be adapted for the local and institutional context. For example, in Lusaka the temperature had to be increased near the end of TS drying time due to swelling of the faecal sludge (Ward *et al.*, 2021). For information on sampling handling and preparation, refer to Chapter 3 and Chapter 8.

2.4.1 Faecal sludge properties

Following is a brief discussion of the chemical and physico-chemical, physical, and biological, properties of faecal sludge and their relevance to the management of faecal sludge.

2.4.1.1 Chemical and physico-chemical properties

Chemical properties refer to properties of materials that change as a result of chemical reactions, for example oxidation state, and whether they are flammable, corrosive, radioactive, or an acid or base. Physico-chemical properties are dependent on both physical (see Section 2.4.1.2) and chemical processes, and are determined by the interactions of components within faecal sludge.

Solids and moisture content

Fractions of TS and moisture content are important for determining appropriate emptying methods for onsite containment technologies, loadings of technologies such as drying beds and settling-thickening tanks, and to evaluate dewatering and drying performance. As defined in Section 2.2, and further explained in Chapter 8, the four defined types of faecal sludge by TS are also used to determine analytical methods, and sample preparation and handling (liquid TS <5%, slurry TS 5-15%, semi-solid TS 15-25%, and solid TS>25%).

The moisture content of faecal sludge is highly variable, resulting in uncertainties when expressing different properties based on the total volume or mass. For more liquid samples, the volumetric method is used because it provides more precision, with concentrations reported as gTS/L total sample volume. For semi-solid or solid samples, the gravimetric method is more precise and concentrations are reported as gTS/g total wet mass of the sample. The density can be used to convert between volumetric and gravimetric for comparison to values in the literature. Total solids can be divided into categories based on organic content (volatile or fixed), and based on physical properties (suspended and dissolved). Total solids can be fractionated into total fixed solids and volatile solids by ignition at 550 °C. Total fixed solids (ash) are the material left behind after ignition, and are the minerals that do not biodegrade over time (*e.g.* inorganic inputs and soil in pit latrine samples). Volatile solids are volatised during ignition at 550 °C and are an indicator of the biodegradability of samples. Care has to be taken not to directly transfer empirical relations from wastewater, as the VS/TS ratio of faecal sludge is heavily influenced by the wide range of inorganic substances in samples. Dissolved and suspended solids are defined by their physical properties. Total solids can be fractionated into total dissolved solids (TDS) and total suspended solids (TSS) through filtration. TDS are defined as being the solids contained in the filtrate that passes through a filter with a pore size of 2.0 μm or less, whereas TSS are not as well defined. Suspended solids are defined as those that do not pass through a filter, but the pore size of filter paper ranges from 0.45 to 2.0 μm due to the clogging of filters with thicker samples. This is why it is especially important to document with clear methods exactly how analysis was carried out.

The moisture content will directly and indirectly affect the biodegradability and viscosity of faecal sludge, the solid-liquid separation and dewaterability potential, pumpability, viscosity, shear thinning, mixing, and drying. Steps for measuring and calculating moisture content of different fractions of solids (TS, VS, TSS) are provided in Chapter 8.

Organic content

Organic matter is important for evaluating the level of stabilisation of faecal sludge, biodegradation potential for biological treatment, and impact on receiving environments. Total organic carbon (TOC) and COD are measurements of the total organic fraction of carbon. COD is measured as the amount of an oxidant (*e.g.* dichromate in acid solution) that reacts with the sample, chemically oxidising it. The results are reported in oxygen equivalents. COD will always be greater than the biodegradable fraction of organic matter, as the strong chemical oxidant can oxidise more organic carbon bonds than biological reactions. The BOD_5 assay is an empirical test to quantify the fraction of organic content that is biodegradable.

Since faecal sludge is stored under predominantly anaerobic conditions, more experimental work needs to be conducted on the best ways to measure stabilisation and potential for biodegradation during treatment. This is important, as the level of stabilisation is related to the dewaterability of faecal sludge, and the potential for biological treatment, as discussed further in Chapter 4. Aggregate methods for concentrations of organic matter are provided in Chapter 8, but not for individual compounds (*e.g.* trace organic contaminants).

Nutrient content

Nutrients in faecal sludge are present in organic or inorganic forms. Nutrients are important to monitor for NH_3 inhibition, adequate nutrients for biological processes, fate in the environment, and potential for valorisation in agriculture as compost or fertiliser. Total Kjeldahl Nitrogen (TKN) is a metric of the sum of organic nitrogen and NH_3. To quantify organic nitrogen, the NH_3 concentration can be measured and subtracted from TKN. Other forms of inorganic nitrogen are nitrate (NO_3^-) and nitrite (NO_2^-). The different forms of nitrogen provide information on the redox potential (*e.g.* aerobic, anaerobic, anoxic) of faecal sludge, and level of stabilisation in biological processes such as compost (Nikiema and Cofie, 2014). Similarly, total phosphorus includes organic and inorganic forms. Ortho-phosphate (PO_4^{3-}) is the inorganic form, which is soluble and bioavailable.

pH, conductivity alkalinity and corrosion

pH is important to measure as it can influence reaction rates, chemical speciation, and biological processes, and also because it can be an indicator of the source of the faecal sludge (*e.g.* industrial contamination). Sample preparation and how the pH is measured is an important factor, as the method can change the pH of the sample. Conductivity is a metric of ions in a solution. Ion concentration is important as high salt concentrations can inhibit biological processes such as in stabilisation ponds. Alkalinity represents the acid-neutralising capacity of water, and is commonly referred to as 'soft' or 'hard' water. Alkalinity is important in many biological processes, such as nitrification, which consumes alkalinity and lowers pH (7.07 $gCaCO_3/gNH_4$-N, plus additional alkalinity to maintain pH) (Tchobanoglous *et al.*, 2014). Corrosion potential (EC, pH, Cl^-, $CaCO_3$, H_2S) is important for tanks and pipes, and can lead to failure.

Metals

Chemical elements are important to quantify, as varying concentrations of metallic elements (*e.g.* macro and micro-nutrients) are necessary for treatment performance (*e.g.* microbial growth) and plant and animal growth (*e.g.* iron, chromium, copper, zinc, and cobalt), but can also be toxic depending on their concentrations. Guidelines for heavy metal concentrations for land application of sludge are summarised in Hanay *et al.* (2008), McGrath *et al.* (1994) and ISO 31800 (2020).

2.4.1.2 Physical properties

Physical properties are characteristics that do not change the chemical composition of a material such as faecal sludge. Examples of physical properties are density, particle size, turbidity, colour, odour, and thermal conductivity.

Settleability and dewaterability

Metrics of settleability and dewaterability are important for the operation of treatment plants, as dewatering is one of the most important steps in the treatment process. Metrics can include general settleability in a settling-thickening tank (Imhoff cone), dewaterability (centrifuge), and time for dewatering on drying beds or geotextiles (*e.g.* capillary suction time (CST)). Settleability and dewaterability can vary significantly depending on sludge characteristics, such as solids concentration and level of stabilisation.

Mechanical properties

Mechanical properties of faecal sludge are important for the design and sizing of emptying technologies (*i.e.* manual and mechanical), collection and transport options, and for the design of onsite sanitation systems and offsite treatment facilities. Measurements such as density, particle size, and rheological properties provide information on the 'pumpability' of materials, or the 'stiffness' versus the ability to 'flow'. The overall tendency of faecal sludge is that it tends to 'flow' - a phenomenon known as 'shear thinning', where the increasing shear rate is expected to ease emptying processes from onsite containments (Septien *et al.*, 2018a).

Thermal properties

Evaluation of thermal properties such as thermal conductivity and diffusivity, specific heat, and calorific value are important for resource recovery implementations with treatment end products, such as combustion as a solid fuel or biofuel. The calorific value of a material is the quantity of heat produced by combustion. Thermal conductivity is the ability of a material to conduct heat and is important for assessment and understanding of faecal sludge end use processes such as combustion and composting. Heat capacity is the quantity of heat energy required to change the temperature of an object by a given amount.

2.4.1.3 Biological properties

Biological examinations of samples are important along the entire service chain. The above chemical and physicochemical, and physical properties create a habitat for many organisms. Some are involved in nutrient and organic cycles, some are pathogens, and others can be associated with environmental impacts and resource recovery. Biological activities related to production and consumption of organic matter, or respiration, are included under the physico-chemical section. Further types of analytical methods for biological examinations include identifying pathogens (*e.g.* virus, bacteria, protozoa, helminths), metrics of toxicity (*e.g.* use of bioassays), enumeration (*e.g.* plate counts, flow cytometry, MPN), and types and functions of organisms (*e.g.* DNA/RNA analysis). The methods presented in Chapter 8 focus on pathogens.

Pathogens

Monitoring of pathogens is essential for the protection of public health, to protect workers handling sludge, to verify treatment efficiency prior to discharge, and for resource recovery. A risk-based approach can be taken to determine the adequate level of pathogen removal depending on the intended end use (WHO, 2015; WHO, 2018). Chapter 8 covers helminth eggs, as they are one of the most resistant pathogens to remove during treatment, and *E.coli*, as it is a type of faecal coliform that is used as an indicator of faecal contamination or of other organisms that can be present.

2.4.2 Selection of appropriate methods for characterisation

After defining the purpose, objectives and the desired properties to determine the characterisation process, the next step is to select appropriate methods for analysis. There are no strict guidelines to adhere to, but general considerations are the TS content, required level of accuracy, available resources, and laboratory capacity, as summarised in Figure 2.14 and explained in the following section. The sampling plan prior to analysis is discussed in Chapter 3.

Type of faecal sludge samples defined by total solids content

The type of faecal sludge samples defined by TS (liquid TS <5%, slurry TS 5-15%, semi-solid TS 15-25%, or solid TS >25%) should always be taken into consideration before designing a plan for characterisation. Some of the methods for sample preparation, chemical analysis, and solids fractionation in Chapter 8 are different for more liquid sludge or for semi-solid or solid sludge. For example, faecal sludge from dry sanitation facilities can require higher dilution during the sample preparation, compared to FS samples coming from wet sanitation facilities. In practice, the easiest way to determine the type of faecal sludge is to conduct a preliminary TS analysis of the faecal sludge that is going to be characterised.

Level of accuracy

The level of accuracy is defined by the purpose of characterisation, the laboratory equipment used for a particular analysis, and the level of competency required to undertake this analysis. For a particular analysis, the level of accuracy could be of high importance. For example, molecular tests to establish pathogenic or other groups of microbial populations require a high level of accuracy and sample preparation using specialised techniques and methods. In other cases, the level of accuracy is not as significant and the priority could be almost immediate data to establish the presence of pathogens, nutrients or TS. In this case, simple test kits can be used, either simple field or laboratory-based techniques. In reality, it is not always possible to obtain the desired level of accuracy as this will be related to the available budget resources. It should also be noted that some parameters have higher degrees of built-in inaccuracy due to the imperfection of analytical and measuring equipment or preparation and handling procedures of a sample.

Cost of analysis

The cost of the analysis is determined by the type of analysis and equipment. Costs of equipment and required laboratory consumables vary enormously, which also needs to be taken into account. Determining the number of samples is discussed in chapters 3 and 5. For example, for a particular project on faecal sludge characterisation, the number of samples to provide statistically significant results could be 300, but in reality, the available budget might only allow for analysis of 100 or even 50 samples. In this case, focus should be placed on the selection of the most representative number of samples from specific areas, together with rigorous quality assurance and control measures (QA/QC). The cost of analysis is one of the main parameters that will determine the scope and duration of a sampling campaign. For more detailed information on data handling, the reader is referred to Von Sperling *et al.* (2020).

Laboratory capacity

The laboratory capacity is defined by the skill level required for a particular analysis, the availability of the desired equipment, and the number of analyses that the laboratory is able to carry out in a certain period. This includes special technical staff in the laboratory to undertake the desired analysis, or whether they could be performed by an employee, researcher, field worker, or student. For example, the TS content method using an oven at 105 °C is a relatively simple method that does not require extensive training, while determining the calorific value with a bomb calorimeter requires a higher level of training to operate the more sophisticated equipment.

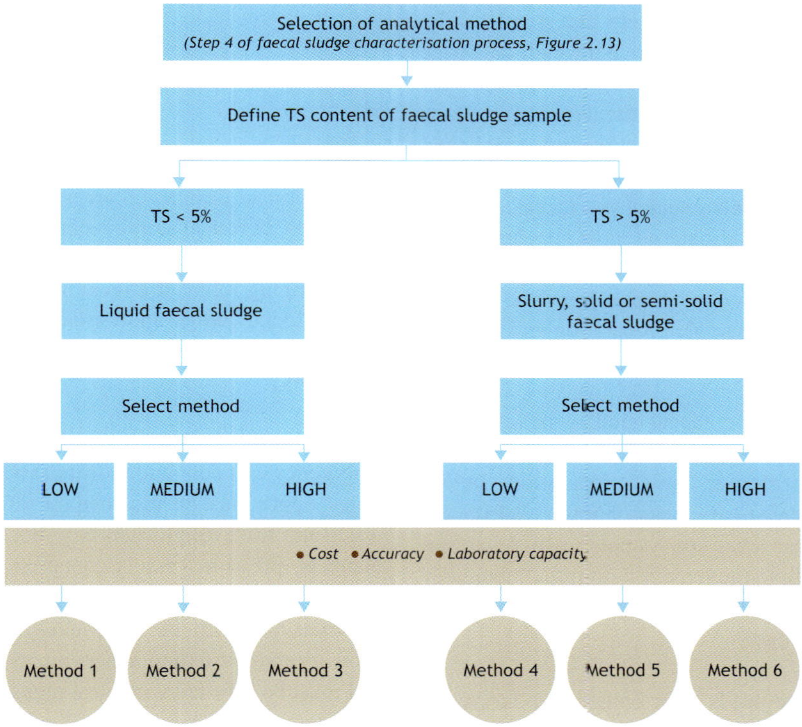

Figure 2.14 An example of a decision-making tree for the selection of a method of analysis depending on the purpose of characterisation. Step 4 refers to Figure 2.13, Steps in the faecal sludge characterisation process. Note: This example is specific to methods that differ for less than or are greater than 5% TS.

In Chapter 8, different methods based on the required level of accuracy (low, medium, or high) are provided (Figure 2.14, Section 8.4). This is based on the assumption that a high level of accuracy will be the most expensive option and will require more specialised laboratory equipment and/or personnel to undertake this kind of analysis. The lower-accuracy methods usually cost less because they require a lower level of laboratory training and less expensive equipment. However, in the end, which method is selected will depend on decisions that must be made based on the specific local context.

2.5 SETTING UP LABORATORIES FOR FAECAL SLUDGE ANALYSIS

Laboratories in many fields of research have essential similarities. However, setting up a faecal sludge laboratory needs special attention to health and safety due to the potential for pathogens. Hence, when working with faecal sludge, health and safety is of the highest priorities. This section considers the importance of a strategic workflow, layout, management system, and best-practice health and safety procedures for setting up a laboratory for faecal sludge analysis. It is followed by case studies of established and operational laboratories that have different objectives in different locations, including research laboratories, international collaborations, and a mobile field laboratory for emergency settings.

2.5.1 Faecal sludge laboratory workflow

A workflow is a systematic pattern that stipulates the order in which a sample will move through the space as it is received, prepared, analysed and disposed, until there is data output from that particular sample. Once the samples arrive at the laboratory, an established workflow needs to take place. When receiving biohazardous materials such as faecal sludge, exposure to this type of material must be restricted as much as possible. Sample collection and transport to laboratories is discussed in Chapter 3 and the specific methods for storage, sample preparation and analysis are provided in Chapter 8.

Once received, the faecal sludge samples pass through a number of steps in designated areas, such as sample intake, storage and preparation before reaching the analytical areas (Figure 2.15). By ensuring that these areas are systematically organised and the bulk sample movement within the laboratory is restricted, this will thereby limit the exposure of personnel to pathogens.

Workflow also needs to be considered during the construction, design, or adaptation of a laboratory for faecal sludge analysis. This allows a dedicated sample-receiving area to be placed adjacent to the storage and sample preparation spaces. 'Clean' rooms can be included to accommodate precision analytical equipment or microbial analysis, and these can be located away from 'dirty' areas where samples are received, stored and prepared. Clean rooms are also required for the preparation of chemical reagents and standard concentration solutions to avoid cross-contamination. Dedicated areas for data capture and analysis adjacent to the analytical rooms prevents cross-contamination as laptops and laboratory notebooks are moved between laboratory and office space.

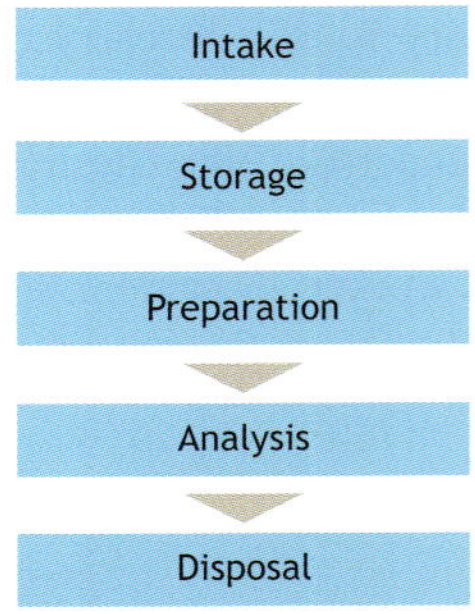

Figure 2.15 Sample processing workflow in a faecal sludge laboratory.

Additional designated areas including an external wash area, chemical storage rooms, equipment storage rooms, personal protective equipment (PPE) storage and changing rooms are recommended. They should be equipped with appropriate handwashing and disinfection facilities for staff prior to leaving the laboratory.

For safety, there must be more than one emergency exit door and they must be accessible at all times. Space must be allocated for safety showers, eye wash, and fire extinguishers (*e.g.* buckets of sand, fire blankets, pressure vessels containing extinguishers), determined by the size of the laboratory and the activities that will be undertaken. Access to safety showers and fire extinguishers must not be obstructed and must be labelled with clear signs.

If a faecal sludge laboratory is being set up as new construction, the systematic workflow will give guidance to the location of required utilities, equipment and designated areas for specialised equipment (Rice *et al.,* 2017). Conversely, if a faecal sludge laboratory is to be retrofitted into an existing space, the laboratory workflow will likely be influenced by the existing layout.

- Taps and sinks should be located in the areas for sample intake, preparation and analysis. They must be located in a safe manner to prevent splashes on nearby electricity power points. A water connection must also be available for safety showers and a basin near the main exit door. If possible, there should be drains on the floors that are linked to the sewerage system.

- Electrical power points must be placed high on the walls, and not at floor level to avoid water leaks, spillages or cause tripping hazards. The number and location of power points will be determined by the analytical equipment required. The switchboard for all the power points must be clearly labelled and should be easily accessible in an emergency.
- Space should be allocated for a gas cylinder storage area that is separated from the main working areas in the laboratory. Gas cylinders must be secured and stored in a ventilated area, with limited exposure to sunlight and ignition sources.
- Odours in a faecal sludge laboratory come from contained faecal sludge, faecal sludge combustion and from chemicals used during analysis. As such, an extraction system that can remove odours for general laboratory users and the public is important. High efficiency particulate air extraction systems are recommended and are coupled to pathogen filters to improve and maintain air quality in the laboratory.

In addition to the utilities discussed, a laboratory needs appropriate workstations and floors - hard, non-porous and chemically resistant. Furniture such as cupboards should be made of materials easy to disinfect.

2.5.2 Health and safety practices

Safe working practices and a written record of these practices are vital to reduce the exposure of personnel to pathogens in faecal sludge and harmful chemicals in a faecal sludge laboratory. The hierarchy of controls shown in Figure 2.16 should be considered when developing safe working practices. The preferred controls are those closer to the top of the pyramid. For example, a manual handling task could make use of a trolley to eliminate the risk of injury from incorrect lifting techniques. Similarly, a test method that uses hazardous chemicals could be substituted for a test method that uses less hazardous chemicals, if appropriate fume hoods are not in place. Fume hoods and ventilation systems are engineering controls which reduce the risks associated with inhalation of fumes and dust. Administrative controls are procedures designed to keep workspaces clean and form a key part of laboratory management systems. The last line of hazard control for laboratory safety is personal protective equipment (PPE). When dealing with pathogenic samples, laboratory coats, closed footwear, nitrile gloves and goggles form part of the necessary safe working wear. PPE might also be required based on the specific task and this can be determined by carrying out a risk assessment.

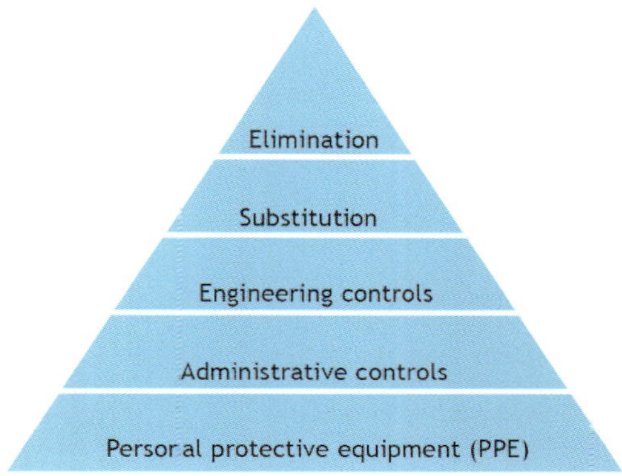

Figure 2.16 Hierarchy of controls for health and safety practices.

All tasks undertaken in the laboratory should have printed copies of standard operating procedures (SOPs) that include risk assessment and management. These documents identify all of the steps needed to carry out a task, the hazards associated with each task, who is at risk from the identified hazards, and the controls that can be implemented to mitigate the risks.

2.5.3 Laboratory management systems

Laboratory management systems establish protocols that govern laboratory processes and maintain a functional system. Laboratory management ensures that proper procedures are adhered to at all times and support is required from all organisational levels in order to ensure safe operation. Laboratories without management systems in place become easily disorganised and cluttered.

Laboratory management systems cover tasks at all levels, as shown in Figure 2.17. Personnel management procedures can vary depending on whether they are suited to staff, students, researchers or visitors. Similarly, facilities management can apply to onsite and offsite laboratories, research test sites and community test sites which can require different procedures.

Personnel

- Human resources induction
- Vaccines
- Laboratory induction
- Laboratory training
- Competency assessment
- Project design
- Project completion and submission

Facilities

- Access control
- First aid provision
- Fire safety
- Chemical storage safety
- Sample storage safety
- Monitoring and control
- Maintenance

Samples

- Site induction
- Sample box and toolkit
- Sampling checklist
- Sample collection forms
- Transportation forms
- Sample observations
- Sample storage logs
- Sample preparation
- Sample testing

Health and safety

- Safe working environment
- Transport management
- Security
- Sample containment
- Safe working practices
- Standard operating procedures
- Risk assessment
- Personal protective equipment
- Waste management

Equipment

- Equipment purchase
- Installation and commissioning
- Method development
- Calibration
- Troubleshooting
- Maintenance logs
- Service repair logs
- Usage records
- Costing for equipment use

Reporting

- Data capture sheets
- Data analysis spreadsheets
- Data reporting
- Data sharing

Figure 2.17 Types of protection provided through laboratory management systems.

When designing laboratory management systems, this should be done in a systematic manner so that processes and procedures are not overlooked. For example, Figure 2.18 shows the laboratory processes from receiving a sample to data distribution. By having procedures written down, there is less confusion about the steps necessary to process samples and where information about existing samples can be found.

Recordkeeping is an important aspect of laboratory management systems. Examples of records are: laboratory induction, laboratory training and competency assessments, sampling field trips, samples received, laboratory daily usage, laboratory analysis, instrument usage, instrument maintenance, and quality controls.

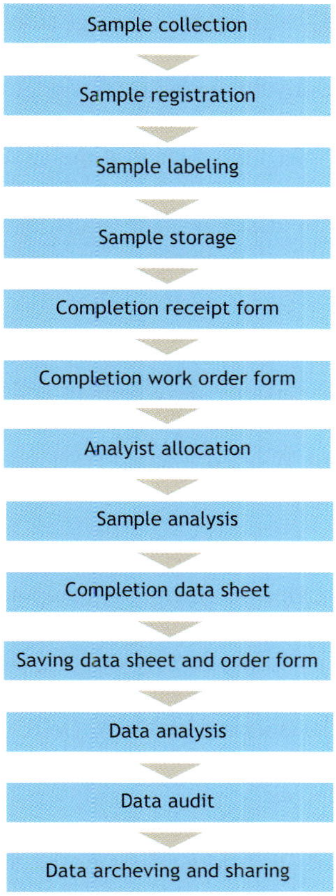

Figure 2.18 A sample management chain from collecting a sample to the results being distributed.

2.5.4 Case studies of global faecal sludge laboratories

Presented here are case studies of established research-based faecal sludge laboratories that are designed to perform analysis and performance evaluation of sanitation systems, also to accommodate teaching, postgraduate research students, local and visiting researchers, and to facilitate trainings. There is also an example of a field-based laboratory that was developed using low-cost alternatives to laboratory equipment, and can be deployed in emergency settings and areas with no laboratory capacity. The final example is of a network of laboratories for knowledge exchange.

Case study 2.1 *UKZN PRG faecal sludge laboratory*

Overview

The Pollution Research Group's (PRG) faecal sludge laboratory is based at the University of KwaZulu-Natal (UKZN), Durban, South Africa and has been operational since the 1970s. The research focus was initially on industrial wastewater and has gradually shifted to water and sanitation with a primary focus on faecal sludge laboratory practices and analysis over the last decade. In 2014 the laboratory undertook a major reconstruction and purchased additional analytical equipment and instruments in order to increase and optimise the laboratory space and management systems.

Focus areas

- Teaching and research of postgraduate students
- Capacity building - training and/or hosting visiting researchers and research students; supporting the development of other sanitation laboratories globally or locally
- Testing and analysis of different faecal sludge samples (Figure 2.19) and developing methods and procedures for faecal sludge analysis and faecal sludge handling procedures
- Testing and evaluating innovative sanitation systems
- Shipping and receiving of faecal sludge samples (Figure 2.20)

Equipment and instruments

The laboratory is fully equipped with analytical instruments used for the purpose of teaching, training, research and capacity building of undergraduate students, postgraduate students, international researchers and practitioners.

Main activities

- Capacity building and collaboration with other laboratories

  An example of the areas of collaboration and support to other laboratories are: improvement of laboratory management systems including health and safety, planning and improvement of laboratory workflows, training and knowledge dissemination of methods and procedures for faecal sludge analysis.

- A collaboration through a Memorandum of Understanding with a local municipality (eThekwini)
  This is a long-term collaboration aiming at a science-based integrated approach, incentives and innovation of the planning activities within the municipality.
- Engineering field testing
  A programme for testing and evaluation of innovative and emerging sanitation prototypes based in the field. The performance is evaluated by researchers and students on a daily basis and the samples are transported, stored and analysed in the UKZN PRG laboratory.

Figure 2.19 Preparation of samples for microwave digestion (photo: UKZN PRG).

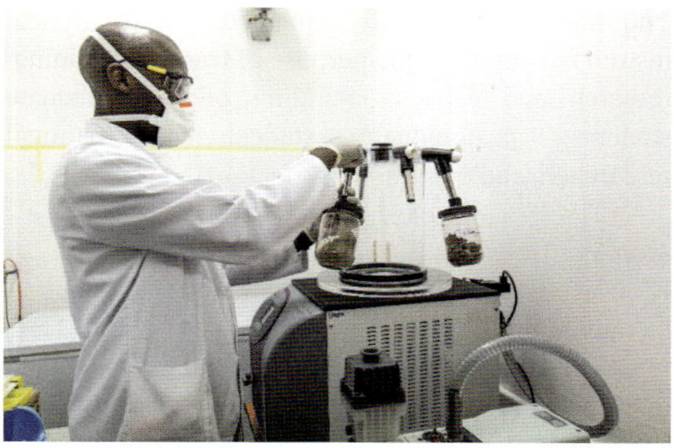

Figure 2.20 Freeze-drying of faecal sludge samples (photo: UKZN PRG).

Case study 2.2 *IHE Delft faecal sludge laboratory*

Overview

This is a relatively new faecal sludge laboratory, constructed in 2018 at the facilities of IHE Delft, The Netherlands. In this state-of-the-art laboratory, sanitation professionals and academics from all over the world can develop their skills and carry out research on the characteristics, use and end use of faecal sludge.

Focus areas

- Teaching, capacity development and tailor-made training
- Support of laboratory-based research at Master's and Doctoral level.

Equipment and instruments

After a thorough assessment, the equipment that was selected for the new laboratory was either new or complementary to the already existing equipment, in order to expand the current teaching and analytical capacity of the laboratories at IHE Delft.

Laboratory layout

Due to exposure to the potentially hazardous materials and pathogenic microorganisms in the faecal sludge laboratory, necessary health and safety requirements have been introduced at this facility (Section 2.5.2 and Chapter 8). These and other standards and requirements were taken into account while designing the laboratory (Figure 2.21) which consisted of five thematic rooms: (*i*) the entrance area, (*ii*) practicum/lecture room, (*iii*) research/analytical section, (*iv*) helminth eggs analysis room and (*v*) preparation room.

The entrance to the faecal sludge laboratory is the point where students and staff enter (or exit) the laboratory; this area has storage facilities for the health and safety equipment and has hand-washing facilities. It is connected with the main practicum section that is also used as a lecture room designed to accommodate up to 15 students at one time, working in parallel in up to four groups (Figure 2.22).

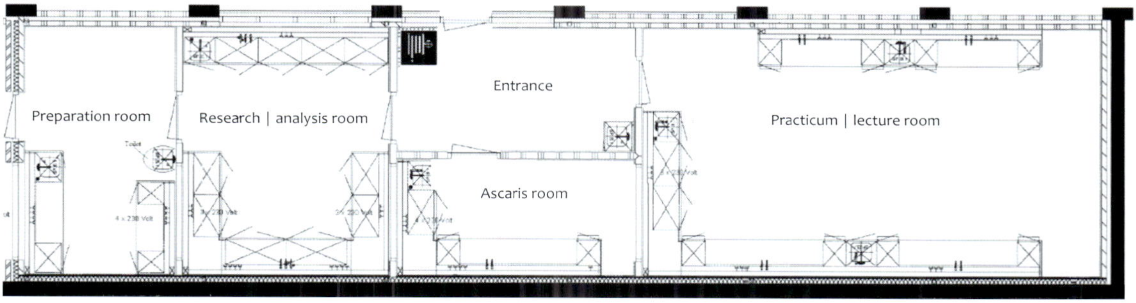

Figure 2.21 Final design of the faecal sludge laboratory at IHE Delft (source: IHE Delft).

Figure 2.22 Practicum/lecture room (photo: IHE Delft).

Figure 2.23 Research section (photo: IHE Delft).

Each group has parallel access to a shared sink, air extraction, electricity and a gas supply connection. This room is equipped with a digital lecture board and the equipment for total and volatile solids analysis. It is designed to be standalone, meaning that teaching can take place while other areas in the laboratory are being used.

Two other rooms can be accessed via the entrance area: the research laboratory and the Ascaris analysis room. The research laboratory is where the analytical equipment such as the analytical balances, Thermal thermogravimetric analyser (TGA) and differential scanning calorimeter (DSC), rheometer, bioreactor and calorimeter with the space and equipment for experimental setups is housed (Figure 2.23).

The Ascaris analysis room is a separate room for helminth eggs and other microbiological analysis. At the back of the faecal sludge laboratory is the sample reception and preparation room, with a separate external entrance for the samples. All samples are handled in this room, before being analysed or used in teaching in other parts of the laboratory.

Main activities

- Teaching and training of students.
  Since opening, the laboratory has been used for teaching the first cohorts of students of the Global Sanitation Graduate School and for the preparation of some of the video materials for the online course that will complement the material presented in this book.

Case study 2.3 *NATS AIT faecal sludge laboratory*

Overview

With more than 15 years' experience of monitoring, sampling and testing of faecal sludge in Southeast Asia, the NATS laboratory was established in 2016 under the Department of Energy, Environment and Climate Change, School of Environment, Resources and Development, Asian Institute of Technology (AIT), Thailand.

Focus areas

- Support of the field research project on 'Sustainable Decentralised Wastewater Management Systems' that covers assessments of faecal sludge management, non-sewered sanitation systems and implementation of reinvented toilet technologies.
- Further field monitoring and assessing the impacts of the toilet interventions on public health and environmental quality, in particular their compliance with national and/or international standards, *i.e.* ISO 30500 (2018).

Main activities

- Accreditation under ISO/IEC 17025
  The NATS lab has established a laboratory quality management system for analysis of high-strength wastewater and faecal sludge in compliance with ISO/IEC 17025. The accreditation process was applied in late 2017 and is expected to be accredited in late 2020, which will improve the quality control and technical competency in calibration and testing of the laboratory. It is envisaged that the knowledge and experience will be shared with other partner laboratories in the region in support of their accreditation (a voluntary process).
- Support of research students

Laboratory management system

Competency assessments have been implemented annually as well as regular laboratory training and proficiency testing to increase the technical skill and experience for laboratory staff. Quality control and quality assurance systems are in place, and equipment and laboratory glassware are calibrated on an annual basis. The working space of the NATS laboratory is organised in a way to provide a systematic laboratory workflow and best practice for analytical processes. There is a sample receiving area, sample storage, sample preparation, analytical area, cleaning areas for laboratory glassware and an external washing area. The analytical equipment area, chemical storage and clean room for microbiological analysis are positioned away from possible cross-contamination zones.

The NATS laboratory plans to upgrade to a 'Proficiency Testing Centre' for faecal sludge, according to ISO 17043 by supporting the testing process of innovative toilet technologies during product development, supporting performance testing of faecal sludge treatment plants and providing a supporting role for the establishment of other faecal sludge laboratories in the region in the form of training, monitoring and knowledge dissemination.

Figure 2.24 Training on faecal sludge analysis (photo: AIT).

Figure 2.25 External audit in NATS faecal sludge laboratory (photo: AIT).

Case study 2.4 *Eawag faecal sludge laboratories*

Overview

Eawag (the Swiss Federal Institute of Aquatic Science and Technology) was founded as a water and wastewater treatment research institute in 1936, with laboratory analysis of faecal sludge starting over 25 years ago. The department Sandec (Sanitation, Water, and Solid Waste for Development) focuses exclusively on development related research, with the mandate to develop and test methods and technologies that help the world's poorest to access sustainable water and sanitation services.

Focus areas and main activities

- Collaborative research:
  Applied research projects are conducted in collaboration with local universities, municipalities, and NGOs. Over the last 10 years research has been conducted in laboratories in Burkina Faso, Cameroon, India, Malawi, Senegal, Tanzania, Thailand, Uganda, Vietnam, and Zambia (Figure 2.26), in addition to the campus in Switzerland, which is well equipped with state-of-the-art laboratory facilities. Research is conducted with PhD and Master's students to develop fundamental knowledge required for integrated management and technology solutions, such as governing mechanisms of solid-liquid separation of faecal sludge and resource recovery.
- Technology innovations:
  Research development with industrial and implementation partners takes place in the Water Hub in the NEST building on the campus in Switzerland (Figure 2.27). NEST is a modular research and innovation site for testing of new technologies, materials and systems and off-grid, closed-loop technology solutions.
- Training/education:
  Training and education is a core tenet of Sandec, including laboratory training on methods for faecal sludge analysis. All of the Sandec educational resources are available free of charge on the Sandec website, including publications, books, online courses, workshops, newsletters and reference materials[5].

Figure 2.26 Collaborative research project on quantities and qualities of faecal sludge in the laboratory at the University of Zambia in Lusaka (photo: Eawag).

Figure 2.27 Dewatering research conducted by PhD students in the NEST building in Dübendorf, Switzerland (photo: Eawag).

Case study 2.5 *Faecal sludge field laboratory (FSFL)-Austrian Red Cross and Eawag*

Overview

In 2017 a consortium of the Austrian Red Cross, the University of Natural Resources and Life Sciences, Vienna (BOKU), WASTE Netherlands and Butyl Products Ltd Group, developed a FSFL that is now further supported by the International Federation of Red Cross and Red Crescent Societies (IFRC), Swiss Humanitarian Aid (SDC/HA) and Eawag. The laboratory can be operated almost entirely off-grid with a solar panel and wind turbine.

[5] www.sandec.ch

Focus areas

- The FSFL was designed as a mobile facility for implementation in emergency settings, and other locations without laboratory capacity.

Main activities

- Methods and equipment have been adapted for these special conditions, and includes analysis of 25 parameters, such as process control parameters (pH, TS, ash, biogas composition, COD), and public health metrics (Helminth eggs, Salmonella, Enterococcus, E. *coli*) (Bousek *et al.*, 2018).
- Selection of cost effective alternatives of laboratory equipment and development of low-cost, low-tech methods for parameters, *e.g.* for COD: using a cooking pot filled with sand as a heating block for the digestion of chemicals in cuvettes.
- The modularity of the FSFL makes it adaptable to many contexts, and the methods will continue to be further refined and tested.

Figure 2.28 A) first deployment of FSFL to Bangladesh in 2019, B) FSFL compactly fits on two pallets for shipping (photos: Eawag and Austrian Red Cross, respectively).

2.5.5 Global Partnership of Laboratories for Faecal Sludge Analysis (GPLFSA)

Experts on faecal sludge analysis recently established the Global Partnership of Laboratories for Faecal Sludge Analysis to address together the challenges and to work towards standardised methods for the characterisation and quantification of faecal sludge from onsite sanitation technologies, including sampling techniques and health and safety procedures for faecal sludge handling. The Partnership also delivers on-campus courses and training and aims to improve communication between sanitation practitioners, provide a comparative faecal sludge database, and improve confidence in the methods and obtained results.

The Partnership currently consists of eleven laboratories: IHE Delft (The Netherlands), UKZN (South Africa), Eawag (Switzerland), CSE and CDD (India), AIT (Thailand), Columbia University (USA), 2iE (Burkina Faso), BITS (India), ENPHO (Nepal) and ITB (Indonesia). More details are provided in Annex 1.

2.6 OUTLOOK

Understanding the purpose of characterisation, the associated faecal sludge properties, and the characterisation process are crucial for both increasing scientific knowledge and making informed decisions for best practices in faecal sludge management. The laboratory methods presented in Chapter 8 are the first step towards establishing standard methods of faecal sludge analysis. However, analytical methods alone are not adequate to provide reliable and repeatable analysis, and must be conducted by adequately trained personal. The background information in this chapter presents material that is necessary prior to conducting analysis of faecal sludge. Four types of faecal sludge, liquid, slurry, semi-solid or solid, are defined, based on total solids content. Their distinction is necessary for implementing the correct steps in the characterisation process, such as appropriate dilutions, and selection of methods (*e.g.* gravimetric or volumetric). However, these types are not reflective of other characteristics such as COD and nutrients, which can also be spread over a wide range of concentrations. An understanding of factors that affect the variability of characteristics along the entire service chain is important in order to understand what analyses are relevant, and must be considered with sampling plans as described in Chapter 3. Selection of appropriate methods for characterisation needs to be based on the available resources, including budget and laboratory capacity. Importantly, all of this must be conducted in an adequately equipped laboratory, with safety measures in place.

As the methods in this book are implemented, and further methods are developed and added to future editions, knowledge of faecal sludge will be greatly improved. Provided in Annex 2 is a link to a database with faecal sludge characteristics reported in the literature, as part of a UKZN PRG study. What is not inherent in the numbers is the innate level of uncertainty and error between the different data sets, due to a lack of standard methods. This highlights the need for development of a global database of characteristics of faecal sludge based on standard methods, so that solutions for faecal sludge management can be pursued with deeper insight, advanced knowledge, and greater confidence.

Figure 2.29 Education and training are key pillars of capacity development in the field of faecal sludge analysis (photo: IHE Delft).

REFERENCES

Ahmed I., Quarshie A.M., Ofori-Amanfo D., Cobbold F., Amofa-Sarkodie E.S. and Awuah E. (2018). Assessment of foreign material load in the management of faecal sludge in the Greater Accra Region of Ghana. *International Journal of Energy and Environmental Science*, 3(1), 27-36.

Anderson C., Malambo D.H., Gonzalez Perez M.E., Nobela H.N., De Pooter L., Spit J., Hooijmans C.M., Van de Vossenberg J., Greya W., Thole B., Van Lier J.B. and Brdjanovic D. (2015). Lactic Acid Fermentation, Urea and Lime Addition: Promising Faecal Sludge Sanitizing Methods for Emergency Sanitation, *International Journal of Environmental Research and Public Health*, 12(11), 13871-13885.

Andriessen N., Ward B.J. and Strande L. (2019). To char or not to char? Review of technologies to produce solid fuels for resource recovery from faecal sludge. *Journal of Water, Sanitation and Hygiene for Development*. 9(2), 210-224.

Appiah-Effah E., Nyarko K.B., Gyasi S.F. and Awuah E. (2014). Faecal sludge management in low income areas: a case study of three districts in the Ashanti region of Ghana. *Journal of Water, Sanitation and Hygiene for Development*, 4(2), 189-199.

Bakare B.F., Foxon K.M., Brouckaert C.J. and Buckley C.A. (2012). Variation in VIP latrine sludge contents. *Water SA*, 38.

Bakare B.F. (2014). Scientific and management support for ventilated improved pit latrines (VIP) sludge content. University of KwaZulu-Natal (doctoral dissertation).

Balasubramanya S., Evans B., Ahmed R., Habib A., Asad N.S.M., Vuong L., Rahman M., Hasan M., Dey D. and Camargo-Valero M. (2016). Pump it up: making single-pit emptying safer in rural Bangladesh. *Journal of Water, Sanitation and Hygiene for Development*, 6, 456-464.

Bassan M., Tchonda T., Yiougo L., Zoellig H., Mahamane I., Mbéguéré M. and Strande L. (2013). Characterization of faecal sludge during dry and rainy seasons in Ouagadougou, Burkina Faso. *36th WEDC International Conference, Nakuru Kenya.*

Bosch A., Schertenleib R. (1985). Emptying On-site Excreta Disposal Systems: Field Tests with Mechanized Equipment in Gaborone (Botswana). IRCWD Report. IRCWD.

Bourgault C. (2019). Characterization and quantification of faecal sludge from pit latrines. Université Laval, Canada (doctoral dissertation).

Bourgault C., Shaw K. and Dorea C.C. (2019). Dominant decomposition pathways in pit latrines: a commentary. *Water Science and Technology*, 80(7), 1392-1394.

Bousek, J., Skodak, M., Bäuerl, M., Ecker, G., Spit, J., Hayes, A. and Fuchs, W. (2018). Development of a Field Laboratory for Monitoring of Faecal Sludge Treatment Plants. *Water*, 10(9), 1153.

Brandberg, B. (2012). Evaluation of the UN-habitat Vacutug Development Project Pit Latrine Exhausting Technology. UN Habitat.

Brdjanovic D., Zakaria F., Mawioo P.M., Garcia H.A., Hooijmans C.M., Pean T.Y. and Setiadi T. (2015). eSOS® - emergency Sanitation Operation System. *Journal of Water, Sanitation and Hygiene for Development*, 5(1), 156-164.

Brouckaert C.J., Foxon K.M. and Wood K. (2013). Modelling the filling rate of pit latrines. *Water SA*, 39.

Buckley C.A., Foxon K.M., Brouckaert C.J., Rodda N., Nwaneri C., Balboni E., Couderc A. and Magagna D. (2008). Scientific Support for the Design and Operation of Ventilated Improved Pit Latrines (VIPs) and The Efficacy of Pit Latrine Additives. WRC Report No TT 357/08: Water Research Commission, South Africa.

Byrne A., Sindall R., Wang L., De Los Reyes F. and Buckley C. (2017). What happens inside a pour-flush pit? Insights from comprehensive characterization. *40th WEDC International Conference, Loughborough, UK.*

Campos, H. M. and Von Sperling, M. (1996). Estimation of domestic wastewater characteristics in a developing country based on socio-economic variables. *Water Science and Technology*, 34(3-4), 71-77.

Chipeta, W. C., Holm, R. H., Kamanula, J. F., Mtonga, W. E. and De Los Reyes F.L. (2017). Designing local solutions for emptying pit latrines in low-income urban settlements (Malawi). *Physics and Chemistry of the Earth, Parts* A/B/C, 100, 336-342.

Chiposa R., Holm R.H., Munthali C., Chidya R.C.G. and De Los Reyes, F. L. (2017). Characterization of pit latrines to support the design and selection of emptying tools in peri-urban Mzuzu, Malawi. *Journal of Water Sanitation and Hygiene for Development,* 7, 151-155.

Chirwa C.F., Hall R.P., Krometis L.H., Vance E.A., Edwards A., Guan T. and Holm R.H. (2017). Pit Latrine Fecal Sludge Resistance Using a Dynamic Cone Penetrometer in Low Income Areas in Mzuzu City, Malawi. *International Journal of Environmental Research and Public Health,* 14.

Colón, J., Forbis-Stokes, A. A. and Deshusses, M. A. (2015). Anaerobic digestion of undiluted simulant human excreta for sanitation and energy recovery in less-developed countries. *Energy for Sustainable Development*, 29, 57-64.

Connelly S., Pussayanavin T., Randle-Boggis R., Wicheansan A., Jampathong S., Keating C., Ijaz U., Sloan W. and Koottatep T. (2019). Solar Septic Tank: Next Generation Sequencing Reveals Effluent Microbial Community Composition as a Useful Index of System Performance. *Water*, 11(12).

Dodane P.H. and Bassan M. (2014). Settling-thickening tanks. *In: Faecal Sludge Management - Systems Approach for Implementation and Operation.* IWA Publishing, London, UK.

Elmitwalli, T. (2013). Sludge accumulation and conversion to methane in a septic tank treating domestic wastewater or black water. *Water Science and Technology*, 68(4), 956-964.

Englund M. and Strande L. (eds.) (2019). Faecal Sludge Management: Highlights and Exercises. ISBN 978-3-906484-70-9, Eawag - Swiss Federal Institute of Aquatic Science and Technology, Dübendorf, Switzerland.

Englund M., Carbajal J.P., Ferré A., Bassan M., Vu T.H.A., Nguyen V.A. and Strande L. (2020). Modelling quantities and qualities (Q&Q) of faecal sludge in Hanoi, Vietnam and Kampala, Uganda for improved management solutions. *Journal of Environmental Management*, 261, 110202.

Fakkaew K., Koottatep T. and Polprasert C. (2018). Faecal sludge treatment and utilization by hydrothermal carbonization. *Journal of Environmental Management,* 216, 421-426.

Gold M., Harada H., Therrien J.-D., Nishida T., Cunningham M., Semiyaga S., Fujii S., Niwagaba C., Dorea C., Nguyen V.-A. and Strande L. (2018). Cross-country analysis of faecal sludge dewatering. *Environmental Technology*, 39(23), 3077-3087.

Gold M., Dayer P., Faye C., Clair G., Seck A., Niang S., Morgenroth E. and Strande L. (2016). Locally produced natural conditioners for dewatering of faecal sludge. *Environmental Technology*. 37(21):2802-2814.

Gray, N. F. (1995). The influence of sludge accumulation rate on septic tank design. *Environmental Technology*, 16(8), 795-800.

Grolle K., Ensink J., Gibson W., Torondel B. and Zeeman G. (2018). Efficiency of additives and internal physical chemical factors for pit latrine lifetime extension. *Waterlines,* 37, 207-228.

Gross A., Maimon A., Alfiya Y. and Friedler E. (2015). Greywater Reuse. IWA Publishing, London, UK.

Gueye A., Mbeguere M., Niang S., Diop C. and Strande L. (2016). Novel plant species for faecal sludge drying beds: Survival, biomass response and forage quality. *Ecological Engineering*, 94, 617-621.

Günther I., Horst A., Lüthi C., Mosler H.J., Niwagaba C.B. and Tumwebaze I.K. (2011). Where do Kampala's poor "go"?-Urban sanitation conditions in Kampala's low-income areas. *Munich Personal RePEc Archive.*

Hanay Ö., Hasar H., Kocer N.N. and Aslan S. (2008). Evaluation for agricultural usage with speciation of heavy metals in a municipal sewage sludge. *Bulletin of Environmental Contamination and Toxicology*, 81(1), 42-46.

Hashemi S. and Han M. (2019). Field evaluation of the fertilizing potential of biologically treated sanitation products. *Science of the Total Environment,* 650, 1591-1598.

Heinss U., Larmie S.A. and Strauss M. (1998). Solids Separation and Pond Systems. *For the Treatment of Faecal Sludges in the Tropics. Lessons Learnt and Recommendations for Preliminary Design*. SANDEC Report.

Hennigs J., Ravndal K.T., Blose T., Toolaram A., Sindall R.C., Barrington D., Collins M., Engineer B., Kolios A.J., McAdam E., Parker A., Williams L. and Tyrrel S. (2019). Field testing of a prototype mechanical dry toilet flush. *Science of the Total Environment*, 668, 419-431.

Hiolski E. (2019). The toilet gets a makeover. How chemistry can help solve the sanitation crisis in low-resource areas. *ACS Central Science,* 5(8), 1303–1306.

Howard, T. L. (2003). Solids accumulation rates for onsite sewage treatment and disposal systems: A focus on Charlotte county, Florida. University of Florida (doctoral dissertation).

ISO (2018). ISO 30500. Non-Sewered Sanitation Systems - Prefabricated integrated treatment units - General safety and performance requirements for design and testing. Switzerland: ISO.

ISO (2020). ISO 31800. Faecal sludge treatment units — Energy independent, prefabricated, community-scale, resource recovery units — Safety and performance requirements. Switzerland: ISO.

Junglen K., Rhodes-Dicker L., Ward B.J., Gitau E., Mwalugongo W., Stradley L. and Thomas E. (2020). Characterization and prediction of fecal sludge parameters and settling behaviour in informal settlements in Nairobi, Kenya. *Sustainability*, 12(21), 9040.

Kemboi E., Van De Vossenberg J., Hooijmans C. and Mamani G. (2018). Impacts of Pit Latrine Additives on Volatile Solids and E. coli in Faecal Sludge. *Water Quality Management.*

Kengne E.S., Kengne I.M., Arsenea L.N.W., Akoa A., Nguyeng-Viet H. and Strande L. (2014) Performance of Vertical Flow Constructed Wetlands for Faecal Sludge Drying Bed Leachate: Effect of Hydraulic Loading. *Ecological Engineering,* 71, 384–393.

Klinger M., Gueye A., Manandhar Sherpa A. and Strande L. (2019). Scoping Study: Faecal Sludge Treatment Plants in South-Asia and Sub-Saharan Africa. eFSTP Project Report. *Gates Open Research* 2019, 3, 1716 (document) (https://doi.org/10.21955/gatesopenres.1116557.1)

Klute A. (ed.) (1986). Methods of soil analysis. Part 1 Physical and mineralogical properties. Madison WI: American Society of Agronomy: Crop Science Society of America: Soil Science Society of America.

Koottatep T., Pussayanavin T. and Polprasert C. (2020). Nouveau design solar septic tank: Reinvented toilet technology for sanitation 4.0. *Environmental Technology and Innovation*, 100933.

Koottatep, T., Surinkul, N., Paochaiyangyuen, R., Suebsao, W., Sherpa, M., Liangwannaphorn, C. and Panuwatvanich, A. (2012). Assessment of faecal sludge rheological properties. Environmental Engineering Program, School of Environment, Resources and Development Asian Institute of Technology.

Krueger B.C., Fowler G.D., Templeton M.R. and Moya B. (2020). Resource recovery and biochar characteristics from full-scale faecal sludge treatment and co-treatment with agricultural waste. *Water Research*, 169, 115253.

Lalander C., Diener S., Magri M.E., Zurbrügg C., Lindström A. and Vinnerås B. (2013). Faecal sludge management with the larvae of the black soldier fly (Hermetia illucens) - From a hygiene aspect. *Science of the Total Environment,* 458, 312-318.

Lugali, Y., Zziwa, A., Banadda, N., Wanyama, J., Kabenge, I., Kambugu, R. and Tumutegyereize, P. (2016). Modeling sludge accumulation rates in lined pit latrines in slum areas of Kampala City, Uganda. *African Journal of Environmental Science and Technology*, 10(8), 253-262.

Mawioo P.M., Garcia H.A., Hooijmans C.M., Velkushanova K., Simonic M., Mijatovic I. and Brdjanovic D. (2017). A pilot-scale microwave technology for sludge sanitization and drying. *Science of the Total Environment,* 601-602, 1437-1448.

McGrath S.P., Chang A.C., Page A.L., and Witter E. (1994). Land application of sewage sludge: scientific perspectives of heavy metal loading limits in Europe and the United States. *Environmental Reviews*, 2(1), 108-118.

Mercer S.J., Sindall R., Cottingham R.S., Buckley C., Alcock N., Zuma L. and Gounden G. (2018). Implementing an engineering field testing platform for sustainable non-sewered sanitation prototypes. *41st WEDC International Conference, Nakuru, Kenya.*

Mikhael G., Robbins D.M., Ramsay J.E. and Mbéguéré M. (2014). Methods and means for collection and transport of faecal sludge, IWA Publishing, London, UK.

Murray Muspratt A., Nakato T., Niwagaba C., Dione H., Baawuah N., Kang J., Stupin L., Regulinski J., Mbéguéré M. and Strande L. (2014). Fuel Potential of Faecal Sludge: Calorific Value Results from Uganda, Ghana, and Senegal. *Journal of Water Sanitation and Hygiene for Development,* 4(2), 223-230.

Nakagiri A., Niwagaba C.B., Nyenje P.M., Kulabako R.N., Tumuhairwe J.B. and Kansiime F. (2015). Are pit latrines in urban areas of Sub-Saharan Africa performing? A review of usage, filling, insects and odour nuisances. *BMC Public Health*, 16(1), 1-16.

Narayana D. (2020). Co-Treatment of Septage and Fecal Sludge in Sewage Treatment Facilities. IWA Publishing, London, UK. ISBN: 9781789061277

Nikiema J. and Cofie O.O. (2014). Technological options for safe resource recovery from fecal sludge. International Water Management Institute (IWMI) and CGIAR Research Program on Water Land and Ecosystems (WLE).

Niwagaba, C. B., Mbéguéré, M. and Strande, L. (2014). Faecal sludge quantification, characterization and treatment objectives. *In*: *Faecal Sludge Management - Systems Approach for Implementation and Operation*. IWA Publishing, London, UK.

Nzouebet, W. A., Kengne, I. M. and Rechenburg, A. (2015). Does Depth And Sanitation Type Affect The Quality Of Faecal Sludge In The Tropics? The Case of Yaoundé, Cameroon. *Open Water Journal*, *3*(1), 15.

Onabanjo T., Patchigolla K., Wagland S.T., Fidalgo B., Kolios A., McAdam E., Parker A., Williams S.T. and Cartmell E. (2016). Energy recovery from human faeces via gasification: a thermodynamic equilibrium modelling approach. *Energy Conversion and Management*, 118, 364-376.

Orner, K. D. and Mihelcic, J. R. (2018). A review of sanitation technologies to achieve multiple sustainable development goals that promote resource recovery. Environmental Science: Water Research and Technology, 4(1), 16-32.

Page A.L. (eds.) 1983. Methods of Soil Analysis Part 2 Microbiological and Biochemical Properties. Madison, WI: American Society of Agronomy: Crop Science Society of America: Soil Science Society of America.

Parker A. (2014). Membrane technology plays key role in waterless hygienic toilet. *Membrane Technology*, 12.

Polprasert, C., and Koottatep, T. (2017). Organic waste recycling: technology, management and sustainability. IWA Publishing, London, UK.

Portiolli, G. F. (2019). Investigating Links in the Sanitation Chain: Collection of Human Waste and Anaerobic Digestion of Food Waste. North Carolina State University (doctoral dissertation).

Prasad P., Andriessen N., Moorthy A., Das A., Coppens K., Pradeep R. and Strande L. (2021). Field-testing of the methods for estimating quantities and qualities (Q&Q) of faecal sludge: field evaluation in Sircilla, India. *Journal of Water, Sanitation and Hygiene for Development.*

Radford J.T. and Fenner R.A. (2013). Characterisation and fluidisation of synthetic pit latrine sludge. *Journal of Water, Sanitation and Hygiene for Development*, 3(3), 375-382.

Radford J.T., Underdown C., Velkushanova K., Byrne A., Smith D.P.K., Fenner R.A., Pietrovito, J. and Whitesell A. (2015). Faecal sludge simulants to aid the development of desludging technologies. *Journal of Water, Sanitation and Hygiene for Development*, 5(3), 456-464.

Reynaert E., Greenwood E.E., Ndwandwe B., Riechmann M.E., Sindall R.C., Udert K.M. and Morgenroth E. (2020). Practical implementation of true on-site water recycling systems for hand washing and toilet flushing. *Water Research*, 100051.

Rice E.W., Baird R.B. and Eaton A.D. (eds.) (2017). Standard Methods for the Examination of Water and Wastewater, 23rd edition. American Public Health Association; American Water Works Association; Water Environment Federation. ISBN: 9780875532875.

Robbins D.M., and Ligon G.C. (2014). *How to Design Wastewater Systems for Local Conditions in Developing Countries*. IWA Publishing, London, UK.

Rogers T.W., De Los Reyes F.L., Beckwith W.J. and Borden R.C. (2014). Power earth auger modification for waste extraction from pit latrines. *Journal of Water, Sanitation and Hygiene for Development,* 4, 72-80.

Rose C., Parker A., Jefferson B. and Cartmell E. (2015). The Characterization of Feces and Urine: A Review of the Literature to Inform Advanced Treatment Technology. *Critical Reviews in Environmental Science and Technology,* 45, 1827-1879.

Roy T., Saroar M. and Haque S. (2019). Use of Co-compost from Faecal Sludge and Municipal Organic Waste in Urban Green Space Plantation of Khulna City: Prospects and Problems. *Waste Valorisation and Recycling.* Springer.

Russel K., Hughes K., Roach M., Auerbach D., Foote A., Kramer S. and Briceño R. (2019). Taking Container-Based Sanitation to Scale: Opportunities and Challenges. *Frontiers in Environmental Science*, *7*, 190.

Sahondo T., Hennessy S., Sindall R.C., Chaudhari H., Teleski S., Lynch B.J., Sellgren K.L., Stoner B.R., Grego S. and Hawkins B.T. (2020). Field testing of a household-scale onsite blackwater treatment system in South Africa. *Science of the Total Environment*, 703, 135469

Schoebitz, L., Andriessen, N., Bollier, S., Bassan, M. and Strande, L. (2016). Market driven approach for selection of faecal sludge treatment products. Eawag - Swiss Federal Institute of Aquatic Science and Technology, Dübendorf, Switzerland. https://www.fsmtoolbox.com/assets/pdf/114.17.29_market_driven_approach.pdf

Seck A., Gold M., Niang S., Mbéguéré M., Diop C. and Strande L. (2015). Technology development of unplanted drying beds for resource recovery from faecal sludge: fuel production in Sub-Sahara Africa. *Journal of Water Sanitation and Hygiene for Development*, 5(1), 72-80.

Semiyaga S., Okure M.A., Niwagaba C.B., Katukiza A.Y. and Kansiime F. (2015). Decentralized options for faecal sludge management in urban slum areas of Sub-Saharan Africa: A review of technologies, practices and end-uses. *Resources, Conservation and Recycling*, 104, 109-119.

Septien S., Pocock J., Teba L., Velkushanova K. and Buckley C.A. (2018a). Rheological characteristics of faecal sludge from VIP latrines and implications on pit emptying. *Journal of Environmental Management*, 228, 149-157.

Septien, S., Singh, A., Mirara, S. W., Teba, L., Velkushanova, K. and Buckley, C. A. (2018b). 'LaDePa' process for the drying and pasteurization of faecal sludge from VIP latrines using infrared radiation. *South African Journal of Chemical Engineering,* 25, 147-158.

Simha P., Lalander C., Vinneras B. and Ganesapillai M. (2017). Farmer attitudes and perceptions to the re-use of fertiliser products from resource-oriented sanitation systems - The case of Vellore, South India. *Science of the Total Environment,* 581-582, 885-896.

Sisco T., Rogers T., Beckwith W., Chipeta W., Holm R., Buckley C.A. and De Los Reyes F.L. (2017). Trash removal methods for improved mechanical emptying of pit latrines using a screw auger. *Journal of Water Sanitation and Hygiene for Development,* 7, 85-91.

Sonko E.H.M., Mbéguéré M., Diop C., Niang S. and Strande L. (2014). Effect of hydraulic loading frequency on performance of planted drying beds for the treatment of faecal sludge. *Journal of Water Sanitation and Hygiene for Development,* 4(4), 633-641.

Sparks D.L., Page A.L., Helmke P.A., Loeppert R.H., Soltanpour P.N., Tabatabai M.A., Johnston C.T. and Sumner M.E. (eds.) (1996). Methods of Soil Analysis Part 3 Chemical methods. Madison, Wis: American Society of Agronomy: Crop Science Society of America: Soil Science Society of America.

Starkl M., Mbatha S., Roma E., Jeffrey P., Stenström T.A., Hawksworth D. and Gounden T. (2010). Evaluation of Community and Household Based Sanitation Systems in Ethekwini Municipality, South Africa: Lessons Learned and Recommendations. *Water Practice and Technology,* 5.

Stasse-Wolthuis M., Albers H.F.F., Van Jeveren, J.G.C., Wil De Jong J., Hautvast J.G.A.J., Hermus R.J.J., Katan M.B., Brydon W.G. and Eastwood M.A. (1980). Influence of dietary fiber from vegetables and fruits, bran or citrus pectin on serum lipids, fecal lipids, and colonic function. *The American Journal of Clinical Nutrition*, 33, 1745-1756.

Stenström T.A. (2004). Guidelines on the safe use of urine and faeces in ecological sanitation systems. EcoSanRes Programme.

Stephen A.M., and Cummings J.H. (1979). Water-holding by dietary fibre in vitro and its relationship to faecal output in man. *Gut*, 20, 722-729.

Strande L., Ronteltap M. and Brdjanovic D. (2014). Faecal Sludge Management: Systems Approach for Implementation and Operation, IWA Publishing, London, UK.

Strande L. (2017). Introduction to faecal sludge management: an online course. Available at: https://www.eawag.ch/en/department/sandec/e-learning/. Eawag: Swiss Federal Institute of Aquatic Science and Technology, Duebendorf, Switzerland.

Strande L., Schoebitz L., Bischoff F., Ddiba D. and Niwagaba C. (2018). Methods to reliably estimate faecal sludge quantities and qualities for the design of treatment technologies and management solutions. *Journal of Environmental Management*, 223(1), 898-907.

Strauss M., Larmie S.A. and Heinss U. (1997). Treatment of sludges from on-site sanitation- Low-cost options. *Water Science and Technology*, 35(6), 129.

Taweesan A., Koottatep T. and Polprasert C. (2015). Effective faecal sludge management measures for on-site sanitation systems. *Journal of Water, Sanitation and Hygiene for Development*, 5(3), 483-492.

Tayler K. (2018). *Faecal Sludge and Septage Treatment,* Rugby, UK, Practical Action Publishing.

Tchobanoglous G., Stensel H.D., Tsuchihashi R. and Burton F. (2014). *Wastewater Engineering: Treatment and Resources Recovery*, 5th edn. New York, NY: McGraw-Hill Education.

Thomas A.R., Rosario A.P., Philip L. and Kranert M. (2019). Performance Evaluation of In-vessel System for Co-composting of Septage. *Waste Management and Resource Efficiency.* Springer.

Tilley E., Ulrich L., Lüthi C., Reymond P. and Zurbrügg C. (2014). Compendium of Sanitation Systems and Technologies. Eawag/IWA/WSS.

Van Eekert M.H., Gibson W.T., Torondel B., Abilahi F., Liseki B., Schuman E., Sumpter C. and Ensink J.H. (2019). Anaerobic digestion is the dominant pathway for pit latrine decomposition and is limited by intrinsic factors. *Water Science and Technology*, 79(12), 2242-2250.

Velkushanova K.V., Zuma L. and Buckley C. (2019). Mechanical properties of faecal sludge. *Gates Open Research*, 3, 1582.

Wagner E. G., Lanoix, J. N. and World Health Organization. (1958). *Excreta disposal for rural areas and small communities.* World Health Organization.

Ward B.J., Gueye A., Diop B., Traber J., Morgenroth M. and Strande L. (2019) Evaluation of conceptual model and predictors of faecal sludge dewatering performance in Senegal and Tanzania. *Water Research*, 167, 115101.

Ward, B.J., Andriessen, N., Tembo, J.M., Kabika, J., Grau, M., Scheidegger, A., Morgenroth, E., Strande, L. (2021). Predictive models using cheap and easy field measurements: Can they fill a gap in faecal sludge monitoring and process control? *Water Research.* https://doi.org/10.25678/00037X.

Welling C.M., Sasidara S., Kachoria, P., Hennessy S., Lynch B.J., Teleski S., Chaudhari H., Sellgren K.L., Stoner B.R., Grego S. and Hawkins B.T. (2020). Field testing of a household-scale onsite blackwater treatment system in Coimbatore, India. *Science of the Total Environment*, 713, 136706.

World Bank (2019). *Evaluating the Potential of Container-Based Sanitation.* World Bank, Washington, DC., US.

World Health Organization (2015). *Sanitation Safety Planning: Manual for Safe Use and Disposal of Wastewater Greywater and Excreta.* World Health Organization

World Health Organization (2018). Guidelines on Sanitation and Health. ISBN 978-92-4-151470-5. WHO, Geneva, Switzerland.

Yadav K.D., Tare V. and Ahammed M.M. (2012). Integrated composting–vermicomposting process for stabilization of human faecal slurry. *Ecological Engineering,* 47, 24-29.

Zakaria F., Harelimana B., Ćurko J., Van de Vossenberg J., Garcia H.A., Hooijmans C.M. and Brdjanovic D. (2016). Effectiveness of UV-C light irradiation on disinfection of an eSOS® smart toilet evaluated in a temporary settlement in the Philippines, *International Journal of Environmental Health Research,* 26(5-6), 536-553.

Zakaria F., Ćurko J., Muratbegovic A., Garcia H.A., Hooijmans C.M. and Brdjanovic D. (2018). Evaluation of a smart toilet in an emergency camp, *International Journal of Disaster Risk Reduction*, 27, 512-523.

Zuma L., Velkushanova K. and Buckley C. (2015). Chemical and thermal properties of VIP latrine sludge. *Water SA*, 41(4), 534-540.

Zziwa, A., Nabulime, M. N., Kiggundu, N., Kambugu, R., Katimbo, A. and Komakech, A. J. (2016). A critical analysis of physiochemical properties influencing pit latrine emptying and feacal sludge disposal in Kampala Slums, Uganda. *African Journal of Environmental Science and Technology*, 10(10), 316-328.

Figure 2.30 Proper sampling is essential prerequisite for successful faecal sludge characterisation. Photo depicts experts of 500B and Eawag conducting a quantities and qualities (Q&Q) study with field testing of the Volaser in Kohalpur, Nepal, as part of the development of a city sanitation plan (photo: S. Renggli).

3

Faecal sludge sample collection and handling

Thammarat Koottatep
Amédé Ferré
Saroj Chapagain
Krailuck Fakkaew
Linda Strande

OBJECTIVES

The objectives of this chapter are to:

- Select different sampling techniques depending on objectives
- Select sampling devices and locations
- Develop appropriate and reliable faecal sludge sampling schemes and plans
- Ensure sample representativeness and integrity
- Protect health and safety of employees and users of onsite sanitation.

3.1 INTRODUCTION

Quantities and qualities (Q&Q) of faecal sludge vary significantly along the entire faecal sludge management (FSM) service chain. Further understanding of factors such as biodegradability, nutrient content, pumpability, dewaterability, resource recovery potential, pathogens, and potential inhibitory compounds are all particularly important for effective faecal sludge planning and management. Sampling is the action or process of taking a subset of a larger volume for characterisation. This process assumes that samples are representative of the larger volume, and there are measures to put in place to help ensure this. Therefore, a proper sampling scheme and subsequent analysis of faecal sludge is paramount for sustainable FSM. As introduced in chapters 5 and 6, the modelling of onsite sanitation will also help to bring a more systematic approach to data collection and sampling, and the number of parameters of interest will continue to grow, resulting in increased demands for sampling and analytical work.

How and where samples are taken, transported, and analysed depends on the specific objectives of the sampling. Examples of sampling objectives include designing a treatment facility, planning for emptying and transport services, evaluating rates of sludge accumulation, selecting and operating treatment processes, evaluating resource recovery options, and complying with regulatory requirements. Sampling and sample handling need to be carried out in such a way that the respective traits being measured (*i.e.* volumes, characteristics) are as similar as possible during the analysis as when the sample was taken. Analysis of samples can be done either *in situ* (*e.g.* within containments), in the field, or in a laboratory after being transported. Proper preservation help ensure that no significant changes in composition occur before the analyses are made. To ensure representativeness of collected data, emphasis is also placed on proper sample collection and tracking. A preliminary site visit, or familiarity with sampling locations, is necessary prior to any survey, sampling, or analysis decisions being made. Furthermore, sampling and sample handling need to be carried out in such a way that is safe for the people collecting and analysing the samples. Examples of safety aspects that need to be considered include collapsing pit latrines, falling or tripping hazards, working in confined spaces, asphyxiation, and hygiene. These topics are all presented in more detail in this chapter.

3.2 SAMPLING OBJECTIVE

The sampling objective is the defined purpose for collecting the data, which analysis of the samples will provide. Identifying the sampling objective is the first step in a sampling campaign. The next step is to develop a sampling plan specifically to answer the question you are asking. Sampling locations, frequency, timing, tools, and methods can greatly affect the outcome. For example, if you are interested in how faecal sludge accumulates within pit latrines, it would not necessarily make sense to sample what collection trucks are discharging at a large-scale treatment facility. However, if you want to improve the operation of a treatment facility, then directly sampling what is being discharged does make sense. Faecal sludge can be sampled for analysis at each step of the sanitation service chain depending on the question, for example, directly from the containment, from the collection vehicles, or during and after treatment. Each location and sampling purpose comes with different considerations. Below are examples of sampling strategies that are relevant to different sampling objectives. These are presented in the sequence of the sanitation service chain.

3.2.1 Containment

Estimating rates of accumulation at the community to city-wide scales

If the objective is to understand actual rates of accumulation that are occurring at a community to citywide scale, then it is logical to measure *in-situ* volumes and estimate time periods for the accumulated sludge. This is in contrast to measuring what is delivered to a treatment plant, which is probably less than the total accumulated amount. Accumulation rates are important for planning purposes, and for designing treatment technologies. For more information on how to develop a sampling campaign to estimate rates of accumulation, refer to Chapter 5.

Evaluating faecal sludge stabilisation with location and time in onsite sanitation systems

If the objective is to understand how faecal sludge changes with location and time within containment in order to improve management, then sampling should take place directly within the containment, for example at various depths and/or times. However, logistically this might be quite difficult. In addition, the *in-situ* environment is altered during an emptying operation, making it very difficult to analyse what is actually occurring underground. Therefore, assumptions might need to be made; for example, that taking samples every 300 mm while the containment is being emptied is representative. This needs to be managed with logical and transparent assumptions.

3.2.2 Collection and transport

Planning of emptying services for a community

If the objective is to design emptying services, then it is important to be able to select an emptying technology that is compatible with the sludge thickness; for example, if the sludge is too thick then a gulper (or manual pump) might need to be used instead of a vacuum pump. It is also important to have adequate volumes for transport, and so estimates need to be made regarding how much faecal sludge accumulates over time. Therefore, samples should be collected within onsite containments, or during emptying operations. If water is added during emptying, then samples should be taken prior to the addition of water.

Designing a technology for emptying or treatment

For the design of different innovative technologies in the sanitation service chain, the steps will be similar. However, there may be specific requirements for the properties, number, frequency, and type of samples that need to be taken. For example, the design of an emptying technology may require information on waste content, viscosity, rheology, ash content, and moisture content. If this data is provided by sampling from trucks or collected at the delivery point of the faecal sludge treatment plant (FSTP), the final design of the technologies for emptying containments may not be applicable for all the containments in this particular area.

3.2.3 Treatment

Designing a new faecal sludge treatment plant

If the objective is to design a new FSTP), part of the design study will include evaluating the characteristics that will arrive at treatment, in order to specify design values. Samples should be taken at an existing FSTP. When there is not an existing FSTP then sludge is frequently dumped in locations around town. Potentially samples could be taken at illegal discharge locations, but this can be difficult to arrange with the emptiers since it is an illegal activity. Illegal dumping is an undesirable practice and the sampling from such locations is only for the purpose of improving the current situation, not to endorse it. For more information on estimating Q&Q at this scale, refer to Chapter 5.

Evaluating operational parameters during the start-up phase of a faecal sludge treatment plant

When commissioning a new FSTP, the start-up period can require months of continuous testing and optimisation to reach the required treatment performance and to optimise treatment capacity. During the ongoing operation, operators will need to adjust operations and loadings on a regular basis; for example, resting time of settling-thickening tanks, and loading rates and residence times on drying beds. Sampling needs to be appropriate for the targeted treatment processes; for example, at the inlet and on the drying beds to determine the optimal drying time on unplanted drying beds.

Monitoring overall faecal sludge treatment plant treatment efficiency

If the objective is to evaluate compliance with environmental regulations for effluent, then sampling should be consistent with the requirements of the regulations (*e.g* effluent prior to discharge). If the objective is to evaluate overall treatment performance, then sampling should be done at the influent (*e.g.* truck discharge), and also effluent and the final treated solids.

3.2.4 End use

Assessing compliance with requirements for end use

If the objective is to assess compliance with requirements for end use or resource recovery, then appropriate sampling should be done on the final product for the characteristics of concern (*e.g.* nutrients, stabilisation, calorific value, pathogens). For example, for concerns specific to use as a dry combustion fuel, see Andriessen *et al.* (2019).

3.3 REPRESENTATIVENESS

Faecal sludge varies temporally and spatially at different scales (*e.g.* within containment, within communities). Due to this high variability, obtaining a representative sample for volumes, properties or characteristics can be very challenging. The goal is to obtain a sample that has a similar composition to the whole substrate that is being sampled. When this is achieved, then the sample can be considered representative of the targeted faecal sludge. It is important to remember that it is highly unlikely to obtain a representative faecal sludge sample if it is taken only at one time and from one sampling point. A single sample will most likely not provide meaningful information to support the sampling objectives.

Factors to consider when determining representativeness include solid or liquid nature, homogeneity or heterogeneity, changes with time, and scale. Various types of containment technologies such as pit latrines, septic tanks, cess pits, and composting toilets will have different sampling requirements that need to be considered (see Example 5.1). If the containment or sampling location is stratified, then the level of stratification needs to be taken into account (*e.g.* septic tanks, wet pit latrines, stabilisation ponds). A representative sample of faecal sludge from a septic tank includes the scum, supernatant and sludge layers, which are not homogenised within the tank. These concepts are applicable to the entire faecal sludge management service chain, from collection, transport, treatment, to final end use or disposal.

3.4 SAMPLING TECHNIQUES

Once the sampling objective has been determined, the resulting sampling locations and substrates can also be identified. According to the degree of variability of the faecal sludge to be sampled, different sampling techniques are suggested.

3.4.1 Grab sampling

A grab sample, also known as a catch sample or individual sample, provides a snapshot of the current situation. This sampling technique refers to the collection of a single sample at a specific sampling location and time or over a short period of time (typically seconds or minutes). The sampling time should always be carefully determined to reduce bias and increase representativeness. Typically, grab samples are not representative of things that change with time, or a flow of heterogeneous substrates. As faecal sludge characteristics can be highly variable, care should be taken that a grab sample is representative of the whole. Discrete grab samples are taken at a selected location, depth, and time. When a source is known to be relatively constant in composition over an extended time or over substantial distances, then a grab sample may represent a larger sampling area or longer time period (Rice *et al.*, 2017). Another possibility is to use a sequence of grab samples to monitor a condition over time. Samples can then be collected at suitable intervals and analysed separately to document the extent, frequency, and duration of these variations (Rice *et al.*, 2017); for example, for typical diurnal or seasonal variations at a FSTP. Similarly, several grab samples across different locations can be used to monitor the condition of a larger space. In faecal sludge treatment processes such as inflow chambers, settling-thickening tanks, or outlet of the FSTP, samples need to be representative of the cross-section of the entire treatment unit. The samples will be individually analysed, and then they cumulatively represent a time series.

Grab samples are most appropriate for:

- Substrate with negligible changes in composition with time
- When other sampling techniques that require more resources would not provide significant improvement in terms of representativeness (see Section 3.4.2 on composite sampling)
- For small FSTPs, decentralised or semi-centralised treatment facilities with low flow and limited capacity and resources for continual sampling (however, it must be taken into account that variations can also be much greater in these cases), and

- For cases where obtaining a composite sample is not feasible because of limited access, for example from pit latrines, leach pits or septic tanks with access only through a small drop hole or access port.

3.4.2 Composite sampling

Composite samples provide a representative sampling of heterogeneous matrices in which the characteristics vary over periods of time and/or space (Rice *et al.*, 2017); for example, the flows arriving from trucks discharging at FSTPs. A composite sample can be obtained by combining portions of multiple grab samples manually over time (Rice *et al.*, 2017). Automatic sampling devices are also available for some situations, and they are often used for the sampling of wastewater in centralised, sewer-based wastewater treatment plants (WWTPs). In many cases for the sampling of faecal sludge, composite grab sampling will be the preferred method. The main advantage is analysing a composite sample instead of analysing a larger number of individual grab samples, and obtaining results that are representative of heterogeneous matrices and flows. An adequate number of grab samples is taken so that the composite is representative.

Composite samples can be prepared in different ways. Sequential (time) composite samples are made up of sub-samples of equal volume taken at specific time intervals. For example, grab samples could be sub-samples taken once an hour, which are then combined to make a single daily sample, whereas flow-proportional sampling is proportional to the flow or loading. They can be taken by mixing equal volumes of substrate collected at time intervals that are inversely proportional to the volume of flow, or by mixing volumes of substrate proportional to the flow collected at regular time intervals (Rice *et al.*, 2017). This can be done manually, or with a purpose-designed sampler. For static heterogeneous substrates, a composite can be made up from randomly taken grab samples distributed throughout the entire substrate source. It should be noted that the composite samples must be comprised of grab samples that have been collected within a short period time: between a few hours and a few days. If the grab samples have been collected in longer time intervals such as a number of weeks or longer, they cannot be mixed as a composite sample and they need to be analysed separately as the characteristics may have changed significantly over this period. It is critical when compiling a composite sample to make a representative sample from the combination of all the grab samples collected. The aliquot of a composite sample needs to be well-mixed and effort must be made to minimise the possibility of sample contamination during the process.

Below are examples of composite grab sampling:

- In the case of sampling a sludge blanket layer in a septic tank, grab samples from multiple chambers and locations may be required to make a representative composite sample of the sludge contained in the tank (see Section 3.5.2).
- In the case of a liquid stream, equal volumes of a sample could be taken at time intervals to create a composite sample. For example, during truck discharge (taking one sample at the beginning, two in the middle, one at the end, see Section 3.5.2), or at the effluent of the FSTP. Another example of making a composite sample is to weight grab samples according to the faecal sludge loading patterns of each unit in a treatment chain at a plant.
- In the case of a completed or stabilised pile of compost as shown in Figure 3.1, a composite grab sample could be taken by grabbing samples distributed throughout the pile and then evenly mixing them into one composite sample. This is based on the assumption that stabilised compost is relatively solid, could be heterogeneous, and does not change with time.
- In the case of monitoring the dewatering of sludge on a drying bed, composite grab samples are taken from throughout the bed, for example using a grid system and taking one sample from each grid. It is important to take a core sample, and not only sample from the surface. Dewatered sludge on a drying bed is also relatively solid (depending on the level of dewatering), is probably heterogeneous, and does not change rapidly with time. A difficulty is if the sludge is not dry enough to walk on, in this case if only the edge of the drying bed can be reached, then the sample would not be as representative.

Figure 3.1 Stabilised pile of compost at the Niayes faecal sludge treatment plant in Dakar, Senegal, 2019 (photo: A. Ferré).

3.5 SAMPLING AND MEASURING DEVICES

Provided in Table 3.1 is an overview of the measuring devices that are described in this chapter, together with the measurements that they are suited for, and the advantages and disadvantages of each device. The devices are then described in more detail including how they can be used along the service chain. Sampling devices must be made of materials that will not contaminate or react with faecal sludge. Polypropylene, polycarbonate, high-density polyethylene (HDPE), polytetrafluoroethylene (Teflon), glass, and stainless steel are relatively inert and are all appropriate for sampling. However, the cost of Teflon and stainless steel equipment might prohibit or restrict their use, and potential for breakage of glass should be considered. If using metal equipment, depending on the analysis, galvanised or zinc-coated items should not be used as these materials will release zinc into the sample.

Table 3.1 Overview of sampling devices for faecal sludge.

Sampling device	Type of measurement	Advantage	Disadvantage
L-stick sludge and scum measuring device	• Depth of containment (septic tank) • Scum and sludge depth	• Affordable • Can be self-constructed	• Lower accuracy • Requires some training • Not suitable for thicker sludge
Core sampling device	• Characterisation of more liquid sludge • Height of scum, supernatant, and sludge layers • Visualisation of the different layers	• Easy to use • Can be self-constructed	• Not suitable for thicker sludge • Needs attention to prevent leakage at the bottom of the device (*e.g.* due to solid waste preventing watertight closure)
Vacuum sludge sampling device	• Characterisation of more liquid sludge	• Collection of sludge at a specific depth • Able to sample thicker sludge at bottom of containment • No mixing of sludge sample with other layers	• Energy required for vacuum pump • Heavy to transport • Not necessarily available on local market • Relatively expensive
Cone-shaped sampling device	• Characterisation of thicker sludge	• Suitable for thicker sludge • Possibility to sample sludge at a specific depth	• Depending on depth and thickness, cannot sample from bottom of containment
Grab sampling device, horizontal	• Characterisation of liquid flow	• Avoids contact with sludge • Easy to use • Affordable • Can be self-constructed	• Limited use (*i.e.* specific to truck discharge, effluent samples) • Reliant on emptying operation • Not suitable for onsite containments
Grab sampling device, vertical	• Characterisation of liquid flow (treatment plant)	• Adequate for homogenous liquid stream • Allows samples to be collected in deep tanks • Can be self-constructed • Affordable	• Representativeness needs to be evaluated • Not suitable for onsite containments
Automatic composite sampler	• Characterisation of liquid flow (treatment plant)	• Consistent sampling • Effective means to collect data for daily operation at treatment plants • Time-saving • Flexible sampling programs	• Energy required • Expensive • Not always locally available • Not applicable for thick sludge
Distance-laser measuring device	• Sludge and containment depth and volume	• Greater precision and accuracy • Obtains quantitative measurement	• Cannot measure extremely large/small containment sizes
Portable penetrometer	• Shear strength of faecal sludge (related to viscosity)	• Rapid estimation of total solids (requires more testing) • No need to collect sample	• Requires trained staff • Measurement takes time • Not locally available • Requires further testing

3.5.1 L-stick sludge and scum measuring device

When sampling *in situ* from septic tanks, cess pits, and 'wet' pit latrines, it is sometimes important to consider the height or depth of the sludge layer, scum layer, and supernatant separately (refer to Example 5.1). An L-stick, shown in Figure 3.2, can be used to measure these layers; it is a long stick similar to a garden or concrete hoe.

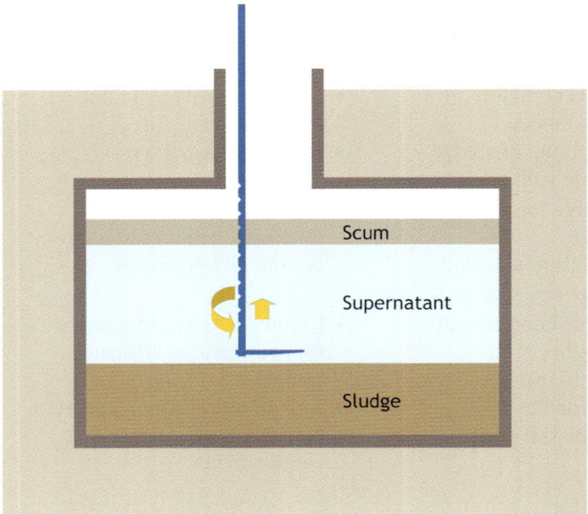

Figure 3.2 L-stick measuring device for depth of layers in a septic tank.

The stick has calibrated notches or nails to measure the depth at which it is inserted. For scum, the layer needs to be firm, with a crust, but not solid. The stick is poked through the scum, rotated 90°, and gently raised until the 'L' touches the bottom of the scum. For the sludge blanket layer, as the hoe is lowered it can be difficult to tell when the hoe first hits the sludge, and requires some practice. In some countries, L-sticks are used by emptiers to determine whether septic tanks should be emptied. The top of the sludge blanket layer is noted, and then the device is lowered to touch the bottom of the tank. One rule of thumb is that if resistance is felt from the top of the sludge blanket to halfway to the bottom, it requires emptying (Khan *et al.*, 2007). The core sampling device described in the following section is an alternative for measuring the depth of layers.

3.5.2 Core sampling device

The core sampling device shown in Figure 3.3 captures a vertical section of the substrate matrix. It is useful for sampling representative sub-samples of different layers in wet containments that have settled for many months or years, such as scum, supernatant, and thickened sludge. It can also be used to take samples from the access port of collection trucks, or tanks at treatment facilities. However, this type of sampling device is difficult to use with thicker sludge or sludge with large amounts of municipal solid waste, because it is difficult to push the device through the layers (Figure 3.3).

Figure 3.3 A) taking a core sample from a septic tank in Lusaka, Zambia, and B) the sampling device becomes clogged if the sludge is too thick. This example shows the collection from a 10-year old septic tank that had never been emptied. The tank was leaking, and so the supernatant leached out into the soil, resulting in a very thick sludge accumulation (photos: Eawag).

Another example of a core sampling device is shown in Figure 3.4. It consists of four transparent tubes (Figure 3.4, 1) that fit together, and four stainless steel rods (Figure 3.4, 2) that screw together inside the tubes. The device can be disassembled for transport, as well as shortened or extended as required. The tubes are graduated to measure volume. In the bottom tube the rod is attached to an airtight cover or plunger (Figure 3.4, 3) to close off the bottom of the sampler. This cover can be constructed from different materials, but it is very important that it can make a watertight seal. An alternative to the rod is a string. During sampling, the tube is inserted in the containment until the cover touches the bottom. Upon reaching the bottom, the cover should be left to settle for 30-60 seconds, allowing for any disturbed solids to settle. The hollow tube is then placed slowly over the cover, which is tightened with the string or rod (Figure 3.4, 4) to ensure a watertight seal so the sample can be removed. It is important not to make the device too large or it will be difficult to remove the sample without spilling.

3.5.3 Vacuum sludge sampling device

The vacuum sludge sampling device shown in Figure 3.5, also called a sampling hand-pump device, was developed by the Asian Institute of Technology (AIT). It was designed to take a sample at a designated depth with minimal disturbance to the surrounding layers. The device consists of a sample collection tank, a vacuum tank, and a hose. When taking a sample, the sample collection tank is evacuated, the vacuum pressure is set, and then the hose is placed in the exact location where the sample is desired. The suction valve of the vacuum tank is then released to collect the sample. The hose is brought back up, and the collected sample is released into a container by opening the discharge and air valves to normalise the pressure. This device is suitable for sampling from onsite containments and treatment technologies, to collect a sample at a specific depth.

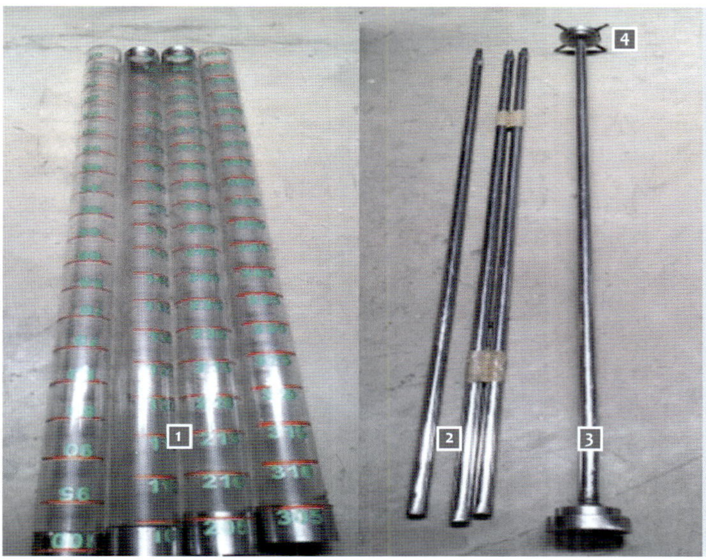

1 Core sampling tubes that fit together
2 Rods that screw together inside the tubes
3 Air-tight seal to close bottom tube
4 Locking system to attach to top-most rod

Specifications (source: CDD, India):

- Self-constructed acrylic tubes in Bangalore, India made with local material.
- Used in Sircilla, Telangana to collect faecal sludge from onsite containment systems (on average total sludge = 25 g/L).
- Internal diameter: 6 mm.
- Price: approximately 85 USD (constructed in 2018 in India).

Figure 3.4 Graduated core-sampling device developed by CDD, India (Prasad *et al.*, submitted, photos: CDD, India).

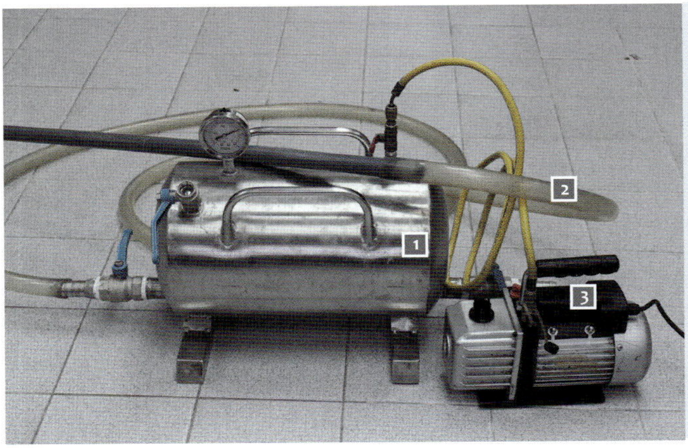

1 Sample collection tank
2 Flexible hose
3 Vacuum pump

Specifications (source: AIT, Thailand):

- Vacuum pump, 0.10-0.15 kW, 220 V.
- Vacuum tank, stainless steel tank, capacity 5-10 L.
- Hose, flexible hose of 1.5 to 2.5 cm.
- Approximately price (in Thailand): 1,500 USD.

Figure 3.5 Vacuum sludge sampling device (photo: AIT).

3.5.4 Cone-shaped sampling device

The cone-shaped sludge sampling device shown in Figure 3.6 can be used to collect samples in relatively 'dry' or less liquid onsite containments. Samples can be taken at a specific depth through a controlled valve that opens to take the sample, and closes to bring the sample out. Solid waste in containments complicates the operation and obtaining a representative sample due to clogging. The cone-shaped sampler in Figure 3.6 is 3 meters high with hinged arms, to allow for sampling of onsite containments within super-structures. The sample size is approximately 1 L. Similar devices have been used in many locations in sub-Saharan Africa, including the University of Zambia (Tembo, 2019), Makerere University (Zziwa, 2019), Egerton University (Muchiri 2019), and Jimma University (Beyene *et al.,* 2019). Modifications include a hinged opening and closing instead of a valve operation. Production of one unit in sub-Saharan Africa is around 300 USD in Kenya and Zambia. An example of sampling in Lusaka is provided in Case study 3.1.

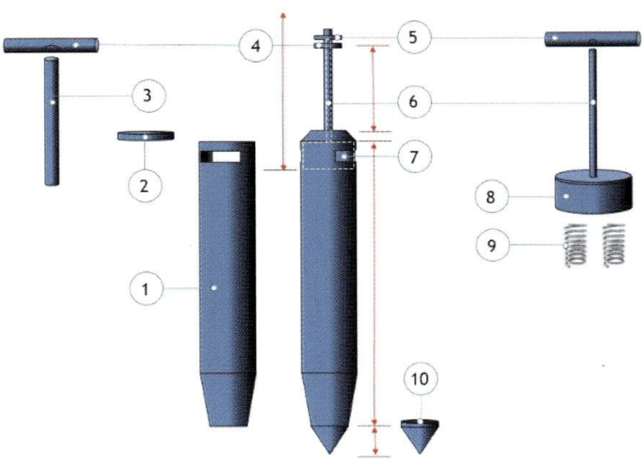

1 Sample-holding tube
2 Joint between the sample-holding tube and the extension pipe
3 Extension pipe connected to the joint
4 Handle of sample-holding tube
5 Handle used to close and open sample inlet door
6 Steel rod to hold the sample inlet closing and opening cup (extendable)
7 Sample inlet door (can be opened and closed at any depth)
8 Sample inlet closing cup
9 Two spring coils inside tube connecting closing cup and sample-holding tube (semi-automatic)
10 Pointed bottom cup - it can be tighten with a screw to facilitate the penetration and can be used to empty the sample

Figure 3.6a Schematic of the cone-shaped pit-sampling device.

Figure 3.6b Cone-shaped pit-sampling device in use in a study in Ethiopia (photo: Beyene *et al.*, 2019).

3.5.5 Grab sampling device - horizontal

The grab sampling device shown in Figure 3.7 consists of a sampling container of a known volume mounted on the end of a bar or rod.

Figure 3.7 A grab sampling device used for sample collection during truck discharge in Kampala, Uganda (photo: Eawag).

This sampling device is suitable for collecting faecal sludge at the discharge valve of the vacuum truck, as well as in some locations in treatment facilities (*e.g.* an FSTP outlet pipe). The sampling container is usually made of rigid plastic or stainless steel with a wide opening and a spout for emptying the sample. The bar or rod needs to be strong enough to avoid deformation or breaking during the sampling, because the flow from the outlet of the vacuum truck can be quite strong, and also long enough to protect the person collecting the sample from being splashed by sludge. The device allows for samples of a known volume of faecal sludge to be taken at a point in time. The sampling container is typically 1 L.

3.5.6 Grab sampling beaker device - vertical

The sampling device shown in Figure 3.8 is similar to the one shown in Figure 3.7, but the sampling container is oriented for samples to be taken vertically at depth of relatively homogenous substrates, such as supernatant in a settling tank. The length of the rod is dependent on the depth at which samples are taken. The sampling container should have a flat bottom, and the rod should be slightly flexible.

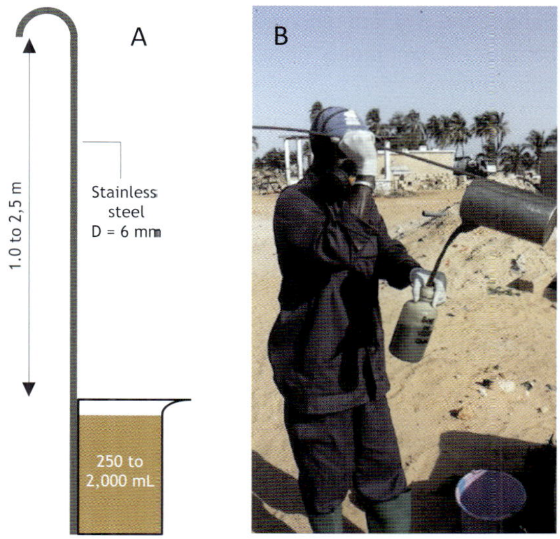

Figure 3.8a A) schematic of the grab sampling beaker device. B) at the outlet of the Cambérène FSTP, ONAS staff, Dakar, Senegal. Note: wide-mouth containers are preferable for sample allocation, as they aid sample collection without spillage (photo: Eawag).

Figure 3.8b Use of the sampling device in the liquid stream at the treatment plant (photo: IHE Delft).

3.5.7 Automatic composite sampler

Automatic composite samplers as shown in Figure 3.9 are commonly used in WWTPs, and can also be used for sampling the effluent of FSTPs. The system requires energy and is equipped with a peristaltic pump. A composite sampler usually includes several modes and sampling methods such as composite sampling (multiple samples are combined in a single large container), or sequential distribution (multiple samples are taken and stored in multiple bottles). Sample interval and time need to be selected, and can be uniform (commonly once an hour for 24 hours), or non-uniform.

Figure 3.9 An automatic composite sampling device (photo: IHE Delft).

Multiple samples are necessary when samples larger than 1,000 mL are required for analysis. As explained in Section 3.4.2, composite samples can be taken as fixed volume or flow proportional. Composite samplers include refrigeration for sample preservation. Single-bottle composite sampling is commonly used for influents and effluents, while multiple-bottle sampling is used to identify operational issues in treatment technologies.

3.5.8 Distance laser measuring device

The Volaser (vol*ume* laser) measuring device shown in Figure 3.10 is being developed by Eawag for measuring *in-situ* volumes of accumulated faecal sludge and volumes of containments (Andriessen and Strande, in preparation). The Volaser can be used to estimate accumulation rates as presented in Chapter 5 and Case study 3.1. The Volaser consists of a distance laser measuring unit, a tripod stand, and a probe to measure depth. The tripod is set up over a vertical access port to a containment. The laser unit is then lowered into the containment, and rotated as it measures the distance to the walls of the containment. Afterwards, the same laser unit is used to measure the distance from the top of the containment to the faecal sludge surface. A collapsible metal probe that is 3 m long is used to physically determine the depth of the containment. These measurements, along with the GPS coordinates, are recorded in a smartphone app which then automatically calculates the required volumes. The measurements take on average less than ten minutes with an accuracy of <10% error (*e.g.* ± 0.2 for a 2 m^3 containment). The Volaser device is not applicable for extreme cases (*e.g.* depth greater than 3m, access ports at an angle, or extremely large storage tanks). The Volaser can be operated by one person, and works well with a team of 2-3 people if sampling also includes characterisation and questionnaires. A version that can be locally assembled for less than 350 USD is expected by 2021 (Andriessen and Strande, in preparation). The tool can be adapted to local needs, and is applicable for all types of onsite containment technologies. Previous attempts at *in-situ* measuring devices include a laser measuring device to measure the 3D surface of sludge in pit latrines; however, further development is required due to light interference (Dahmani, 2010).

Figure 3.10 A) schematic of the Volaser with a laser measuring head that enters the containment to measure the area and distance to the sludge, with a smartphone mounted on top. Photos are from Lusaka, Zambia of (B) the prototype version used in 2019, with the Volaser placed over an access to a pit latrine (C) (photos: Eawag).

3.5.9 Portable penetrometer

The portable penetrometer shown in Figure 3.11 is intended as a relatively simple and quick *in-situ* test for shear strength of faecal sludge (related to viscosity) (Radford and Sugden, 2014). The penetrometer gives a continuous profile of how sludge varies throughout the depth of a containment. The device still requires further development, but the goal is to predict TS based on the *in-situ* penetrometer measurements, for rapid estimates at community to citywide scales. One measurement takes approximately twenty minutes with a skilled team of two to three operators.

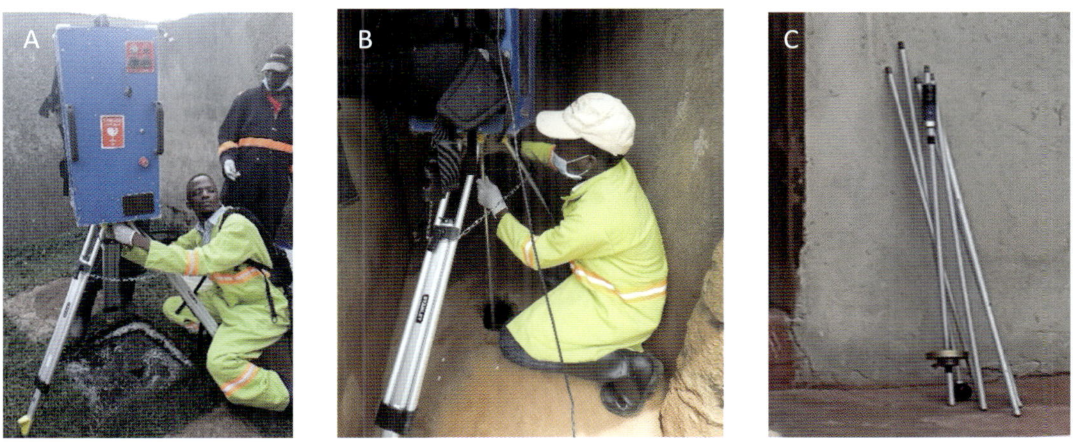

Figure 3.11 A and B) the portable penetrometer in use in Kampala, Uganda; C) the new 'P-lite' model for easier mobility in the field which is under development (photos: J. Radford).

3.6 SAMPLING METHODS AND LOCATION

Once the sampling objective has been determined, sampling locations in the faecal sludge management service chain and the sampling methods and devices can be selected. There are specific concerns for each step in the faecal sludge management service chain, including type and usage of onsite containment, collection and transport, type of treatment processes, and final end use or disposal. The reality is that obtaining representative sampling from containments can be difficult, as they are closed, underground systems, and samples cannot always be taken exactly where preferred. When selecting the sampling location, if the preferred location is not possible, then the closest representative alternative should be selected. The decision process should be documented, and evaluated for bias. For example, if the objective is to determine *in-situ* total loadings of accumulated faecal sludge, and sampling takes place during discharge at treatment plants, it will not necessarily be reflective of the total accumulated sludge if containments are not fully emptied. Another example is if sampling can only be done when they are full and require emptying (Strande *et al.,* 2018), because as illustrated in Figure 3.12, accumulation rates of total volumes of faecal sludge in containment do not accumulate linearly due to biological, physical and chemical properties (see Chapter 5). What triggers the emptying event is typically a nuisance event such as backing up or overflowing.

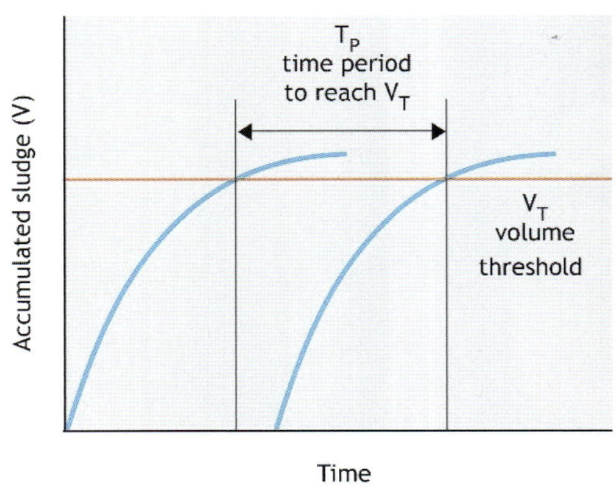

Figure 3.12 The blue lines illustrate change in the faecal sludge accumulation rate and T_P is the amount of time to reach V_T, the volume threshold where emptying is triggered.

The reality is that sampling will be dependent on the available resources. Assumptions will have to be made when designing a sampling campaign, which can be validated during implementation from different sampling locations. This is further discussed in Case Study 3.3 and Chapter 5. The following section presents examples of sampling along the service chain.

3.6.1 Sampling *in situ* from onsite containment technologies

In Chapter 2, faecal sludge is classified as liquid (total solids content <5%), a 'pumpable' slurry (total solids 5-15%), a 'spadable' semi-solid (total solids 15-25%), or a 'solid' (total solids >25%). Sampling methods are classified for more 'wet' or 'dry' faecal sludge, but in reality, in many systems or locations the faecal sludge will be a combination of types, and what is most appropriate for each situation will be context-specific, as illustrated in Case study 3.1.

In-situ sampling of a wet toilet system (faecal sludge < 5% TS)

This category can include many types of containment, including pit latrines or septic tanks, lined, unlined, or partially lined, one or multiple chambers, with or without overflows, and with soakaways or drain fields. The sampling location depends on the objective, and also on accessibility (Figure 3.14).

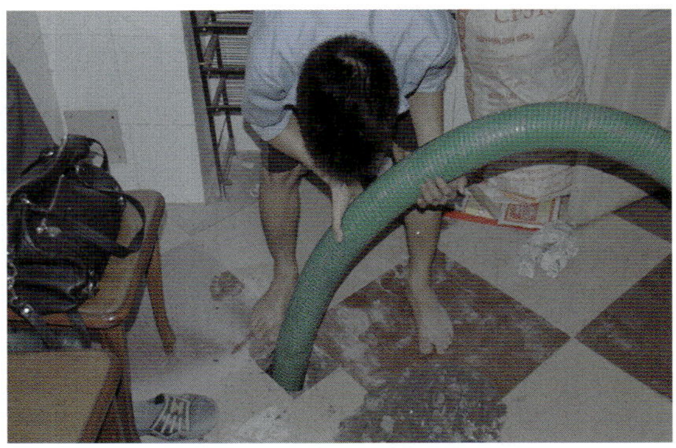

Figure 3.13 A septic tank located below a house, requiring the floor to be broken for emptying or sampling, Hanoi, Vietnam (photo: Eawag).

Sampling from septic tanks can be done via access ports, but they are also frequently sealed, covered over, or even located under buildings, as shown in Figure 3.13. In the latter case it can be difficult to know which part of the septic tank is being sampled.

Samples are frequently collected as core samples to collect a representative sample of all accumulated sludge layers. Grab samples of the effluent from the septic tank can also be collected to evaluate settling performance/solids removal. Examples of sampling locations in a two-chamber septic tank are provided in Figure 3.14. According to the sampling objective and strategy, a composite sample may be made from core samples from the different chambers of the wet toilet system or from grab samples collected at regular time intervals. Sampling could be also done directly through the toilet access hole in 'wet' pit latrines. In other cases, the depth of the sludge layer, supernatant, and scum layer can be measured with an L-stick

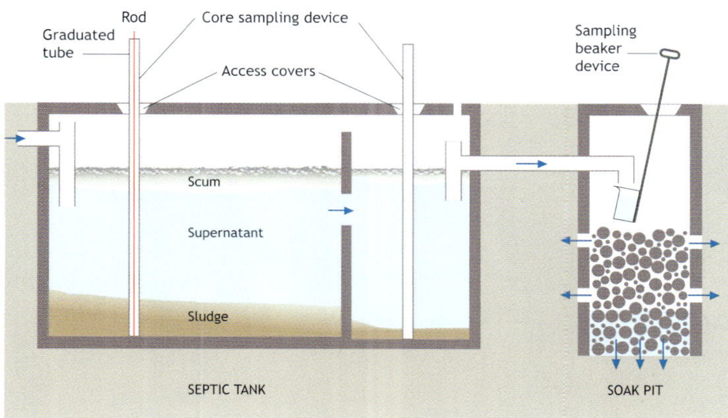

Figure 3.14 Sampling points in a septic tank: on the left is a vertical core sample in the first chamber, in the middle is a vertical core sample in the second chamber and on the right is a grab sample of the septic tank effluent (depending on configuration, e.g a distribution box or open drain).

In-situ sampling of dry containments (faecal sludge > 15% TS)

In-situ sampling of dry containments for characterisation can be done with the cone-shaped sampling device (Section 3.5.4) through the toilet access hole. Sludge volumes and depths can be measured with the Volaser measuring device (Section 3.5.8). Examples of in situ sampling are found in Case Study 3.1.

Case study 3.1 *In-situ sampling to estimate quantities and qualities (Q&Q) of faecal sludge in Lusaka, Zambia*

Eawag and UNZA conducted a study from September to December 2019 in Lusaka, Zambia to estimate quantities and qualities (Q&Q) of faecal sludge, specifically characteristics and accumulation rates (see Chapter 5). 82% of Lusaka relies on onsite sanitation, with 55-70% being pit latrines and 10-20% septic tanks (GFA Consulting Group GmbH, 2018). Observed total solids concentrations of 420 collected faecal sludge samples ranged from 0.1 to 40% measured gravimetrically, illustrating the wide range of characteristics that are present. To account for this diversity, different sampling devices were needed for *in-situ* sampling from septic tanks and pit latrines. For the design of the sampling plan see Case study 5.1.

Upon arrival at the sampling site, the containment was inspected to see if it could be sampled. A collapsible metal probe (3 m length) was used to measure the depth of the containment. For pit latrines, a cone-shaped sampler with a hinged arm was used to collect samples, as shown in Figure 3.15. Faecal sludge up to 40% total solids could be sampled with the cone-shaped sampler, and the minimum required sludge for sampling was 50 cm. The pit latrine samples were collected directly through the opening in the toilet. A core sampling device was used for septic tanks (Figure 3.15). The core sampler was graduated, to simultaneously measure the depth of the total sludge level and the sludge blanket layer. Faecal sludge from septic tanks was sampled in the first chamber of the tank. The cone-shaped sampler and the core sampler were both 3 m long.

Figure 3.15 A) a core sampling device, B and C) a cone-shaped sampler, D) the Volaser measuring device (photos: Eawag).

To measure the total volume of the containment, the Volaser measuring device was used (Ward *et al.*, 2021). The measurement was started through the smartphone app, and the Volaser was rotated while the laser was measuring the distance to the walls, angle of rotation, and calculating the area of the containment (Figure 3.10). The distance to the sludge surface was also measured. Based on collected data including time since last emptied, it was possible to estimate the sludge accumulation rates.

Samples collected for characterisation were poured into a bucket, stirred for homogenisation, and 0.9 L was transferred to a plastic container. Samples were stored in a cooler box with ice packs during transportation and delivered to the laboratory at the end of the sampling day, where they were immediately stored in a refrigerator. Analysis included TS, VS, COD, electrical conductivity, pH, NH_4-N, capillary suction time (CST), colour, odour, foam and C/N ratio. Duplicate sampling was conducted every 5 samples and triplicate sampling every 20 samples to test the replicability of the sampling method. Following this procedure, 6-7 samples could be collected per sampling team in one day.

3.6.2 Sampling during emptying of onsite containment technologies

As discussed, *in-situ* sampling is often not possible, and so sampling is frequently conducted during emptying operations.

Sampling of dry toilet containment during manual emptying

Manual emptying occurs with all types of faecal sludge in areas where vacuum trucks cannot access due to narrow lanes or paths, where faecal sludge is too thick for vacuum pumps, or where vacuum trucks are not available. Faecal sludge is commonly emptied into barrels, which can then be transported by cart or small trucks to a treatment plant or transfer station. Figure 3.16 shows examples of manual emptying operations in Lusaka, Zambia, and in Durban, South Africa. If the sampling objective is to determine average characteristics, grab samples could be taken from the barrels, and combined into a composite sample. Examples of dry toilet systems are urine diverting dry toilets (UDDT) and dry pit latrines with total solids > 15%.

Figure 3.16 A) sampling during a manual emptying operation in Lusaka, Zambia, and B) Durban, South Africa (photos: Eawag).

If the sampling objective is to evaluate how sludge degrades over time and with depth inside a pit latrine, samples can be taken from different vertical layers during emptying. Buckley *et al.* (2008) propose that faecal sludge in dry toilet systems can be classified in four layers as: (*i*) fresh stools, (*ii*) a partially degraded aerobic surface layer, (*iii*) a partially degraded anaerobic layer beneath the surface, and (*iv*) a completely stabilised anaerobic layer. Velkushanova (2019) and Zuma *et al.* (2015) developed their sampling methodology based on Buckley *et al.* (2008) and proposed that a dry toilet system can be further divided into two sub-sections: a back section and a front section (under the pedestal) as presented in Case study 3.2. Faecal sludge sampling should be done at different depths at the front and back of the pit, as containment of sludge in dry ventilated improved pit (VIP) latrines is not evenly distributed. In contrast to wetter sludges, it is possible to have a higher heap of sludge accumulate directly underneath the pedestal. Similarly, faecal sludge samples can be selected from both active and standing vaults of the UDDT toilets and other dry containment systems, outlined in Case study 3.2. These separations or distinctions should be considered during sampling to ensure an overall representative sample of the entire containment system, and are represented by the numbers in Figure 3.17.

Case study 3.2 *Sampling methods and locations of different dry onsite sanitation systems in Durban*

The Pollution Research Group at the University of KwaZulu-Natal (UKZN PRG), South Africa carried out a study into the properties of faecal sludge from onsite sanitation facilities in the Durban metro area, including: wet and dry household VIP latrines, household UDDTs, household unimproved pit latrines, community ablution block (CAB) VIP latrines, and school VIP toilet blocks. The goals were to provide a better understanding of the potential use of faecal sludge as a biofuel or fertiliser, to support the design and sizing of mechanical pit-emptying devices, transportation and processing systems for the excavated sludge, and the design of future onsite sanitation facilities. The study took place during 2012 and 2013.

Pit emptying

The first phase of the project involved a sampling program (Table 3.2) to obtain faecal sludge samples from selected onsite sanitation facilities in peri-urban and rural areas of Durban that are serviced by the eThekwini Municipality.

Table 3.2 Distribution of 45 samples in peri-urban and rural areas of Durban.

Facility type	Characteristics	Usage level	Number of onsite sanitation systems sampled	Location
Household VIP latrine	Dry	Low usage (<5 users/onsite system)	5	Besters
		High usage (>5 users/onsite system)	5	
	Wet	Low usage	5	Besters
		High usage	5	
Household UDDT toilet		Low usage	5	Mzinyathi
		High usage	5	
Household unimproved pit latrine	Dry	Low to high usage	2	Ocean Drive
Community ablution block VIP	Dry	High usage	9	Malacca Road
School VIP toilet block	Wet and dry	High usage	4	Mzinyathi
Total			45	

Sludge sampling

The faecal sludge in pit latrines varies widely, which makes the comparison between samples collected from the different onsite sanitation facilities challenging. In order to provide a uniform data comparison, a sampling method was developed and applied for selection of samples from different depth levels at the front and back of the pit for all dry VIPs (Figure 3.17, top left). Sample 1 represents a fresh deposit and is usually right beneath the pedestal, sample 5 is partially degraded aerobic faecal sludge but some of the fresh material may have fallen there, samples 2 and 6 are partially degraded aerobic faecal sludge, samples 3 and 7 are partially degraded anaerobic faecal sludge, and samples 4 and 8 are at the bottom of the containment and are completely stabilised and anaerobic faecal sludge.

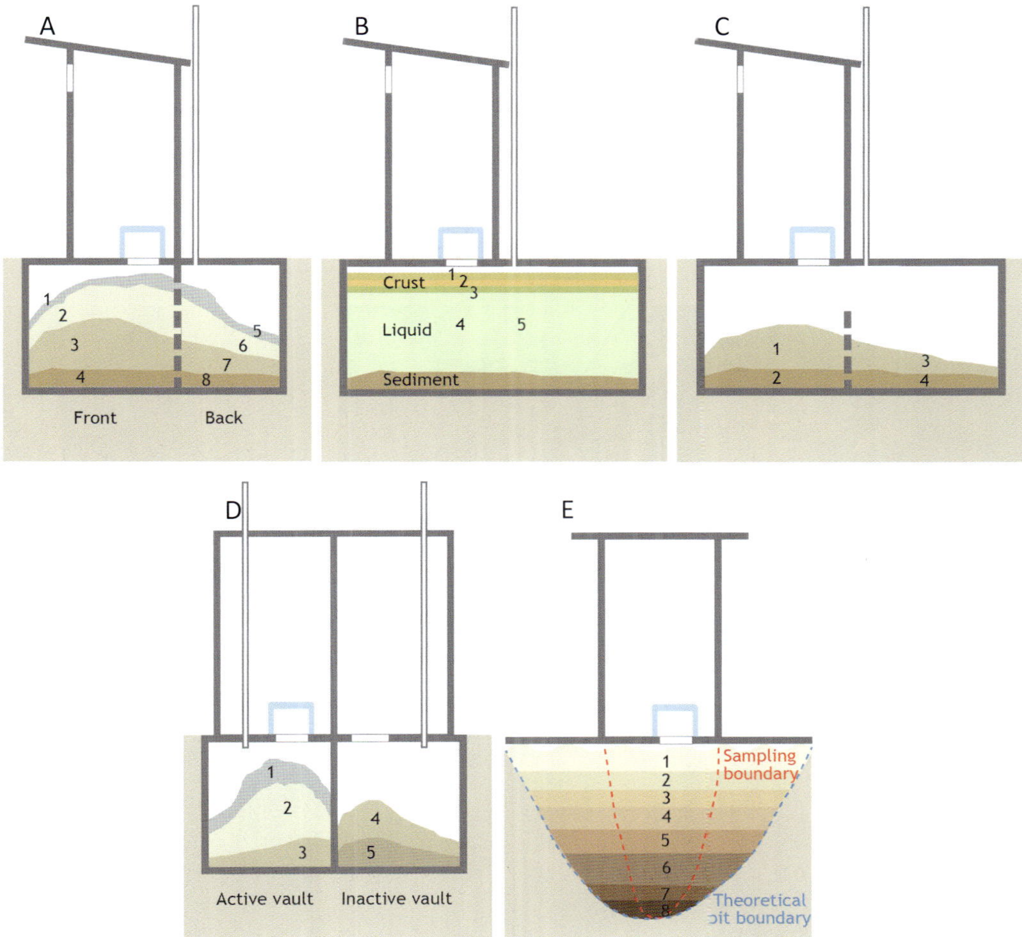

Figure 3.17 The selection of faecal sludge analytical samples used in Case study 3.1 is from: A) dry ventilated improved pit latrines (VIPs), B) wet VIPs, C) school VIPs, D) UDDT toilets, and E) unimproved pit latrines. The numbers illustrate where samples were taken.

A similar approach was followed for the UDDT toilets, where samples were selected from both active and standing vaults (Figure 3.17, bottom left). For the wet VIPs, samples were selected from the sludge crust concentrated at the top of the pit and from the liquid beneath the sludge layer but no distinction was made between the front and the back of the pit (Figure 3.17, top middle). The community ablution block VIPs did not allow for structured sampling, because of the limited accessibility for pit emptiers due to the large size of the containments and large amounts of solid waste. Samples were selected from the top sludge layer and the liquid beneath, similarly to the wet household VIPs. For the school VIP toilets, the sampling procedure was similar to the one followed for dry VIPs. Due to the shallower sludge layers, only four samples were selected from each pit (two from the front and two from the back (Figure 3.17, top right), except for one where six samples were selected in total. For the unimproved pit latrines, seven to eight samples per pit were selected as indicated in Figure 3.17 (bottom right). This procedure was followed as there was no superstructure as for the VIP toilets, hence there were no clear boundaries between the faecal sludge disposed in the pit and the surrounding soil.

On average, eight samples were selected from each dry VIP, between four and six samples from each wet VIP, two to six from each UDDT toilet, two from each CAB VIP, four from each school toilet VIP, and eight samples from each unimproved pit latrine over a period of 18 months, where 211 samples were collected in total. The selected samples had a capacity of approximately 1 litre and were stored in plastic containers at 4°C in a cold room in the UKZN PRG laboratory for further analytical tests.

Figure 3.18 Photographs of faecal sludge samples in Case study 3.2 taken from: A) a dry ventilated improved pit latrine (VIP), B) a wet VIP, C and F) a school VIP, D) a UDDT toilet, and E) an unimproved pit latrine (photos: UKZN PRG).

Sampling from collection and transport vehicles

Sampling from collection and transport trucks is another possibility, and fits the sampling objective of knowing what will be delivered to treatment. Depending on the type of truck, samples can be taken directly from the access port on the top of the truck tank or during discharge from the discharge valve (Bassan *et al.*, 2016). In the first option a core sampling device can be used, while in the second option a composite of grab samples is collected (Figure 3.19).

Figure 3.19 A) collecting grab samples from the truck discharge valve to make a composite sample, and B) collecting a core sample from a truck access port with a 180 cm length PVC core sampling device with a 5 cm internal diameter, Hanoi, Vietnam (photos: Eawag).

The composite sample usually consists of taking one sample at the beginning of discharge, two in the middle, and one at the end (Bassan *et al.*, 2013). When possible, a volume gauge on the back of the truck can be used measure volumes, and to determine when to take samples. Samples should be collected from the truck immediately after emptying, or from the discharge valve immediately upon arrival at the discharge facility. If trucks are left standing for even a short period of time, solids will rapidly start to settle out in the tank. A comparison of sampling methods is provided in Case study 3.3.

Case study 3.3 *Comparison of four sampling methods in Hanoi, Vietnam*

This case study is based on a Master's thesis by Amédé Ferré (2014), a collaborative project between Eawag and the Institute of Environmental Science and Engineering at Hanoi University of Civil Engineering. Sampling methods were evaluated during a characterisation study that took place between September 2013 and June 2014. More than 90% of households in Hanoi have septic tanks, with the overflow going directly to rainwater drains or sewer systems. Samples were taken from six different septic tanks with the number of chambers varying from two to three, and for each of the six septic tanks four different sampling locations were compared. Core samples were taken with a 1.8 m high PVC core sampler with an internal diameter of 5 cm. Grab samples were taken with a 1 L grab sampling device (bucket mounted on the end of a 1 m long bar).

1. Septic tank: samples were taken *in situ* from septic tanks with a core sampling device. This included from the bottom to the liquid surface (*i.e.* a core sample of sludge layer, supernatant and scum layers). However, the specific location in the septic tank where the sampling occurred could not be identified.
2. Truck access port: samples were taken with a core sampling device *in situ* from the access port on the top of the vacuum trucks, immediately following collection of septic tank sludge from households.
3. Beginning discharge: a single grab sample of 2 L taken from the truck valve at the beginning of the discharge.
4. Composite discharge: a composite sample comprised of four grab samples of 1 L each, taken from the truck valve at the beginning, middle and end of the discharge in a ratio of 1:2:1.

Presented in Figure 3.20 is a comparison of the TS and COD results for each of the sampling methods. The results illustrate the importance of sampling location depending on the objective and evaluating bias. The septic tank is more relevant if the objective is to determine sludge accumulation rates in the septic tank, whereas either the truck access port or the composite discharge is preferable for constituents of faecal sludge being delivered to treatment. In the case of thick faecal sludge (septic tanks 1 and 3), the composite discharge may be more representative than the truck access port, whereas for more liquid sludge (*e.g.* septic tank 6) the truck access port may be more suitable (*i.e.* larger supernatant volume). The beginning discharge appears to be biased to solids that settle out in the truck, and are washed out at the beginning of discharge (*e.g.* septic tanks 1, 2 and 3). Further analysis is needed to fully understand the effect of sampling location. Samples were taken from trucks, as service providers were reticent to allow sampling during discharge. There is no legal discharge location in Hanoi, and sampling would draw attention to their illegal discharge (although the businesses are legally registered). Samples were also analysed for total suspended solids (TSS), nutrients, volatile fatty acids (VFA), and proteins, and the raw data is available for download using the link provided in Englund *et al.* (2020).

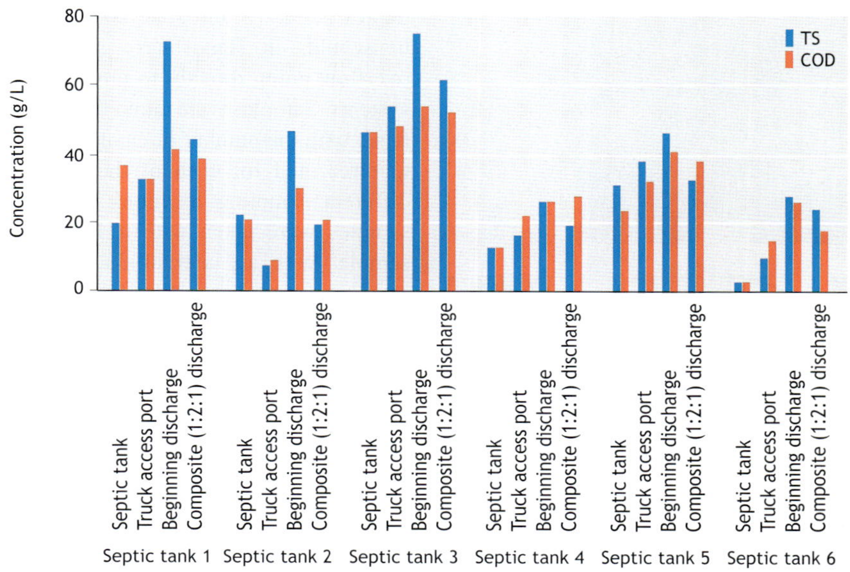

Figure 3.20 Total solids (TS) and organic matter (COD) results for six septic tank samples that were collected with four different methods.

3.6.3 Sampling at faecal sludge treatment plants

FSTPs can have combinations of various technologies such as settling-thickening tanks, drying beds, waste stabilisation ponds, and mechanical dewatering. Sampling locations and strategies will depend on the objective, for example, treatment performance, operational concerns, monitoring, resource recovery, and optimisation of loadings. In general, liquid and solid streams require different approaches to collection and analysis. For the sampling of liquid streams with similar characteristics to wastewater, refer to Meijer and Brdjanovic (2012), and the USEPA operating procedure for wastewater sampling (2017). Below are two examples of sampling at FSTPs; more information on dewatering and drying is available in Chapter 4 and Ward *et al.* (2019).

Case study 3.4 *Assessing FSTP performance*

A hypothetical FSTP in South East Asia consists of two settling tanks in parallel, planted drying beds, and vertical flow constructed wetlands. The effluent is discharged by gravity into a river. The FSTP opening hours are from 8 am to 6 pm on Monday to Saturday. The FSTP operator has defined the sampling objectives as evaluating the FSTP performance to assess future investment needs, and defined the sampling plan summarised in Table 3.3.

Table 3.3 Sampling plan to evaluate a FSTP performance in South East Asia.

Item	Sampling plan	Observation
Sampling location	• Discharge channel right after screening (laminar flow). • Manhole at the outlet of the vertical flow constructed wetlands.	• Flow and turbulence are high in the channel before screening.
Sampling technique	• Composite of 6 grab samples of equal volume taken every 2 hours. • Grab samples.	• Single daily composite. Due to limited human resources and time, interval time between 2 sampling is set at 2 hours (ideally every hour). • The outlet flow composition is assumed to be constant.
Sampling equipment	• Grab sampling device, vertical, 1 L volume with a 1 m rod to collect samples at half depth of the sludge flow in the discharge channel. • Grab sampling device, vertical, with 2 L volume and 3 m rod to access the bottom of the manhole.	• Sampling devices and containers were first rinsed 3 times with the targeted substrates (*i.e.* untreated faecal sludge or effluent). • The beaker is lowered to a depth of around 50 cm into the channel and then inclined to face the flow. It is assumed that the collected sample is representative of the flow.
Storage containers	• 6 x 500 mL PTFE plastic containers. • 2 x 2 L PTFE plastic containers. • 2 x 250 mL sterilised glass containers.	• Sterilised glass containers for further microbiological analyses. • First rinsed 3 times with the targeted substrates (i.e. untreated faecal sludge or effluent).
Sample preservation technique	• The six grab samples are immediately stored in a cool box with ice. • The effluent grab sample is transported to the lab in a cool box together with the two glass containers.	• Since microbiological parameters must be analysed within 6 hours, a single grab sample is taken specifically for these parameters.
Protective equipment	• Rubber boots, protective gown, protective glasses, active carbon filter mask, and rubber gloves.	

Case study 3.5 *Planning for measures to reduce exposure to contamination risk*

A hypothetical FSTP in West Africa consists of unplanted drying beds, each equipped with a discharge channel with a screening grid, a buffering storage tank for treated effluent reuse and a dried sludge storage area. The effluent, if not used, is infiltrated. After being removed from the drying beds, sludge is stored for one year (Figure 3.2.1). The FSTP operator has defined the sampling objective of evaluating compliance of dried sludge with agriculture reuse requirements to reduce farmers' exposure to faecal contamination risk. In order to fulfil this objective, the FSTP operator will assess the pathogen content of the dried and stored sludge, as described in Table 3.4.

Figure 3.21 Dried sludge at a faecal sludge treatment plant in West Africa (photo: A. Ferré).

Table 3.4 Sampling plan to evaluate the compliance of dried sludge with agricultural reuse requirements in West Africa.

Item	Sampling plan	Observation
Sampling objective	• Verify compliance of dried sludge with agricultural reuse requirements	• To reduce farmers' exposure to faecal contamination risk.
Sampling location	• Storage area: stabilised sludge after 1 year of storage	• See example in Figure 3.21
Sampling technique	• Composite of five random single grab samples distributed throughout the stabilised sludge pile.	• Stabilised sludge composition may vary throughout the pile.
Sampling equipment	• Grab device: tongs, spoon, gloves, etc depending on size	• Stabilised sludge is relatively inert, reaction with a plastic container has low probability.
Storage containers	• 1 L PVC container with wide opening	
Composite	• The sub-grab samples will be gently crushed in a mortar and the resulting powder will be mixed.	
Sample preservation technique	• The sample will be transported to the soil laboratory in a cool box with ice.	• No preservative required for microbiological parameters.
Protective equipment	• Rubber boots, protective gown, rubber gloves.	• Risk of ingestion is low.

3.7 SAMPLE SIZE

Guidelines on how to develop sampling and analytical plans taking into account the adequate number of duplicate samples to ensure accuracy and precision are presented in Chapter 8. A detailed plan for quality assurance and quality control (QA/QC) needs to be developed in advance of sampling to take into account the increased number of samples for duplicates and controls. In reality, there are no hard and fast guidelines for determining the 'right' number of samples, and frequently the selected sample number will come down to available time and resources. Even with a limited number of samples, by taking them in a logical fashion with defined objectives and QA/QC procedures in place, the results will still be more meaningful than if collected without these controls in place. In Example 3.1 are sample sizes based on a normal distribution. However, as presented in Chapter 1, faecal sludge does not follow a normal distribution and a statistically valid number of samples cannot be determined until a distribution is known. This means that in reality, the samples actually have to be taken before these assumptions can be validated. It is important to keep in mind that with more samples there is increased accuracy, but the increase is not linear. How to calculate the effect of sample size on uncertainty is discussed in Chapter 5, along with further information and examples of developing sampling plans for community to citywide scales, and statistical relationships that can be used to reduce the required time and resources for analysis.

Example 3.1 *Sample sizes for normal distributions*

If the probability distribution of a sampling population is known, equations exist to determine a statistically significant number of random and independent samples. The number of samples will depend on the selected confidence interval (margin of error) and confidence level. For example, as shown in the table for a normal distribution, if a city has a population of 2,000,000, served by 70% onsite sanitation with an average of 10 users per containment, this would mean 140,000 onsite containments. Based on the values in Table 3.5 with a 90% confidence interval and 5% margin of error, this would mean 270 samples. However, Q&Q of faecal sludge will probably not follow a normal distribution, and a much lower number of samples could logically be selected with a transparent explanation of how the number and type of samples were selected.

Table 3.5 Required sample size to fulfil a confidence interval of 90% and 95% with a margin of error of 5% and 2% for normally distributed data.

Population	Confidence interval = 90%		Confidence interval = 95%	
	Margin of error		Margin of error	
	5%	2%	5%	2%
100	74	95	80	97
200	115	179	132	185
500	176	386	217	414
1,000	213	629	278	706
10,000	264	1,447	370	1,936
100,000	270	1,663	383	2,345
1,000,000	271	1,689	384	2,396
2,000,000	271	1,690	385	2,398

3.8 HEALTH AND SAFETY

It is important to have a health and safety plan in place for sample collection and transport. Personal protective equipment (PPE), as shown in Figure 3.22, must be worn to ensure protection from pathogens and other potentially harmful constituents in faecal sludge, including appropriate handling and cleaning of contaminated clothing. Other safety considerations include working in confined and dangerous spaces, toxic gasses that build up during anaerobic digestion of faecal sludge, water, and electricity, and moving components at FSTPs. The sampling area must also be kept clean to protect the general population from risk of exposure to faecal contamination. Any faecal sludge that is spilled during sampling must be immediately cleaned, and waste matter properly disposed of. For more information, refer to Chapter 8, and for a detailed overview of recommendations, refer to Health, Safety and Dignity of Sanitation Workers An Initial Assessment (World Bank 2019), and for hygiene practices to Louton *et al.* (2018).

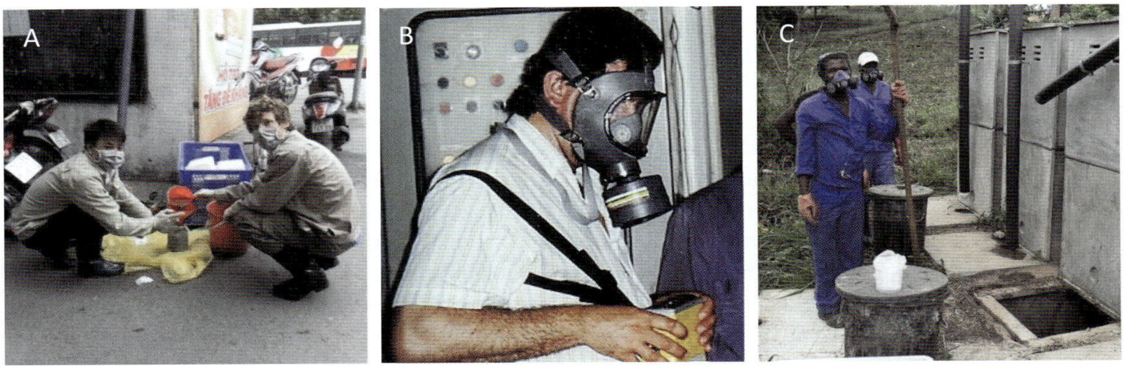

Figure 3.22 Personal protective equipment. A) transferring homogenised samples to sample containers with a plastic funnel and ground protection. B) a mask and meter/alarm for H_2S. C) collecting samples from pit latrines (photos: Eawag, M. Henze, UKZN PRG, respectively).

3.9 SAMPLE COLLECTION

Prior to sampling, arrangements need to be made with the laboratory carrying out the analysis regarding sample volume and laboratory capacity. It is important to consider transportation times, working hours, weekends, and available staff. The minimum required sample volume needs to be determined based on the number and type of analytical procedures to be carried out. An example of calculating the required volume based on planned analysis is presented in Figure 3.23. Extra sample volume should be added to account for potential spillage and other unforeseen needs during processing.

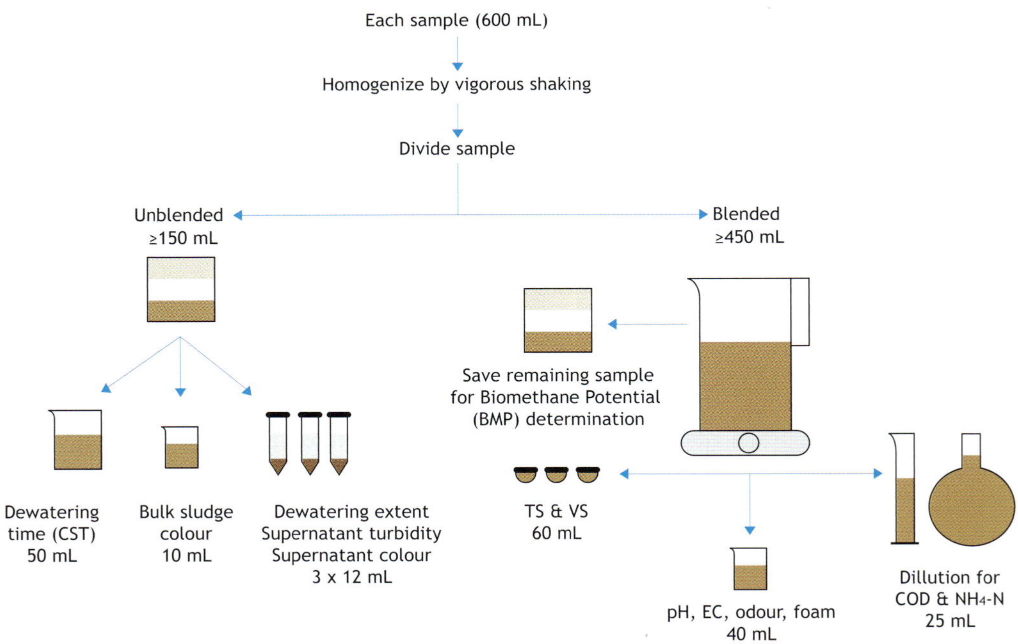

Figure 3.23 Example of how to calculate the required sample volume based on the planned laboratory analyses.

When samples will be sub-divided the sample must be homogenised. This can be done by stirring rapidly with a ladle to get all the particles in suspension and then immediately distributing to sub-sample containers. Whichever method is used, it is important to record the method, and to evaluate the accuracy and replicability. Wide-mouth sampling containers are preferred, and the use of a funnel is recommended for transfer of samples.

If making a time-related composite, all the grab samples must be stored at 4 °C until the entire sampling process is completed. If a refrigerator is not available at the sampling site, then samples should be stored in a cooler box with ice packs. In this case the composite is often prepared at the laboratory.

All the containers used to store the samples should be labelled prior to sample collection to prevent sample misidentification. Labels must be water resistant, and include at a minimum a unique sample number or code, the sampling date, nature of the sample for health and safety, and name of the laboratory where the samples will be delivered. All the sampling equipment and material has to be cleaned immediately after sampling to avoid contamination of future samples and ensure the health and safety of workers.

Equipment used to collect samples should be cleaned in the field with water and detergent. Detergent should be a standard brand of phosphate-free laboratory detergent. Under extenuating circumstances where cleaning in the field is not feasible, equipment can be containerised, bagged or sealed and cleaned upon return to the laboratory. Sampling containers must be properly cleaned prior to use or reuse and, if needed, sterilised in an autoclave. For more information on the specific procedures, methods and considerations to be used and observed when cleaning and decontaminating sampling equipment during the course of field investigations, the reader is referred to USEPA (2015).

3.10 RECORDING OF SAMPLE COLLECTION

Details on each sampling event need to be documented in a logbook immediately at the time of sampling. This documentation is useful for troubleshooting if the laboratory results are atypical or suspect; it serves to demonstrate that the proper sampling protocols were used, and is useful to interpret and compare analytical results. It is good practice to record sufficient information that the sampling procedure can be reconstructed from the logbook alone. Recorded information should include at a minimum:

- sample identification code (specific to sampling event *i.e.* type, location, date, treatment process and condition, etc.)
- number of samples and volume of sample taken,
- type of sample (*e.g.* grab, 24-hour composite), sampling equipment and a brief description of sampling procedures
- volume of sample
- date and time
- sample location, GPS coordinates
- preservatives
- analytical parameters
- name of person who performed the sampling or measurement
- special conditions or remarks, *i.e.* weather conditions at the time of sampling and other observations which could potentially impact the laboratory analytical results
- brief description of the sludge collected, *e.g.* colour, odour, viscosity, consistency.

A chain-of-custody document is required to provide a record of sample transfer from person to person including everyone involved from taking the sample until delivery at the laboratory, and at what time they had the samples. All the personnel need to sign the form with the date and time of day, along with the sample ID code (see also Chapter 8).

3.11 TRANSPORT

When analysis will be performed away from the sampling location, the faecal sludge samples must be packaged and transported. Samples should be delivered to the laboratory as soon as possible following collection, and the travel time and conditions need to be recorded. Samples typically need to be transported in a cooler with ice packs to maintain a sample temperature of 4 °C for the duration of the collection and transport. Faecal sludge sample containers must be packaged in order to protect them and reduce the risk of leakage. Containers should be held upright and cushioned from shock. For more details on samples handling reader is referred to Chapter 8 and reference literature (*e.g.* Rice *et al.*, 2017 and Van Loosdrecht *et al.*, 2016).

3.12 STORAGE AND PRESERVATION

Preservation of samples is crucial to allow reliable analytical results. Sludge composition changes over time, depending on factors such as light, oxygen, temperature and microbial activity, and therefore preservation techniques are required to slow down or stop/inhibit these processes. Analyses should only be done on well-preserved samples, and within the period in which the results will be representative of the initial sludge composition as stated in methods presented in Chapter 8. Samples should always be stored at a temperature of 4 °C to limit biologically induced changes. When several grab samples are collected with the purpose of making a composite, all the grab samples must be stored and preserved at 4 °C during the whole sampling process. Some microbial analysis requires storage preservation at -20 °C or -80 °C for storage longer than 24 hours, whereas samples can be dried and stored for later analysis with acid digestion (*e.g.* heavy metals) or combustion (*e.g.* calorific value, carbon, carbon, hydrogen and nitrogen elemental concentrations). For biologically active samples, it is important to label with an appropriate warning, and to allow gases to vent to avoid explosion.

The same considerations for sample containers need to be considered as discussed in Section 3.5; sampling and storage containers must be made of materials that will not contaminate or react with the faecal sludge. Polypropylene, polycarbonate, HDPE, Teflon, glass, and stainless steel are relatively inert and are all appropriate for sampling. The cost of Teflon and stainless steel equipment might prohibit or restrict their use, and potential for breakage should be considered with glass. If using steel equipment, depending on the analysis, galvanised or zinc-coated items should not be used because these materials will release zinc into the sample. Other considerations for interaction include silica, sodium, and boron which may be leached from soft glass but not plastic, and trace levels of organics and metals may sorb onto the walls of containers. In all cases, opaque containers are recommended to protect the sample from the light.

The addition of preservatives to the sample container can increase the preservation time of the sample from a few days to a few weeks. However, preservatives also change the composition of the sample and can affect the properties, so their usage has to be carefully evaluated. In this case, it is recommended to only use preservative in a sub-sample of the original sample. Chemical preservatives should only be used when there is no interference with the analyses that are still to be made. However, all methods of preservation may be inadequate when applied to suspended matter. Preservatives should not be added if analysis of volatile, semi-volatile or microbial contaminants are to be done, unless specified methods. For solid sludge samples ('cake' with total solids >25%), adding a chemical preservative is generally not useful since the preservative does not usually penetrate the sludge matrix.

3.13 EXAMPLE OF SAMPLING KIT

An example of a check list for a typical sampling kit is presented in Figure 3.24. For more information on the associated paperwork and health and safety forms, the reader is referred to Chapter 8.

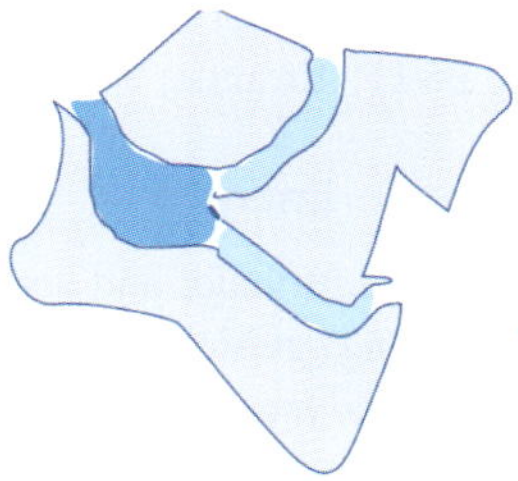

Figure 4.1 Representation of the different forms of water in a sludge floc (adapted from Bassan *et al.*, 2014)

This chapter first discusses the general scales of experimental work and introduces a methodology for experimental design and how to apply these concepts to faecal sludge treatment processes. This is followed by background information that is necessary to apply the concepts for scaling-up dewatering and drying, together with five case studies: two for conditioning for improved dewatering, and three for thermal drying for energy recovery. The background and case studies provide examples of how to implement the methods presented in Chapter 8. Prior to conducting experiments at any scale, preliminary research must first be completed. This includes a literature review to learn from experience, and to ensure efforts are not unnecessarily replicated.

In this era of virtual communication, open access to many materials and online communities of researchers and practitioners has made it easier to share findings and obtain feedback and support. To put this advantage to good use, the sharing of research results and raw data is strongly encouraged.

4.2 EXPERIMENTATION IN FAECAL SLUDGE MANAGEMENT

Before starting experiments, it is important to become familiar with the key elements used in setting up experimental work. Experiments are a way to understand cause-and-effect relationships in a system, by deliberately changing conditions in a controlled fashion, and observing the changes in the system that are produced as a result of what has been altered. An experiment can be defined as '*a series of runs in which purposeful changes are made to the input variables of a process or system so that we may observe and identify the reasons for changes that may be observed in the output response*' (Montgomery, 2019). A run is one component of an experiment, conducted with a specific set of input variables. Tests are measurements of specific faecal sludge characteristics or properties, as described in chapters 2 and 8. Tests can be part of an experiment (*e.g.* measuring pathogen levels in treated faecal sludge), but can also be conducted outside of a planned experiment. For example, laboratory tests are used for routine characterisation to monitor the performance of existing treatment systems.

4.2.1 Scales of experiments

In the development and scaling-up of treatment technologies, reasons for experimentation will include developing fundamental knowledge (*e.g.* mechanisms controlling dewaterability), designing and developing new processes or technologies (*e.g.* LaDePa), and transferring and optimizing existing technologies (*e.g.* geotextiles, pelletisers, and conditioners). To accomplish this, the different levels of laboratory-, pilot-, and full-scale experimentation are employed depending on the stage of development and specifically defined objectives.

Laboratory-scale experiments

Laboratory-scale experiments are conducted in a laboratory, often using existing conventional analytical equipment and Standard Operating Procedures (SOPs). This is the smallest, bench-scale for experimentation, typically using low volumes of faecal sludge (*i.e.* mL to several L). Laboratory-scale experiments allow for controlled conditions when the experimenter wants to investigate the isolated effects of specific process parameters. This scale of experimentation lends itself well to comparisons with results from other researchers, as they should be replicable in other laboratories with the same setup. One caveat is variability in faecal sludge, which can be addressed through experiments that include simulant faecal sludge, as presented in Chapter 7. Laboratory-scale experimentation is also often used for establishing proof-of-concept for a new technology, and answering questions about fundamentals and mechanisms involved with faecal sludge treatment.

Pilot-scale experiments

Pilot-scale experiments are regarded as a necessary step on the way from laboratory-scale research to full-scale process optimisation and implementation (Wood-Black, 2014). Typical pilot-scale experiments operate at capacities between 50-2,000 L of faecal sludge per day. Pilot-scale experiments help to answer questions about practical operation and feasibility of the process. Reasons for piloting a treatment process can be predicting costs and energy requirements, establishing the needs for process control, understanding practical operating conditions, and anticipating any potential unforeseen impacts of adopting a new technology or process unit on the rest of a faecal sludge treatment plant (FSTP). Pilot-scale experiments can ultimately be used to determine whether it is feasible to implement a new technology at full-scale.

Full-scale experiments

Full-scale experiments are conducted at existing FSTPs that have been designed for treatment capacities ranging from 1,000 - 800,000 L of faecal sludge per day (Klinger *et al.*, 2019). Experiments that take place at full-scale are used to optimise FSTP performance. The FSM sector is undergoing rapid change, so transitions from pilot- to full-scale application for many treatment processes is expected to happen with increasing frequency in the near future. However, while full-scale experiments are necessary to make faecal sludge treatment as effective, efficient, robust, and sustainable as possible, they must always be balanced with the responsibility of maintaining certain standards of treatment for the protection of public and environmental health.

Working with faecal sludge and with transferring or innovative treatment technologies inherently includes uncertainties and risks that need to be managed. The transitions between laboratory-, pilot-, and full-scale experiments may be iterative and will require time and dedication to achieve high-quality experimental design and execution. It is critically important to incorporate a research component into any faecal sludge treatment project from its inception. Risks can be mitigated by forming partnerships between municipalities and universities/research institutes, which can help guide experimentation from the start of the project to the optimisation and monitoring of a full-scale FSTP (Strande, 2017).

4.2.2 Designing an experiment

After identifying the purpose, rationale, and scale of experimentation, the following guidelines adapted from the book Design and Analysis of Experiments (Montgomery, 2019) can be used to design experiments. Montgomery (2019) can also be consulted for detailed information about experimental design and statistical analysis for process engineering. In addition, information on experimental methods is available in Van Loosdrecht *et al.*, 2016 and on experimental data handling and analysis in Von Sperling *et al.*, 2020.

The experimental design guidelines are presented here, together with examples specific to helminth inactivation during drying (in italics):

1. Specify the research question.
   What is the optimum retention time for drying of faecal sludge in an infrared dryer to achieve complete helminth inactivation?
2. Select the response variable to measure.
   Mean percentage viability of helminths after drying.
3. Identify relevant design factors, levels, and ranges over which the experiment should operate.
   Infrared drying technology can operate over the range of 105-125 °C, so the retention time will be evaluated at 105, 115, and 125 °C. Retention times of 10, 30, 60, and 120 seconds will be evaluated at each temperature.
4. Identify factors that could influence the response variable, and evaluate if they can be kept constant during the experiment.
   Moisture content of air, characteristics of faecal sludge.
5. Identify laboratory methods and SOPs to measure the response variable, influencing factors, and operating conditions.
   See Section 8.9.3 Helminth Method.
6. Determine how many replicates to run to determine the uncertainty in your response variable.
   Triplicate runs for each combination of temperature and residence time.

7. Develop a QA/QC protocol to ensure meaningful results (*e.g.* standards, blanks, duplicates).
   Use Ascaris suum egg standards with known egg count and percentage viability as a positive control, and sludge simulant as a negative control. Prepare 3 positive controls by spiking sludge simulant with Ascaris suum egg standards. Once the drying experiments have been carried out, test the negative control and positive controls along with the test samples, as per Section 8.9.3 Helminth Method.
8. Perform the experiment.
   Carry out experiments as previously described; write down any deviations from the original plan.
9. Interpret the results.
   Visual inspection of data, graphs, statistical interactions, and empirical models.
10. Define the next steps based on conclusions and recommendations from interpretation of the results.
    All residence times tested at 125 °C yield complete helminth egg inactivation; 10 seconds is the recommended residence time based on these results. Conduct further tests on a broader range of sludges, and a cost-benefit analysis of the operating temperatures.

Presented in this chapter are five case studies for dewatering and drying of faecal sludge, together with adequate background information for understanding of the case studies.

4.3 TRANSFERRING TECHNOLOGY: CONDITIONING TO IMPROVE DEWATERING

Presented in this section is background information on the use of conditioners to improve dewatering of sludge, followed by two real-life case studies of experimental design for faecal sludge conditioning processes.

4.3.1 Introduction to faecal sludge dewatering with conditioners

Prior to dewatering, faecal sludge can be up to 99% water by weight. Separation of solids and liquids is required in order to fully treat the liquid fraction before end use or discharge into the environment. It is also required before treatment of the solids fraction for disposal or end use, and enables more efficient transportation of the solids fractions.

Separating the solids and liquids in faecal sludge can be achieved through settling (*e.g.* settling-thickening tanks), filtration (*e.g.* drying beds or geotextiles), or mechanical methods (*e.g.* screw presses, filter presses, or centrifuges). Settling-thickening tanks (Figure 4.2) and drying beds (Figure 4.3) are widely-used, established technologies for faecal sludge treatment, however they require large areas of land and long residence times to sufficiently dewater sludge.

Figure 4.2 Settling-thickening tanks at Lubigi FSTP in Kampala, Uganda (photo: Eawag).

Figure 4.3 Drying beds at Niayes FSTP, Dakar, Senegal (photo: Eawag).

Depending on its specific properties, sometimes sludge dewaters more quickly and thoroughly, and other times dewatering performance is quite poor. To address this, transferring technologies from wastewater treatment, such as geotextiles (Figure 4.10, Case study 4.2) or mechanical presses are being considered to increase throughput and treatment performance, and reduce footprint. However, these transferring technologies do not reliably or predictably perform without the addition of dewatering aids called 'conditioners'.

Conditioners are chemicals that are used to improve dewatering and settling performance. They are well-established in wastewater and water treatment, food processing, and the pulp and paper industry, which have relatively more homogenous waste streams than faecal sludge. Empirical and observational knowledge is starting to be gathered about conditioning of faecal sludge at the laboratory- and pilot-scale, but very little fundamental knowledge is available. Further experimentation at all scales will be necessary to scale up the use of conditioners.

Conditioners are mixed into a slurry or suspension, and added to sludge during treatment at optimal 'doses'. Selection of the optimal conditioner and dose of that conditioner are based on physical and chemical characteristics of the sludge to be dewatered. Accurate dosing is required, as both under-dosing and over-dosing result in poor flocculation, which results in quickly clogged filters and slow or incomplete dewatering performance (*i.e.* increased organic loadings in the filtrate, clogged or blocked drying beds or geotextiles, and higher residual moisture in dewatered faecal sludge). Dosing needs to be frequently reassessed and varied in response to changes in influent characteristics, and is based on online monitoring, making the high variability in quantities and qualities of influent faecal sludge currently a barrier to implementing them at scale.

Recent research on the optimal dosing of conditioners for faecal sludge has been based on laboratory testing, which is too time- and labour-intensive to be scaled up. However, it indicates that when the right conditioner and dose are applied, significant improvements to faecal sludge dewatering performance are possible; for example, faster dewatering on drying beds, and cleaner effluent from drying beds, settling-thickening tanks, and geotextiles (Gold *et al.*, 2016). Research needs to be directed at developing methods to rapidly characterise influent faecal sludge quantities and qualities (Q&Q, see Chapter 5) to determine the conditioner dose (Gold *et al.*, 2018, Ward *et al.*, 2019). In addition, considerations such as cost, availability, supply chain, chemical safety, and possible requirement of additional infrastructure (storage tank, dosing device, mixing tank) need to be taken into account when designing experiments and selecting conditioners and dosing processes to apply at pilot- and full-scale.

4.3.2 Types and mechanisms of conditioners

The following section has been adapted from Chapter 5 (Section 5.2) of the book Faecal Sludge Management: Highlights and Exercises (Ward and Strande, 2019), and provides additional background information on the use of conditioners to improve dewatering performance of faecal sludge, and methods for measuring performance.

Conditioners can be inorganic chemicals such as lime, ferric chloride or aluminium sulphate, or they can be charged polymers ('polyelectrolytes'). Polymers can be locally produced from natural materials, such as chitosan or *Moringa oleifera*, or can be proprietary materials sourced from chemical companies. It is expected that cationic (positively charged) polyelectrolytes will work best with faecal sludge, as they will be more likely to interact with organic particles, which are negatively charged. Conditioners work by destabilising small suspended particles to form larger aggregates (shown in Figure 4.4). This happens through coagulation, which is the initial destabilisation and aggregation of colloidal particles. This is followed by flocculation, which is the formation of larger particles, or 'flocs', from smaller particles.

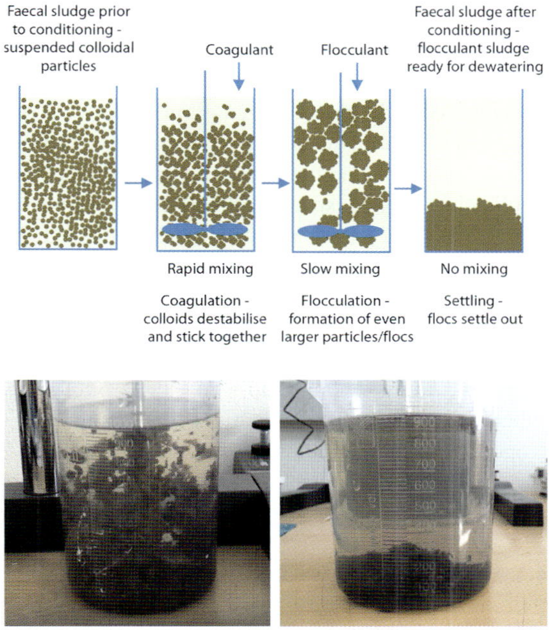

Figure 4.4 Above: steps in faecal sludge conditioning, coagulation, flocculation, and sedimentation; below left: flocculation of faecal sludge; below right: settling of faecal sludge flocs (figure: adapted from Ward and Strande, 2019, photos: IHE Delft).

4.3.3 Key parameters for selection of conditioners and optimal dose

The selection of conditioners and the optimal dosage is specific to the faecal sludge properties, the dewatering technology, and the mixing conditions of the chemicals with the sludge.

- *Faecal sludge properties*: conditioners are commonly dosed as a function of total suspended solids, or with faecal sludge often total solids is used in the absence of total suspended solids measurements. Other factors such as the electrical conductivity or the degree of stabilisation may influence which type of conditioners work best.
- *Dewatering technology*: conditioners need to be compatible with technologies used for dewatering. For example, centrifuge dewatering requires conditioning with polymers that produce flocs that are resistant to high shear (*i.e.* very high molecular weight, and usually branched or structured polymers).
- *Mixing conditions*: complete mixing of faecal sludge with conditioners is necessary to make the particles collide and stick together (coagulate) and grow into flocs (flocculate); however, mixing speeds need to be selected to avoid floc destruction. Mixing for coagulation needs to be vigorous in order to cause many particle collisions. However, mixing for flocculation needs to be gentle to keep flocs from breaking up. This should also be considered during the selection of pumps for example, for the transfer of conditioned faecal sludge from a settling-thickening tank to a drying bed.

Use of conditioners will also impact the properties of the dewatered faecal sludge, which needs to be taken into account when designing further process steps. Conditioners can increase total solids production, and affect the rheology and water-binding behaviour of the conditioned sludge.

4.3.4 Laboratory- and pilot-scale testing

The following methods used to evaluate the suitability of conditioners in faecal sludge are included in the Chapter 8:

- *Jar test*: a common method for testing conditioner performance at different doses. Faecal sludge is mixed with different doses or different types of conditioner. After mixing, the settling and/or dewatering performance of the conditioned faecal sludge is compared to unconditioned faecal sludge (Figure 4.5).
- *Sludge volume index (SVI)*: a metric for settling performance using Imhoff cones (Figure 4.6).
- *Chemical oxygen demand (COD)*: a metric for organic loading in the supernatant after settling, or in the filtrate after filtering.
- *Total suspended solids (TSS)*: a metric for particulate loading in the supernatant after settling, or in the filtrate after filtering.
- *Capillary suction time (CST)*: a metric for dewatering time (Figure 4.7).
- *Dewatered cake dryness*: a metric for dewaterability, determined by dewatering using a centrifuge or a lab-scale filter press. Dry solids fractions in the dewatered sludge cake are measured and compared.

Figure 4.5 Example of a jar test setup to test suitability of a conditioner (photo: Eawag).

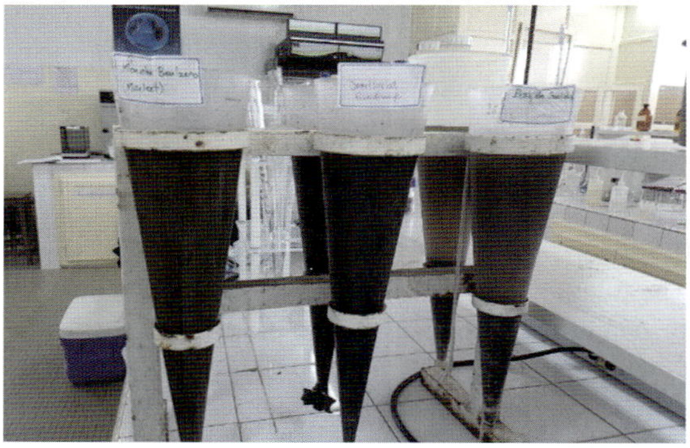

Figure 4.6 Example of an SVI settling test setup with graduated Imhoff cones (photo: Eawag).

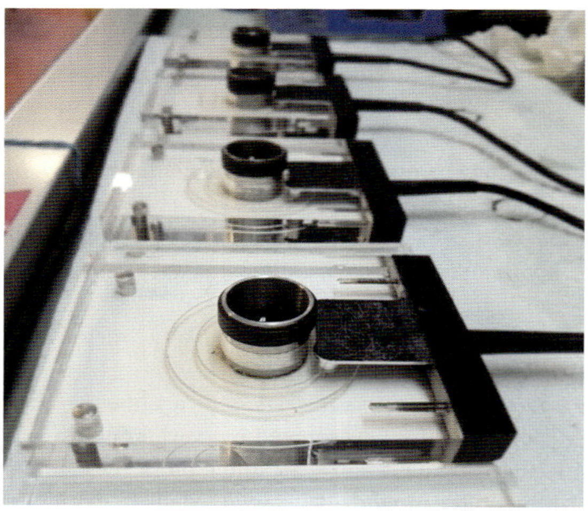

Figure 4.7 Example of replicates being measured in a CST test to determine sludge dewatering time (photo: Eawag).

At the pilot-scale, similar experiments can be conducted with settling-thickening columns, pilot-scale drying beds, or pilot-scale mechanical presses. Specific considerations when transitioning from laboratory experiments to pilot-scale conditioner trials include mixing conditions and sampling protocols. Replicating mixing speed and turbulence achieved during laboratory-scale jar tests is often difficult at pilot-scale. The shape and power of the mixer, and shape and aspect ratio of the mixing tank influence the completeness of mixing, and may therefore alter the optimal dose. Sampling protocols are another point to consider when scaling-up. If the pilot-scale experiments require a comparison of faecal sludge properties before and after conditioning, mixing, and dewatering, the pilot facility should be designed to accommodate this.

4.3.5 Case studies – conditioning for improved dewatering

In the following case studies, examples are provided of (*i*) a laboratory-scale comparison of different conditioners followed by discussion of how to implement pilot-scale testing on drying beds, and (*ii*) an account of a pilot-scale study of online conditioner dosing combined with geotextile dewatering, with lessons learned for full-scale implementation.

Case study 4.1 *Evaluating conditioners produced from locally-available materials for improved faecal sludge dewatering in Dar es Salaam, Tanzania*

This case study is based on a two-year Master´s project by Nuhu Moto at the University of Dar es Salaam (UDSM), a collaborative project between Eawag and UDSM in Dar es Salaam, Tanzania (Moto *et al.*, 2018). This project was motivated by the desire to increase the capacity of unplanted drying beds at an FSTP. Laboratory-scale experiments were conducted to find out whether conditioners could be a possible treatment option for faecal sludge in Dar es Salaam. Two conditioners, which could be produced from locally-available materials, were compared using jar tests, and conclusions were drawn about which conditioners and which doses to select for pilot-scale drying bed trials.

Research question

Which locally-available conditioners and at which doses should be selected for pilot-scale trials?

Response variables

- CST was used to quantify filtration time.
- TSS of the supernatant after settling was used to quantify particulate removal.

Factors, levels, and ranges

- Type of conditioner tested.

  Two types of conditioners that could be manufactured from locally-available natural materials were tested: chitosan and *Moringa oleifera.*

- Conditioner dose
  0, 1, 2, 3, 5, and 8 mg/gTS for chitosan and 0, 10, 50, 100, 250, 500, 750, 1,000 mg/gTS for *M. oleifera*

Factors that might influence the response variables

- Mixing speeds and durations and beaker size/shape can influence results of a jar test. To avoid interference from these factors, consistent mixing speeds, mixing durations, and beakers were used for all of the jar tests.
- Physical-chemical characteristics of faecal sludge (TS, TSS, pH, conductivity) can affect how well a conditioner works. To account for this, one large faecal sludge sample was used for every jar test, and care was taken to homogenise the sample well so that all the beakers contained representative sludge. To make sure that they were not selecting the best conditioner and dose for just one specific batch of sludge, jar tests were run with multiple faecal sludge samples.
- Faecal sludge processing procedures (*e.g.* homogenising with a blender) can change the dewatering performance of a sludge. Blending can disrupt particles and flocs, which can change dewatering behaviour, so homogenisation was done by hand mixing so as to not destroy particles.

Experimental design details

The number of replicates was based on suggestions in standard methods for specific SOPs. An optimal conditioner dose was defined as the lowest dose that achieves > 75% reduction of CST (based on literature, explained in Ward and Strande, 2019).

Interpreting the results

To determine the optimal doses of chitosan and *M. oleifera*, jar tests were performed with the following concentrations of conditioners, and the CST and TSS of supernatant were measured. Results for CST are shown in Table 4.1 and Figure 4.8. Trends in TSS were similar to trends in CST, and are not shown.

Table 4.1 Results of jar tests to determine the effect of different doses of conditioners chitosan and *M. oleifera* on CST reduction.

Conditioner	Dose (mg/gTS)	Reduction in CST (%)
Chitosan	0	0
	0.5	45
	1	60
	2	79
	3	88
	5	90
	8	92
M. oleifera	0	0
	10	13
	50	25
	100	33
	250	68
	500	83
	750	87
	1,000	80

The results indicated that for this sludge, the optimal dose of Chitosan is approximately 2-3 mg/gTS, and the optimal dose of *M. oleifera* is approximately 250-500 mg/gTS (the red dots in Figure 4.8).

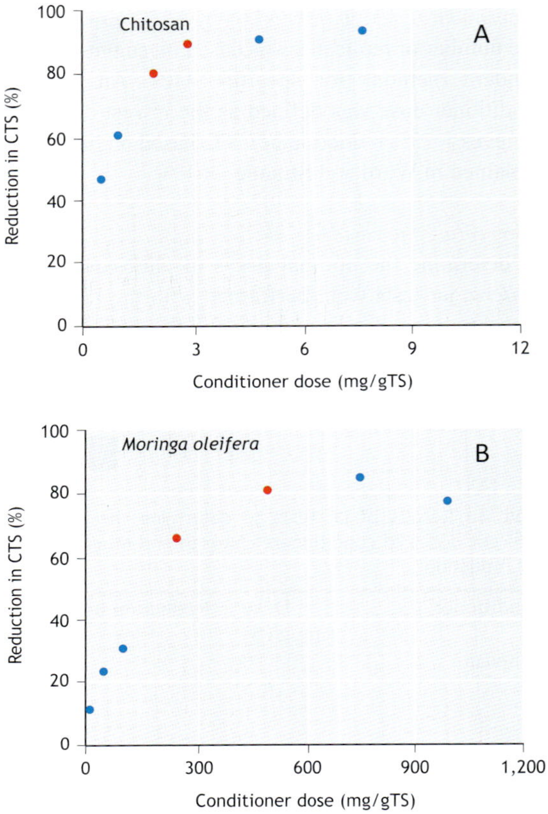

Figure 4.8 A) Results of jar tests with chitosan. B) Results of jar tests with *M. oleifera*. The red dots indicate the optimal dose of each conditioner.

Figure 4.9 A) mixing chitosan conditioner for pilot-scale trials. B) the pilot-scale dewatering research facility at the University of Dar es Salaam, including the settling-thickening tanks, conditioner mixing tank, and six sand drying beds (photos: Eawag).

Scaling-up from laboratory to pilot-scale

Both conditioners that were tested achieved similar performance in terms of CST and TSS reduction, but the optimal doses for each were very different. In Dar es Salaam, chitosan was estimated to cost 15 US$/kg and *M. oleifera* 30 US$/kg. The cost of each conditioner at optimal dose for 1 tonne of faecal sludge (with TS of 10 g/L) would be 0.38 US$ for chitosan and 112 US$ for *M. oleifera* (see Ward and Strande, 2019 for full details). Because *M. oleifera* was prohibitively expensive at the optimal dose, only chitosan was chosen to proceed to the pilot-scale trials (Figure 4.9).

New research questions were developed for the pilot-scale experimentation, including:

- Does chitosan decrease residence time on unplanted drying beds?
- Can chitosan be used to condition every batch of incoming faecal sludge, or does it only work for sludge with certain physical and chemical characteristics?

- Does the benefit of reduced residence time on drying beds justify the cost of conditioners?

For more information on the results, refer to Moto *et al.*, 2018.

Case study 4.2 Scaling-up conditioner dosing for full-scale faecal sludge dewatering

This case study is based on research by Naomi Korir, Jonathan Wilcox, and Catherine Berner at Sanivation in Naivasha, Kenya. This pilot-scale research was done to inform the design of a full-scale dewatering process for a new FSTP in Naivasha, Kenya (capacity 4,000 tonnes faecal sludge per month, delivered by vacuum trucks from pit latrines and septic tanks). Requirements for the plant included a small treatment footprint for the dewatering step, and economic viability. Previous laboratory-scale research characterised hundreds of samples of faecal sludge from Naivasha and established the selection of polymer conditioner and the optimal dose for flocculation. Sanivation wanted to scale up dewatering with geotextiles. To do this requires experimentation for the online dosing, as presented in Section 4.3.1. Because of the iterative experimental approach, questions should be answered one at a time. Therefore the following experiments were carried out on the assumption that geotextiles would work. The pilot-scale setup was sized to process sludge from one vacuum truck at a time, and was designed to test different online conditioner dosing and mixing configurations followed by a subsequent dewatering step using geotextile skips suspended on metal supports (Figure 4.10).

Research question

What is the optimal configuration for online dosing and mixing of conditioners?

Response variables

Sanivation defined the 'optimal' dosing configuration as one that yields fast dewatering and low solids loading in the filtrate while requiring the lowest possible conditioner cost.

- Dewatering time was the amount of time it took for sludge to dewater in geotextile skips (residence time); sludge was considered 'dewatered' when it reached 15-20% TS (80-85% moisture). This benchmark was chosen as it is the required input dryness for Sanivation's heat treatment method, the next step in the treatment process.
- Filtration efficiency was used to quantify how well the geotextiles filtered solids from the incoming faecal sludge. Filtration efficiency was calculated using measured values of TSS of the influent faecal sludge (TSS_{FS}) and of the filtrate leaving geotextile skips ($TSS_{filtrate}$), using the following equation:

Figure 4.10 A) a geotextile skip setup at the pilot facility; B) a geotextile skip being loaded with conditioned faecal sludge; C) dewatered sludge ready to be unloaded from a geotextile skip (photos: Sanivation).

$$\text{Filtration efficiency} = \frac{TSS_{FS} - TSS_{filtrate}}{TSS_{FS}} \quad (4.1)$$

Every batch of filtrate was also characterised for TS, COD, BOD, ammonia and nitrates, to understand the removal of different pollutants by the geotextiles, and the type of treatment that would be required to treat the liquid effluent to required standards (NEMA Standards).

- Cost of polymer per tonne faecal sludge was used to predict material costs for a full-scale process.

Factors, levels, and ranges

- Dosing configurations: different numbers of dosing ports (one or multiple dosing ports) and different mixing conditions (no mixing, mixing with baffles, mixing with a mechanical stirrer) were tested (Figure 4.11). Figure 4.12 shows the actual setup.

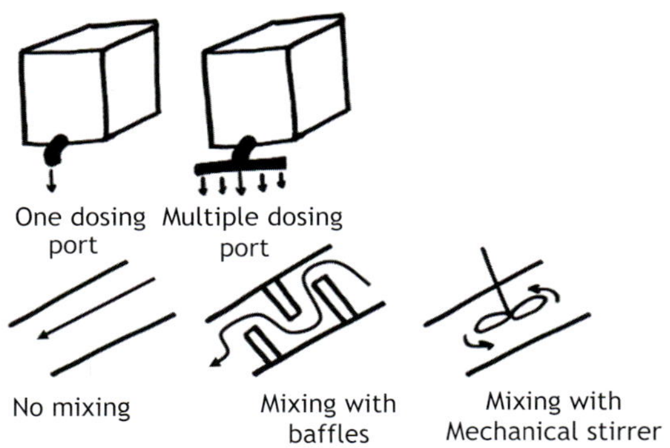

Figure 4.11 Diagram representation of the different conditioner dosing and mixing configurations evaluated by Sanivation.

- Conditioner doses: the laboratory-scale conditioner experiments indicated that the optimal polymer conditioner dose was 2 g polymer per kg faecal sludge; however, the Sanivation team suspected that due to different mixing conditions at the pilot-scale, the optimal dose for the scaled up process could be different. Doses of 2-60 g polymer per kg faecal sludge were tested at the pilot-scale.
- Geotextile cleaning methods: geotextiles were cleaned to determine whether their lifetime could be extending by cleaning between receiving batches of faecal sludge. Three cleaning methods were investigated detergent, detergent + salt, detergent + salt + high-pressure water rinse (Figure 4.12).

Figure 4.12 A) a Sanivation employee washes detergent, salt, and particulate residue from a geotextile using a high-pressure water rinse and B) example of a conditioner dosing configuration: one dosing port followed by mixing with baffles (photos: Sanivation).

Factors that might influence the response variables

- Age of the geotextile/frequency of cleaning can affect dewatering time. New geotextiles dewater quickly (several minutes for septic tank faecal sludge, several hours for pit latrine faecal sludge), but older geotextiles require more time. To account for this, trials were carried out with three geotextile skips that were the same age and had undergone the same cleaning regimen.
- Weather: rain and humidity can affect how long it takes sludge to dewater, since geotextile boxes were open to the air and could gather rainwater. To account for this, the Sanivation team set up a tray that was exposed to the same conditions as the geotextiles. At the end of the study, no rainwater had accumulated in the tray. Physico-chemical characteristics of sludge can change the optimal dose and dewatering speed. Every batch of incoming faecal sludge was characterised for TS, TSS, COD, BOD, ammonia and nitrates. Sanivation engineers designed different conditioner dosing flow rates for pit latrine sludge and septic tank sludge to account for higher levels of observed TS in sludge from pit latrines and lower levels observed in septic tanks.

Experimental design details

Each dosing configuration and each geotextile cleaning method were typically trialled with at least one batch of pit latrine sludge, and one batch of septic tank sludge. If the first repetition was not successful, then further replicates were not completed. For promising configurations, more replicate testing was performed to determine the reproducibility and variability of performance.

Interpreting the results

The optimal conditioner dose was not directly transferable from lab-scale studies to pilot-scale. Different, less ideal mixing conditions at the pilot-scale called for increased doses of polymer to be used to account for incomplete mixing with sludge particles. Multiple dosing ports performed better than a single port, and the addition of both baffles and mechanical mixing led to the most thorough mixing of conditioner and subsequently the shortest dewatering times in the geotextile skips (less than 5 days compared to 14 days with not optimal conditioner dose) and highest filtration efficiency. With the optimal setup, polymer doses from 2-8 g/kg produced the best results. Overdosing occurred at doses over 8 g/kg, resulting in immediate clogging of the geotextiles and a prolonged dewatering time. The team continued to experience issues with achieving precise dosing with respect to TS. Because of this, it was difficult to avoid overdosing even when doses < 8 g/kg were targeted.

Geotextiles were able to be reused after employing the optimal cleaning method: detergent + salt + high-pressure washing. After cleaning, geotextiles were restored to about 30% of the performance of original unused geotextile at negligible material cost increase. However, cleaning was labour-intensive and required 1.5 hours of work to clean every bag after every loading/unloading cycle.

Scaling-up from pilot to larger-scale FSTP

Based on their performance at pilot-scale, the Sanivation team decided not to scale up geotextile skips. This decision was based on the estimated land area required for dewatering using performance data from optimised dosing, mixing, and geotextile cleaning processes in place (with mechanical mixing, multiple dose ports, and cleaning between every load cycle). The average residence time in the geotextile skips at optimal conditions was 5 days per truckload. The full-scale FSTP is designed for a capacity of 20-25 truckloads per day, and the footprint of a geotextile skip is 8 m^2. In the best-case scenario involving constant operation 7 days/week and just one day to unload and clean a geotextile skip, 150 geotextile skips would be required, which means 8 $m^2 \cdot 150 = 1{,}200$ m^2 or 0.12 hectares of land would be required for dewatering (10% of the entire land allotment for the new FSTP). Labour costs were also a significant factor in the decision not to scale up geotextiles. Sanivation also identified that geotextiles can be reused for dewatering up to 10 times with washing in between loadings.

Sanivation is moving forward with the design and implementation of their full-scale FSTP, and will proceed with their optimal polymer dosing configuration. However, the team will switch to a screw press as an alternative, lower-footprint technology. The screw press technology is more resilient to overdosing and the team hopes it will not clog as easily as geotextiles. Screw presses operate continuously instead of being batch processes, allowing for a higher throughput of 20 m^3 sludge per hour. The allotted footprint of the full-scale dewatering process is 120 m^2, an order of magnitude lower than geotextiles would have allowed. Piloting experiments with screw presses are now planned in order to inform the FSTP design. New research questions can be asked, for example, 'What are the optimal operation conditions of the screw press (hydraulic loading rate, conditioner dose, wash water flow rate)?'.

Fast, easy, and reliable methods for online measurements to adjust conditioner doses are still lacking. This is one of the key research topics that needs to be addressed in order to avoid overdosing and reduce conditioner costs. Research is actively being pursued to advance this knowledge (Ward *et al.*, 2021). When accurate methods for online dosing have been adequately developed, the use of geotextiles will be more readily transferable to faecal sludge. However, there are other cases where geotextiles are currently being successfully employed for dewatering, for example, the Dumaguete FSTP in the Philippines (Strande, 2017).

4.4 TRANSFERRING TECHNOLOGY: THERMAL DRYING FOR RESOURCE RECOVERY OF DRIED SLUDGE FOR ENERGY

Presented in this section is background information on thermal drying of sludge, followed by three real-life case studies of experimental design for thermal drying processes.

4.4.1 Introduction to resource recovery of faecal sludge as solid fuel

Producing value-added end products from faecal sludge can be an incentive for appropriate management and treatment. Revenue from resource recovery can be used to offset operational and maintenance costs at FSTPs, which can incentivise adequate collection and delivery of sludge to treatment plants and achievement of consistent treatment targets (Diener *et al.*, 2014). A market-driven approach should be used to determine the revenue potential from possible end products of faecal sludge treatment (Schoebitz *et al.*, 2016). In Accra, Ghana and in Kampala, Uganda, use as a solid fuel for manufacturing industries (*e.g.* brick and cement factories) was identified as a high-demand end product of faecal sludge (Diener *et al.*, 2014). Many industries in these cities typically rely on wood and waste biomass, and struggle when availability of these fuels fluctuates. Solid fuels produced from faecal sludge can have comparable energy densities to these traditionally used fuels (Andriessen *et al.*, 2019; Gold *et al.*, 2017; Murray Muspratt *et al.*, 2014). The decision to target resource recovery allows FSTP designers to set treatment targets based on the requirements set by the consumers (*e.g.* moisture content, energy density, pathogens), and select appropriate treatment technologies accordingly.

4.4.2 Introduction to faecal sludge drying

Drying is a requirement for producing solid fuels from faecal sludge. In addition to increasing net energy gains (Murray Muspratt *et al.*, 2014; Septien *et al.*, 2020), drying also reduces the mass, making it easier to handle and decreasing transportation costs. Drying can be achieved passively, for example with drying beds, but this requires a large footprint and long residence times (weeks to months). Hence, researchers are pursuing heat drying of dewatered faecal sludge as a transferring technology from the food processing industry. One example is the LaDePa process, developed by the eThewkini municipality and Particle Separation Systems (Durban, South Africa). The LaDePa can be used at a full-scale treatment plant to dry and pasteurise sludge from ventilated improved pit latrines (VIPs) (see Case study 4.4 and Septien *et al.*, 2018a). Another example is the Tehno Sanitizer® (also known as The Shit Killer®), based on microwave technology that has been used for food drying for years (*e.g.* pasta, fruit etc., see Case study 4.5). Requirements for how much moisture needs to be removed are dictated by the treatment process design and by the end-user requirements. Different technologies require different input moisture contents, and further drying may be necessary after sludge has been processed (Figure 4.13).

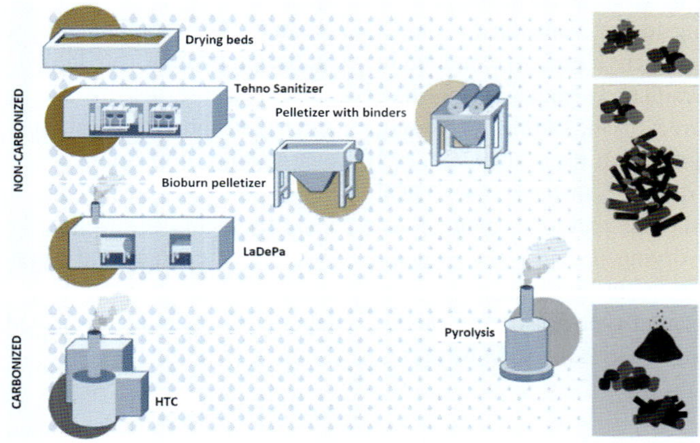

Figure 4.13 An overview of technology options for producing solid fuel, starting from dewatered faecal sludge at 80% moisture and ending at non-carbonised or carbonised solid-fuel end products. The position of the technology icons from left to right indicates the required dryness of the input sludge for each technology, as indicated by the size of the droplets, ranging from 80% moisture on the left to 10% moisture on the right (modified from Andriessen *et al.*, 2019).

In general, solid fuels do not perform well if they contain too much moisture, but this needs to be balanced with higher energy inputs or longer drying times.

It has been difficult to adapt and scale up drying technologies to full-scale faecal sludge treatment processes. Drying technologies face many of the same challenges as any faecal sludge treatment process, for example high variability in Q&Q of the influent sludge. However, drying faecal sludge presents its own specific technical challenges as well. These include the high energy demand, the release of strong odours during drying, and the stickiness acquired by faecal sludge during the drying process. As with conditioning, more research on the mechanics of faecal sludge drying is required to generate a fundamental understanding of the process and to inform the development and adaptation of well-functioning drying processes.

4.4.3 Types and mechanisms of thermal drying (technical background)

Understanding the underlying physical, chemical, and biological processes supporting a technology is crucial for making informed decisions about adapting it to work with faecal sludge. During thermal drying, heat is transferred to the sludge from a heating source (*e.g.* hot fluid, heated wall, infrared radiation) or generated internally after conversion of another form of energy (*e.g.* microwave, dielectric radiation), leading to the movement of moisture to the sludge surface where it evaporates. The rate of drying depends on the temperature or irradiance from the heat source, humidity, flow rate, and pressure of the ambient air, and on the area, thickness, and thermal properties (*i.e.* heat capacity and thermal conductivity) of the exposed sludge surface. A schematic representation of heat drying of faecal sludge is shown in Figure 4.14.

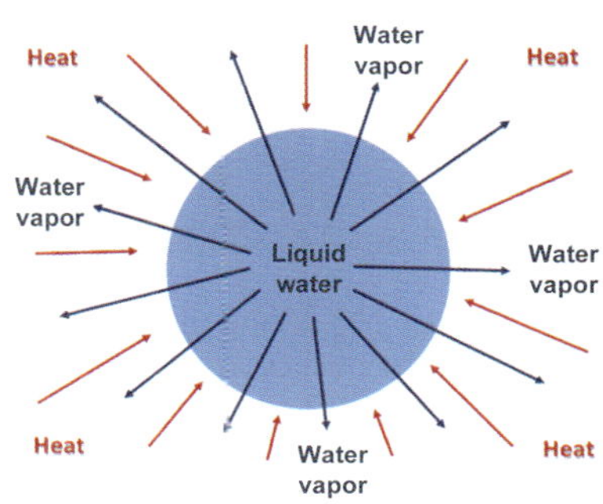

Figure 4.14 Schematic representation of drying faecal sludge. Red arrows represent heat transfer, blue arrows represent mass transfer of water (H_2O).

The most common way to classify thermal drying technologies is according to the heat transfer mode, which are convection (convective drying), conduction (contact drying), and radiation (radiative drying). Convective drying (or direct drying) works by passing hot air or gases directly through the sludge. Contact drying (or indirect drying) instead uses heat exchangers to heat a surface that the sludge is in contact with. Radiative drying provides the heat for moisture evaporation by solar, infrared, microwave, or dielectric radiation. Different types of drying modes can be combined for a given technology. Most drying systems include a ventilation or vacuum system to evacuate the evaporated moisture and avoid saturation of the air, which can inhibit the drying process. In passive drying systems such as drying beds, the sun and wind provide heat and air flow to promote evaporation. For more detailed information about drying mechanisms or types of industrial dryers used in other fields see Mujumdar (2014). Examples of convective, contact, and radiative drying technologies that have been used with faecal sludge at pilot- and laboratory-scales are presented in figures 4.15 to 4.18.

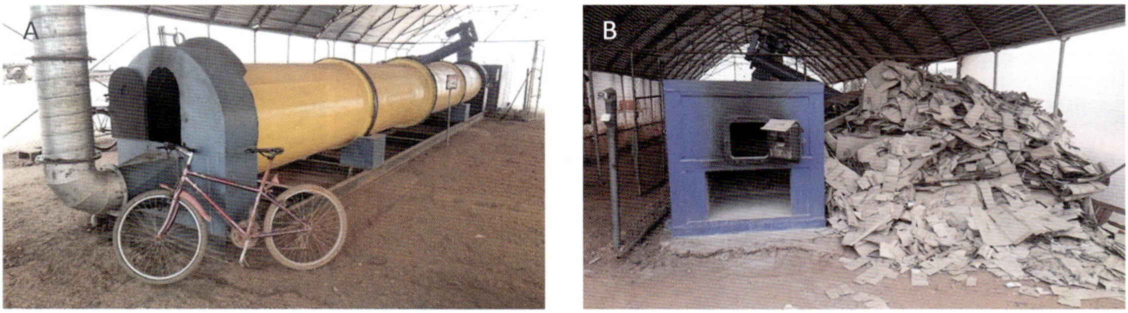

Figure 4.15 A) a rotary convection dryer operated by Pivot, in Kigali, Rwanda and B) waste cardboard is burned in a boiler and the hot gases produced are used as the heat source for drying. The sludge is pre-dried by a solar dryer before entering the convective dryer pictured here (photos: UKZN PRG).

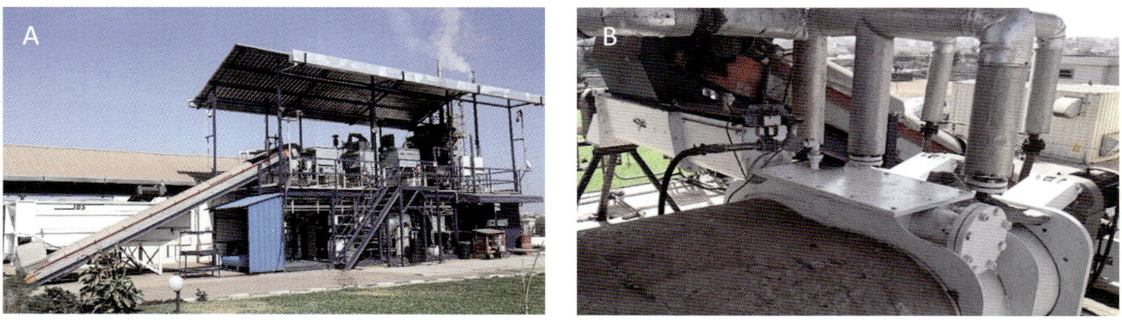

Figure 4.16 A) a contact dryer at the Omni-processor pilot plant at Niayes FSTP in Dakar, Senegal. In this plant, the sludge is incinerated leading to the generation of heat and electricity. B) part of the heat from combustion is recirculated in the process for the drying of the sludge with this contact dryer unit (photos: UKZN PRG).

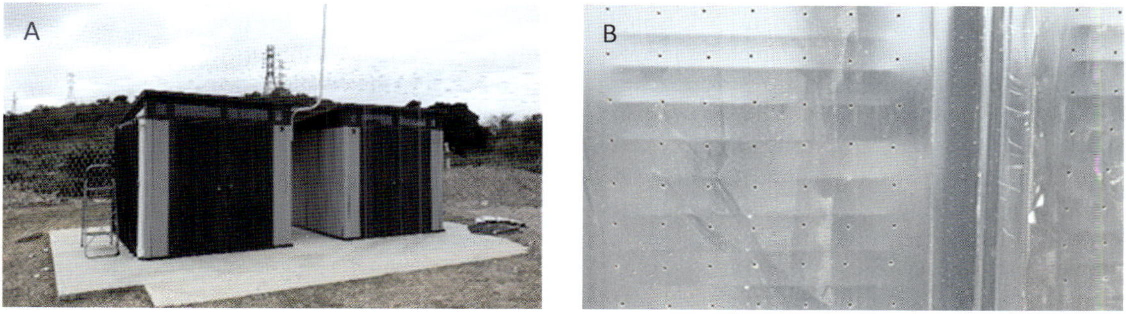

Figure 4.17 A pilot-scale solar dryer developed by Swansea University and tested in Durban, South Africa. A) sludge is placed inside these sheds to dry. B) the walls of the sheds absorb solar energy and transfer it to the sludge inside through a ventilation system (photos: UKZN PRG).

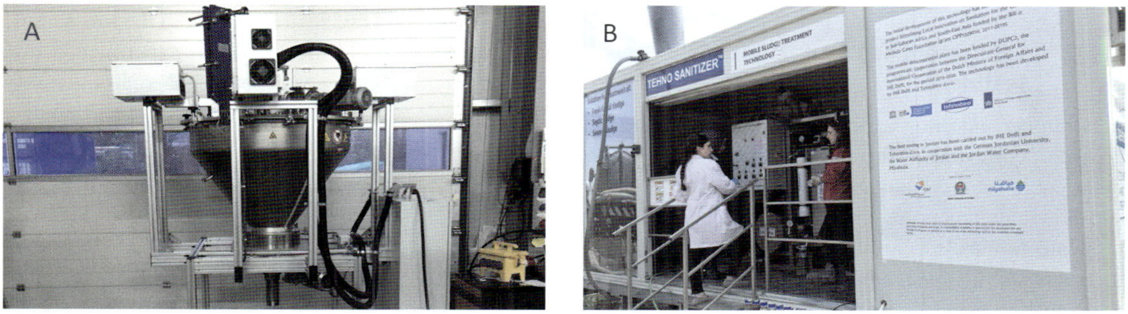

Figure 4.18 A) the bench-scale Shit Killer microwave-based technology for sludge treatment developed by IHE Delft and Fricke und Mallah Microwave Technology GmbH, and B) the follow-up prototype – the pilot-scale Tehno Sanitizer developed by IHE Delft and Tehnobiro d.o.o. (photos: IHE Delft).

4.4.4 Key parameters when implementing thermal-drying technologies

When designing or implementing a drying technology, first the amount of time it takes the sludge to dry to the desired moisture content (*i.e.* the drying rate) needs to be determined, along with the amount of required energy. Optimal combinations of key process parameters will yield dry sludge with the desired moisture content at the lowest energy cost. Methods such as pre-treatment of the sludge with stirring, or techniques to enhance the heat and mass transfer such as mechanical vibrations or ultrasound can also be investigated to improve drying performance. The following factors will influence the drying rate and energy consumption of the faecal sludge drying process, and need to be taken into account during technology transfer and process optimisation (Septien *et al.*, 2018a):

- Intensity of the energy source used to heat the sludge influences the evaporation, heat and mass transfer rates, and energy consumption. Examples of how this is measured are air temperature for convective drying; temperature of the heated surface for contact drying; and irradiance of the radiation source for radiative drying.
- Residence time of sludge in the dryer influences the energy consumption and the final moisture content of the treated sludge. Optimal residence time is used to design for capacity of specific treatment technologies.
- Relative humidity and flow rate of the air stream influence the heat and mass transfer kinetics.

Faster air flow and lower relative humidity promote faster and more complete evaporation, but also often require higher energy input.
- Physical, chemical and physico-chemical characteristics of the faecal sludge influence how much moisture needs to be removed, for example different starting moisture contents and water-binding characteristics, which influence the required energy to remove moisture.
- Sludge volume and geometry influence the rates of heat and mass transfer during drying, for example pellets, bulk sludge, thin or thick layers. Configurations with a higher sludge surface area to volume ratio, such as pellets, promote faster drying, whereas thick layers of bulk sludge require more time.

4.4.5 Laboratory-scale and pilot-scale testing

Laboratory- and pilot-scale testing of drying needs to consider comparable drying temperatures, air-flow rates, energy sources, and humidity ranges to the pilot- and full-scale technologies. Pilot-scale testing should replicate full-scale conditions as closely as possible, using knowledge of scientific mechanisms to evaluate performance for scaling-up. For example, a pilot-scale drying technology should produce pellets of the same size and aspect ratio as the full-scale system.

As mentioned in the previous section, the performance of the drying process at any scale is measured through the evolution of the faecal sludge moisture content as a function of time. In an experimental setup, this can be done through different methods:

- Online or intermittent measurement of the mass of the sample over time, assuming that the mass loss is exclusively due to moisture removal.
- Measurement of the moisture content after sampling a fraction of the sludge at a given time interval.
- Online or intermittent measurement of the humidity at the air-flow outlet, assuming that the gain of humidity in the air is due to moisture evaporation.

The determination of the drying rate under different conditions enables a better understanding of the process, and facilitates the development of kinetic models that can be used as tools for the design, operation and optimisation of drying technologies. Drying kinetics can be characterised at the laboratory-scale using commercially available instruments or custom-designed drying rigs. The commercially available thermogravimetric analyser (TGA) offers high-precision mass measurement during the thermal decomposition of materials, under controlled conditions. It can be coupled to a differential thermal analysis (DTA) or differential scanning calorimetry (DSC) unit, in order to determine the heat released or consumed during the transformation of the material. The main drawbacks of this method are the high cost of the TGA and DSC instruments, and the low sample weight that has to be used in experiments (*i.e.* milligrams), which can lead to reproducibility problems due to the heterogeneity of faecal sludge. The moisture analyser is a more affordable commercial instrument that can record the loss of mass of the sludge during drying. In this device, the sludge is heated by an infrared radiator and a ventilation system evacuates the evaporated moisture. A larger amount of sample can also be used (*i.e.* grams). However, the drying conditions cannot be controlled as well as in the TGA. Custom-designed drying rigs can be adjusted in size and complexity according to the needs and means of the experimenter, and can give a more tailored representation of the drying kinetics of a specific technology. Custom rigs can be as simple as a conventional oven where sludge is occasionally removed to track the mass loss, or a sophisticated experimental rig with high levels of instrumentation and an interface to log the data. Provided in Case study 4.3 is an account of the use of a custom experimental rig to measure faecal sludge drying kinetics under variable process settings.

The physical and chemical changes that the sludge undergoes during drying must be characterised in order to have a deeper understanding of the drying process. Periodic characterisation of the sludge properties during drying also helps researchers to target drying processes to produce suitable end products. The properties of the dried sludge can be quantified with the methods described in Chapter 8:

- Total solids of dried sludge; measured gravimetrically by sludge weight before and after complete drying in a 105 °C oven.
- Calorific value is a measure of energy density, and is measured using a bomb calorimeter.
- Ash and volatile solids content of the sludge are measured gravimetrically with a 550 °C muffle furnace.
- Rheological properties, such as shear stress and viscosity under different shear rates, are measured with a rheometer or viscometer.
- *E. coli* or Helminth eggs can be monitored as indicator organisms for pathogen inactivation, if the end product is required to be pathogen-free.

4.4.6 Case studies - thermal drying for energy recovery

The following three case studies provide examples of (*i*) how to get useful kinetics data from laboratory-scale devices for the design and development of pilot-scale and full-scale dryers, (*ii*) how to optimise the performance of a full-scale drying process using experiments conducted with a laboratory-scale apparatus, and (*iii*) how to optimise the performance of a full-scale drying process using experiments conducted at full-scale.

Case study 4.3 *Determination of faecal sludge drying kinetics with a custom-designed experimental rig*

This case study presents an example of how to determine faecal sludge drying kinetics in a laboratory-scale custom-designed experimental rig. This investigation was carried out by the Pollution Research Group (PRG) at the University of KwaZulu-Natal (UKZN) in Durban, South Africa. It was part of a MScEng project to learn about the rate at which pit latrine faecal sludge dries under different operating conditions (Makununika, 2017). A rig was custom-designed to study drying rates under different operational conditions in a convective dryer. In this rig, faecal sludge pellets were dried with hot air while their mass loss due to evaporation was measured in real time. The determination of the drying rates will aid in the development of drying technologies suitable for faecal sludge. Determination of kinetic data is an important step towards the design, development, optimisation and scaling-up of drying technologies. It provides information that is used to size the dryer, to determine the optimum operating conditions, to fix the residence time (continuous mode) or holding time (batch mode), and to estimate the power consumption of the process.

Research question

What is the rate of faecal sludge drying with varying temperature, humidity, air velocity, and pellet diameter?

Response variable

Change in moisture content over time was characterised gravimetrically by a custom-designed convective drying rig. A photograph and schematic representation of the convective drying rig are presented in Figure 4.19.

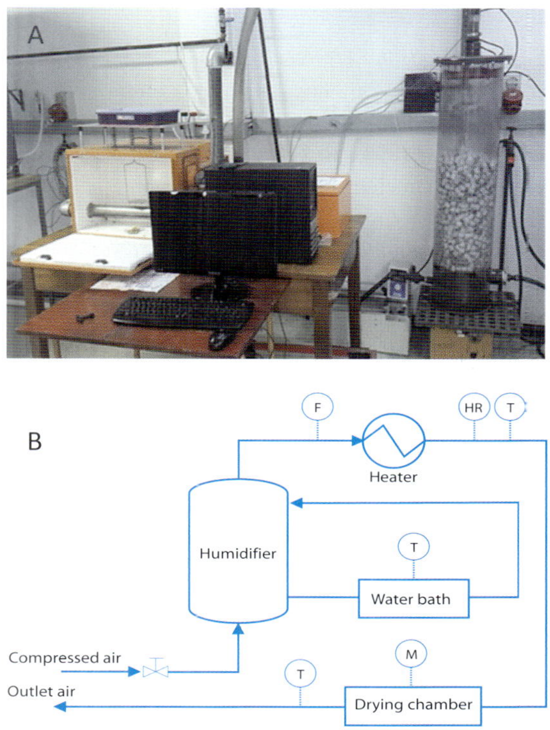

Figure 4.19 A) the custom-designed convective drying rig (photo: UKZN PRG) and B) a schematic representation of the convective drying rig, F: air-flow measurement; T: temperature measurement; M: mass measurement; HR: relative humidity measurement.

During the experiments, dehumidified compressed air was fed into the drying rig. The airflow rate was measured by a differential pressure measurement device and was controlled by a globe valve. The air stream was humidified in a packed column by counter-current contact with a water flow. The relative humidity of the air was adjusted by controlling the water temperature. The humidified air then passed through an electric heater to raise its temperature to the set value. The hot air stream was then introduced into the drying chamber where the faecal sludge sample was placed on a sample holder linked to a precision weighing strain gauge load cell with an accuracy of 0.01 g. The sample mass was measured online to track the change in mass with time. The air temperature and relative humidity were monitored at the inlet and outlet of the drying chamber. All the measurements were continually logged on a computer.

Factors, levels, and ranges

- Temperature: 40, 60, and 80 °C
- Relative humidity: 5, 15, and 25%
- Air velocity: 0.1, 0.2, and 0.4 m/h
- Pellet diameter: 8, 10, 12, and 14 mm

Factors that might influence the response variable

- Presence of solid waste: faecal sludge can contain considerable amounts of rubbish that can cause interferences and clogging during the drying experiments. In order to avoid this, the sludge samples were screened prior to the experiments, and large pieces of rubbish were removed.
- Heterogeneity of faecal sludge: faecal sludge is highly heterogeneous, which can lead to inconsistent experimental results. In order to reduce heterogeneity and ensure repeatability, the sludge samples were thoroughly mixed prior to the experiments.

Experimental design details

Each run was performed in triplicate. Table 4.2 displays the runs performed in this study from all the possible runs. If all the possible combinations of the selected factors, levels, and ranges had been tested, 108 different runs would have been required. However, this was not feasible in terms of time and resources, therefore the most appropriate combination of runs was selected in order to study the influence of each variable. This was done by varying the value of a single variable while keeping the others constant at a reference value.

Table 4.2 Matrix with the different runs performed (marked with the symbol ■) out of all the possible combinations.

Temperature (°C)	Relative humidity (%)	Air velocity (m/h)	Pellet diameter (mm)			
			8	10	12	14
		0.1	■	■		
	5	0.2		■		
		0.4		■		
		0.1		■		
40	15	0.2				
		0.4				
		0.1		■		
	25	0.2				
		0.4				
		0.1	■		■	■
	5	0.2				
		0.4				
		0.1				
60	15	0.2				
		0.4				
		0.1				
	25	0.2				
		0.4				
		0.1	■			
	5	0.2				
		0.4				
		0.1				
80	15	0.2				
		0.4				
		0.1				
	25	0.2				
		0.4				

Interpreting the results

The results of the experiment are presented in Figure 4.20.

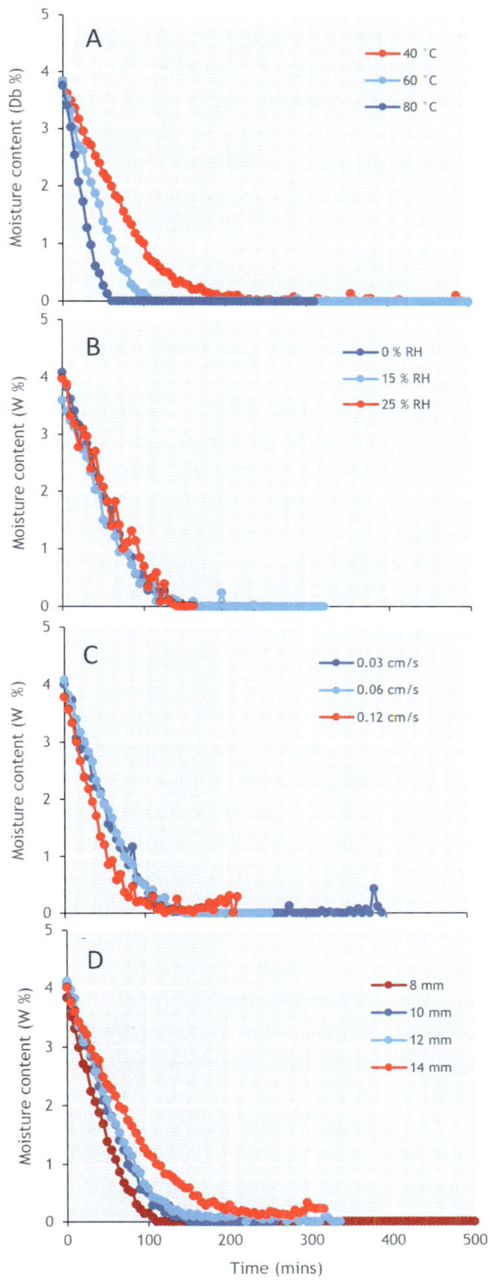

Figure 4.20 A) Drying rate as a function of air temperature, B) relative humidity, C) air velocity, and D) pellet diameter.

The main findings of this study were:

- Air temperature has a major influence on the drying rate. Increasing the temperature from 40 to 80 °C decreased the drying time from 3 hours to 1 hour.
- The diameter of the sludge pellets also has an important influence on the drying rate. The 8 mm pellets were completely dried within 100 minutes, whereas the 14 mm pellets required drying times greater than 200 minutes.
- The relative humidity and air velocity had low or negligible influence on the drying kinetics under the explored conditions.

Scaling-up from laboratory to pilot-scale

According to the experimental results in this case study, the most critical parameters to optimise during drying are the air temperature and diameter of the sludge pellets.

The experimental data from this work was used to develop a mathematical model that could be inserted into reactor models as a tool for simulation to design new dryers, and can be used in process control for scaled up systems (Makuninika, 2017). ■

Case study 4.4 *Optimising the LaDePa process for infrared faecal sludge drying*

This case study is based on a Master's thesis by Simon Mirara (Mirara, 2017). Further information can be found in Septien *et al.*, 2018a, 2018b, and Septien *et al.*, 2020. The motivation for this research project was to optimise the existing full-scale Latrine Dehydration Pasteurisation (LaDePa) process. The LaDePa process was implemented in the eThewkini municipality in Durban, South Africa to treat the faecal sludge from ventilated improved pit (VIP) latrines through infrared drying, to produce dry, pathogen-free pellets for use as a soil conditioner or solid fuel. The LaDePa process was developed by the eThewkini municipality and Particle Separation Systems as a transferring technology from the mining industry where it was where it was applied for drying of minerals. Based on the treatment performance of the full-scale LaDePa, the municipality decided that it needed to be optimised to minimise energy consumption while maximising the drying rate, pasteurisation performance, and end-use potential in the treated sludge. In order to optimise drying in the LaDePa process and to develop a deeper understanding of the drying process, a 1:10 laboratory-scale replica of the full-scale LaDePa was constructed (Figure 4.2.1).

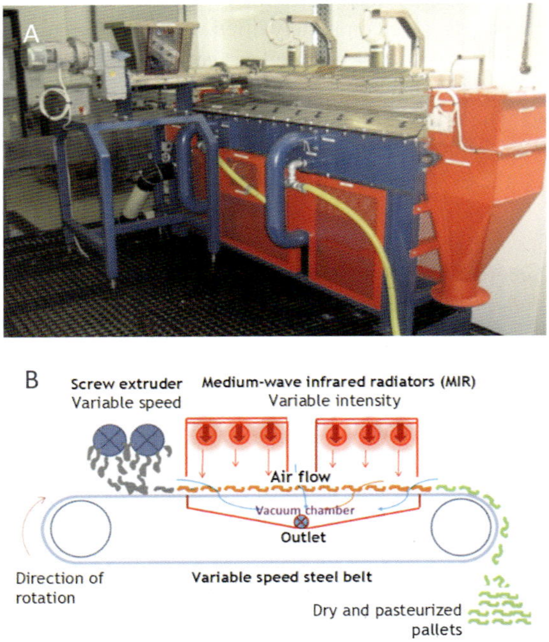

Figure 4.21 A) the laboratory-scale LaDePa, and B) a corresponding schematic representation of the process (photo and schematic: UKZN PRG).

Research question

What process settings for faecal sludge drying with the LaDePa infrared dryer minimise energy consumption and maximise sludge drying rate?

Response variables

The laboratory-scale LaDePa was used to characterise the moisture content of the dried pellets, and energy consumption of the process, at different conditions (see factors, levels and ranges). The sludge was fed into the machine as pellets formed with a screw extruder, which were conveyed by a moving belt under two successive infrared emitters (providing heat for drying). An air stream was induced in the drying zone through an air suction box system installed below the belt to keep humidity low (Figure 4.21).

The dried pellets after processing were analysed to determine physical, chemical and biological properties, such as moisture content, volatile solids content, nutrient content, calorific value, thermal properties and helminth eggs. The drying and pasteurisation performance of the process were measured through the moisture content evolution and helminth egg viability. The end-use potential of the dried sludge was evaluated through the measurement of their properties.

Factors, levels, and ranges

- Emitter intensity (infrared irradiance): 6, 24, and 34 kW/m^2.
- Residence time: 4, 8, 12, 17, 26, and 39 minutes (varied by adjusting the speed of the belt).
- Distance between the belt and infrared emitters: 50, 80 and 115 mm (varied by adjusting the belt height).
- Suction air-flow rate: 11.1 and 18.3 m^3/s.
- Pellet diameter: 8, 10, 12 and 14 mm.

Factors that might influence the response variables

- Heterogeneity of sludge and presence of solid waste: as in Case Study 4.3, large pieces of solid waste were screened and removed from the sludge, and screened sludge samples were thoroughly homogenised prior to experimentation.
- Ambient temperature and humidity: ambient air is used for ventilation in the LaDePa, thus, the temperature and humidity of the suction air stream is dependent on ambient conditions. As the laboratory is climate-controlled, the ambient conditions are quite steady throughout the year and it was assumed that these parameters did not significantly change throughout the course of the study.
- Loading density of the pellets on the belt: this could have an influence on the performance of the process, as it could be expected that the drying of large sample loads would require a higher heat input. To address this, the loading density on the belt was kept consistently low in this investigation. Prior to scaling-up, higher loadings will be investigated.

Experimental design details

Due to available time and resources, the following runs, indicated with a ■ in Table 4.3, were determined to be the most relevant.

Table 4.3 Matrix with the different runs performed (the most relevant marked with ■) out of all the possible combinations.

Emitter irradiance (kW/m^2)	Height emitter (mm)	Air-flow rate (m^3/h)	Pellet diameter (mm)			
			8	10	12	14
6	50	11.1				
		18.1	■			
	80	11.1				
		18.1				
	115	11.1				
		18.1				
24	50	11.1				
		18.1	■	■	■	■
	80	11.1				
		18.1	■			
	115	11.1	■			
		18.1	■			
34	50	11.1				
		18.1	■			
	80	11.1				
		18.1				
	115	11.1				
		18.1				

Interpreting the results

Results of the experiment are presented in Figure 4.22.

As expected, the rate of drying increased as the intensity of the infrared radiation increased and the distance between the pellets and the heating source decreased. Drying was faster for pellets with a smaller diameter. Increasing the suction air-flow rate caused a cooling effect on the sludge (negative for the process) but also enhanced the evacuation of moisture from the surface of the pellets (positive for the process). Under the explored conditions, these opposing effects counteracted each other and the overall drying rate was not affected by changing the air-flow rate. The pre-treatment of the sludge also did not affect the drying rate.

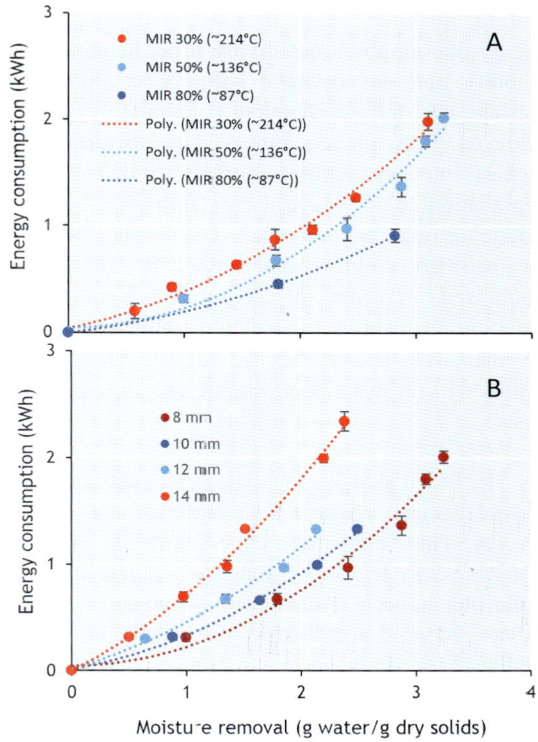

Figure 4.22 A) Plot of moisture removal vs energy consumption at varied medium-wave infrared intensities (MIR). MIR of 30, 50 and 80% equals infrared irradiance of 6, 24, and 34 kW/m^2, respectively, and B) plot of moisture removal vs energy consumption at varied pellet diameters.

The energy consumption for moisture removal was determined from the kinetic data. Depicted in Figure 4.22, the drying process consumes less energy to remove a given amount of moisture when operating at higher infrared heating intensity and with smaller diameter pellets. However, it was observed that drying at too high a heating flux could induce thermal degradation of the sludge, which could lead to charring or burning. During the trials, drying at the highest infrared intensity (34 kW/m^2) resulted in the pellets starting to smoke.

Process optimisation from laboratory- to full-scale
Based on the results of the laboratory-scale experiments, it is recommended to operate the LaDePa at the highest possible infrared radiation intensity that does not cause thermal degradation. During laboratory tests, in addition to monitoring energy consumption and drying time, helminth egg viability and net calorific value were measured. During tests, full deactivation of helminth eggs was achieved. It is not recommended to operate at the highest intensity setting, as the resulting thermal degradation could reduce the suitability of the dried sludge for reuse as a solid fuel. The distance between the infrared emitters and the belts should be minimised, in order to maximise the amount of radiation received by the pellets without the need of an increased power supply. Implementing these results will result in lower energy use and operating costs.

The faecal sludge should also be pelletised at the lowest diameter possible for a more efficient drying process. This will require experimentation with the full-scale extruder to determine the smallest diameter achievable at scale. After process changes are made, pellets produced at full-scale will need to be further evaluated for pathogens to ensure protection of public health during the end use. ■

4.5 TRANSFERRING TECHNOLOGY: MICROWAVE DRYING FOR RESOURCE RECOVERY OF DRIED SLUDGE FOR ENERGY

Microwave drying is a type of radiative drying where microwave radiation is used to heat the sludge. In the microwave drying process, microwave radiation heats the core of the sludge particles promoting the transport of water molecules from the inside to the surface; this results in a large amount of water molecules at the surface of the sludge that can be more easily evaporated compared to the water bound deeper within sludge particles. Due in part to this mechanism, microwave drying can offer energy savings compared to other thermal drying technologies.

Case study 4.5 *Optimising the Tehno Sanitizer technology for microwave faecal sludge sanitisation and drying*

This case study is based on two PhD and several MSc studies carried out at IHE Delft Institute for Water Education in The Netherlands (IHE Delft). It concerns the development of a novel microwave-based technology for sludge sanitisation and drying. The new technology is an example of a development that has passed through all the Technology Readiness Levels (TRLs) (Héder M. 2017), starting from a small laboratory-scale setup using an adapted kitchen microwave (Mawioo *et al.,* 2016a; Mawioo *et al.,* 2016b), to a bench-scale unit (Mawioo *et al.,* 2017) and finally, to a full-scale prototype (Kocbek *et al.,* 2020, in preparation). This technology, called the Shit Killer, was initially developed for decentralised faecal sludge treatment in emergency sanitation (Brdjanovic *et al.,* 2015) and has evolved into a robust and efficient technology known nowadays as the Tehno Sanitizer (Figure 4.23). The Tehno Sanitizer prototype, recently tested in Jordan, is equipped with four technologically-independent but inter-connected functional components, namely: (*i*) microwave-based sludge treatment, (*ii*) liquid stream treatment, (*iii*) air treatment, and (*iv*) an energy-recovery system (Figure 4.23).

The bench-scale Shit Killer unit was successfully tested for pathogen removal and sludge drying in Slovenia. At that time, the specific energy consumption (SEC) (energy consumed per liter of evaporated water) was not the primary objective and thus was, as expected, sub-optimal. The main reasons for this were: (*i*) lack of thermal insulation, (*ii*) inefficient use of microwave energy, (*iii*) less efficient mixing at higher sludge densities, (*iv*) cold ambient temperature (5 °C), (*v*) poor extraction of the condensate from the cavity, (*vi*) unnecessary heating of the cavity, and (*vii*) absence of energy recovery features.

All of these shortcomings have been addressed and mitigated in the next generation full-scale prototype: the Tehno Sanitizer. This system is a semi-decentralised and containerised mobile full-scale prototype designed for the treatment (drying, pathogen inactivation, and resource recovery) of diverse types of sludges such as fresh faecal sludge and waste activated sludge, with different water and dry solids contents. This mobile unit has the capacity to process 300 kg of wet sludge per day. The integration of the different technologies provides an attractive approach for treating sludge and wastewater streams generated while producing valuable resources that can be utilised in agricultural and domestic applications, with up to 95% DS. The initial results obtained from studies focusing on pathogen indicator organisms (Mawioo *et al.*, 2016a and 2016b), carried out at laboratory- and bench-scale setups, suggest that the Tehno Sanitizer could be an effective technology for sanitisation of sludge.

A

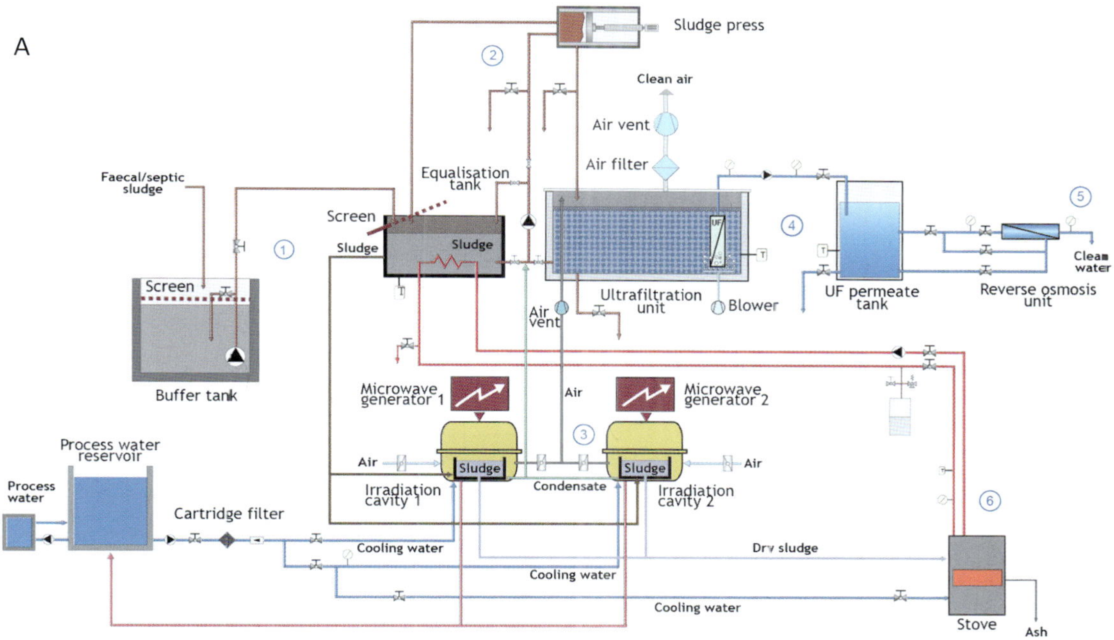

B

Figure 4.23 A) simplified process flow diagram of the Tehno Sanitizer: (1) sludge intake, (2) sludge pre-treatment, (3) the sludge sterilisation and drying unit, (4) microfiltration, (5) reverse osmosis, and (6) the sludge energy recovery unit, and B) a full-scale Tehno Sanitizer prototype (source: Tehnobiro d.o.o.).

The main challenge addressed in the development of the full-scale prototype was how to minimise the specific energy consumption (SEC) of the system from the value initially observed in the bench-scale unit of 4.0 kWh/L of evaporated water, to the target level of below 1.0 kWh/L of evaporated water.

Research question

Which microwave power output settings on the full-scale prototype achieve the target dryness (85% DS) while minimising the SEC to below 1.0 kWh/L of evaporated water?

Response variables

The experimental setup was designed to measure the SEC (kWh/L) of the system. The SEC was calculated using the power output setting of the microwave generator, set at the desired value (kW). This value was multiplied by the time of the exposure and divided by the mass of water that had evaporated at that exposure time.

The mass of the sludge in the microwave cavity was continuously measured and the moisture content and the DS were calculated from the TS measurement of the sludge sample taken just before the start of the test. Also the sludge temperature was continuously measured by a sensor installed inside the cavity.

Factors, levels, and ranges

Microwave power output: 1.0, 1.5, 3.0, 3.25, 4.5 and 6.0 kW (adjusted manually)

Factors that influence the response variable

The factors that influence the SEC include:

- Energy losses due to the lack of thermal insulation.
- Frequencies at which the microwave energy is delivered.
- Mixing conditions at the irradiation cavity.
- Condensation of the evaporated water in the microwave cavity.
- The microwave energy absorption capacity of the sludge (power density) at the evaluated microwave power outputs.

Experimental design details

Experiments were conducted using the full-scale prototype. The experimental setup (Figure 4.24) consisted of two stainless steel microwave cavities equipped with a rotating polypropylene turntable and an oval sludge-holding vessel, two microwave power supply units, and two microwave generators with a combined power output of 12.0 kW operated at a frequency of 2,450 GHz. An electromotor was used to rotate the sludge samples at a speed of 1 rpm to alleviate the effect of non-uniform sludge heating. Ancillary equipment included an air extraction and treatment unit and a microwave generator-cooling water-based system. In total six identical tests were executed (each at different power level) because only one cavity was equipped with a load cell to continuously measure the mass of the sludge. Each test had a different duration (the shortest was 21 minutes at power output of 6 kW) and lasted until the target DS of 85% was achieved.

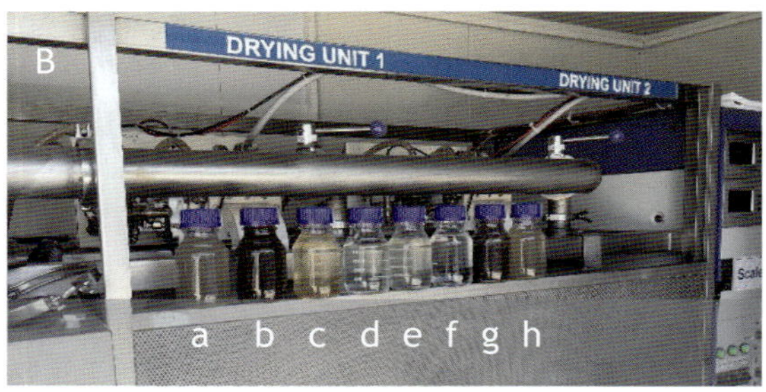

Figure 4.24 A) an experimental microwave-based faecal sludge drying unit, and B) samples taken at different points in the process: a) filtrate from the sludge press, b) concentrated sludge from the sludge press, c) ultrafiltration concentrate, d) ultrafiltration permeate, e) reverse osmosis concentrate, f) reverse osmosis permeate, g) dry sludge, and h) condensate (photo: IHE Delft).

Interpreting the results

Figure 4.25 depicts the drying rate as a function of dry solids content at different power outputs of the microwave generators. As expected, the higher the power output, the higher the drying rate. At the start of the drying process the drying rate increased at all the evaluated power outputs until it reached a maximum and constant drying rate value. This constant drying phase was dominant and extended through almost the entire drying process; this is a positive characteristic of the microwave drying process and introduces a competitive advantage compared to thermal drying technologies where such constant drying phases are not commonly observed. Such a constant drying phase is associated with the removal of unbound (free) water from the surface of the sludge which demands much less energy to be evaporated than other types of water contained in the sludge (Figure 4.1).

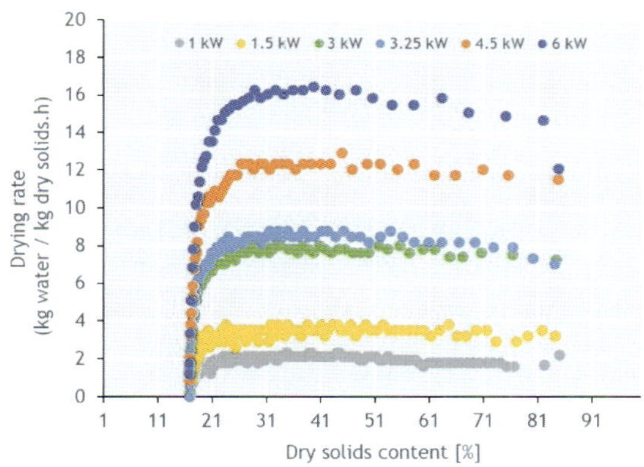

Figure 4.25 The sludge drying rate as a function of sludge dry-solids content at different power outputs of the microwave generators (Kocbek *et al.*, 2020).

Figure 4.26 shows the SEC of the system during the period of drying sludge from 17% to 85% DS at the evaluated microwave generator power output range. It has been observed that increase in power output lowers the SEC. The lowest SEC of approximately 1 kWh/L of evaporated water was reported at power outputs higher than 3 kW. The observed changes in the SEC were due to the microwave radiation generation efficiency which was between 50% (at power below 3 kW) and 70% (at the highest power outputs).

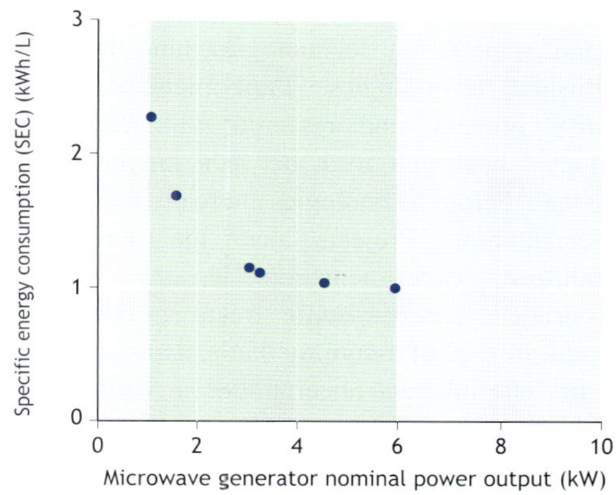

Figure 4.26 Effect of microwave generator power output on the specific energy consumption (SEC) (Kocbek *et al.*, 2020).

Implications of scaling-up

The SEC results obtained in this research provided the evidence that the modifications and innovations built in the Tehno Sanitizer mitigated the early development issues experienced with the Shit Killer, largely reducing the energy requirement resulting in achieving the target SEC of 1 kWh/L. Such results bring Tehno Sanitizer into the mix with conventional thermal drying (convective and conductive) technologies (Bennamoun *et al.*, 2013). Given the fact that in commercial-scale applications a more efficient microwave generator will be used (with an efficiency rate of up to 90%), the SEC is expected to decrease by an additional 10 to 20%. Furthermore, the energy recovery unit in the Techno Sanitizer in this study was not turned on. With the additional heat becoming available from co-incineration of dry sludge (energetic value of obtained dry sludge was 20 MJ/kg or 5.6 kWh/kg) for pre-heating of the incoming sludge, and when the system starts to be continuously used, the calculated SEC will further decrease. If less stringent requirements for water treatment are applicable, an SEC of below 0.8 kWh/L can be achieved. Such results are promising and make this new technology a viable alternative for faecal sludge management.

4.5 OUTLOOK

Faecal sludge management is a rapidly evolving sector. The information described in this chapter is important for developing new technologies, scaling-up and transferring technologies, and optimising established technologies. Experimentation is an iterative process, and research will need to be conducted back and forth between laboratory- and pilot-scale before technologies are ready for full-scale implementation. Projects that incorporate well thought-out experimentation ensure that an appropriate, context-specific treatment solution is selected, instead of assuming that a standard solution will fit. The inherent uncertainties in working with faecal sludge, and with innovative and transferring technologies, make risk management an essential focus in the development and scaling-up of any treatment technology. Risks can be mitigated through dedication to quality experimental design and execution, and through partnerships between municipalities and research institutions, which can help guide experimentation from the start of a project to the optimisation and monitoring of a full-scale FSTP.

Future research needs for scaling-up dewatering and drying technologies will be driven by requirements to optimise treatment technologies that work for faecal sludge. The next advances in dewatering research will include establishing how to more rapidly and cost effectively monitor faecal sludge such that optimal conditioner dosing can be achieved. Another step will be acquiring a fundamental understanding of the processes occurring during stabilisation that affect dewaterability. Future focuses in thermal drying research will address the need for a more holistic understanding of the drying process of faecal sludge, for example morphological changes that occur such as stickiness, and a better understanding of how moisture is bound to faecal sludge.

REFERENCES

Andriessen N., Ward B.J. and Strande, L. (2019). To char or not to char? Review of technologies to produce solid fuels for resource recovery from faecal sludge. *Journal of Water, Sanitation and Hygiene for Development*, 9(2), 210-224.

Bassan M., Dodane P.H. and Strande L. (2014). Treatment Mechanisms. In *Faecal sludge management: systems approach for implementation and operation*, 45-66. IWA Publishing, London, UK.

Bennamoun L., Arlabosse P. and Léonard A. (2013). Review on fundamental aspect of application of drying process to wastewater sludge. *Renewable and Sustainable Energy Reviews*, 28, 29-43.

Brdjanovic D., Zakaria F., Mawioo P.M., Garcia H.A., Hooijmans C.M., Pean T.Y. and Setiadi T., (2015). eSOS® - emergency Sanitation Operation System. *Journal of Water, Sanitation and Hygiene for Development*, 5(1), 156-164.

Diener S., Semiyaga S., Niwagaba C.B., Muspratt A.M., Gning J.B., Mbéguéré M., Ennin J.E., Zurbrugg C. and Strande L. (2014). A value proposition: Resource recovery from faecal sludge - Can it be the driver for improved sanitation?. *Resources, Conservation and Recycling*, 88, 32-38.

Gold M., Harada H., Therrien J.D., Nishida T., Cunningham M., Semiyaga S., Fujii S., Dorea C., Nguyen V.A. and Strande, L. (2018). Cross-country analysis of faecal sludge dewatering. *Environmental Technology*, 39(23), 3077-3087.

Gold M., Ddiba D.I.W., Seck A., Sekigongo P., Diene A., Diaw S., Niang S., Niwagaba C. and Strande L. (2017). Faecal sludge as a solid industrial fuel: a pilot-scale study. *Journal of Water, Sanitation and Hygiene for Development*, 7(2), 243-251.

Gold M., Dayer P., Faye M.C.A.S., Clair G., Seck A., Niang S., Morgenroth E. and Strande L. (2016). Locally produced natural conditioners for dewatering of faecal sludge. *Environmental Technology*, 37(21), 2802-2814.

Héder M. (2017). From NASA to EU: the evolution of the TRL scale in Public Sector Innovation. *The Innovation Journal: The Public Sector Innovation Journal,* 22(2), 2017, article 3.

Klinger M., Gueye A., Manandhar Sherpa A. and Strande L. (2019). Scoping Study: Faecal Sludge Treatment Plants in South-Asia and Sub-Saharan Africa. eFSTP Project Report. *Gates Open Research* 2019, 3, 1716 (document) (https://doi.org/10.21955/gatesopenres.1116557.1)

Makununika B. (2017). Thermal Drying of Faecal Sludge from VIP latrines and Charcaterisation of Dried Faecal Material. MSc thesis, University of KwaZulu-Natal (South Africa).

Kocbek E., Garcia H.A., Hooijmans C.M., Mijatović I., Lah B. and Brdjanovic D. (2020). Microwave treatment of municipal sewage sludge: Evaluation of the drying performance and energy demand of a pilot-scale microwave drying system. *Science of the Total Environment,* 742, 140541.

Kocbek E., Garcia H.A., Hooijmans C.M., Mijatović I. and Brdjanovic D. (in preparation). Microwave treatment of municipal sewage sludge: effects of sludge thickness and mass/power output ratio on the drying performance and energy demand.

Kocbek E., Garcia H.A., Hooijmans C.M., Mijatović I., Zakariya D. and Brdjanovic D. (in preparation). Pilot-scale evaluation of a microwave sanitisation system treating septic sludge at an experimental site in the Jordan Valley.

Mawioo P.M., Hooijmans C.M., Garcia H.A. and Brdjanovic D. (2016a). Microwave treatment of faecal sludge from intensively used toilets in the slums of Nairobi, Kenya. *Journal of Environmental Management,* 184, 575-584.

Mawioo P.M., Rweyemamua A., Garcia H.A., Hooijmans C.M. and Brdjanovic D. (2016b). Evaluation of a microwave-based reactor for the treatment of blackwater sludge. *Science of the Total Environment*, 548–549:72–8.

Mawioo P.M., Garcia H.A., Hooijmans C.M., Velkushanova K., Simonič M., Mijatović I. and Brdjanovic D. (2017). A pilot-scale microwave technology for sludge sanitization and drying. *Science of the Total Environment*, 601–602, 1437-1448.

Mirara S.W. (2017). Drying and pasteurisation of VIP latrine faecal sludge using a bench scale medium infrared machine. Msc thesis, University of KwaZulu-Natal, South Africa.

Montgomery D.C. (2019). *Design and analysis of experiments*. John Wiley & Sons.

Moto N., Esanju M., Andriessen N., Kimwaga R. and Strande L. (2018). Use of chitosan and Moringa oleifera as conditioners for improved dewatering of faecal sludge. In *Transformation towards sustainable and resilient WASH services: Proceedings of the 41st WEDC International Conference, Nakuru, Kenya.*

Mujumdar A.S. (2014) Handbook of Industrial Drying. *CRC Press.*

Murray Muspratt A., Nakato T., Niwagaba C., Dione H., Kang J., Stupin L., Regulinski J., Mbéguéré M. and Strande L. (2014). Fuel potential of faecal sludge: calorific value results from Uganda, Ghana and Senegal. *Journal of Water, Sanitation and Hygiene for Development*, 4(2), 223-230.

Niwagaba C.B., Mbéguéré M. and Strande L. (2014). Faecal sludge quantification, characterisation and treatment objectives. In *Faecal Sludge Management - Systems Approach for Implementation and Operation.* IWA Publishing, London, UK.

Schoebitz L., Andriessen N., Bollier S. and Strande L. (2016). Market Driven Approach for Selection of Faecal Sludge Treatment Products. Eawag, Duebendorf, Switzerland. https://www.fsmtoolbox.com/assets/pdf/114.17.29_market_driven_approach.pdf

Septien S., Mirara S.W., Makununika B.S.N., Singh A., Pocock J., Velkushanova K. and Buckley C.A. (2020). Effect of drying on the physical and chemical properties of faecal sludge for its reuse. *Journal of Environmental Chemical Engineering*, 8(1), 103652.

Septien S., Singh A., Mirara S.W., Teba L., Velkushanova K. and Buckley C.A. (2018a). 'LaDePa' process for the drying and pasteurization of faecal sludge from VIP latrines using infrared radiation. *South African Journal of Chemical Engineering*, 25, 147-158.

Septien S., Mirara S., Singh A., Velkushanova K. and Buckley C. (2018b). Characterisation of On-Site Sanitation Material and Products: VIP Latrines and Pour-Flush Toilets. WRC project final report. Pretoria, South Africa.

Stensel H.D. and Makinia J. (2014). Activated sludge process development. *Activated Sludge–100 Years and Counting.* Ed. by Jenkins D. and Wanner J., IWA Publishing, 33-51.

Strande L. (2017). Introduction to faecal sludge management: an online course. Available at: https://www.eawag.ch/en/department/sandec/e-learning/. Sandec: Department of Sanitation, Water and Solid Waste for Development, Eawag: Swiss Federal Institute of Aquatic Science and Technology, Duebendorf, Switzerland.

Van Loosdrecht M.C.M., Nielsen P.H, Lopez-Vazquez C.M. and Brdjanovic D. (eds.) (2016). *Experimental Methods in Wastewater Treatment*, IWA Publishing, London, UK.

Von Sperling M., Verbyla M.E. and Oliveira S.M.A.C. (2020). Assessment of Treatment Plant Performance and Water Quality Data: A Guide for Students, Researchers and Practitioners, IWA Publishing, London, UK.

Ward B.J., Traber J., Gueye A., Diop B., Morgenroth E. and Strande L. (2019). Evaluation of conceptual model and predictors of faecal sludge dewatering performance in Senegal and Tanzania. *Water Research*, 167, 115101.

Ward B.J. and Strande L. (2019). Chapter 5.2 Conditioning in Faecal Sludge Management: Highlights and Exercises. Eawag: Swiss Federal Institute of Aquatic Science and Technology, Duebendorf, Switzerland. ISBN: 978-3-906484-70-9.

Ward, B.J., Andriessen, N., Tembo, J.M., Kabika, J., Grau, M., Scheidegger, A., Morgenroth, E., Strande, L. (2021) Predictive models using "cheap and easy" field measurements: Can they fill a gap in planning, monitoring, and implementing faecal sludge management solutions? *Water Research.* https://doi.org/10.25678/00037X.

Wood-Black F. (2014). Considerations for scale-up – Moving from the bench to the pilot plant to full production. In *Academia and Industrial Pilot Plant Operations and Safety*. American Chemical Society.

World Health Organization (2018). Guidelines on sanitation and health.

Figure 4.27 Settling tests following faecal sludge conditioning experiments at Niayes FSTP, Dakar, Senegal (photo: Eawag).

5

Estimating quantities and qualities (Q&Q) of faecal sludge at community to city-wide scales

Linda Strande
Miriam Englund
Nienke Andriessen
Juan Pablo Carbajal
Andreas Scheidegger

OBJECTIVES

The objectives of this chapter are to:

- Explain the importance of being able to reasonably estimate Q&Q of faecal sludge
- Define the six stages in the faecal sludge service chain where Q&Q of faecal sludge can be estimated
- Summarise the existing state of knowledge and future prospects for making projections of Q&Q of faecal sludge
- Provide an overview of a methodology to estimate Q&Q of faecal sludge on a scale relevant for the planning of management and treatment solutions, from community scale to city-wide planning.

 Methods for Faecal Sludge Analysis. Edited by Velkushanova K., Strande L., Ronteltap M., Koottatep T., Brdjanovic D. and Buckley C. ISBN: 9781780409115. Published by IWA Publishing, London, UK.

5.1 INTRODUCTION

The goal of this chapter is to present steps for collecting and analysing data to make reasonable projections for faecal sludge loadings at larger-scales that are relevant for planning of city-wide inclusive sanitation. Reasonable projections for quantities and qualities (Q&Q) of faecal sludge that accumulate in given areas, are fundamental for the design of appropriate and sustainable management and treatment solutions. The methodology is based on the hypothesis that **d**emographic, **e**nvironmental, and **t**echnical forms of data that can be referenced or presented in **spa**tial formats (SPA-DET) can be used in planning as predictors of Q&Q of faecal sludge. SPA-DET data can come from existing sources, and also from collection of information with questionnaires during a sampling campaign. The methodology is designed to make adequate estimates for planning, with a reasonable amount of resources. A simple analysis of the data collected in this fashion, can provide projections and trends of Q&Q of faecal sludge. Additional possibilities for analysis of the collected data are numerous, and include sophisticated and advanced modelling approaches. The required steps are identical for any scale, from small communities to entire cities, and are applicable anywhere. The methodology has been continually evolving, from the ideas for sludge production or sludge collection estimates as presented in Strande *et al.*, 2014, to what is presented here. The method will continue to be refined over time to meet the rapidly growing demand for implementing faecal sludge management systems.

This chapter does not consider the complexities of what is fundamentally occurring with physical, chemical, and biological transformations at the micro-level inside individual onsite containments, but rather levels out these complexities to determine total amounts of faecal sludge that need to be managed on a larger scale. With the current state of knowledge, trying to make community to citywide estimates based on the perspective of what is happening within each individual containment would not be sensible due to time, financial and other practical constraints. However, in the future, as more is known at both the macro- and micro-levels, large-scale projections could also be reinforced by insights obtained by the use of models at the individual containment level. As presented in Chapter 6, models at the level of onsite containment will also be useful to describe processes that influence individual rates of sludge accumulation.

What is needed for planning are coarser, larger-scale estimates. This is similar to considering entire populations or community dynamics in ecology. Analogously, to learn about the movement of a population of crickets through an agricultural area, it would not be helpful to inspect one cricket in the laboratory under a microscope. It would instead require zooming out to consider the entire population. As presented in Figure 5.1, in centralised, sewer-based wastewater treatment the design of wastewater treatment plants is based on relatively more homogenised values for entire communities, with less fluctuation due to mixing during transportation in the sewer. In faecal sludge management however, the complexity of what is occurring at the level of individual household or containment is transmitted to the treatment plant. Projections for loadings of faecal sludge are therefore more complicated, due to the unknown nature of the underground containments, together with the widely varying Q&Q of faecal sludge. The methodology presented here, has been developed to address these complex needs.

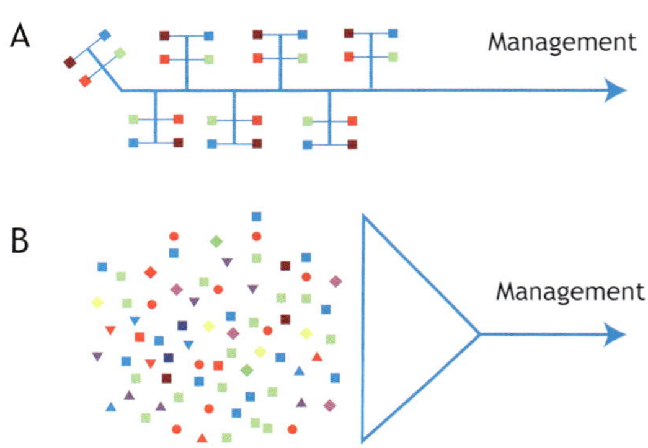

Figure 5.1 A) schematic of wastewater transported through sewer to treatment plant, where it is somewhat homogenised during transport. Squares represent level of each individual connection (*e.g.* household, business). B) schematic of faecal sludge, which is collected, transported, and delivered to treatment plant at the level of individual, onsite containment, without homogenisation. Shapes also represent level of individual onsite containments.

This chapter presents relevant background information, followed by an implementation section for practitioners with guidelines on how to apply this methodology in the field, and a section on future possibilities of how the methodology can continue to advance with future developments. This chapter focuses on projections for faecal sludge loadings at large-scales, and does not address treatment processes or effluent quality, as other mass balance-based methods already exist for that purpose.

5.2 BACKGROUND

Urban areas of low- and middle-income countries are experiencing rapid growth, creating a constant demand for upgrading faecal sludge collection and transport services and treatment infrastructure. As illustrated in Figure 5.2, in addition to planning for total population growth, adaptive management plans are necessary that take the complex and dynamic citywide sanitation context into account.

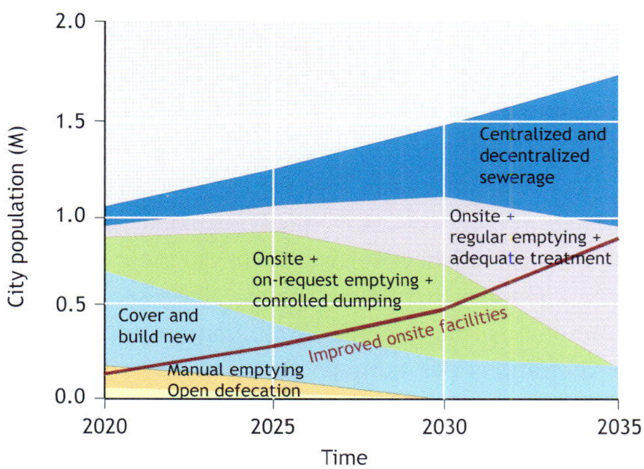

Figure 5.2 SanMix (time-technology diagram) example of adaptive, long-term planning (source: Eawag, 2020).

This requires informed projections along the entire sanitation service chain. An example of the dynamic nature of citywide sanitation, is planning for a new faecal sludge treatment plant to be built in an area where previously there were no legal options available for the discharge of faecal sludge. If projections for the new treatment plant design were based on existing collection and transport practices, the treatment plant would most likely be at capacity or overloaded within a short period after commissioning. This is because once the treatment plant is commissioned, for the first time collection and transport companies will have a legal and affordable place to discharge sludge. With a legal option, formal service providers will likely start collecting and transporting sludge, leading to competition and lower prices. This could subsequently increase the demand for emptying at the household level, creating a much different, higher loading than what was previously projected. Projections are important, as sub-optimal design (both under- and over-sizing) results in risks to public and environmental health, and waste of financial resources.

5.2.1 Scenario projections for planning and management

Implementing adaptive management for complex and dynamic citywide inclusive sanitation requires appropriate projections. In the case of faecal sludge management, this entails characteristics or qualities of the faecal sludge, together with the rates of accumulation. Qualities of faecal sludge include properties, and are often measured as concentrations. Examples of quality parameters include organic matter, solids, nutrients, and dewaterability (Ward *et al.*, 2019, Gold *et al.*, 2018). These parameters are useful for the design of treatment technologies, collection and transport technologies (*e.g.* pumpability as solids or rheological properties), and estimating public and environmental health impacts (*e.g.* pathogens, degradable organic matter, nutrients). Quantities of faecal sludge are expressed as flows, or volumes per time (*e.g.* L/cap.yr). Q&Q together represent loadings (M). For the design of treatment and handling facilities loadings are needed, not quantities or qualities alone. Figure 5.3 is a schematic of how projections for loadings estimated with this methodology would fit into overall planning strategies and projections. Additional examples are developing citywide sanitation plans that include infrastructure plans for faecal sludge treatment plants; community planning for a regularly scheduled desludging program; design of an interim transfer station; designing and sizing a faecal sludge treatment plant; or considering different treatment options based on sludge loadings.

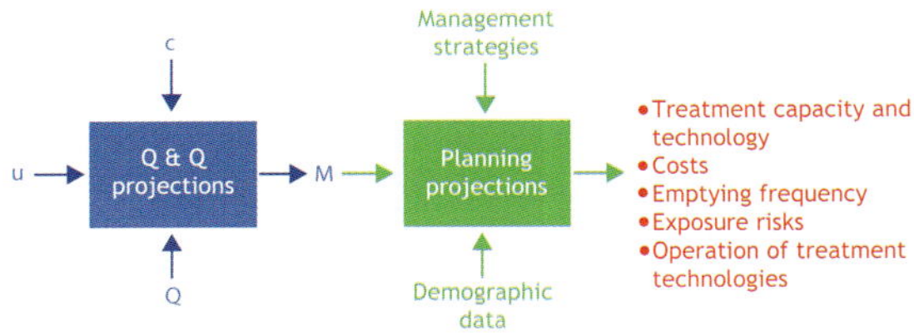

Figure 5.3 Examples of how inputs for scenario projections for Q&Q of faecal sludge can be used in citywide sanitation planning. Q = loading rate, c = concentrations, u = total number of units, M = total loadings, and are further defined in the following section.

It is important to keep in mind that scenario projections provide rough estimates, not exact numbers. This is partly because they are based on accumulation rates and concentrations obtained from field sampling, which are themselves widely variable, but also due to the inherent uncertainty of future scenarios. Furthermore, scenarios are based on the assumption that accumulation rates and concentrations for given categories are stable over time, which will not always be true. For example, assuming that faecal sludge characteristics remain constant during different seasons. This assumption is especially important to consider with changes in infrastructure that will affect accumulation rates and concentrations. For example, if water was provided by stand pipes in an informal settlement during data collection, and then later piped water is delivered, the projections will most likely no longer be valid. In this case, assumptions, data collection and the projection scenarios would have to be revisited.

As is further described in Section 5.3, to make simple projection models the objectives of the study must first be specified , which will then shape the data collection. This includes: (*i*) the defined region boundaries (*e.g.* neigborhood or city or district); (*ii*) accumulation rates and characteristics of interest; (*iii*) categories of SPA-DET data; and (*iv*) estimated future growth of total units (*i.e.* containments). The following values that are required for projections are then collected during sampling:

x_i the categories of data around which the sampling plan and data analysis will be developed (*e.g.* a combination of containment type and income level)

$\bar{Q}(x_i)$ the average accumulation rate of sludge per unit category x_i

$\bar{c}(x_i)$ the average concentration of parameters of interest in sludge of category x_i

$u(x_i)$ the number of units in category x_i that results will be extrapolated to (*e.g.* total number of pit latrines and septic tanks)

Once these values are obtained, projections for total loads are calculated in two steps, first the load that a single unit produces is calculated for every category of data x_i:

$$M(x_i) = \bar{Q}(x_i) \cdot \bar{c}(x_i) \qquad (5.1)$$

Second, the total load is then calculated with the total number of units of containments estimated for the defined area:

$$M_{total} = \sum_{i=1}^{K} u(x_i) \cdot M(x_i) \qquad (5.2)$$

As shown in Figure 5.4, the calculations of these loading projections are easily carried out with common spreadsheet software. For example, to size a faecal sludge treatment plant that serves two communities, the average accumulation rate and TS concentrations per sampled data categories could be used to extrapolate the total TS loading generated by the communities. These loadings could then be used with further information on collection and transport services, to estimate the loading that will actually be delivered to the treatment plant in order to size it. Further details on how to obtain average accumulation rates and concentrations are provided in Section 5.3. The following section presents locations along the service chain where Q&Q of faecal sludge could be calculated.

Figure 5.4 Example calculation of total solids (TS) loading projections for total accumulated faecal sludge in two communities.

Category x_i		Accumulation rate Q(x_i) (L/cap.yr)	Concentration c(x_i) (gTS/L)	Community A: Number of units u(x_i) (-)	Community A: Loading M(x_i) (gTS/cap.yr)	Community B: Number of units u(x_i) (-)	Community B: Loading M(x_i) (gTS/cap.yr)
Type	Income						
(1)	(2)	(3)	(4)	(5)	(6)=(3)·(4)·(5)	(7)	(8)=(3)·(4)·(7)
Pit latrine	Low	50	23	200	230,000	5,000	5,750,000
Pit latrine	Medium	70	19	2,000	2,660,000	3,000	3,990,000
Pit latrine	High	95	12	1,500	1,710,000	2,000	2,280,000
Septic tank	Low	100	8	300	240,000	900	720,000
Septic tank	Medium	180	6	1,000	1,080,000	400	432,000
Septic tank	High	200	2	2,000	800,000	200	80,000
Total					6,720,000		13,252,000

5.2.2 Mass balance: quantifying loadings of faecal sludge

From a mass balance perspective, there are six stages along the faecal sludge management service chain where it is logical to estimate loadings (M), as illustrated in Figure 5.5.

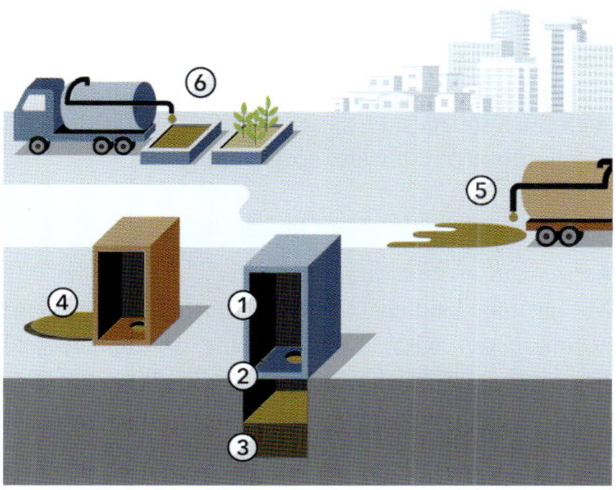

Figure 5.5 Illustration of six stages for mass balance calculations: 1. excreta production; 2 faecal sludge production; 3. faecal sludge accumulation; 4. faecal sludge emptied, not collected; 5. faecal sludge collected, not delivered to treatment; and 6. faecal sludge collected, delivered to legal discharge/treatment (image: Strande *et al.*, 2018).

It is important to distinguish the six stages and estimate them separately for management purposes. Although they are interrelated, they measure very different accumulation rates, concentrations, and environmental fates. Hence, values for the same parameter will vary significantly between them. Pit latrines, mechanical emptying with trucks, and treatment with drying beds are depicted in the figure, but the stages and concept are the same for all arrangements of the faecal sludge management service chain, including manual emptying, all types of onsite containment and treatment technologies, and all methods of collection and transport.

As illustrated in Figure 5.6, stages one and two represent production of excreta and faecal sludge, stage three the accumulation of faecal sludge, and stages four, five and six together the fate of accumulated faecal sludge. When planning for the total amount of faecal sludge that will need to be managed in a community or city, it is most important to consider stage three, the total amount of faecal sludge that is accumulating (*i.e.* total latent demand). However, it is also the most difficult to estimate, as net accumulation rates depend on a large number of factors that are too complex to account for individually. Hence, the estimation of what is actually accumulating in onsite containment is the focus of the methodology in Section 5.3. The following is an overview of each of the six stages.

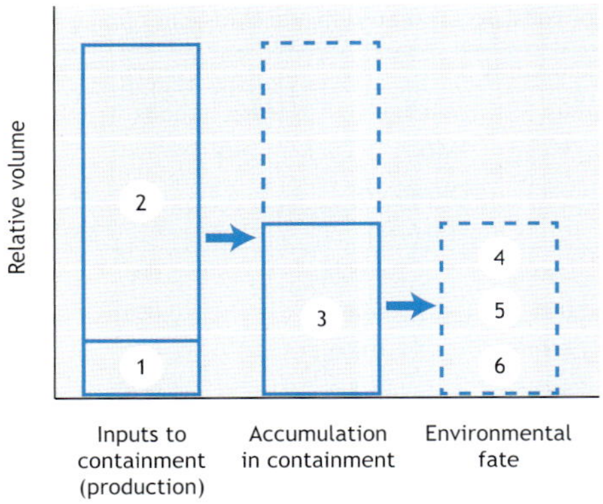

Figure 5.6 Comparison of the relative volume of the six stages, stages one to three represent production and accumulation of excreta and faecal sludge, whereas stages four to six are the fate of faecal following emptying. As illustrated by the dashed line, the total volume of excreta and faecal sludge produced is not the same as the accumulated amount due to biological, physical, and chemical factors that result in a change in the volume of faecal sludge. 4, 5, and 6 cumulatively add up to 3, but volumes of each depend on the local context.

5.2.2.1 Production of excreta and faecal sludge

Excreta production (M_1)

The total load of excreta production (M_1) is the sum of the loads from urine and faeces production from all users of a facility, as represented by equation 5.3.

$$M_1 = M_{urine} + M_{faeces} \quad (5.3)$$

M_1 is not particularly useful for faecal sludge management, other than potentially for the design of container-based sanitation, because as explained in the following sections, excreta alone does not represent faecal sludge. Reasonable estimates for Q_1 and c_1 for excreta could be made based on literature, with adaption for the local context. Further details of ranges of characterics and volumes of produced excreta are provided in Chapter 7 (Penn *et al.*, 2018).

Faecal sludge production (M_2)

The total load of faecal sludge production (M_2) is the sum of the loading from excreta production (M_1) in addition to anything else that is going into the containment (M_{in}), as represented by equation 5.4.

$$M_2 = M_1 + M_{in} = M_{urine} + M_{faeces} + M_{in} \quad (5.4)$$

The total Q&Q of faecal sludge that are produced is dependent on technical factors such as existence and type of flush systems and water connections, and social, economic and political factors, such as available municipal solid waste services and cleansing materials, as explained in more detail in Chapter 2. Estimations for Q_2 could start with existing municipal information on water usage and solid waste, if it is available, together with data from literature, field visits and questionnaires, whereas c_2 would need to be determined through a sampling campaign.

Although the amount of solid waste or garbage in onsite containment can be significant, total amounts will be very context specific. Economic and political factors will play a role, for example in informal settlements in Kampala, Uganda faecal sludge emptying services are paid for by residents, whereas in eThekwini in Durban, South Africa, emptying services are paid for by the municipality. The indirect result is that there is much greater solid waste accumulation in eThekwini than in Kampala where solid waste tends to be dumped outside of pit latrines (Nakagiri *et al.*, 2015, Buckley *et al.*, 2008). Technical factors also play a role, for example there will in general be less solid waste in containment associated with flush toilets such as septic tanks, as it is difficult to pass through the water seal syphon.

5.2.2.2 Accumulation of faecal sludge

Accumulation of faecal sludge (M_3)

M_3 is the load of the total faecal sludge that accumulates with time. From a fundamental perspective, to be able to calculate loadings for total faecal sludge accumulation (M_3) would require knowing total faecal sludge production (M_2), in addition to rates of degradation and accumulation for the biological, physical, and chemical (M_{BPC}) factors that result in reduction of volumes of faecal sludge, as represented by equation 5.5.

$$M_3 = M_2 - M_{BPC} = M_{urine} + M_{faeces} + M_{in} - M_{BPC} \quad (5.5)$$

As a result, every onsite system has different values for M_3, which is why the developed methodology for averaging out complexities is required. Biological factors affecting accumulation include degradation of organic matter, growth of microorganisms, and nutrient cycling, which are affected by many parameters including varying levels of oxygen, water content, and temperature. Physical processes include infiltration and inflow of groundwater or the liquid fraction in containment, and infiltration of soil and sand, which can be affected by construction, soil type and groundwater level. Other factors explained in Example 5.1 that affect the variability of accumulation include how the containment is designed, constructed, used, and maintained, and sludge age and hydraulic retention time. It is important to recognise that loadings from total faecal sludge production (M_2) are not equivalent to loadings from faecal sludge accumulation (M_3), since M_3 is what remains in containment over time (storage) and in most cases the volume, and hence Q_3, will be much smaller (see Figure 5.6). Using instead estimations from any of the other five stages would greatly over- or under-estimate the total faecal sludge that currently needs to be managed. To illustrate the effect that the different volumes have on accumulation rate, excreta production (Q_1), total faecal sludge production (Q_2) and faecal sludge accumulation (Q_3), estimates based on examples from the literature are presented in Table 5.1.

Table 5.1 Estimates based on values in literature for rates of accumulation of excreta production (Q1), faecal sludge production (Q2), and faecal sludge accumulation (Q3) for Kampala, Uganda; Hanoi, Vietnam; and Durban, South Africa.

Location	Excreta production (Q_1)	Faecal sludge production (Q_2)	Faecal sludge accumulation (Q_3)
Kampala (Uganda)	600 L/cap.yr[1,2]	24,480 L/cap.yr[1,2,3,4]	270-280 L/cap.yr[5]
Hanoi (Vietnam)	600 L/cap.yr[1,2]	34,070 L/cap.yr([1,2,5,7]	30 L/cap.yr[8]
Durban (S. Africa)	600 L/cap.yr[1,2]	31,260 L/cap.yr[1,2,9]	21-200 L/cap.yr[10,11,12]

[1]Rose *et al.*, 2015; [2]Brown *et al.*, 1996; [3]Fichtner, 2015; [4]Ojok *et al.*, 2012; [5]Strande *et al.*, 2018, [6]De Bercegol *et al.*, 2017; [7]Otaki *et al.*, 2013; [8]Englund *et al.*, 2020; [9]Van Zyl *et al.*, 2007; [10]Brouckaert *et al.*, 2013; [11]Still and Foxon, 2012; [12]Still *et al.*, 2005[1].

In addition, to illustrate the large variability for values of Q_3, rates reported in the literature for Q_3 from different cities throughout the world are presented in Figure 5.7 (left).

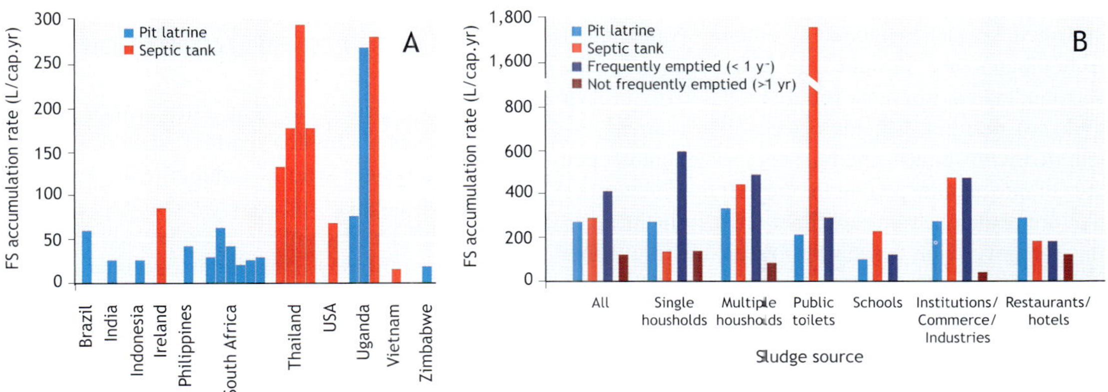

Figure 5.7 Reported diversity of accumulation rates between different cities, and within one city, reproduced from Strande et al. (2018). A) Reported accumulation rates in the literature categorised by country in alphabetical order (Brazil and India (Wagner *et al.*, 1958), Indonesia (Milles *et al.*, 2014), Ireland (Gray, 1995), Philippines (Wagner *et al.*, 1958), South Africa (Brouckaert *et al.*, 2013; Stills and Foxon, 2012; Still *et al.*, 2005), Thailand (including cesspits) (Koottatep *et al.*, 2012), Uganda (Lugali *et al.*, 2016; Strande *et al.*, 2018), USA (Howard, 2003), Vietnam (Harada *et al.*, 2014), Zimbabwe (Morgan *et al.*, 1982). B) Estimated accumulation rates for Kampala, Uganda, by containment type, emptying frequency, and usage, raw data fully available in Englund *et al.* (2020).

The values in Figure 5.7 (left) range from 15 to 300 L/cap.yr. In addition, a study of 30 cities in Asia and Africa reported rates from 36 to 959 L/cap.yr (Chowdhry and Koné 2012) and a recent study in Accra, Ghana reported accumulation rates up to 4,137 L/cap.yr (Sagoe *et al.*, 2019). Also presented in Figure 5.7 (right), are projected values for Q_3 for different types of land usage, all within Kampala, Uganda, to illustrate the high variability of Q_3 even on a citywide scale. Also important to note, is the relation between greater emptying frequency and Q_3. The reported variability of two orders of magnitude for Q_3, illustrates the importance of looking at Q_3 for the specific context, and the need for a standardised approach for determining total amounts of faecal sludge that need to be managed.

5.2.2.3 Fate of faecal sludge

Faecal sludge emptied, but not collected (M_4)
Faecal sludge collected, not delivered to treatment (M_5)
Faecal sludge collected, and delivered to treatment (M_6)
M_4, M_5 and M_6, cumulatively represent the fate of M_3, and will have different values depending on the local context, as represented by Eq. 5.6, and depicted in Figure 5.6.

$$\Sigma M_3 = M_4 + M_5 + M_6 \qquad (5.6)$$

Examples of faecal sludge that is emptied but not collected (M_4), include when containment technologies are designed to drain out into the surrounding environment (or are intentionally broken to do so), or when difficult to access containments are emptied with shovels and buckets into the immediate area or into another pit dug for the purpose. Pit latrines that are abandoned or backfilled are also included in this category, as in dense urban areas this results in a similar fate in the environment. M_4 is difficult to quantify, as it is typically an illegal activity. A rough estimate can be developed through observational site visits, key informant interviews with emptiers and households, and questionnaires. The most important reason to estimate M_4 is for advocacy purposes. The focus should be put on eradication, as it is never an acceptable form of faecal sludge management.

Faecal sludge that is collected but not delivered to treatment (M_5) typically occurs when there is no legal discharge location available, or costs associated with travel and discharge make illegal dumping for emptiers more attractive. Estimates for M_5 can be useful for managing the current situation, for example setting up intermediate transfer or receiving stations until longer-term solutions are implemented. M_5 is also difficult to quantify due to its illegal nature, and is also never an acceptable form of faecal sludge management (Bassan *et al.*, 2013a,b, 2014).

Loadings of faecal sludge that is collected and delivered to legal discharge or treatment facilities (M_6) can be more straightforward to estimate based on existing operating records. However, in reality frequently records do not exist, and there are in general inadequate laboratory resources (Schoebitz *et al.*, 2014). If reports are available, whether there is an incentive to under- or overestimate the amount being discharged should be considered, for example in the case where fees are charged per volume discharged. Instituting a manifest or ledger-based system at treatment plants that includes information such as truck volumes, sludge volumes, emptying frequency, and origin or source of sludge is important for proper design and operation of treatment plants, and could also provide very valuable information for estimating citywide rates of accumulation.

5.3 STEPS FOR IMPLEMENTATION

The first step prior to any implementation is to build a qualified team. Implementation should include a sanitary engineer who is familiar with both faecal sludge management and sewered sanitation solutions[1]. The overall approach of this methodology for making projections of Q&Q of faecal sludge is presented in Figure 5.8. Limited resources should not result in skipping any of the steps, rather the depth of analysis should be adjusted. In this way, the steps can be applied iteratively as new resources become available. In general, it is recommended that projections are revisited in iterations of the approach with progressively deeper rounds of data collection as more information becomes available about the status of

[1] https://sanitationeducation.org/alumni-community/

sanitation within a city. Knowledge of previous sampling campaigns can be used to further tailor sampling plans to increase accuracy, and projections can be gradually refined bringing in additional statistical relationships as they are developed.

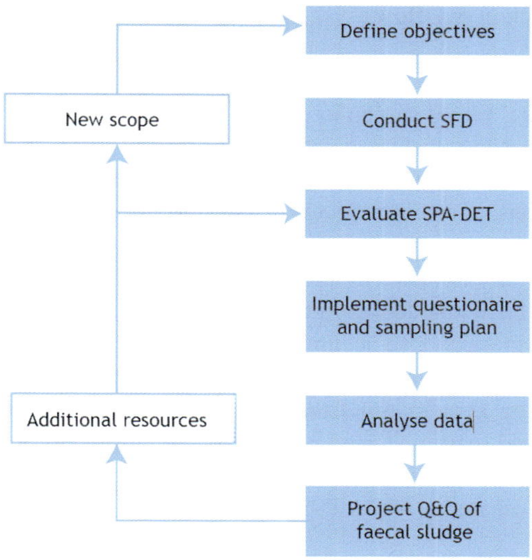

Figure 5.8 Flow diagram including the six steps (blue boxes) of the Q&Q methodology presented in Section 5.3, for data collection and analysis for projections of Q&Q of faecal sludge.

Step 1. Define objectives and region of interest

Planning for different technical and management solutions requires different forms of data collection, so it is necessary to define clear regional boundaries, and objectives for how the Q&Q data is to be used. Based on the defined objectives and local context, how rates of accumulated faecal sludge will be defined and measured is a very important distinction. Refer to Example 5.1 for a discussion of defining accumulation rates. At this initial step, the types of laboratory analysis and analytical data that will be needed to fulfil the objectives should also already be defined.

Example 5.1 *Defining accumulated faecal sludge*

Defining boundaries for values of sludge accumulation will depend on the objective of the study, as further discussed in Chapter 3. Objectives could include knowing what will be delivered to treatment, or researching *in situ* sludge accumulation, or recommending emptying frequency for septic tanks. Regardless of the objective, it is important to keep in mind that evidence suggests that accumulation rates in urban areas are much greater than the historic design filling rates for pit latrines of 42 L/cap.yr that were based on use in rural areas, with five users and an emptying frequency of 10-15 years (Wagner and Lanoix, 1958). This is because onsite containments in dense urban areas have much different usage patterns, a much greater number of users per toilet, and more frequent emptying (refer to Table 5.1 and Figure 5.7). In addition, typically the current reality in low-income cities is little to no level of standardisation for construction of onsite containments. This translates into a wide variety of types of containments, ranging from properly to inappropriately and haphazardly constructed. Most likely, it will not be entirely known beforehand what can be expected, or will be encountered while sampling. Therefore, assumptions about containment type and construction quality will have to be made and then validated during sampling. Prior to making these assumptions and determining sampling locations, it is important to consider how faecal sludge is actually expected to accumulate within the containments.

Septic tanks

Theoretically, the total volume of faecal sludge in septic tanks with an outflow is fixed, with a sludge blanket layer that accumulates as solids settle out, a supernatant zone, and a scum layer (Figure 5.9). Hence, historically the sludge blanket accumulation rate was most commonly estimated as the faecal sludge accumulation rate. Although this is accepted practice, there is a lack of detailed, evidence-based information on actual in-field operating conditions, and in reality, most septic tanks do not operate as intended. They are frequently only emptied upon emergency events such as clogging, extreme odor, or backing up into the house or drains. This means that distinctly different layers cannot necessarily be expected. Therefore, in some cases, it is more relevant to consider the total (fixed) volume when estimating the accumulation rate based on what is emptied over time (L/cap.yr), together with concentrations, to be able to predict loadings that arrive at treatment plants. This is an example of managing the current (not ideal) situation, *versus* improved future solutions that are desired. In areas where septic tanks are properly maintained and operated as designed, it could be more useful to determine rates of sludge blanket accumulation in order to be able to recommend emptying frequencies. However, sludge blanket accumulation is difficult to measure, and can vary a lot over time depending on the operating conditions of the septic tank. In Sircilla, India, no distinguishable change could be measured based on monitoring of sludge blankets in new septic tanks conducted six times over eight months (Prasad *et al.,* 2021). Containments with outflows provide a clear example of how total faecal sludge production is many times greater than actual accumulation within the tank.

Fully lined tanks

In some cities, fully lined tanks emptied at frequent intervals are common for containment in industrial areas, for example for employees working at a factory, or large-volume generators such as hotels, or hospitals (Figure 5.9). This can result in very high accumulation rates, as nothing is leaching out into the surrounding area, and in this case can be as high as total production (Figure 5.6). However, in other cities, industrial areas have been observed to have lower accumulation rates than households (Prasad *et al.* 2021), illustrating the importance of considering the local context. For these types of tanks, accumulation of the total volume of faecal sludge is relevant, as that is what is accumulating and needs to be emptied and treated. It is important to consider non-household types of faecal sludge in any Q&Q study, as they can represent a significant proportion of total flows. In Kampala, Uganda, non-household sources were observed to be up to 50% of the total flow delivered to treatment, and the population of the city doubles during the day due to people commuting in for work (Strande *et al.,* 2018). Fully lined tanks are also sometimes used in flood-prone areas at the household level, with or without overflows.

Partially lined pit latrines

'Dry' faecal sludge in partially lined pit latrines may not have such distinct layers of solids and liquid fractions, but as discussed in Chapter 3, could have layers of different levels of stabilisation (Figure 5.9). In this case, it is relevant to estimate the total volume that accumulates in the pit, or the total volume that is emptied and delivered to treatment. Partially lined pit latrines can also accumulate a very dense layer at the bottom that will never be emptied. However, it needs to be kept in mind that in many cities around the world, partially lined pit latrines are commonly used for all types of faecal sludge, including very 'liquid' faecal sludge (<5% TS).

Cesspits

Cesspits, leach pits, and leaking septic tanks are also very common in urban areas (Figure 5.9). Operating conditions can be assumed to be somewhere between septic tanks and partially lined pit latrines, although in general they have not been studied, and represent an enormous range of possible conditions. Due to a wide range of local terminology, they are also frequently referred to as septic tanks. For management purposes, as there is no way of knowing what processes are occurring inside, accumulation rate of the total volume of faecal sludge is probably most interesting.

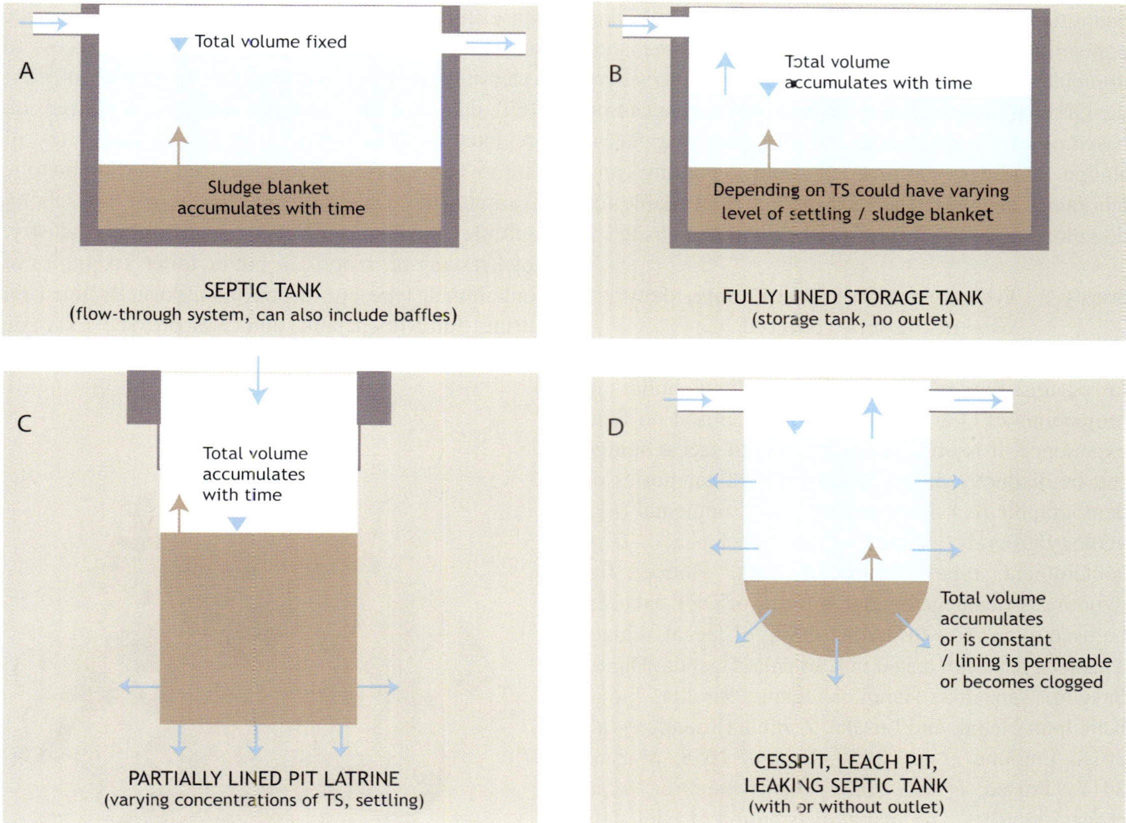

Figure 5.9 Schematic of faecal sludge accumulating in various types of onsite containments: A) septic tank, B) fully lined storage tanks, C) partially lined pit latrines, and D) cesspits (or leach pit, or leaking septic tank).

Step 2. Excreta or Shit Flow Diagram (SFD)

To be able to make reasonable assumptions for sampling plans, data collection, and scenario models, a certain level of expert knowledge is needed. The SFD methodology can be implemented to obtain background information. The SFD is a standardised methodology to collect adequate information to obtain a holistic view of the existing sanitation situation in a city, and producing a report with a diagram for dissemination (Peal *et al.*, 2020). The methodology includes assessing the enabling environment, analysing the sanitation service chain, engaging with stakeholders, and evaluating the credibility of data sources. Through this process, one will become familiar with the types of information that are available for a city.

The SFD approach provides a standardised method to track and document the fate of safely and unsafely managed fractions of total excreta produced by the population through faecal sludge management or sewer-based sanitation, also including open defecation. The SFD diagram itself is meant to be a communication tool that provides an overview of the current sanitation situation in a simple and non-technical fashion. The width of each arrow on an SFD diagram is proportional to the percentage of the population whose excreta contribute to that flow as a proxy for pathogen flows and therefore public health hazard. It is very useful for communicating to decision makers the need for sanitation policy and infrastructure to protect public health. However, it is important to note that the SFD does not estimate quantities of faecal sludge, but rather contributing populations. Depending on the level of implementation, the SFD requires less resources than the Q&Q approach, as the fractions of excreta can be based on expert knowledge, while quantifying faecal sludge loadings requires in field sampling and laboratory analysis. The SFD method is available for download at the SFD Promotion Initiative website[2].

Step 3 Evaluate available SPA-DET data, identify what needs to be collected

SPA-DET data, as defined in the introduction, is used to design the sampling plan, and to build up projections of Q&Q of faecal sludge. Based on field experience, it is observed that Q&Q of faecal sludge can be distinctly different for different categories of demographic (*e.g.* income level), environmental (*e.g.* geology/ground water) and technical (*e.g.* containment type) forms of data. Hence, the hypothesis was developed that forms of DET data can be used as proxies to predict Q&Q of faecal sludge. This idea has been tested in Kampala, Uganda; Dar es Salaam, Tanzania; Hanoi, Vietnam; Sircilla, India; Kohalpur, Nepal; and Lusaka, Zambia (Strande *et al.*, 2018; Englund *et al.*, 2020; Esanju, 2018; Marwa, 2017; Prasad *et al.*, 2021, Andriessen *et al.*,in preparation (b)). The spatial distribution of DET data is important when designing the sampling plan, and when used for scenario planning projections to identify trends and patterns, to identify different infrastructure or interventions needs, and to know the locations and transport distances of existing infrastructures. Because the data is spatially analysable, it can be used to derive citywide projections for Q&Q of faecal sludge, or break them out by community or neighbourhood. An example of SPA-DET data is presented in Figure 5.10, with a spatial distribution of income category and access to sewer network in Kampala, Uganda.

SPA-DET data do not necessarily require a direct cause-effect relationship on Q&Q of faecal sludge to serve as predictors, as long as consistent statistical relationships are observed. For example, significant differences with Q&Q of faecal sludge based on income level were observed in Kampala, Uganda (Case study 5.2). Income level is not the direct cause, but could be explained by factors such as access to water and quality of construction. Examples of SPA-DET data are provided in Table 5.2. Based on previous implementation experience, categories of data in Tier 1 of the table have been good predictors. Examples of building types or usage are: household, multiple household, institution/industry, hotel/restaurant, school, or public toilet. Examples of containment type are: septic tank, partially lined pit latrine, fully lined tank, and cess pit (see Example 5.1).

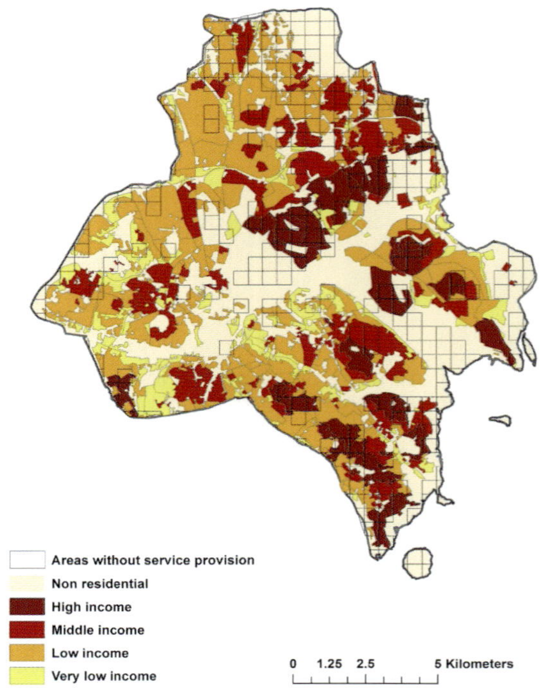

Figure 5.10 Spatial distribution of DET data in Kampala, Uganda. Income categories and non-residential areas shown by color, and areas unserved by sewers shown by outlined grids (image: Schoebitz *et al.*, 2017).

Tier 2 of the table is categories of data that specifically need to be collected to make loading projections based on accumulation rates and characteristics. Tier 2 data is collected during field implementation together with GPS points, so that the

[2] http://sfd.susana.org/

data is spatially analysable and can be evaluated for statistical relations to Tier 1 (and Tier 3). Methods for taking *in situ* samples for characteristics of faecal sludge include the core sampler and cone shaped sampling device, and for *in situ* volumes of faecal sludge include the Volaser measuring device (Andriessen *et al.*, in preparation a). Samples can also be taken during emptying operations, or at delivery to treatment plants. Obtaining reasonable estimates for the sludge age or time since last emptied, are very important in estimating accumulation rates, but is most likely one of the most difficult values to obtain accurate values for, as official records typically do not exist. Until there is better recording, this information will have to be obtained through a questionnaire (refer to Step 4). Relevant details for sampling plans, techniques and methods are covered in detail in Chapter 3.

In Tier 3 of the table are categories of data that have not yet been tested or are in the process of being tested, and based on intuition also seem like potential candidates. Further information on which are the best predictors, and potentially new categories that have not yet been considered, will continue to be developed with future implementations. SPA-DET data that is used in each study will depend on what can be obtained in each specific city, together with what is deemed relevant based on expert knowledge. For example, in Case study 5.2 in Kampala, ground water or soil type were not considered because it was simply not available. In Sircilla, household connection to water was not considered, as all households had water connections (Prasad *et al.*, 2021). In addition, under the umbrella of the 'Swachh Bharat' mission, many new containments have been constructed in Sircilla in the past few years, and are documented in an online database owned by the municipality. Information was available for the sampling team on type of containment, GPS location, and a picture from before, during and after construction. This was useful in designing a sampling plan and analysing the data.

Table 5.2 Categories of SPA-DET data grouped by whether they have been tested, are required for projections of accumulation rates and loadings, or are currently being tested / of potential interest.

SPA-DET		
Demographic	Environmental	Technical
Tier 1. Have been tested		
• Building type/usage • Income level • Number of users	• Geology • Seasonal flooding	• Age of system • Containment type • Water connection • Emptying frequency • Types of wastewater (grey/black)
Tier 2. Required for projections of accumulation rates and loadings		
		• Volume of accumulated sludge • Time since last emptied • Sample for laboratory analysis
Tier 3. Currently being tested / of potential interest		
• Employment rate • Family size • Housing density • Land usage • Population density • Property value	• Elevation • Groundwater • Hydrology • Soil characteristics • Proximity to water • Topography	• Flush • Emptying frequency • Emptying method • Overflow pipe • Piped water • Truck volume • Truck full following emptying • Containment fully emptied • Water added during emptying • Containment fully lined/water tight • Volume of containment • Number of chambers

Tier 1 and Tier 3 SPA-DET data can be collected prior to sampling through desk-based methods, and during sampling through the questionnaire (Step 4). Presented in Table 5.3 are examples of where SPA-DET data can be found.

Table 5.3 Potential sources of SPA-DET data

- Academic institutions (*e.g.* civil engineering department, urban planning department)
- Geographical tools (*e.g.* Google Maps satellite view[3], BORDA City Sanitation Planning[4])
- Census data (*e.g.* population, housing, land use)
- International non-government organisations (NGOs) (*e.g.* UN, WHO, World Bank, JMP SDG reporting)
- Communities of practice (*e.g.* SuSanA, local WASH networks)
- Local NGOs (*e.g.* national WASH missions)
- Contractors (*e.g.* construction and installation of containment)
- Ministries (*e.g.* housing and urban affairs[5], economics, sanitation)
- Call centers (*e.g.* desludging, latrine contractors, plumbing)
- Municipality offices (*e.g.* local assembly, district offices)
- Desludging businesses (*e.g.* trade associations, call centers)
- National bureau of statistics (*e.g.* statistical year books)
- Environmental protection authorities or agencies (*e.g.* soil, elevation, groundwater maps)
- Private sector players (*e.g.* environmental consultancy firms)
- Faecal sludge treatment plants (FSTPs)
- Public water and sanitation utilities

The first step in evaluating SPA-DET data, is to determine whether access to the categories listed in Table 5.2 is easily available . If they are not accessible, evaluate if they can be obtained through the possible sources listed in Table 5.3. If they cannot be obtained, then they will need to be included in the questionnaire-based data collection (Step 4) together with the field sampling.

Based on expert knowledge, and insight gained during the SFD process, a list can then be made of other relevant and interesting categories of SPA-DET data. The list should contain clear links or reasons as to why they might be predictors of Q&Q. For example, 'size of building' is probably interesting because it could be related to accumulation rates, whereas "color of building" is probably not. The listed categories can then be evaluated as to whether they should be included in the study, based on whether or not they are already available, can be easily obtained, or can be readily collected using a questionnaire. Increasing the number and type of SPA-DET data should not significantly increase the cost of data collection, however it can increase the complexity of data analysis. Selecting how many categories of SPA-DET data are feasible to analyse, will be a tradeoff between available time and resources, and more detailed or insightful results. Information that is available by neighbourhood or community can be entered into GIS database during data collection (eg. QGIS[6], or other similar open-source software programs).

Step 4. Location-specific questionnaire

Following collection of available SPA-DET data, a context specific questionnaire-based data acquisition plan needs to be developed based on the study objectives and taking account of available information. Questionnaires can be used to interview customers, service providers during emptying operations or sludge delivery, and treatment plant operators. The person conducting the survey in the field needs to be adequately trained, with an appropriate level of expertise in faecal sludge management to be able to evaluate the validity of answers, fact-check collected information, and to make field observations (refer to Chapter 3 for information on data validation). To reduce costs, if a larger water, sanitation and hygiene (WASH) scoping study will be implemented, a carefully thought out questionnaire could be used to 'piggy-back' onto existing studies, and improve estimates for Q&Q. However, questionnaires have to be conducted at the same location and time point as measurements for Q&Q. Further ideas for reducing costs are presented in Section 5.4.3 and Case study 5.3.

Examples of questionnaires and scoping studies that can serve as a starting place are available online, such as the World Bank's FSM Tools[7] and the Joint

[3] www.google.com/maps
[4] http://citysanitationplanning.org/
[5] http://www.smartcities.gov.in/content/
[6] https://www.qgis.org/en/site/
[7] https://www.worldbank.org/en/topic/sanitation/brief/fecal-sludge-management-tools

Monitoring Program's (JMP) Core questions on water, sanitation and hygiene for household surveys[8]. It is important to consider data resolution when adapting questionnaires to the specific context. It is better to have boxes that the interviewer can check or insert numbers, *versus* qualitative observations. Except for truly categorical variables (*e.g.* septic tank *versus* pit latrine, household *versus* non-household), it is usually recommended to ask for actual numbers. Numbers can be grouped later in categories if desired for the analysis, but not the other way around. 'Slider' responses are one way to ask for continuous response variables, that let respondents rate an item on a numerical scale by indicating values on an interactive slider.

Waypoints (GPS points) need to be recorded during data collection so that the data can be represented spatially. The most efficient way of carrying out surveys is with the help of mobile-based applications on smartphones and tablets (Figure 5.11). There is a wide array of free to use software that is available for mobile data collection (*e.g.* KoboToolbox[9], Akvoflow[10], Open Data Kit[11]). Advantages compared to traditional paper based questionnaires are that data is available immediately, constraints help to ensure the quality of collected data, and coordinates can be obtained automatically via GPS. Factors to be specifically addressed in a Q&Q approach are outlined in Table 5.4.

Figure 5.11 Implementation of a questionnaire in Karnali Province, Nepal. Trained enumerators conducted interviews using smart phones loaded with Open Data Kit (ODK) software to assess households' access to and perceptions of basic services. In the photo, a study participant discusses her feelings of ownership for local communal infrastructure using a 5-point visual scale (photo: M. Vogel).

8 https://washdata.org/monitoring/methods/core-questions
9 https://www.kobotoolbox.org/
10 https://akvo.org/
11 https://opendatakit.org/

Table 5.4 Factors to be included in context specific questionnaires developed for Q&Q studies

Factors	Description
User level	
• Type of onsite containment	Examples of types of onsite containment include septic tanks, pit latrines, and cess pits. It is important to capture as realistic a picture, or sense, as possible of what is existing, as common usage for these terminologies vary widely. Important points to capture include is the containment fully lined (watertight), partially lined, or unlined? For this section, refer to the SFD manual (SFD Promotion Initiative 2018), and the sanitation technology compendium (Tilley *et al.,* 2014).
• Fate of faecal sludge in local environment	If faecal sludge is not transported to treatment, what is its fate following emptying? For example, dumped in local proximity of emptying operation, or transported away?
• Volume of onsite containment (m^3)	The validity of this answer will depend on the context. In general, users of onsite sanitation do not necessarily have any idea of the volume. However, if the respondent was responsible for paying for construction, they most likely have a very good idea of the volume. This can be validated and/or collected during emptying operations, and with tools such as the Volaser measuring device (Chapter 3).
• Outlet	How do liquids leave the containment, is there an outlet pipe, are there multiple containment in series, is there a leach pit, does it go to an open drain?
• Land usage	Possibilities include household, school, industry, commercial (e.g. hotel, restaurant, shop), place of worship, or public toilet. Q&Q will vary depending on land use. For example, industries frequently have larger containment volumes, high number of users, more frequent emptying and different characteristics.
• Number of users (population equivalent)	This can be difficult to evaluate, as records most likely will not exist for number of users of non-single household toilets. Inadequate access to sanitation, and large commuting populations, can result in very high values. Techniques like counters on doors can also be used to validate results. Do not use default values.
• Income	To obtain accurate values for income data, proxy indicators of wealth are potentially more accurate than asking households for income data, but need to be adapted for the local context. See for example Filmer and Pritchett (2001), and the World Bank's tools for measurements of living standards[12].
• Water availability	Increased water access will in general increase volumes of faecal sludge produced. Evaluate whether houses have piped water, stand pipes, or no access.
• Wastewater streams	Water streams connected to the containment will also increase volumes produced. Evaluate whether there are flush or no-flush toilets, if users cleanse with water or paper, and if greywater sources are connected to containment.
• Solid waste	Solid waste in containment can increase the volume produced, but even if it is not a contributor, it could correlate to Q&Q. Flush toilets will tend to have less solid waste in containment due to the water seal.
• Quality of construction	What types of materials are used (*e.g.* concrete, plastic, fiberglass), is it self-constructed or standardised?
Environmental	
• Fact check SPA-DET data	Environmental data will be difficult to collect during a survey, but important to fact check or ground truth in the field. Some data such as percent sand, silt or clay for soil characteristics, or proximity to surface water is feasible to be collected.
Emptying operation	
• Emptying method	Is collection of faecal sludge conducted manually, mechanically, or mechanically assisted. By whom? Is water added during collection, and how much?
• Time since last emptied	The time since last emptied is required for estimating rates of accumulation. Could be measured in days, weeks, months or years.
• Typical emptying interval	This can provide useful information on the management of the containment. Could also be measured in days, weeks, months or years.
• Volume emptied (m^3)	The volume is also important in estimating rates of accumulation. It is important to have multiple ways to evaluate this to ensure accuracy. For example, check for a gauge on the truck, or barrels of standard size.
• Fully emptied	This is important to validate whether the volume emptied is equivalent to the volume of containment.
• Truck volume	This can correlate to Q&Q, as different types of trucks tend to empty containment for different types of land uses, and can also be used to validate containment volume.
• Truck full	This is also important to validate the size of containment.
• Number of containments	Did the truck empty more than one containment? Commonly, operation will be optimised for costs, meaning operators will empty one containment per trip, with a truck that is a similar size to the containment. But will depend on local context, and could include one truck emptying multiple containments, or multiple truck loads for one containment.

[12] http://surveys.worldbank.org/lsms

Step 5. Sampling plan

Once all of the above decisions have been made on the categories of data that need to be collected, and how best to collect it, then the required number and distribution of samples needs to be determined. The recommendations given here are based on theoretical considerations and are only intended as guidelines, as in reality decisions will have to be made based on available resources and practical constraints, an example is provided in Case Study 5.1. The sampling plan is derived through the following steps:

a) Defining categories of SPA-DET data to sample
b) Determining the number of samples
c) Allocating the distribution of samples
d) Building data validation into sampling plan

a) Defining categories of SPA-DET data to sample

a1) Based on expert knowledge, identify the most relevant categories of SPA-DET data (x_i). A category may also be defined by a combination of multiple variables, for example, a given containment type and income level. A larger number of categories allows for finer variation in the scenarios used for projections, but also requires a larger number of samples. Therefore it is recommended to limit the number of categories.

a2) Allocate the number of samples to take in each category. In the simplest case the same number is used for every category. However, in some cases this can be further optimised, as explained in the following section, *Allocating the distribution of samples*

a3) Identify the units to sample from by first identifying all units of a category, and then *randomly* selecting the ones for sampling. However, in many cases a category is defined by information that is not known prior to sampling. For example, if a category is "single-story building + septic tank", depending on the situation, it might be possible to obtain information about the building type prior to sampling, but not information on the containment type. In this case, it is best to randomise over unknown factors. For this example, that would mean simply sampling randomly over all single-story buildings. Randomisation is a very important technique to avoid biases. For example, imagine that the sludge quality in pit latrines varies across the city due to groundwater influence, but this influence is not yet known, and groundwater maps are not available. If samples are randomly selected across the entire area, the sampled average would still be correct. However, if all samples were taken in a region that had a similar groundwater influence, the results would be biased.

b) Determining the number of samples

It is not possible to provide a hard or simple rule on how many samples are required. The number of samples collected during a study will be dictated by the objective of the study, knowledge gaps to be covered, the desired level of accuracy, and available resources. The more samples that are taken, the greater the accuracy of the results. However, this relationship is not linear, meaning that there are diminishing returns with increased sample number, as discussed in Example 5.2. For Q&Q of faecal sludge, the variability between sampled units will typically be quite large, but results will depend on the specific local conditions and number of categories. The absolute minimum number of samples should be selected to reduce the uncertainty to at least 25 %. However, regardless of the sample size, a carefully designed sampling plan is needed to obtain meaningful results that are not biased to increase the reliability of estimates for Q&Q of faecal sludge. In most cases, it is therefore recommended to decide on as many samples as possible depending on the available budget, and then distribute them optimally over the categories. This is another example of the value of implementing incremental studies in the Q&Q approach. In the future, as more implementations of the Q&Q approach are conducted, and distributions of data are better understood, estimates for sample size and distribution should become more straightforward. To facilitate this learning process, open sharing of raw data sets should be encouraged. Sharing of raw research data will benefit the design of future studies, development of statistical methods, and reproducibility, transferability, and learning from results.

Example 5.2 *Influence of sample size*

The loading calculations are based on the *average* accumulation rates $\bar{Q}(x_i)$ and concentrations $\bar{c}(x_i)$. The uncertainty of these averages depends on the standard deviation, or variability, of the accumulation rates sd(Q(x_i)) and of the concentrations sd(c(x_i)) *of the individual samples*, and the number of samples taken, n. More samples will reduce the uncertainty. However, this effect is not linear, meaning taking twice as many samples will reduce the uncertainty by less than half (Figure 5.1.1). The exact relationship for accumulation is represented by equation 5.7. The same equation can be applied for concentrations.

$$sd\left(\bar{Q}(x_i)\right) = \frac{sd\left(Q(x_i)\right)}{\sqrt{n}} \quad (5.7)$$

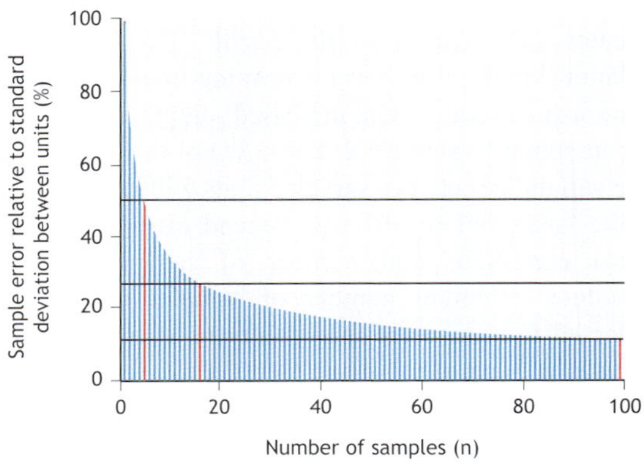

Figure 5.11 Associated reduction of uncertainty with increasing number of samples.

c) Allocating the distribution of samples

If the distribution of samples is going to be adjusted, it can be done based on educated guesses (that are ideally based on data) for the expected averages $\bar{Q}(x_i)$, $\bar{c}(x_i)$ of each category, and the total number of units $u(x_i)$. How much variability is expected between the accumulation rates $Q(x_i)$ of individual categories is also estimated, expressed as $\varepsilon\left(Q(x_i)\right)$, the relative standard deviation of $Q(x_i)$. The same estimations are also made for concentrations. Based on these assumptions, an optimal fraction of samples to be allocated to each category x_i can be determined with equation 5.8.

$$w_i = \frac{u(x_i)\bar{Q}(x_i)\bar{c}(x_i)\sqrt{\varepsilon\left(\bar{Q}(x_i)\right)+\varepsilon\left(\bar{c}(x_i)\right)}}{\sum_{j=1}^{K} u(x_j)\bar{Q}(x_j)\bar{c}(x_j)\sqrt{\varepsilon\left(\bar{Q}(x_j)\right)+\varepsilon\left(\bar{c}(x_j)\right)}}$$

(5.8)

Not every quantity in equation 5.8 of the scenario projection will have the same influence on the result. Intuitively, it is clear that an error in the accumulation rate for a category with a small number of units is less relevant than if this number is large. In addition, there will be more variability within some categories than others. Therefore, it is sensible to take more samples from units of the influential categories and/or categories with more variables. However, this technique is typically not suitable for the first iteration of a sampling campaign, unless there is reliable existing expert knowledge about the variability and the average of the Q&Q. However, if in doubt, an equal number of samples per category should be used. In subsequent iterations of sampling, the necessary information to make these decisions will then be more readily available.

d) Building data validation into sampling plan

Any sampling plan requires data validation to verify the accuracy and precision of obtained values, but this is especially important with the high intrinsic variability of faecal sludge. The accuracy of the overall estimation will only be as good as the least accurate parameter. Therefore, the number of samples in the sampling plan and laboratory analysis will have to be increased (or reduced total number of sampling sites) to validate the collected data and assumptions, and the results. An example of sampling for Q&Q of faecal sludge from *in situ* containments *versus* during collection and transport is presented in Case study 3.3. Guidelines on quality assurance and quality control (QA/QC) for how to develop sampling and analytical plans taking into account the adequate number of duplicate samples to ensure accuracy and precision are presented in Chapter 8, section 2.2: Quality assurance. Another example is determining the number of users of a public toilet in an informal settlement, a large factory, or a toilet at a public market is difficult to assess with a questionnaire. Records might not exist, and the number of users will also need to be converted to daily population equivalents if per capita flows are going to be estimated. One example could be to place a counter on the toilet door to validate questionnaire data (Zakaria *et al.*, 2018). However, any effort (*i.e.* time and money) spent on improving measurements must take into account specifically the required level of accuracy of collected Q&Q data.

Case study 5.1 Development of Q&Q sampling plan in Lusaka, Zambia

The University of Zambia (UNZA) and Eawag implemented a Q&Q study in Lusaka, Zambia between September and December 2019. In total, samples were collected from 421 onsite containments together with a questionnaire and laboratory analysis (Ward *et al.* 2021). The following steps were taken to develop the sampling plan.

For households:

- ArcMap was used to develop the sampling plan, as shown in Figure 5.12. The boundaries of the study were set as the official Lusaka city boundaries, in addition to any areas served by the Lusaka Water and Sewerage Company (LWSC) outside of these boundaries. Areas where service is provided through the sewer network were excluded, in addition to the airport.
- A layer was added with information on geological formations, and the area was separated by the three different rock formations that are present in Lusaka (Cheta limestone, Dolomite and Schitst/Quartzite). It is known that risk of disease from groundwater varies by these locations (Museteka *et al.*, 2019) so sample locations were assigned from all three.
- A one square kilometer grid layer was added to the area. Sampling locations for the field team were randomly selected by assigning one point to each grid with ArcMap. Quadrants with no, or only a few, households were excluded, in addition to the industrial area, and the area served by the sewer.
- High density areas were identified based on expert knowledge and visual inspection. They are highlighted in green on the map. In these areas, two sampling locations were randomly selected per quadrant.
- During implementation, the field team always went to one of the randomly selected points. If for some reason it was inaccessible this was documented, and then the sample was always taken at the next location *to the right* if facing the building. In this way, the randomness of the sampling was maintained. During sampling, the sampling locations were marked in Google Maps.

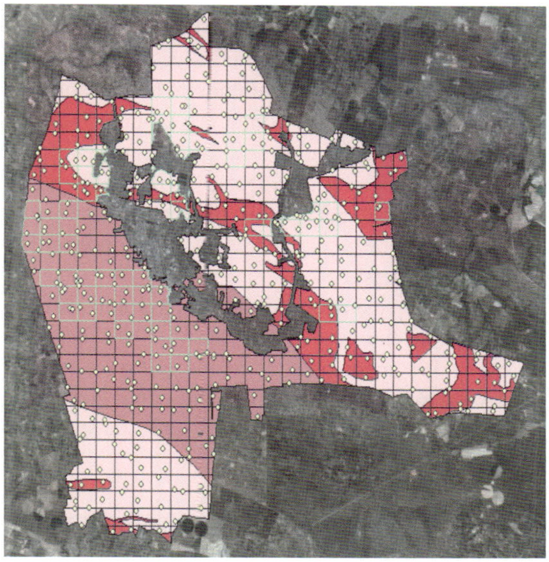

Figure 5.12 Sampling plan for Lusaka, Zambia developed in ArcMap.

For non-households:

- Samples for commercial areas were separated into four categories: public toilets, office buildings, schools, and malls. These were selected because they were determined to be the most relevant for Lusaka based on local expert knowledge.
- For each of these categories, the goal was to obtain 15 samples spread evenly throughout the boundaries.
- Non-household sampling points were selected based on local expert knowledge. Malls and schools could be identified on Google maps, public toilets and office buildings were identified by the sampling teams and local knowledge (*e.g.* sampling team drivers, city council members, community leaders).
- The spread of the commercial sampled points was monitored during the sampling campaign.

Step 6. Data analysis

Once data collection is completed, then the planned projections can be made. Recommended steps for data analysis include first a visual examination of the data, and verifying whether all of the results seem reasonable based on expert knowledge. This includes:

- Identifying minimum and maximum values, and evaluating them as to whether they are feasible.
- Visually identifying extreme values, and checking to see if any recorded data points look suspicious (*e.g.* missing decimal points, wrong units, number entered in wrong field).
- Visually checking if expected correlations can be found in the data (*e.g.* higher income level expected to be associated with a higher proportion of flush toilets).
- Excluding suspicious data from further analysis based on the inspections above. This requires expert judgment, as no hard and fast rules exist to decide what is an outlier and what not. This process needs to be clearly and transparently documented and reported.

Data analysis and reporting of projections will depend on the defined objective(s) for the Q&Q study. Recommended steps for evaluating categories of SPA-DET data to use in projections include the following:

- Evaluate whether there are relevant differences between categories of SPA-DET data (*e.g.* type of containments, income levels, building type). For an example, see Case study 5.2.
- Investigate what combinations of categories of SPA-DET data make sense to combine for the specific study region and objectives, and evaluate different scenarios. Depending on the amount of collected data, and relevant differences, it is recommended to select at most a few significant categories and to avoid cross-correlations. For example, if "water connection" and "containment type" exhibit significantly different relations to measured parameters, but all buildings with water connections have septic tanks, it does not make sense to include both "water connection" and "containment type". However, if there are also pit latrines with water connections, then it could make sense. Summarise in a table the most relevant combinations of categories of SPA-DET data (Table 5.5).
- Evaluate if there are differences in loadings for different regions of a city. This is important, to identify potential indicators of loadings that were not considered or known during study implementation. For example, as discussed in Step 5, this could help to identify areas of groundwater intrusion when groundwater maps are not available.

Table 5.5 Example of breaking down loadings based on categories of SPA-DET data.

	Pit latrine	Septic tank
Households	$M_{PL,HH}$	$M_{ST,HH}$
Non-households	$M_{PL,NHH}$	$M_{ST,NHH}$

As discussed in the introduction, the projections for total loadings at the community to city-wide scale can now be used along with management strategies and demographic data for planning projections, such as selecting treatment capacity and technologies (Figure 5.3). Further options for data analysis are introduced in Section 5.4.

Case study 5.2 *Evaluating Q&Q of faecal sludge in Kampala, Uganda*

This case study is described in detail in Strande *et al.* (2018) and the complete raw data set is available for download at https://doi.org/10.25678/0000tt. From December 2013 to March 2014, in total 180 faecal sludge samples were collected in Kampala, Uganda, spanning both the dry and (short) rainy season. Categories of SPA-DET data were found to be significantly different for Q&Q of faecal sludge in Kampala, Uganda. Presented in Figure 5.13 are results for TS concentrations for the categories of collected data. Differences were determined by evaluating the confidence interval around the median (notches in the boxplots), calculated by equation 5.9, where IQR is the interquartile range and n is the sample size.

$$\text{Confidence interval} = \pm\frac{1.58 \cdot \text{IQR}}{\sqrt{n}} \qquad (5.9)$$

For each set of potential indicators, if confidence intervals of the median did not overlap they were considered to be statistically different. High-income areas had lower median TS concentration (7 gTS/L faecal sludge) than low-income areas (29 gTS/L faecal sludge). Other observed predictors were black water only, solid waste, number of users, containment volume, emptying frequency, and truck size. The average accumulated faecal sludge for the entire city was projected as 270-280 L/cap.yr (Figure 5.7).

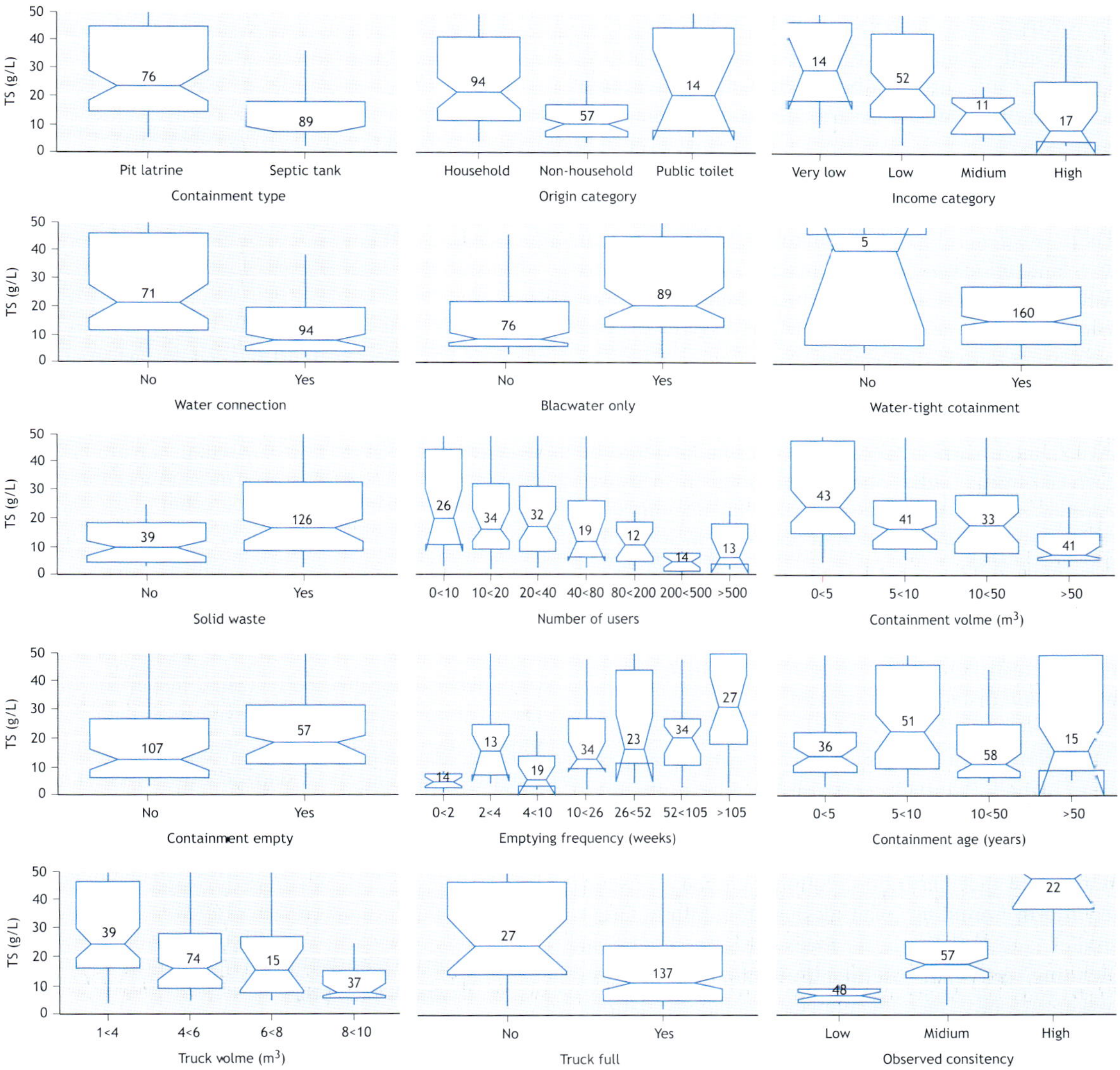

Figure 5.13 Total solids (TS) concentration of faecal sludge, based on collected categories of SPA-DET data in Kampala, Uganda. Number on box plots is number of samples in category.

5.4 FURTHER RESEARCH AND ANALYTICAL POSSIBILITIES

The methodology presented in this chapter includes steps for data collection that can then be used to build up estimations or projections of Q&Q of faecal sludge with straightforward and non-complicated calculations. As implementations and experience with the methodology increase, it will also continue to evolve and become more refined and sophisticated. Tools that are currently being evaluated in research activities of Eawag are described in this section.

5.4.1 Remote sensing

There is a general lack of available SPA-DET data in low- and middle-income countries. To help fill this gap, the use of Earth observation data and remote sensing-based indicators are being explored as a strategy to derive such missing information (Baud *et al.,* 2010; Kohli, 2015). Eawag and the Department of Geoinformatics (Z_GIS) University of Salzburg investigated whether SPA-DET information could be derived from Earth observation data in Lusaka, Zambia, and evaluated it for statistical relations with Q&Q data collected in the field (Nödel 2020). Presented in Figure 5.14 is the example of building density, based on building footprints extracted from satellite imagery. Data was also collected for land use, roof type, distance to green space, distance to water bodies, and distance to treatment.

Referring back to Figure 5.12, it can be seen that areas of high building density are similar to the high density areas designated on the sampling plan. Based on the results, the main findings from this exploratory study were that Earth observation data can be useful to inform sampling plan design for future Q&Q studies, and could indicate focus areas for sanitation planning, providing useful information for decision makers. None of the indicators had statistical relations to quantities, however, building density, building size, street condition, and building use were predictors of TS (Nödel 2020).

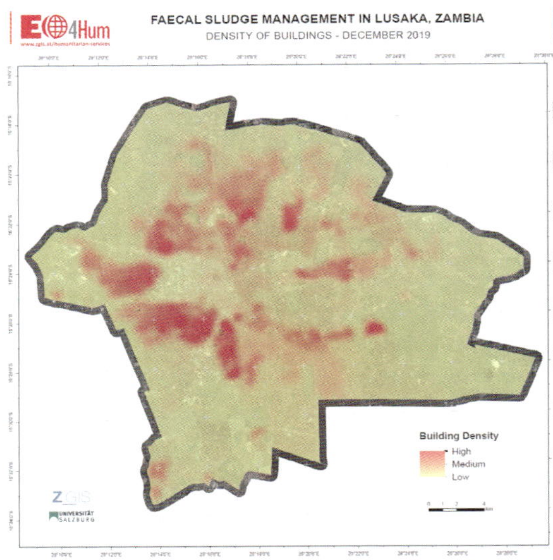

Figure 5.14 Building density in Lusaka, Zambia (map generated by Johannes Nödel and Barbara Riedler). Density calculations were based on data integration of building footprints derived through (*i*) semi-automated, object-based extraction using a very high resolution Pleiades satellite imagery from 2018, (*ii*) OpenStreetMap, and (*iii*) field data provided by GIZ. Values of building density range from low (0 buildings/ km²) to high (4,800 buildings/km²).

5.4.2 Additional spatial analysis

Plotting of results from Q&Q studies in GIS software provides another method of visual inspection of data. By evaluating the results visually it can help to identify further relationships that affect loadings, or identify areas that have significantly different results than expected. For example, as discussed in Step 5 of the sampling plan, this could help to identify areas of groundwater intrusion when groundwater maps are not available.

5.4.3 Interrelationships between sludge characteristics

In wastewater treatment, ratios of constituents in wastewater have been empirically established and are commonly used as rough guidelines during design and selection of treatment technologies. For example, untreated wastewater with a VSS/TSS ratio of 0.85, or a BOD/COD ratio of 0.5 or higher can be considered treatable by activated sludge (Tchobanoglous *et al.,*

2014; Henze and Comeau, 2008). These types of relationships have not yet been empirically established for faecal sludge, due to the relative lack of experience and data. Potentially, as more and more Q&Q studies are conducted, these types of relationships could also be established for faecal sludge. However, with the current state of knowledge, empirical relationships for specific types of faecal sludge, or for specific regions, cannot be transferred to other scenarios, as is also recommended for wastewater with correlations of TOC and COD (Rice *et al.,* 2017).

One potential application for established empirical correlations, could be to reduce the cost of characterisation studies, which are quite resource intensive. For example, if consistent COD/TS ratios are observed in an area, all samples could be analysed for TS (which does not require chemicals) and only a fraction measured for COD (see Case study 5.3). Such approaches could also lead to the development of lower cost qualitative methods for rough estimations. For example developing a color chart or smart phone app that indicates the level of stabilisation of faecal sludge (Ward *et al.,* 2021), or an in-field portable penetrometer that could predict TS as a metric of viscosity (see Chapter 3, section 3.5.9).

Case study 5.3 *Further analysis of statistical relationships within data, COD/TS*

Over the past five years, researchers from Eawag have collected 1,000 samples during implementation of Q&Q studies in six cities[13]. This data is currently being analysed to evaluate trends and relationships within cities, across multiple cities, and for categories of data such as pit latrine or septic tank. The example of COD/TS is presented in Figure 5.15a and Figure 5.15b. Relatively good correlations for COD/TS were observed in Dar es Salaam, Hanoi, Kampala, Ougadougou, and Sircilla, but *the relationships were different in each city*. This pattern was also observed in a study employing the Q&Q methodology in an informal settlement in Nairobi, Kenya (COD = 0.86·TS, with R^2 = 0.93) (Junglen, *et al.* in preparation). In contrast, observed correlations in Lusaka were relatively weak, and were slightly improved by breaking down correlations by categories of collected SPA-DET data (Ward *et al.,* 2021). These examples illustrate that even if empirical relationships are established within cities, the results from the different cities are not necessarily transferable. Note: TS for Lusaka is reported as % TS determined gravimetrically, whereas the TS for the other cities is reported as concentration (g/L).

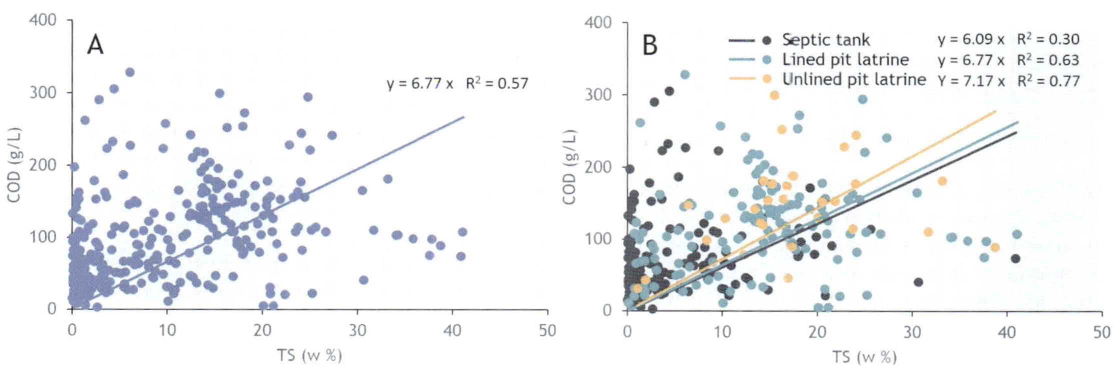

Figure 5.15a A) COD/TS for Lusaka (n=360). B) COD/TS for Lusaka based on type of containment.

[13] For complete data sets see: Englund *et al.* (2020), Strande *et al.* (2018), Ward *et al.* (2021), Prasad *et al.* (2021), and Andriessen *et al.* in preparation.

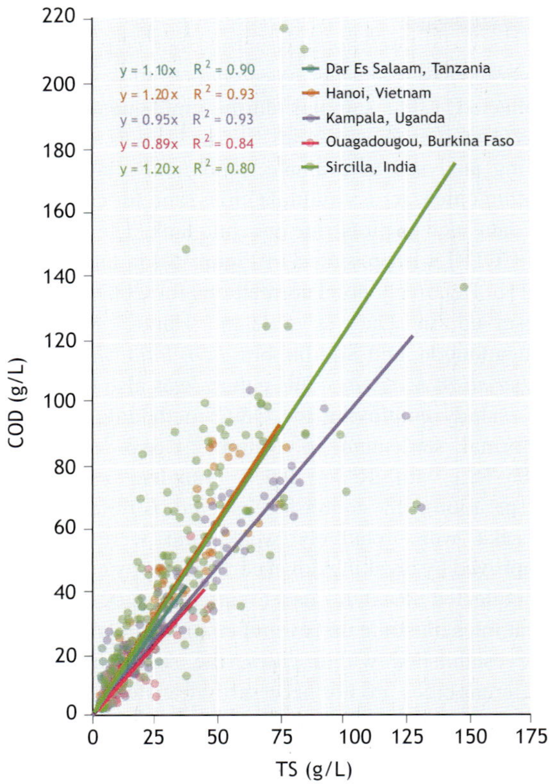

Figure 5.15b City-specific correlations for COD/TS for Kampala (n=180), Dar es Salaam (n=76), Sircilla (n=180), Ouagadougou (n=53), and Hanoi (n=60).

5.4.4 Evaluating categories of data to evaluate separately

Decision trees are models with the main advantage that they are very easy to visualise (Figure 5.16) as they consist of a series 'if' statements separating data that follows different patterns (Safavian and Landgrebe, 1991). The resulting trees should always be compared with expert knowledge for validation. Decision trees could be used to automatically define categories of data that are relevant to analyse separately, instead of only relying on observational experiences and expertise in the field, as was done in the presented methodology. Data analysis can include investigating where and how to break out results for large areas or sample sizes for separate analysis based on categories of SPA-DET data, such as household – non-household and septic tank – pit latrine (Figure 5.16). If the differences are distinct enough, and the sample size large enough, then the SPA-DET data could be analysed separately among these two types of data categories to increase the power and accuracy of predictions.

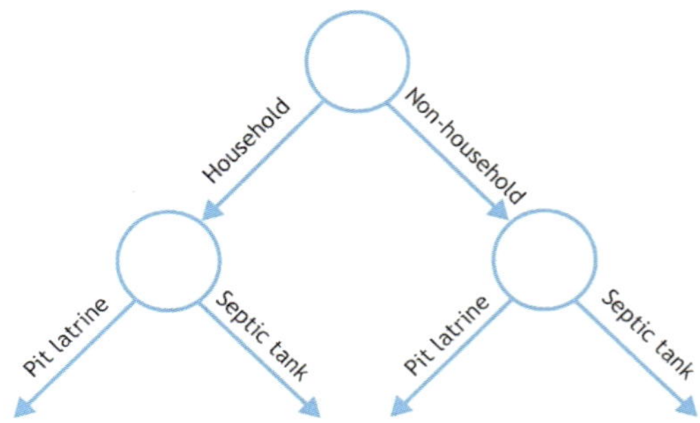

Figure 5.16 Decision tree based on land use patterns (household and non-household) and containment technology (septic tank and pit latrine).

Like any model, the use of decision trees requires adequate input variables. Attempts to train decision trees should start with input variables that are most readily and affordably available. For example, satellite or aerial image analysis can readily be used to distinguish residential and non-residential land use for an entire area. If differences among Q&Q of faecal sludge are expected based on land use, land use should be used as input variable to increase the accuracy of the model. Other variables such as the containment technology are relatively straightforward to obtain, but could require primary data collection depending on the level of information that is available for a city.

The use of decision trees is also useful to test and inform knowledge. Much of the current state of knowledge in faecal sludge management is based purely on observations in the field and it is often not clear which categories are important to analyse separately and which should not be disaggregated. For example, 'public toilet' is frequently grouped as being characteristic of one type of faecal sludge, but analysis shows this is not necessarily valid, for example, in Kampala the type of containment technology was more relevant (Strande *et al.*, 2018).

5.4.5 Predictive models

Statistical models with the aim of predicting a quantity with given inputs can range from simple linear regressions (Case study 5.3) to complex non-linear machine learning models (Case study 5.4). The construction and calibration of such models requires a certain level of expertise, especially since the data collected for Q&Q of faecal sludge can be quite 'noisy'. However, financially the hurdle for such data analysis is quite low when compared to laboratory analysis, as free software, tutorials and online courses are available[14]. The main advantage of machine learning algorithms is that they can identify statistical relationships that are not always noticed by visual inspection. Since relationships can be noisy, care needs to be taken to avoid 'overfitting', to avoid creating a model that just describes the noise of the data. The usefulness of predictive models are application and data dependent. A basic decision tree model can be useful when estimates do not require a high-level of precision. Where higher-resolution predictions are needed, other tools such as machine learning can be used to improve accuracy and reduce error in predictions (Ward *et al.*, 2021) However, the precision will still depend on the available data (Case study 5.4). Stochastic models could be advantageous to predict the loading of faecal sludge at treatment plants, as they also describe peak loadings (a similar application for urine collection is presented in Rossboth, 2013). Using predictive models for data exploration can also lead to deeper learning from results, which can in turn lead to the development of mechanistic models. Mechanistic models are discussed further in Chapter 6.

5.4.6 Sensitivity analysis and error propagation

Sensitivity analysis aims to identify the most critical input of a scenario analysis. Various techniques exist, from simply changing one input at a time to more sophisticated approaches that also reveal interactions (*e.g.* Saltelli, 2004). Error propagation can be applied in cases where the uncertainty of the inputs can be quantified (or guessed) to investigate how these uncertainties influence the model outputs. A common and easy to apply technique is Monte Carlo simulations.

Case study 5.4 *Predictive Models for Hanoi, Vietnam, and Kampala, Uganda*

This case study is based on Englund *et al.* (2020) and the complete raw data set is available for download at https://doi.org/10.25678/0000tt. This study was conducted to evaluate whether SPA-DET data could be used to build predictive models for faecal sludge management. Two data sets from Hanoi and Kampala were used. The data includes 60 field samples and questionnaires from Hanoi and 180 from Kampala, results of the characterisation from Hanoi are presented in Figure 5.17.

Software tools were used in an iterative process to predict TS and emptying frequency in both cities. City-specific data could be predicted with types of SPA-DET data as input variables, and model performance was improved by analysing septic tanks and pit latrines separately. Individual city models were built for TS concentrations and emptying frequency. In addition, a model was built across both cities for emptying frequency of septic tanks based on number of users and containment volume (Figure 5.18). The data appears to be consistent across the two cities, despite the fact that the range of input variables is quite different, indicating that in the future predictive models could potentially be relevant for multiple cities. However, it is important to note that these two cities only represent two data points, and general assumptions for other cities cannot be drawn without validation. Number of users, containment volume, truck volume and income level were identified as the most common variables for the correction function. Results confirm the high intrinsic variability of faecal sludge characteristics, and illustrate the value of moving beyond simple reporting of city-wide average values for estimations of Q&Q.

[14] e.g. https://www.r-project.org/, https://scikit-learn.org/stable/

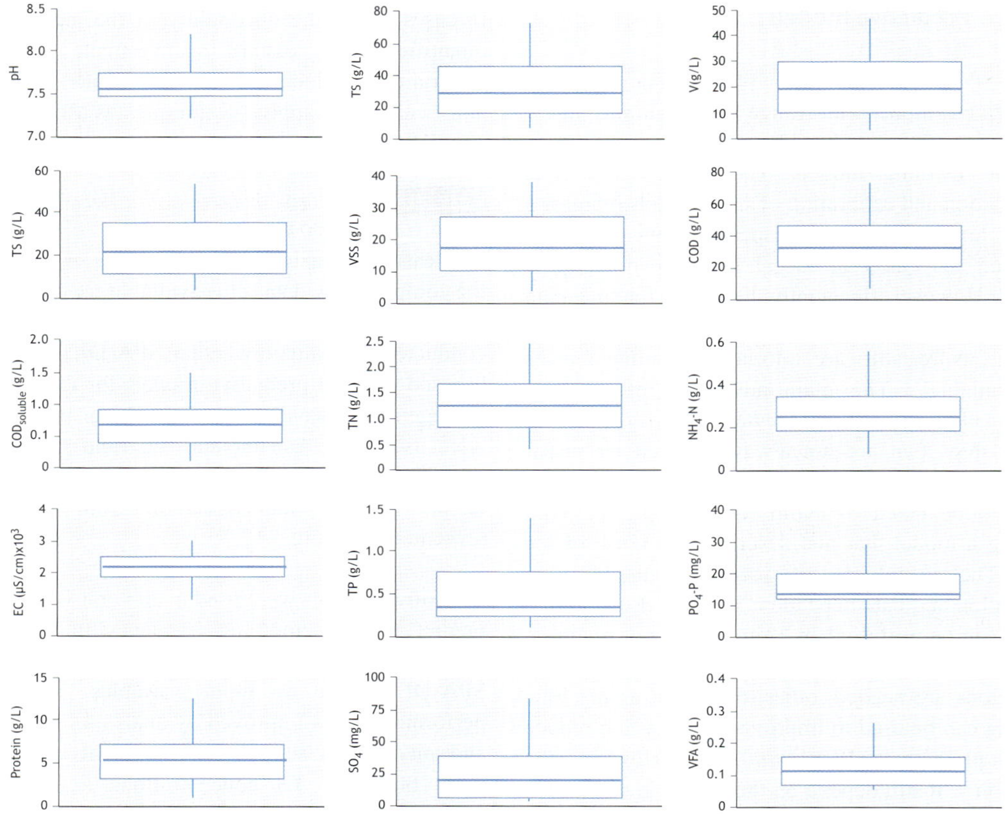

Figure 5.17 Characterisation results for 60 samples taken from household septic tanks in Hanoi, Vietnam (Englund *et al.* 2020).

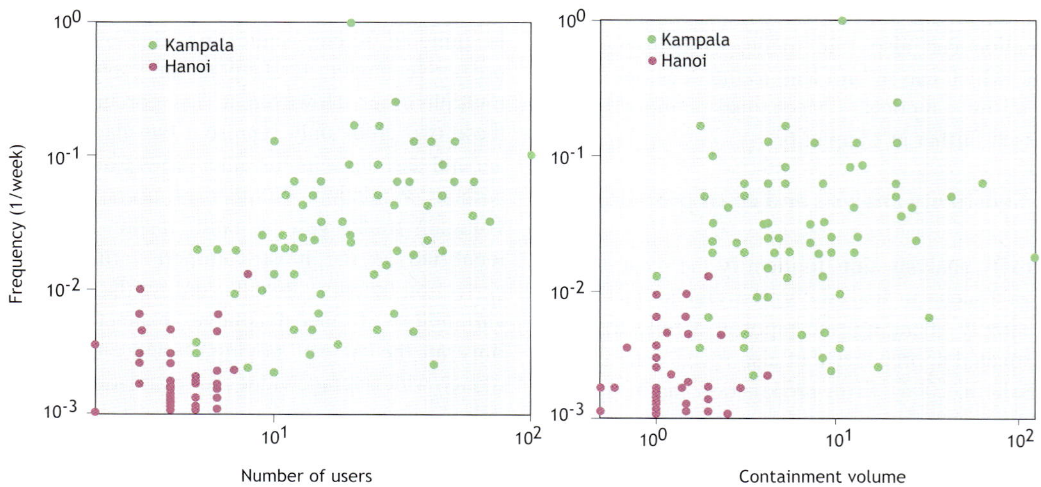

Figure 5.18 Predicted emptying frequency of septic tanks from single and multiple households, showing fit to data from Hanoi and Kampala, the plots are in log-log scale (Englund *et al.* 2020).

5.5 OUTLOOK

The management of faecal sludge is dynamic and complex. Sustainable long-term management requires adaptive planning for population growth and changing infrastructure. The methodology presented in this chapter for the projection of faecal sludge loadings at the community to citywide scale, is a structured, iterative process. The methodology can be implemented with available resources, and revisited with progressively deeper and more data-rich campaign rounds as resources become available. In this way, projections can be improved with time, and additional statistical relationships can be established. Data collected in this fashion will be representative for making projections of Q&Q of faecal sludge, and as more data becomes globally available, that is collected in a logical, replicable, comparable, and transparent fashion, it will allow for greater transferability and learning among cities, countries, and regions.

Important lessons learned include:

- Use of historical accumulation rates intended for the design of pit latrines in rural areas are not transferable to dense urban areas (Strande et *al.* 2018).
- Faecal sludge Q&Q data do not follow a normal distribution (Chapter 1), hence, only reporting values for averages and standard deviations is not adequate. Summary statistics should include at a minimum averages, standard deviations, medians and interquartile ranges, and ideally, complete raw data sets should be shared (Andriessen *et al.* in preparation(b)).
- It is important to clearly identify the goal of a Q&Q study prior to defining system boundaries of onsite containment technologies. Resulting metrics should be determined based on these definitions, together with the availability of resources (Prasad *et al.* 2021).
- The resolution of planning projections only needs to be as precise as the decision-making process requires. City-wide inclusive sanitation planning does not require the same level of precision as process control or optimisation of treatment plants (Ward et *al.*, 2021, Englund et *al.* 2020).
- When designing faecal sludge treatment plants, it is crucial to keep in mind, that even with more reliable predictions for loadings, daily operation still needs to be able to adapt to highly variable influent loadings (Klinger et *al.* 2019).

REFERENCES

Andriessen N. and Strande L. (in preparation a). Manual for assembling and operating Volaser measuring device.

Andriessen N., Scheidegger A. and Strande L. (in preparation b). Projections of quantities and qualities (Q&Q) of faecal sludge for management and treatment: comparison of data from six cities in Africa and Asia.

Bassan M., Mbéguéré M., Zabsonre F. and Strande L. (2013a). Integrated faecal sludge management scheme for the cities of Burkina Faso. *Journal of Water, Sanitation and Hygiene for Development,* 3, 216-221.

Bassan M., Tchonda, T., Yiougo L., Zoellig H., Mahamane I., Mbéguéré M. and Strande, L. (2013b). Characterisation of faecal sludge during dry and rainy seasons in Ouagadougou, Burkina Faso, *Proceedings of the 36th WEDC International Conference*, Nakuru, Kenya, 2013.

Bassan M., Dao N., Nguyen V.A., Holliger C. and Strande L. (2014). Technologies for sanitation: how to determine appropriate sludge treatment strategies in Vietnam. *37th WEDC International Conference,* Hanoi Vietnam, 2014.

Brouckaert C., Foxon K. and Wood K. (2013). Modelling the filling rate of pit latrines. *Water SA*, 39.

Brown D., Butler, D., Orman, N. and Davies, J. (1996). Gross solids transport in small diameter sewers. *Water Science and Technology*, 33(9), 25-30.

Buckley C.A., Foxon K.M., Brouckaert C. J., Rodda N., Nwaneri C.F. and Balboni E. (2008). Scientific support for the design and operation of ventilated improved pit latrines (VIPS) and the efficacy of pit latrine additives, *Water Research Commission,* South Africa.

Chowdhry S. and Kone,D. (2012). Business Analysis of Fecal Sludge Management: Emptying and Transportation Services in Africa and Asia. *Bill & Melinda Gates Foundation*, Seattle, USA.

De Bercegol R., Cave J. and Huyen A.N.T. (2017). Waste Municipal Service and Informal Recycling Sector in Fast-Growing Asian Cities: Co-Existence, Opposition or Integration? *Resources-Basel,* 6(4), 10.

Eawag-Sandec (2020). Training course on Consultant Capacity Development (ConCaD) for City-wide Inclusive Urban Sanitation, Module 3.3.

Englund M., Carbajal J.P., Ferré A., Bassan M., Vu T.H.A., Nguyen V.A. and Strande L. (2020). Modelling quantities and qualities (Q&Q) of faecal sludge in Hanoi, Vietnam and Kampala, Uganda for improved management solutions. *Journal of Environmental Management*, 261, 110202.

Esanju M.M. (2018). Assessment of the applicability of data collection methodology for estimating quantities and qualities of faecal sludge (Q&Q model) in Dar es Salaam. Master's thesis, *University of Dar es Salaam.*

Fichtner Associates GmbH., (2015). Kampala Sanitation Master Plan Update. Volume 1: Report, JV Fichtner *Water & Transportation.*

Filmer D. and Pritchett L.H. (2001). Estimating wealth effects without expenditure data-or tears: an application to educational enrollments in states of India. *Demography,* 38(1), 115-132.

Gold M., Harada H., Therrien J.-D., Nishida T., Cunningham M., Semiyaga S., Fujii S., Niwagaba C., Dorea C., Nguyen V.-A. and Strande L. (2018). Cross-country analysis of faecal sludge dewatering. *Environmental Technology.* 39(23), 3077-3087.

Gray N. F. (1995). The Influence of sludge accumulation rate on septic tank design. *Environmental Technology,* 16(8), 795-800.

Harada H., Nguyet P.-A., Nguyen V.-A. and Fujii S. (2014). Influential factors on septage composition: A case study in Hanoi, *In proceedings*: *FSM3 - International Faecal Sludge Management Conference*, Hanoi, 2014.

Henze M. and Comeau Y. (2008). Wastewater Characterisation. Book chapter in Henze M., Van Loosdrecht M.C.M, Ekama G.A., Brdjanovic D. (eds.), Biological Wastewater Treatment: Principles, Modelling and Design. IWA Publishing, London, UK.

Howard T.L. (2003). Solids accumulation rates for onsite sewage treatment and disposal systems: A Focus on Charlotte County, Florida. *University of Florida*, Master's Thesis.

Junglen K., Rhodes-Dicker L., Ward B.J., Gitau E., Mwalugongo W., Stradley L. and Thomas E. (2020). Characterisation and prediction of fecal sludge parameters and settling behavior in informal settlements in Nairobi, Kenya. *Sustainability*, 12(21), 9040.

Klinger M., Gueye A., Manandhar Sherpa A. and Strande L. (2019). Scoping Study: Faecal Sludge Treatment Plants in South-Asia and Sub-Saharan Africa. eFSTP Project Report. *Gates Open Research* 2019, 3, 1716 (document) (https://doi.org/10.21955/gatesopenres.1116557.1)

Koottatep T., Surinkul N., Paochaiyangyuen R., Suebsao W., Sherpa M., Liangwannaphorn C. and Panuwatvanich A. (2012). Assessment of faecal sludge rheological properties, Report.

Lugali Y., Zziwa A., Banadda N., Wanyama J., Kabenge I., Kambugu R. and Tumutegyereize P. (2016). Modeling sludge accumulation rates in lined pit latrines in slum areas of Kampala City, Uganda. *African Journal of Environmental Science and Technology,* 10(8), 253-262.

Marwa L.I. (2017) Development of potential predictors for faecal sludge characterisation on a community scale, the case study of Mlalakua Community, Dar es Salaam. University of Dar es Salaam (MSc thesis).

Mills F., Blackett I. and Tayler K. (2014). Assessing on-site systems and sludge accumulation rates to understand pit emptying in Indonesia. In *WEDC International Conference*, Hanoi, Vietnam, 2014.

Morgan P.R., Mara D.D. and Mundial B. (1982). Ventilated improved pit-latrines: recent development in Zimbabwe. *World Bank*, Vol. 3.

Museteka L., Petulo P., Karen M., Fahle M. and Godau T. (2019). Groundwater Vulnerability Map of Lusaka Province / Zambia. https://www.whymap.org/EN/Themen/Wasser/Projekte/laufend/TZ/Zambia/map_hygeo_zambia_lusaka-vm.html

Nakagiri A., Kulabako R.N., Nyenje P.M., Tumuhairwe J.B., Niwagaba C.B. and Kansiime F. (2015). Performance of pit latrines in urban poor areas: A case of Kampala, Uganda. *Habitat international*, 49, 529-537.

Nödel, J.M. (2020). Indicator-based city zoning using Earth observation for sanitation management in Lusaka, Zambia. Master's thesis, Interfaculty Department of Geoinformatics-Z_GIS, *Paris-Lodron-University of Salzburg.*

Ojok, Koech M. and Okot-Okumu J. (2012). Rate and quantities of household solid waste generated in Kampala City, Uganda. Science *Journal of Environmental Engineering Science*. 2013, **doi:** 10.7237/sjeer/237.

Otaki Y., Otaki M., Bao P.N., Nga T.T.V. and Aramaki T. (2013). Micro-component survey of residential water consumption in Hanoi. *Water Science and Technology - Water Supply*, 13(2), 469-478.

Peal A., Evans B., Sangaralingam A., Ban R., Blackett I., Hawkins P., Schoebitz L., Scott R., Sleigh A., Strande L. and Roda O. (2020). Estimating safely managed sanitation in urban areas, lessons learned from a global implementation of excreta-flow diagrams. *Frontiers*, https://doi.org/10.3389/fenvs.2020.00001.

Prasad P., Andriessen N., Moorthy A., Das A., Coppens K., Pradeep R. and Strande L. (2021). Field-testing of the methods for estimating quantities and qualities (Q&Q) of faecal sludge: field evaluation in Sircilla, India. *Journal of Water, Sanitation and Hygiene for Development.*

Penn R., Ward B.J. Strande L. and Maurer M. (2018). Review of synthetic human faeces and faecal sludge for sanitation and wastewater research. *Water Research*, 132, 222-240.

Rice E.W., Baird R.B. and Eaton A.D. (eds.) (2017). Standard Methods for the Examination of Water and Wastewater, 23rd edition. American Public Health Association; American Water Works Association; Water Environment Federation. ISBN: 9780875532875.

Rose C., Parker A., Jefferson B. and Cartmell E. (2015). The characterisation of faeces and urine: A Review of the literature to inform advanced treatment technology. *Critical Reviews in Environmental Science and Technology*, 45(17), 1827-1879.

Rossboth T. (2013). Model-based systems analysis of the collection management of source-separated urine in Ethekwini Municipality/South Africa. University of Natural Resources and Life Sciences, Vienna, Austria. (MSc thesis).

Safavian S.R. and Landgrebe D. (1991). A survey of decision tree classifier methodology. IEEE *Transactions on Systems, Man, and Cybernetics*. 21(3), 660-674.

Sagoe G., Danquah F.S., Amofa-Sarkodie E.S., Appiah-Effah E., Ekumah E., Mensah E.K. and Karikari K.S. (2019). GIS-aided optimisation of faecal sludge management in developing countries: the case of the Greater Accra Metropolitan Area, Ghana. *Heliyon*, 5(9).

Saltelli A., Tarantola S., Campolongo F. and Ratto R. (2004). Sensitivity analysis in practice a guide to assessing scientific models, *John Wiley*, Chichester, England.

Schoebitz L., Niwagaba C., Francis O., Bischoff F. and Strande L. (2014). FAQ: Faecal sludge quantification and characterisation - Kampala. *Sandec News*.

Schoebitz L., Bischoff F., Lohri C.R., Niwagaba C.B., Siber R. and Strande L. (2017). GIS analysis and optimisation of faecal sludge logistics at city-wide scale in Kampala, Uganda. *Sustainability,* 9(2), 194.

SFD Promotion Initiative *SFD Manual Volume 1 and 2*, 2018. http://sfd.susana.org/

Strande L., Ronteltap M. and Brdjanovic D. (2014). Faecal Sludge Management - Systems Approach for Implementation and Operation. IWA Publishing: London, UK.

Strande L., Schoebitz L., Bischoff F., Ddiba D. and Niwagaba C. (2018). Methods to reliably estimate faecal sludge quantities and qualities for the design of treatment technologies and management solutions. *Journal of Environmental Management,* 223(1), 898-907.

Still D. and Foxon K. (2012). Tackling the challenges of full pit latrines, *Water Research Commission South Africa*, 1745/2/12.

Still D., Salisbury R., Foxon K., Buckley C. and Bhagwan J. (2005). The challenges of dealing with full VIP latrines, *Proceedings WISA Biennial Conference & Exhibition*, Durban ICC, South Africa.

Tchobanoglous G., Burton F.L. and Stensel H.D. (2014). Wastewater Engineering: Treatment and Reuse. *Metcalf & Eddy* (Boston) 5th ed, McGraw-Hill Higher Education, New York.

Tilley E., Ulrich L., Lüthi C., Reymond P. and Zurbrügg C. (2014). Compendium of sanitation systems and technologies. Swiss Federal Institute of Aquatic Science and Technology (Eawag). Dübendorf, Switzerland, 2nd revised edition.

Van Zyl H., Van Zyl J., Geustyn L., Ilemobade A. and Buckle J. (2007). Water consumption levels in selected South African cities, *WRC* Report No 1536/1/06.

Wagner E.G. and Lanoix J.N. (1958). Excreta disposal for rural areas and small communities, *World Health Organisation*.

Ward, B.J., Andriessen, N., Tembo, J.M., Kabika, J., Grau, M., Scheidegger, A., Morgenroth, E., Strande, L. (2021) Predictive models using "cheap and easy" field measurements: Can they fill a gap in planning, monitoring, and implementing faecal sludge management solutions? *Water Research.* https://doi.org/10.25678/00037X

Ward B.J., Gueye A., Diop B., Traber J., Morgenroth M. and Strande L. (2019). Evaluation of conceptual model and predictors of faecal sludge dewatering performance in Senegal and Tanzania. *Water Research*, 167, 115101. https://doi.org/10.1016/j.watres.2019.115101

Zakaria F., Ćurko J., Muratbegovic A., Garcia H.A., Hooijmans C.M. and Brdjanovic D. (2018). Evaluation of a smart toilet in an emergency camp. *International journal of disaster risk reduction,* 27, 512-523.

Figure 5.19 Faecal sludge collection and transport by a Tanzanian entrepreneur in Kigamboni, Dar es Salaam (photo: Eawag).

6

Towards city-wide inclusive sanitation (CWIS) modelling: modelling of faecal sludge containment/treatment processes

Carlos M. Lopez Vazquez
Francisco J. Rubio Rincon
Damir Brdjanovic

OBJECTIVES

The objectives of this chapter are to:

- Promote modelling of onsite sanitation
- Familiarise readers with the basic principles of established modelling approaches applied in sewered sanitation
- Introduce ideas on how faecal sludge containment/treatment processes can be modelled using the analogy with modelling practices in sewered sanitation
- Bring sewered and onsite sanitation closer together through the integrated approach of community city-wide inclusive sanitation modelling.

6.1 BACKGROUND

The approach presented in this book is to bring urban sanitation modelling closer to city-wide inclusive sanitation (CWIS) modelling. This chapter focuses on modelling the mechanistic microbial and physico-chemical processes that take place inside a single sanitation system (to predict the faecal sludge (FS) degradation and characteristics), while an empirical approach to estimating the quantities and qualities generated in onsite sanitation systems at community or city-wide level is presented in Chapter 5.

In general there is a consensus that developments regarding urban drainage and sewerage, urban flooding, and wastewater/sewage treatment modelling have advanced to the stage that they are considered valuable and standard tools in wastewater practice. However, comparable advances in onsite (non-sewered) sanitation are lagging behind and have only made advances in the last decade. Therefore the approach in this chapter is to make as much use as possible of existing and readily accessible modelling knowledge in the wastewater and sludge treatment field (well-established modelling principles, approaches and protocols) and relate and refer, wherever meaningful, to existing modelling practices as stepping stones for the development of a roadmap for modelling onsite sanitation systems. The ultimate objective is to reach the development of a modelling tool able to predict the biological, chemical and physical characteristics of faecal sludge as a function of local and environmental factors, depending on the timescale and typical characteristics, operation and use of onsite containment/treatment technologies. For this reason, the next sections in this chapter elaborate on the basic concepts of wastewater and sludge treatment models, approaches and protocols (Henze *et al.*, 2008; Rieger *et al.*, 2012; Brdjanovic *et al.*, 2015) to extrapolate for modelling onsite sanitation systems. Thus, three basic potential approaches are suggested to illustrate the modelling of three types of containment systems (*e.g.* a portable toilet, a pit latrine and a septic tank) either in contained or un-contained versions. This is considered the first and an essential step towards true CWIS modelling. Modelling of other CWIS components beyond these selected FSM containment/treatment technologies falls outside the scope of this chapter, as described in its concluding section.

Only recently have efforts been made to improve the understanding and description of the composition and biodegradability of faecal sludge in onsite containment and treatment systems (Elmitwalli *et al.*, 2011; Lopez-Vazquez *et al.*, 2013). Therefore, there is still major uncertainty about what the main biological and physico-chemical processes are that take place during the containment, emptying, transport and treatment components of the entire onsite sanitation chain, as well as what the principal underlying mechanisms are in terms of transformation processes and compounds involved, from both a spatial and a temporal perspective. In contrast, over the last three decades mathematical models have become a mature and reliable tool to support the design, optimisation, retrofit and upgrade of (activated sludge) WWTPs (Brdjanovic *et al.*, 2015). However, despite numerous well-documented and published examples of successfully modelled WWTPs, examples of the application of mathematical modelling to onsite sanitation and treatment systems are rare.

Nevertheless, it should be noted that in general, human excreta (faeces and urine) are the main 'raw materials' of concern in both onsite and sewered sanitation systems. The main difference is that the fate of human excreta in sewered systems is different to that prevailing in onsite sanitation systems. As a consequence, the type of sanitation infrastructure and conditions thereof determine to a large extent the type and speed of conversion of the compounds of interest present in faeces and urine. The specific conditions characterising onsite and sewered sanitation systems have a major influence on both the composition and the quantity of sewage and faecal sludge, resulting in a different 'strength' of such streams. The characterisation and quantification of faecal sludge from septic tanks and pit latrines are elaborated in detail in chapters 2 and 5, respectively. More information on the characteristics of sewage can be found in standard sanitary engineering literature (*e.g.* Metcalf and Eddy *et al.*, 2014; Henze *et al.*, 2008; Chen *et al.*, 2020).

It is common knowledge that the strength of sanitary flows in general is affected by the degree of dilution and the extent of the transformation process in a sanitation system as a consequence of various technical, cultural and socio-economic factors. These include: water usage and consumption, type of toilets, type of containment systems, degree of water infiltration and percolation, discharges of garbage and non-degradable materials, discharges of non-domestic waste streams, the ratio between onsite and sewered sanitation coverage, type of sewage and drainage system, management of rainwater and grey water in onsite sanitation systems, environmental conditions (*e.g.* redox, pH, temperature), hydraulics in the storage and transport components of the system, sewage and waste sludge retention time, and faecal and septic sludge retention in the containment. The dilution effect in a sewerage system is considerable, even in the case of sanitary sewers (either with or without any contribution of rainwater).

As presented in Chapter 5, the average daily production of faeces and urine of a person are 180 g and 1,500 mL, respectively, which are diluted by a large amount of relatively clean water to up to 300 or more times (Rose *et al.*, 2015). Clean water is largely used for toilet flushing, evacuation of sewage from households and transport to the point of treatment and/or discharge. This decreases the concentration of the compounds of interest from the perspective of public health and environmental protection. The situation worsens in the case of combined sewers and sewers with a high infiltration rate. Other important factors are mixing and the hydraulic regimes as they play an important role in defining the environmental conditions in both sewered (*e.g.* on the flow regime in pipes and channels) and onsite sanitation settings (*e.g.* on the degree of sludge stratification in containment units and the degree of mixing and homogenisation during emptying and transport and consequent disposal), affecting the bio-chemical conversions in the system.

It is remarkable that the sanitary engineering community has been investigating activated sludge systems for more than 100 years (Jenkins *et al.*, 2014) and that biological nitrification is the most studied process in wastewater treatment but that, in contrast, interest in more fundamental research on onsite sanitation systems has only gained momentum in recent years. This is even more surprising given that the initial 'raw material' of concern in both onsite and sewered sanitation systems is essentially the same. The failure to distinguish the principal differences between raw sewage, faecal sludge and septic sludge has, on occasion, led to the collapse of existing wastewater treatment plants because they have not been designed to receive high(er) faecal sludge loads (Lopez-Vazquez *et al.*, 2013). Without doubt, there is still a lack of fundamental understanding and scientific evidence of the complex processes taking place in onsite sanitation systems across the world, including latrines as the most common onsite containment unit. However, thanks to the fact that sanitation has had a prominent focus in both UN Millennium Development Goals (MDGs) and is increasingly prominent in the current UN Sustainable Development Goals (SDGs, United Nations, 2015), the interest of both the academic and professional communities in sanitation has increased tremendously over the last decade, in recognition of the fact that onsite sanitation has to be approached with deeper insight, advanced knowledge and greater confidence. Therefore, the authors believe that this chapter will improve the understanding of the dominant microbial and physico-chemical processes that take place in onsite sanitation systems. This understanding should be based on the principles, fundamentals and proven practice documented by researchers, modellers and practitioners dealing with sewers, activated sludge systems and sewage sludge, that can be used as a basis to define the approaches and steps required for modelling onsite containment systems in order to estimate the volumes and characteristics of the faecal sludge generated in different sanitation systems. It will ultimately contribute to the development of a modelling framework that could potentially be used to improve the design and exploitation of onsite and also sewered sanitation systems in the future.

The expected benefits of setting out the basis for modelling onsite sanitation systems are: (*i*) to improve practical understanding of onsite sanitation systems, (*ii*) to increase confidence in the determination of the main faecal sludge characteristics and fractions, (*iii*) to deepen fundamental understanding of the dominant/prevailing biological and physico-chemical processes that take place in onsite sanitation systems, and (*iv*) to help to initiate a community of practice on onsite sanitation modelling.

It is important to highlight that although there are several mathematical models capable of satisfactorily describing carbon, nitrogen and phosphorus removal processes from sewage, pathogen removal has been overlooked by the mainstream modelling community. However, since most onsite sanitation systems are located in low and middle income countries where billions of people have no basic sanitation provision, it is essential to contribute to the prevention of the spread of waterborne diseases and therefore to prevent contact of people with pathogens through the control of contamination pathways and pathogen-removal mechanisms. This is an obvious reason and an important challenge to develop and promote enhanced pathogen removal (or inactivation) practices and approaches supported by mathematical modelling and linked to the transformations of other compounds (*e.g.* organics, nitrogen and phosphorus). Two-directional synergy between the two sanitation fields, in a spirit of CWIS, is useful and recommended given the fact that, for instance, by promoting the generation of inactivation agents during faecal sludge treatment, pathogen reduction and inactivation can be achieved (Nordin *et al.*, 2009; Fidjeland *et al.*, 2013; Anderson *et al.*, 2015).

Similarly to sewage-based modelling, onsite sanitation modelling can have the potential to become a basis or a tool to improve the management and operation of sanitation facilities in onsite settings because, for example, the actual removal capacity, volume and solids accumulation in onsite systems could be better predicted and improved, also enabling better emptying practices (Bhagwan *et al.*, 2008). Recent large faecal sludge characterisation efforts in Sub-Saharan Africa and South Asia and approaches to track material flows (well-established in the wastewater field in the form of mass balances) and represented by Shit Flow Diagrams (SFDs, Peal *et al.*, 2020) are clearly important building blocks of the foundation needed for onsite sanitation modelling. Similarly to the latest trends in wastewater treatment, the quantification and prediction of the transformation processes of faecal sludge may make it possible to replicate developments such as 'WWTP - an energy factory' and 'energy-neutral WWTP' in some way within the onsite sanitation field. As such, despite the intrinsic complexities and drawbacks, it becomes very important to promote modelling of sanitation systems within the framework of a CWIS approach (World Bank, 2019; Löthi and Narayan, 2019), to contribute to the development of the sanitation value and service chain management in an integrated and holistic way.

6.2 INTRODUCTION TO MODELLING – LEARNING FROM ACTIVATED SLUDGE MODELS

6.2.1 What is a model?

A model can be defined as a purposeful representation or description (often simplified) of a system of interest (Ubisi *et al.*, 1997). This consequently means that a model never exactly reflects the reality. So, the question 'Can (does) this model describe a process occurring in an onsite containment system?' is pointless without a definition of what (which) part(s) of an onsite containment system the model should describe. One never develops a model that describes every detail of the process. Models are a simplification of reality that describe that part of reality that is relevant to understand and to deal with (Van Loosdrecht *et al.*, 2015). It is also important to note that a mathematical model can only be successful if it fulfils the expectations that people have of it. From the perspective of time, a model can be developed to describe frozen-state, dynamic-state or steady-state conditions. Frozen-state conditions are those that do change over time, but not in the time interval that one is interested in or dealing with. Often, dynamic-state conditions are the ones that deserve special attention to describe the variations that occur as a function of time. For instance, the concentrations of organic matter, nitrogenous and phosphorus compounds in the influent will vary during the day, the concentration of ammonia in the effluent will vary over time, concentration of nitrate will vary in the activated sludge recators etc. Nevertheless, the concentrations of these compounds in anaerobic sludge digesters (which nowadays with an increased interest in energy and resources recovery are often found as intrinsic components in sewage treatment plants) (Batstone *et al.*, 2014) scarcely vary within a day. One of the reasons is that the hydraulic retention time (HRT) of sludge digesters is usually around 20-30 days and, thus, the characteristics change in time intervals of two to three weeks. As a consequence, the variations or fluctuations in sludge digesters, with regard to the daily dynamics of interest, are therefore assumed to be in a kind of frozen state. The analogy can be drawn with some onsite containment systems that are also less sensitive to daily variations in the load and are based on anaerobic digestion (*e.g.* septic tanks or pit latrines that are not often emptied). Moreover, some other processes occur so fast that they are assumed to be, under the usually applied timescales of a study, under steady-state or equilibrium conditions. An example of such processes are the precipitation processes that occur almost instantaneously (in a few seconds). The speed at which these processes occur is so fast that they do not have to be described in a dynamic way, so they are assumed to be in equilibrium or completed. As such, one of the first considerations is to define what the processes of interest are, the relevant timeframe for their description, an assessment of their dynamics, and an accurate description of those processes that are time-variable within the timeframe of concern. Therefore, the aspect of time is the first major issue in trying to simplify the reality. The recommended approach is to consider the time constants and select those processes that have the dynamics in the order of time constants that one is interested in.

The second relevant issue for modelling is space resolution. One can theoretically make a model that describes every square inch of the process tanks, reactors or section of a sanitation system. However, one needs to realise and define whether such a detailed description is strictly necessary. The answer can be found, once again, in the purpose of the model. In order to describe the concentration gradients of the relevant components in the process tanks, units or reactors, one should determine the scale size that is most appropriate. On a different scale, there is a gradient of concentrations inside the bacterial agglomerations, biofilms, and accumulation of solids that theoretically can also be described by a model. Again, the situation may be different in onsite containment systems (such as pit latrines or septic tanks) where stratification, water content of the sludge, and limited or no mixing may all have a major influence on the choices made. Therefore, one needs to assess whether they are sufficiently relevant to be taken into account.

The next step in modelling is the relevant level of detail in a microbial model. In activated sludge modelling, the closest modelling parallel, there are basically three approaches (Van Loosdrecht *et al.*, 2007): *(i)* the traditional 'black-box' approach, *(ii)* the 'grey-box' approach, and *(iii)* the 'glass-box' approach. Over the years, the black-box approach has been shown to be reliable enough for design purposes, even though it does not provide information about the sludge composition. If one is interested in refining the design and operation of the plant, grey-box models (such as Activated Sludge Model No. 1 - ASM1) split the sludge into relevant fractions composed of the compounds of interest (such as biodegradable and unbiodegradable, soluble and particulate fractions) and microbial biomass (such as ordinary heterotrophic organisms, nitrifying organisms, phosphate-removing organisms, among others). This approach allows modeller to take into consideration different functional aspects of the microbial communities present in the sludge and incorporate them in the model. 'Glass-box' models, such as the metabolic models initially developed for enhanced biological phosphorus removal (EBPR) by Smolders *et al.* (1995), Kuba *et al.* (1996), Murnleitner *et al.* (1997) and for the first time applied at a full-scale WWTP by Van Veldhuizen *et al.* (1999) and Brdjanovic *et al.* (2000), provide a good description of the metabolic routes that take place inside the organisms, almost reaching a 'glass-box' modelling approach. This more complex and detailed level has been shown to be necessary to secure a satisfactory description of phosphorus-removing systems, but it is by no means essential to describe all the biological processes. Therefore, the preference for a black-, grey- or glass-box modelling approach depends on the purpose and application of the model, also in the context of onsite sanitation systems.

Furthermore, two types of mathematical models exist: empirical and mechanistic models. An empirical model is based on the recognition of the parameters that seem to be essential to describe the behavioural patterns of interest, and linking these through empirical relationships established by observation (*e.g.* mathematical regressions to find any dependence between the effluent characteristics and the influent concentrations or environmental conditions such as temperature). As such, in empirical models, the mechanisms and/or processes operating and governing the conversions that occur in the system are not known and are often ignored. Empirical models can be considered to be an example of a classical black-box modelling approach. In contrast, a mechanistic model is based on a particular conceptualisation of the biological/physical mechanisms governing the system. The degree and level of understanding of the biological and chemical processes occurring in the system will determine the complexity of a mechanistic model. As such, since mechanistic models have a conceptual basis, they tend to be more reliable than empirical models. Moreover, empirical models are naturally restricted by the boundaries used to develop the model itself (such as the wastewater or faecal sludge characteristics and system parameters), allowing only certain interpolation. On the other hand, because mechanistic models are conceptually-based, they can be not only interpolated but also extrapolated. Nevertheless, one should not forget that all models need to be rigorously and properly calibrated and verified. In addition, the boundary conditions of application of every model should also be firmly delineated. Historically, and based on how they have been developed and evolved, mechanistic models have been shown to have a greater potential for application in the sanitary engineering field, deserving special attention and interest compared to empirical models.

To set up a mechanistic model, a conceptual model needs to be defined describing the processes of interest occurring within a system and the compounds subject to the transformations and conversions to be described by the processes. Furthermore, the interactions and interlinks between the processes and compounds should also be delineated. Thereafter, a mechanistic model can be developed by formulating the mathematical expressions that describe the stoichiometric relationships and kinetic rates of the processes and their compounds. Strictly speaking, the model should not include all the processes that take place within a system but only those that are significant to meet the expectations raised by the model. To develop a model that includes all the possible processes and their interactions is not feasible, since it would lead to a very complex model that would not completely describe the phenomenon. An example of such a practice is the level of organisation: rather than model every microbial population (for which microbial identification and enumeration techniques may not even be fully and reliably developed) microorganisms are grouped as single entities or groups of 'surrogate' organisms that fulfil or perform a defined function, namely: ordinary heterotrophic organisms (OHO) that carry out the aerobic removal of organics on the upper layers of an onsite containment system that are exposed to air, or anaerobic organisms (ANO) that perform the removal of organics in the deeper layers of the same onsite sanitation system where oxygen is absent. The single entities or surrogate groups of organisms are modelled with a defined set of characteristics and behaviour to describe their prime function within the system. These characteristics will not reflect the particular or specific characteristics of each individual microorganism, but their main function or process of interest that, as a whole, will provide a satisfactory description of the main role of the group in the system. Consequently, the actual overall effect of modelling the group reflects the cumulative net effect of the individual contribution of each microorganism. The advantage of this approach is that it decreases the level of complexity since less information is required for the development, calibration and validation of the model. Usually, most of the information and parameters that are incorporated are of a biochemical or microbiological nature. Also, the more complete, the better the description. Nevertheless, this additional information should be incorporated to the minimum required level where the key processes that govern or describe the response of a system are identified. This is also because detailed microbiological and biochemical information is usually needed (Ubisi *et al.*, 1997) and, even more importantly, data from onsite sanitation systems is often subject to considerable fluctuations and levels of uncertainty (Brouckaert *et al.*, 2013). In this regard, more methods are needed to quantify uncertainty and its sources (Sin *et al.*, 2005; Belia *et al.*, 2009; Benedetti *et al.*, 2010; Flores-Alsina *et al.*, 2012) in onsite sanitation systems. The adaptation to onsite sanitation systems of the findings and developments of the IWA Task Group on Design and Operations Uncertainty (DOUT) (Sin *et al.*, 2005; Belia *et al.*, 2009; Flores-Alsina *et al.*, 2012) can be used to carry out uncertainty evaluations and contribute to defining the minimum levels of complexity and data required to describe the operation and performance of faecal sludge technologies.

The objectives that the model needs to fulfil will determine the parameters that need to be considered based on the defined level of organisation. Generally, two different types of models are developed: steady-state and dynamic models. Steady-state models are simpler since they usually have constant or steady constant flows and loads. Dynamic models are more complex because they tend to have variable or varying flows and loads. Steady-state models are oriented to determining the most important design parameters and therefore are good for design. Dynamic models are useful to predict the time-dependent response of a treatment system.

6.2.2 Modelling basics

6.2.2.1 Model building

Mathematical models can provide a quantitative description of the systems of study and, therefore, are widely applied. Mathematical expressions are used to describe the stoichiometric reactions and the kinetic rates at which the conversions of the parameters occur (usually as a function of time). To provide the required predictions, the mathematical formulations are included with the procedures needed to find their solutions within the boundaries defined by the structure of the model and that of the system (such as temperature and mixing conditions). Mathematical models are not developed in isolation but evolve in close interaction with conceptual and physical models (*e.g.* laboratory-scale or pilot-scale reactors) (Ubisi *et al.*, 1997).

For example, to develop a mathematical model that describes the wastewater (or faecal sludge) conversion processes that take place in sewered or onsite systems, at least four components are needed: (*i*) influent or input characteristics, (*ii*) balance equations, (*iii*) kinetic process rates, and (*iv*) transport processes, as described below.

Influent or input characteristics

An adequate and reliable determination of the influent or input characteristics is vital in order to obtain a satisfactory description of the process conversions and of the actual impact and response of the system. Bearing in mind the objectives to be met, the level that existing models have reached implies carrying out not only a thorough characterisation during a representative period of time, but also a fractionation of the compounds of interest. The characterisation should look into those parameters that better illustrate the strength of the medium (*e.g.* BOD, total COD, soluble COD, total nitrogen, ammonia, among others). Also, the characterisation must include the determination of limiting compounds (whose absence can limit the conversion processes) and inhibiting or toxic compounds (whose presence can slow down or even prevent the (bio-) degradation or conversion processes). An example of limiting compounds can be oxygen for the aerobic removal of organic matter, and an example of inhibiting or toxic compounds can be ammonia or hydrogen sulphide for the anaerobic removal of organic matter. Regarding the fractionation of the main compounds of interest (at least of organic matter in terms of COD), this should be carried out in relation to the potential conversions that are closely related to their physico-chemical and (bio-) degradation properties, under the prevailing redox conditions (generally, aerobic or anaerobic). In this regard, most available protocols focus on the determination of the soluble and particulate fractions, and to what extent these soluble and particulate fractions are (bio-)degradable or not, within the boundaries of the conversion process in question (Hulsbeek *et al.*, 2002; Van Loosdrecht *et al.*, 2015). The determination of the characteristics and fractionation(s) is in general an essential modelling step (also in the cases of faecal and septic sludge), since it contains the main input data of the model and, as expected, will define the success of the description of the conversion processes. Moreover, its correct determination will ease the calibration and validation process (Brdjanovic *et al.*, 2015).

Balance equations

Balance equations are necessary to describe the biological and chemical conversion processes of interest. These processes lead to the consumption of reactants and the generation of products. Often, a product generated by one reaction or conversion process can be the reactant of one or more subsequent processes. Consequently, the concentrations of certain compounds, or parameters in a reactor or system, will change over time. However, when a system or model reaches steady-state conditions, the concentrations are stable and therefore no longer change.

Kinetic process rates

Each reaction has its own rate equation. The rate equations specify the rate at which certain reactants are converted into their products. The kinetic rate expressions can be either substrate-based or growth-based. They range from zero order to second order equations (*e.g.* $r = k$, $r = kC$ to $r = kC^2$, where r is the reaction, k is the kinetic rate and C is the concentration of the component converted during the reaction) (Metcalf and Eddy *et al.*, 2014). In waste(water) conversion models, the most common process rate equations used are the saturation equations defined by, for example, the widely used empirical Monod-type expressions (*e.g.* $r = kC/(K+C)$ where K is the half-saturation concentration of the component converted during the reaction. Such an expression allows us to describe the process rates as a function of the availability of substrate in the systems and reactors (in this case, C). It is also common to use Monod-type equations as switching functions (in the form of $r = K/K + C_I$) to describe the inhibitory effects caused by a toxic or inhibiting compound (C_I) that slows down a reaction process and at high concentrations can even stop it.

Transport processes

Together with the stoichiometric equations and kinetic process rates, transport processes also affect the changes in concentrations in a reactor over time, because the local concentrations observed in a process unit or reactor (besides being affected by the conversion processes whose rates are usually dependent on the local concentrations themselves) are also subject to the transport of reactive compounds between process units. The transport processes can be convective or diffusive. Convective transport processes are commonly used to describe the transport of liquids (directly linked to the hydraulic behaviour of a plant, such as the conduction of a wastewater stream from one tank to another), whereas diffusive transport processes are used to describe the transfer of gases between phases (for instance, to describe the diffusion of oxygen from the atmosphere into a liquid contained in an open reservoir). Thus, transport processes are another key component of a model of a physical nature and must also be carefully determined.

6.2.2.2 General activated sludge model set-up

The different influent or characteristics of the inputs, balance equations, kinetic process rates, and transport processes are the main components of a model. They need to be grouped following a defined framework to provide an adequate representation of the onsite containment system dependent on the objectives pursued by the modelling study. First, the stoichiometric equations that define the main conversion processes of interest need to be incorporated. From a conservation perspective, they need to be mass-balanced to comply with the conservation principles (all inputs should equal all outputs) in terms of loads (*e.g.* carbon, phosphorus) and charges (*e.g.* for nitrogen compounds, alkalinity). Together with their correspondent kinetic process rates, these balanced equations are the main core of the conversion models. Over the years, different research groups and groups of practice have developed extensive aerobic and anaerobic models that present, in a structured manner, the main stoichiometric and balanced conversion processes, as well as their corresponding stoichiometric and kinetic rates and parameters. Among others, with regard to the conversions of organic matter, examples of such developments are the activated sludge and anaerobic digestion models (ASM1 and ADM1, respectively) to describe aerobic- and anaerobic-driven organic matter conversion processes, respectively. The use and application of certain models (for instance, either an ASM-type or ADM1) depends on the objective of the modelling study. Consequently, the most suitable model(s) need(s) to be selected to model either the system or certain process units with one model type (*e.g.* aerobic phases with an ASM-type model) or with another (*e.g.* anaerobic phases with ADM1).

Once the model has been selected, measurable input parameters and fractionations need to be determined as a function of the selected model. In this regard, ASM1 requires very basic characterisations and fractionations (composed of only four COD parameters as a function of their complexity and biodegradability to describe the COD loads) (Henze *et al.*, 1987). On the other hand, ADM1 demands a very thorough and extensive characterisation and fractionation that requires the determination of carbohydrate and lipid concentrations (among other compounds) in the influent (Batstone *et al.,* 2002). After the determination of the corresponding wastewater characteristics and fractionations, they are transformed into an influent vector, becoming the main input of the model.

The transport processes in the FSM unit need to be defined based on the transport (flow or flux) of the main streams or discharges through the treatment system. Initially, the system can be modelled hydraulically, describing the main zones/reactor compartments of the system. An approach is recommended in which each process unit is modelled individually considering its hydraulic behaviour (whether it is a completely-stirred tank or a plug-flow reactor) and redox conditions (aerobic or anaerobic). The process units may be further split or divided into compartments to mimic the dominant or prevailing conditions. For instance, a process unit with plug-flow hydraulic behaviour can be represented by a defined number of completely-stirred tank reactors (CSTR) in series. This practice is common to ease the modelling process (Volcke *et al.*, 2006). Also, one process unit can be split into different compartments to represent the existence or generation of different redox conditions (such as anaerobic or anoxic dead zones in aerobic units due to uneven mixing or aeration conditions). In all the aforementioned conditions, the transport of flows and the concentrations of the compounds of interest between process units and their compartments can be described with convective transport expressions based on the actual hydraulic configuration of the treatment system. With regard to the transport of gases, diffusive transport expressions are commonly applied. This enables the diffusion of oxygen into the process units to be assessed as well as the gas emissions from the conversion processes.

It should be noted that neither the ASM nor the ADM families of models include pathogen removal. Therefore, pathogen removal/inactivation modelling and its integration with other models is addressed in section 6.3.7.3.

The overall model of a system can be generated by compiling the influent characterisation model (or influent vector), the process conversion model (containing the stoichiometric and kinetic components) and the process flow model. The process flow model can be composed of individual units and their phases or sub-units connected by a state vector that includes the corresponding convective and diffusive transport expressions, as required. The overall model is usually solved numerically to compute the concentration of each compound included in the model as a function of time. Every compound entering into the treatment system and consequently into each process unit, reactor or compartment should be converted, exchanged with the gas phase, or leave with the effluent. For example, a schematic representation of the model of a sewer-based system, an activated sludge wastewater treatment plant, is presented in Figure 6.1. It is composed of four units or phases modelled as a continuous stirred-tank reactor

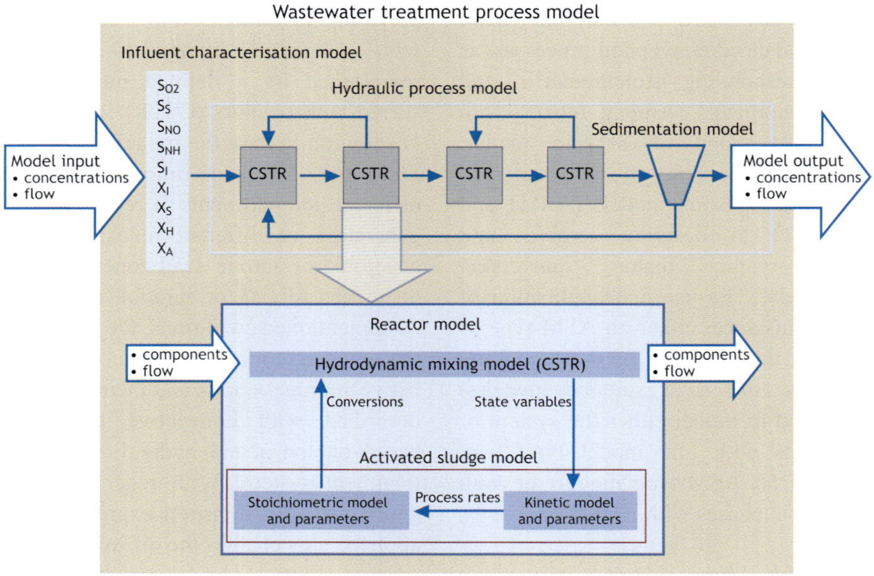

Figure 6.1 Schematic representation of an activated sludge wastewater treatment plant (modified from Meijer, 2004).

(CSTR), interconnected to simulate the potential recycle and return of flows between the tanks. The feed or influent is received or discharged into the first unit before continuing to flow to the next units. The retention and/or accumulation of solids (the solid-liquid separation process) is modelled as a sedimentation model or settling unit before discharging the effluent. In particular, every single unit has both a hydrodynamic model and a process conversion model to describe the transport and conversion processes of the compounds of interest.

6.2.2.3 The matrix notation

Balance equations are used to describe the conversions of the individual compounds of interest depending on the objectives and purpose of the modelling study. Due to the number of relevant compounds, their associated conversions and the dependencies between balance equations (in which the product of a balance equation can become the reactant of other equations whose products can be the reactants of other processes, and even of the previous processes, and so forth), in 1987 the IAWQ Task Group on 'Mathematical modelling of wastewater treatment' (Henze *et al.*, 1987) recommended and adopted the Peterson matrix notation (Peterson *et al.*, 1965), afterwards renamed the 'Gujer matrix', for model presentation. This format facilitates a clear and unambiguous presentation of the compounds and processes and their interactions in a simple and compact manner. Moreover, this format allows a direct comparison between different models, and facilitates the transfer of the expressions into a computer program or modelling simulator. The matrix is presented by a number of columns and rows, in which the columns are used to display the compounds of interest and the rows the processes to which the compounds are subject to, either as reactants or products. Table 6.1 presents an example of a simplified stoichiometric matrix that describes the aerobic removal of readily biodegradable organics (S_S with the stoichiometric coefficient $-1/Y_H$) by the aerobic growth of ordinary heterotrophic organisms (X_H with the coefficient +1) linked to oxygen consumption (S_O with the stoichiometric coefficient $-1/Y_H+1$). A negative coefficient indicates that a component is consumed whereas a positive coefficient indicates that the component is produced or generated. The process rate of the aerobic growth reaction is $\mu_H^{MAX} \cdot (S_S/(K_S+S_S)) \cdot X_H$). The example also includes the lysis or decay process of the ordinary

Table 6.1 Example of a simplified stoichiometric matrix for activated sludge modelling (Henze *et al.*, 1987).

Components i	1: S_O	2: S_S	3: X_H	Process rate equation ρ_j
List of processes j				
Aerobic growth	$-\frac{1}{Y_H}+1$	$-\frac{1}{Y_H}$	+1	$\mu_H^{max} \cdot \frac{S_S}{K_S+S_S} \cdot X_H$
Lysis		+1	−1	$b_H \cdot X_H$
Observed transformation rate r_i	$r_i=\sum_j v_{j,i} \cdot \rho_j$ $[M_iL^{-3}T^{-1}]$			
Definition of stoichiometric parameters: Y_H Heterotrophic yield coefficient $[M_H\ M_S^{-1}]$	Dissolved oxygen (O_2)	Dissolved organic substrate (COD)	Heterotrophic biomass (COD)	Definition of kinetic parameters: μ_H^{max} Maximum specific growth rate $[T^{-1}]$; K_S Saturation coefficient for substrate $[M_{COD}L^{-3}]$; b_H Rate constant for decay $[T^{-1}]$

heterotrophic organism (OHO) biomass (negative coefficient -1) that results in the generation of readily biodegradable organics (S_S with the positive coefficient +1) with a process kinetic rate $b_H \cdot X_H$ (Henze *et al.*, 1987). This example also illustrates the potential interconnections between components in which the product of the first reaction (the heterotrophic biomass, X_H, generated during the aerobic growth process) becomes the reactant of the second reaction (in the lysis process) and, consequently, the product of the second reaction (S_S) is the reactant of the first reaction.

In the previous example, all the units are expressed in terms of COD equivalents and the continuity and, therefore, conservation principles need to be met. These can be assessed by moving across any row in the matrix, summing up all the coefficients whose net sum should be zero. The previous example illustrates how a matrix can be used to summarise and represent complex interactions between compounds and processes in a relatively simplified manner, justifying why the matrix notation is commonly used in mathematical modelling of wastewater treatment systems. It is strongly recommended that matrix notation is used in modelling of onsite sanitation systems, following any necessary adaptation.

6.2.2.4 Wastewater treatment models

As previously described, different extensive aerobic and anaerobic models have been developed over the years to model sewered sanitation systems, and in particular activated sludge systems. The family of mathematical models developed under the leadership of the International Water Association (IWA) includes the most applied models in the field of wastewater treatment. These include the ASM models nos. 1, 2, 2d and 3 (Henze *et al.*, 2000) and ADM1 (Batstone *et al.*, 2002. Also, previous versions that have contributed to the development of the IWA models can be found, such as the UCTOLD or the UCTPHO models (Dold *et al.*, 1981,; Wentzel *et al.*, 1988, 1989a, 1989b), models with a similar basis developed in parallel (Barker and Dold, 1997) or modified or expanded versions of the IWA models (such as the TUDelft model, or the ASM3-Bio-P model) (Meijer, 2004; Rieger *et al.*, 2001). However, in spite of the development of different anaerobic models since the late 1970s (Donoso-Bravo *et al.*, 2011), IWA ADM1 (Batstone *et al.*, 2002) is still the most commonly applied anaerobic treatment model. One important reason is that its core model structure with different adaptations, modifications and extensions (Donoso-Bravo *et al.*, 2011) has proven capable of describing several wastewater and solid waste conversion processes (Kythreotou *et al.*, 2014; Batstone *et al.*, 2015). Furthermore, with the use of suitable interfaces, coupling ASM-types with the ADM1 model has become possible for plant-wide modelling purposes (Mithaiwala *et al.*, 2005; Rosen *et al.*, 2006; Volcke *et al.*, 2006; Alex *et al.*, 2008; Nopens *et al.*, 2009) with the aim of optimising the operation of wastewater treatment plants and for resource recovery purposes. For the implementation of the models, different general purpose simulators are available ranging from open-access simulators such as Aquasim, ASIM[1] or STOAT[2] to proprietary software simulators such as MatLab™/Simulink™[3]. In parallel, different initiatives have led to the development of more comprehensive models that couple aerobic and anaerobic processes. They often belong to more advanced commercial software packages and include BioWin[4], GPS-X[5], SIMBA[6], SUMO[7], and WEST[8]. Some of these comprehensive models have been incorporated in simulators that bring additional advantages. For example, they offer user-friendly interfaces to build process-flow diagrams of sewered sanitation systems, to describe more easily the key chemical and precipitation processes, or to estimate specific operating conditions that can lead to process inhibition due to the presence or accumulation of certain compounds (*e.g.* sulphide, excessive ammonia or nitrite accumulation). All the aforementioned models have defined model structures to describe certain conversion processes and therefore meet specific modelling objectives. Thus, a key decision in the modelling process is to select the model that is most suitable for the required modelling needs. This selection is usually carried out by considering the main conversion processes that take place in the system to be modelled and those that each model can describe. Consequently, the model whose conversion processes are identical or the closest to those governing the system under study can be selected. Excluding models that belong to or are part of proprietary simulators or software packages, Table 6.2 presents an overview of some selected (open-access) models developed for wastewater treatment with specific emphasis on the main conversion processes that they can describe. For modelling onsite sanitation systems, certain processes can probably be excluded (such as nitrification, denitrification and enhanced biological phosphorus removal (EBPR), which require the presence of oxygen prior to, during, or after each of these processes) bearing in mind that most of the conditions prevailing in onsite containment units tend to be anaerobic (due to the absence of aeration systems) or that they are micro-aerophilic (in the upper layers of the systems) (Bakare *et al.*, 2012). As such, to describe the conversion processes occurring in onsite sanitation systems, ADM1 appears to be an essential model coupled with ASM1 or ASM3 to describe the marginal aerobic processes.

6.2.2.5 Modelling protocols

As described previously, different mathematical models have been developed and extensively applied to model several types of aerobic and anaerobic wastewater treatment systems. For this purpose, each model requires to be calibrated for each case study. As such and since different research groups, groups of practice and experts, companies and institutions have been involved in the implementation of modelling studies in different regions, several calibration models have been developed involving different methodologies and approaches (Hulsbeek *et al.*, 2002; Vanrolleghem *et al.*, 2003; Sin *et al.*, 2005). Among them, four calibration protocols have become most popular (Sin *et al.*, 2005): (*i*) the BIOMATH calibration protocol (Vanrolleghem *et al.*, 2003), (*ii*) the HSG guidelines (Langergraber *et al.*, 2004), (*iii*) the WERF protocol for modelling calibration (Melcer *et al.*, 2003) and, (*iv*) the STOWA calibration protocol (Hulsbeek *et al.*, 2002; Roeleveld *et al.*, 2002). Despite the advantages and disadvantages of each protocol, all of them have a similar structure.

[1] www.eawag.ch
[2] www.wrcplc.co.uk
[3] www.mathworks.com
[4] www.envirosim.com
[5] www.hydromantis.com
[6] www.ifak.eu/content/simba-sharp-water
[7] www.dynamita.com
[8] www.mikepoweredbydhi.com/products/west

Table 6.2 Overview of selected mathematical models commonly applied to model sewered treatment systems (modified from Gernaey *et al.*, 2004).

Model	Aerobic organic matter removal	Nitrification	Denitrification	EBPR	Chemical P removal	Hydrolysis	Fermentation	Acetogenesis	Methanogenesis	Reference
ASM1	•	•	•			•				Henze *et al.* (1987)
UCTOLD	•	•	•			•				Dold *et al.* (1981,)
ASM3	•	•	•			•				Gujer *et al.* (1999)
UCTPHO	•	•	•	•		•	•			Wentzel *et al.* (1988, 1989a, 1989b)
ASM2	•	•	•	•	•	•	•			Gujer *et al.* (1995)
ASM2d	•	•	•	•	•	•	•			Henze *et al.* (1999)
B&D	•	•	•	•		•	•			Barker and Dold (1997)
TUDP	•	•	•	•		•	•			Meijer (2004)
ASM3-BioP	•	•	•	•		•				Rieger *et al.* (2001)
ADM1						•	•	•	•	Batstone *et al.* (2002)

Sin *et al.* (2005) carried out a thorough SWOT (Strengths, Weaknesses, Opportunities and Threats) analysis of the calibration protocols previously listed (BIOMATH, WERF, HSG and STOWA). Overall, they concluded that all of them are suitable and reliable; the BIOMATH calibration protocol is the most sophisticated (with regard to its level of detail and thorough characterisation and calibration procedures), the HSG is the most systematic (concerning the calibration steps), the WERF is the most detailed with regard to the experimental methods needed for influent characterisation and fractionation (including a summarised number of calibration studies, which is attractive for inexperienced modellers and consultants), and the STOWA calibration protocol, which is the most straightforward, practical and easy to implement. In particular, the STOWA protocol can be useful for inexperienced modellers and practitioners, since it also gathers and summarises the experience earned through several modelling studies (Roeleveld *et al.*, 2002). Therefore, since the most commonly applied modelling protocols share and follow, to some extent, similar concepts and principles. The STOWA calibration protocol will be briefly presented in this section and the main steps discussed from an faecal sludge modelling perspective. Figure 6.2 presents a flow diagram illustrating the main steps of the STOWA calibration protocol and their inter-relations (Hulsbeek *et al.*, 2002). These are discussed in more detail below.

Formulation of objectives

The definition of the main purpose and objectives is essential to define the scope of the study, its relevance, and also its boundaries. The objectives define whether the modelling study will be carried out to select a (future) design, to optimise an existing design or to develop (improved) strategies to operate existing or future sanitation systems. This will influence the model extension and complexity, and also the required modelling activities, such as the length and frequency of the sampling campaigns and the type and number of operating and analytical parameters to be determined and analysed.

Process description

Depending on the objectives of the study, the process can be described by defining the process components of relevance and identifying the general plant layout and configuration. It is essential to include and define all the inflows, internal flows and outflows from the system (*e.g.* influent, feeds, internal recirculations, effluent, infiltration and percolation flows, whenever applicable).

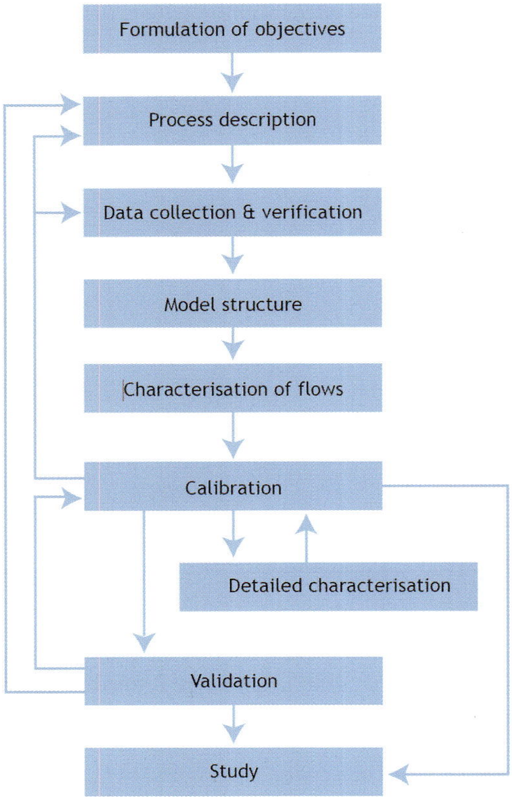

Figure 6.2 Main structure of the STOWA calibration protocol (Hulsbeek *et al.*, 2002).

Data collection and verification

The collection of data is essential in conducting a comprehensive survey of the system under study. In this step, the composition and volume of all the flows going through all the process components need to be defined. If available, data can be collected from (previous) periodic sampling and monitoring programs. This data can be useful to start to define the characteristics and composition of the flows. Furthermore, this preliminary information can be used to run preliminary simulations (after selecting a model) and use them to design an appropriate and more detailed sampling and monitoring campaign to complete the data required for modelling. The concentrations that show the highest variations at certain points may need to be evaluated in detail. It is highly recommended to evaluate the quality of the data collected to find potential gaps and to correct any potential inconsistency. For this purpose, it is strongly advised to conduct water and mass balances on the suspended solids, COD, nitrogen and phosphorus (Meijer, 2004). Depending on the outcomes of the data quality assurance (*e.g.* if the mass balances do not close), additional sampling and monitoring campaigns will be needed to take this into account to complete and/or correct the required data.

Model structure

The structure of the model will be initially defined based on the process description. First, the model will need to be set up based on the hydraulics or transportation processes of the FSM unit, defining each process component. This means that the number of tanks, the compartments of the tanks, redox conditions, and solid-liquid separation compartments will need to be defined. The redox conditions will not only indicate whether a tank or stage is anaerobic or aerobic but also if the redox gradients prevailing in the system may indicate that one single tank should be modelled as a series of aerobic or anaerobic compartments. To set up a proper compartmentation, it is recommended to measure the dissolved oxygen concentration and redox conditions in a vertical and horizontal direction in all the tanks and their compartments. Based on the prevailing or dominant processes conversions, a process model needs to be selected among those available in literature (*e.g.* ASM-type, ADM1).

Characterisation of flows

First (if available), using historical data or specific measurements, the main inputs and flows can be characterised. Depending on the configuration of the system, these need to include the influent, effluent, and the internal and recirculation flows. If there is no data available or certain data points are missing, a sampling campaign needs to be conducted. If the model will be used to select a design, daily average concentrations for three days and the variations in the flow patterns may be enough. However, for process optimisation and control strategies, samples may need to be collected periodically every 2-4 hours over a period of three to seven days at several critical points along the system (*e.g.* not only at the feed or influent and effluent but also at the interfaces between the tanks and compartments). All the data gathered and collected needs to be checked for consistency (*e.g.* performing water and mass balances).

Calibration

Once the data have been checked for consistency and quality assurance, the first simulations can be executed and the model calibrated using the available data. If the description of the performance of the plant shows that a major adjustment is needed (*e.g.* if in order to describe the data or measurements a large adjustment of the kinetic parameters is required), the model structure will probably need to be revised as well as the mass balances and data collection. Based on the experience drawn from modelling activated sludge systems, it is recommended to first model and calibrate the sludge production, followed by the process which is kinetically most sensitive, and afterwards the rest of the kinetic processes. If the process performance and effluent quality are not well predicted, a sensitivity analysis can be conducted to assess which parameters have the strongest impact. In this regard and at this stage, different approaches can be applied to quantify the level of uncertainty and its sources and to assess in more detail their impact to define additional sampling and monitoring criteria (Belia *et al.*, 2009; Flores-Alsina *et al.*, 2012). Following an iterative step-wise process, the model could be calibrated by adjusting the least possible number of kinetic parameters until it provides a satisfactory description of the performance of the containment unit.

Detailed characterisation

The results of the first simulations, calibration and a sensitivity analysis can be used to define an additional (more thorough) sampling campaign with a more detailed influent characterisation (in relevant points along the system), and lab-scale tests for the determination of the key modelling parameters. The needs and characteristics of such a detailed sampling campaign can also be defined based on the uncertainty analysis.

Validation

The calibrated model needs to be validated by assessing its capacity to predict the performance of the plant using operational and environmental data from a different period than that used for the model calibration. If it fails the validation step, the model will need to be re-calibrated iteratively until a satisfactory validation is reached.

Study

A validated model can then be used to assess the scenarios of concern in accordance with the purpose and objectives of the modelling study.

Because of its practical nature and satisfactory application for model wastewater treatment plants, the steps of the STOWA calibration protocol will be reviewed from a faecal sludge modelling perspective, suggesting how they could be extrapolated and adapted to the particular characteristics and features of the most common onsite sanitation systems. This will be used to suggest the required steps towards developing a framework to model sanitation systems whose aim is to describe the dominant processes that take place inside the sanitation systems, in order to estimate the volumes and characteristics of the faecal sludge generated. However, one should bear in mind that while this framework describes different considerations and assumptions that need to be followed, but that also need to be proven and validated by applying and testing the framework and its outcomes in different sanitation systems. Ultimately, a structured and continuous application of the framework could lead over the years to a robust and solid protocol that could be applied with confidence and reliability, as has been observed in the wastewater field (Henze *et al.*, 2008; Van Loosdrecht *et al.*, 2016).

6.3 TOWARDS AN ONSITE SANITATION MODELLING FRAMEWORK

6.3.1 Onsite sanitation modelling: formulation of objectives

The first step is to define the main objectives of carrying out an onsite sanitation modelling study. Considering the prime purpose of sanitation, the main initial objectives should focus on (*i*) providing a tool to describe the accumulation of solids in onsite containment and treatment systems (as a function of the feeding rates and sludge disintegration) and to assess potential strategies to minimise the volumes of sludge, (*ii*) studying pathogen inactivation mechanisms, and evaluating different approaches to enhance and maximise the inactivation of pathogens, (*iii*) improving the prediction of the characteristics of the sludge contained, accumulated and emptied (as a function of the operating and environmental conditions of the sanitation systems) as a tool to contribute to improving the decision-making process in the sanitation chain, and (*iv*) evaluating the potential recovery of resources by maximising biogas production and enhancing nutrient recovery.

Different modelling studies have already been conducted (*i*) to describe the accumulation of solids (Brouckaert *et al.*, 2013; Todman *et al.*, 2015; Lugali *et al.*, 2016; Strande *et al.*, 2018); (*ii*) to model pathogen inactivation by pH, temperature or high ammonia concentrations in containment and treatment sanitation systems (Lübken *et al.*, 2007; Fidjeland *et al.*, 2013; Koottatep *et al.*, 2014; Magri *et al .*, 2015); (*iii*) to model the anaerobic degradation of faecal sludge with special emphasis on biogas production (Elmitwalli *et al.*, 2006, 2013; Wendland, 2008); and (*iv*) to study the aerobic degradation of faecal sludge (Lopez-Zavala *et al.*, 2004a, 2004b). Most of these studies were conducted following empirical approaches and black-box models to achieve a satisfactory description of the accumulation of solids (Brouckaert *et al.*, 2013; Todman *et al.*, 2015; Lugali *et al.*, 2016; Strande *et al.*, 2018).

However, to include and consider additional and intermediate (biological and chemical) conversion processes could provide additional advantages that improve the operation of such systems. For instance, the hydrolysis and fermentation processes involved in the degradation of organic matter are often neglected, but these processes and their by-products can have an important influence on pathogen inactivation (Fidjeland *et al.*, 2013; Magri *et al.*, 2015; Anderson *et al.*, 2015). There are also other models available and applied to describe the degradation of faecal sludge in lab-scale systems operated under well controlled conditions to forecast degradation efficiencies and performance (Lopez-Zavala *et al.*, 2004a, 2004b; Wendland, 2008; Elmitwalli *et al.*, 2006, 2013). These models need to be validated under actual operating and environmental conditions with real data measurements.

Last but not least, the pathogen inactivation models available so far tend to be stand-alone expressions (Lübken *et al.*, 2007; Fidjeland *et al.*, 2013; Koottatep *et al.*, 2014; Magri *et al.*, 2015) that need to be incorporated into mechanistic faecal sludge conversion and degradation models in order to explore different practical alternatives to enhance pathogen inactivation. Overall, the information and knowledge generated and provided by existing models are very valuable and can be combined and used to propose a basis to develop an expanded and structured mechanistic (glass-box) model for onsite containment and treatment sanitation systems that can be used to achieve the aforementioned objectives.

6.3.2 Onsite sanitation modelling: process description

There is a need to conceptually describe the activities and processes that take place in onsite containment and treatment systems. In this regard, onsite containment and treatment units can range from portable toilets (only used for containment prior to emptying, transportation and treatment) to borehole and pit latrines, septic tanks, and anaerobic baffled reactors. In order to define potential modelling approaches that reasonably represent the broader range of onsite sanitation systems, three commonly used technologies will be assessed in detail in this chapter: a portable toilet, a single pit latrine, and a septic tank. Because of the large variations in nature, other onsite containment sanitation systems require different modelling approaches which fall outside the scope of this chapter. Nevertheless, for the sake of completeness, an overview of different models that can be applied to describe different onsite sanitation systems is presented later in this chapter (Table 6.3).

6.3.2.1 Portable toilets

A portable toilet is placed in a defined location to provide a temporary service that can range from a few days to months, and sometimes much longer, for example under emergency situations (Brdjanovic *et al.*, 2015). It is usually made of light, yet durable, material (plastic, PVC, wood, among others) to facilitate its transportation and has no large compartments to store high volumes of solids or liquids. It may have separated compartments (urine diversion toilet - UDT) to collect urine and faeces. There are three types: dry, pour flush, and flush. The latest generation can have three compartments, one each for urine, faeces and grey water, and can even include an extra compartment for internal storage for service water (as a source of grey water). An example of such a toilet, which was recently used in peri-urban areas of Nairobi, is shown in Figure 6.3.

Since it can be used frequently (*e.g.* up to 300-400 times a day during public events or under emergency situations), these containment units fill up rapidly and require emptying. Some may need to be emptied every day whereas other toilets with larger storage volumes may operate for up to 7-10 days without being emptied (Zakaria *et al.*, 2017). When several single toilets are clustered (*e.g.* four or more), it is common to find larger containers, which makes the emptying periods less frequent. Taking into account that portable toilets do not have large compartments and that (consequently) they are emptied frequently, the faecal sludge and urine contained are usually fresh and of high strength (Lopez-Vazquez *et al.*, 2013; Zakaria *et al.*, 2018). Moreover, if the containment units are dry toilets made of impermeable materials (such as plastic) and often located above the ground (raised latrines), they are not subject to infiltration or seepage. Thus, the only input is the filling rate at which they are subject to by the users and the only output is due to emptying. The relatively high filling rates and emptying frequencies that result in short retention times allow little anaerobic or aerobic degradation of the faecal sludge. Arguably, this unit resembles a 'fill and draw' batch type of system (Henze *et al.*, 2007) (Figure 6.4).

Figure 6.3 Portable eSOS Smart Toilet with storage for urine, faeces and grey water, and a service water reservoir as the roof (image: Flex/design).

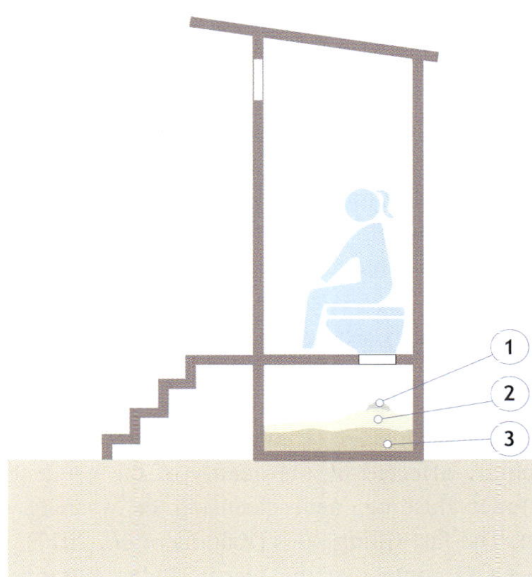

Figure 6.4 Schematic representation of a portable toilet without separated collection of urine (adapted from Bakare et al., 2012).

In this system (and following the approach defined by Bakare *et al.* (2012)), three zones can be identified (Figure 6.4):

[1] Zone 1: the upper part where fresh faecal sludge and urine accumulate and are distributed over the cross-sectional area of the system.
[2] Zone 2: where the fresh faecal sludge and urine are already distributed. They remain in contact with the atmosphere, creating (micro-)aerobic conditions. In this zone the biological and chemical conversion processes start to take place.
[3] Zone 3: due to the accumulation of faecal sludge and urine and the consumption of oxygen in zone 2 (where the biological conversions under micro-aerophilic conditions start), zone 3 starts where the dissolved oxygen can no longer penetrate. As such, zone 3 is anaerobic and it triggers the occurrence of anaerobic conversion processes.

In zone 1, fresh faecal sludge accumulates depending upon the feeding rates in accordance with the number of users (Brouckaert *et al.*, 2013; Todman *et al.*, 2015; Lugali *et al.*, 2016; Strande *et al.*, 2018). This fresh faecal sludge from zone 1 is probably exposed to micro-aerophilic conditions in the exterior and possibly anaerobic in the interior. However, any biological or chemically-induced activity will only be driven by the microorganisms present in the fresh faecal sludge itself and, consequently, the biological conversions (if any) may be negligible. Overall, in zone 1, it can be assumed that the characteristics of fresh faecal sludge and urine will remain practically unchanged. Then, these components will only be distributed over the cross-sectional area of the unit as a function of the rheology of the sludge.

Zone 2 starts where the biological and chemical processes also start. Chemical conversions may begin (such as the hydrolysis of urine, depending on the presence of urease) (Rubio-Rincón *et al.*, 2014), which are affected by the quality of the water used for toilet flushing, anal cleansing or washing the toilet. The fast filling rates (Zakaria *et al.*, 2017), the high COD content of the faecal sludge (Strande *et al.*, 2014; Chapter 2), and a potentially minimal diffusion of oxygen due to the physical characteristics of the faecal sludge (and merely driven by the atmospheric pressure) (Allaire *et al.*, 2008) probably limit the availability of dissolved oxygen down to only a few millimetres in the solids layers. This suggests that zone 2 may be only a thin micro-aerophilic layer of a few millimetres that goes from the exterior layer up to where dissolved oxygen penetrates. Due to the limited availability and diffusion of oxygen, only some of the aerobic hydrolysis processes take place in zone 2 and a full aerobic conversion of organics cannot be expected. This is also because the relatively short retention times (as a consequence of the extremely frequent filling and emptying rates) will limit the accumulation of sludge and organisms.

Zone 3 starts where dissolved oxygen is no longer detected. Therefore, zone 3 is anaerobic and will trigger the anaerobic conversion of compounds. However, the short retention times (unless certain biomass/sludge is unwantedly retained after emptying inside the containment unit) will limit the growth of anaerobic bacteria (in particular the growth of methanogens) (Jabłoński *et al.*, 2015) suggesting that hydrolysis, some fermentation and (as much) a marginal acetogenesis process may be the dominant (biological) mechanisms. Thus, a full anaerobic conversion of the organics may not be expected. Therefore, inert or unbiodegradable compounds will not be excessively generated and accumulated in these systems unless the retention time is extended for some weeks or months. In zone 3 chemical processes are also expected to take place after the hydrolysis of urine and of other organic compounds has occured. This, in combination with the particular quality of used water or the addition of external compounds (*e.g.* magnesium or iron salts) (Zhang *et al.*, 2008), may lead to the formation of certain crystals (*e.g.* calcium phosphate, and struvit) (Udert *et al.*, (2003).

6.3.2.2 Single pit latrines

Another widely used onsite sanitation system is the single pit latrine. Excreta, along with anal cleansing materials or water, are deposited into the pit. They are emptied with a frequency that ranges from a few months (4-6 months) or a few years (1-2 years) to several years (even longer than 10 years) depending on the faecal sludge accumulation rate, which is the function of percolation, degradation and consolidation of collected sludge (Broukaert *et al.*, 2013; Todman *et al.*, 2015; Zziwa *et al.*, 2016). In one of the first efforts made to describe the accumulation of faecal sludge in these systems in more detail, Bakare *et al.* (2012) provided a conceptual description of the main processes that take place in a pit latrine.

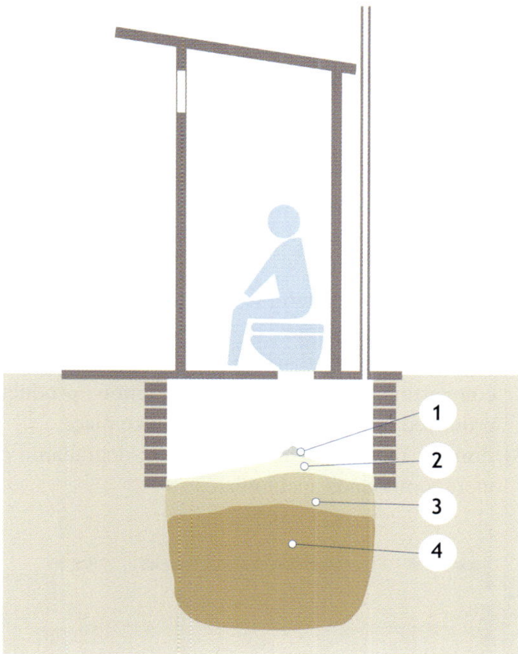

Figure 6.5 Schematic diagram illustrating the different theoretical layers within a pit latrine (adapted from Bakare *et al.*, 2012).

They identified four zones (Figure 6.5):

[1] Zone 1: the upper part of where fresh faecal sludge and urine will only accumulate and be distributed over the cross-sectional area of the system.

[2] Zone 2: in this zone the fresh faecal sludge and urine are already distributed in the system. They remain in contact with the atmosphere, creating (micro-) aerophilic conditions where the biological and chemical conversion processes start.

[3] Zone 3: due to the accumulation of faecal sludge and the consumption of oxygen in zone 2, the third zone starts when the dissolved oxygen can no longer penetrate, creating anaerobic conditions and therefore triggering the occurrence of anaerobic conversion processes.

[4] Zone 4: in the fourth zone, located at the bottom of the faecal sludge system, biological activity is minimal or no longer observed and only non-degradable or inert compounds accumulate.

A latrine is a larger, permanent system, which fills up and gets emptied less frequently than a portable toilet. Thus, the retention times are longer. This allows: (*i*) the retention of biomass, (*ii*) aerobic but mostly anaerobic conversions that can lead to the removal of organics and the accumulation of inert and non-degradable components, (*iii*) a substantial generation of gases (such as methane, carbon dioxide and also hydrogen sulphide), (*iv*) the occurrence of chemical processes and, (*v*) the infiltration of groundwater and the percolation/leakage of soluble components into the ground if the pit latrine is not well lined (sealed). Thus, zones 1 and 2 will be similar to those found in a portable toilet, zone 3 will allow the full anaerobic conversion of organics, and zone 4 will appear where most of the inert and non-degradable products from the conversion processes will accumulate. However, since it is a system with underground storage which is often an unlined pit, it is subject to the influence of the groundwater level, a particular problem in flood-prone areas. As such, it may suffer from groundwater infiltration that not only affects the biological and chemical conversions (*e.g.* due to the dilution effect as well as an increasing generation of gases if, for instance, sulphate-rich water intrudes into the latrine) but also allows the percolation of water and soluble compounds from the pit latrine into the ground.

The rheology of faecal sludge (Forster, 2002; Woolley *et al.*, 2014a, 2014b; Liu *et al.*, 2016) in combination with the impact of the infiltration of percolation processes will determine the way the faeces, urine and water are distributed and percolate between the different zones and will also affect the consumption and production of soluble and particulate products of the dominant conversion process(es) prevailing in each phase. The longer retention times will allow an extended conversion of organic matter in zone 3 that will lead to the generation of gases (such as methane, carbon dioxide and possibly also hydrogen sulphide) that will mostly diffuse into zones 2 and 1. The diffusion of such gases into zone 2, in combination with the high organic loads present in the faecal sludge discharged into the latrine, will decrease the volume and thickness in zone 2. While the start of zone 3 can be determined based on the profile of dissolved oxygen, its depth and thickness cannot be easily determined. This is mostly because, as pointed out by Nwaneri *et al.* (2008), this phase finishes at a depth where the accumulation of inert and non-degradable compounds is dominant, meaning that zone 3 finishes where the anaerobic biological conversions become negligible or are no longer observed. Zone 4 starts where zone 3 finishes and in this layer mostly unbiodegradable or non-degradable organic and inorganic compounds accumulate.

6.3.2.3 Septic tanks

Septic tanks are a common onsite sanitation system. They can be relatively simple and made of concrete, fibreglass, vynil or plastic. They are composed of at least two compartments divided by one baffle (Figure 6.6). Excreta and anal cleansing materials are deposited into the septic tank. They are emptied with a frequency that ranges from a few (1-2 years) to several years (even longer than 10 years) (Broukhaert *et al.*, 2013; Todman *et al.*, 2015; Zziwa *et al.*, 2016) depending on faecal sludge accumulation, but the hydraulic retention time can be as short as a few hours (12-24 hours) especially when the tank is full. Settleable solids accumulate at the bottom of the system whereas floating material accumulates at the top. Mostly anaerobic conversion processes contribute to the removal and reduction of the organic matter. From a process description perspective, the two (or more) compartments can be divided into different zones (Figure 6.6) as explained below.

Compartment 1:

[1] Zone 1: the upper part where the wastewater is received and settleable and non-settleable matter is split.

[2] Zone 2: a small (micro-)aerobic zone where some dissolved oxygen may be present, either from the influent or due to oxygen diffusion. Thus, some aerobic processes may take place.

[3] Zone 3: the anaerobic zone. This zone can be further divided into two sub-zones where the soluble compounds (3a) and the particulate compounds (3b) can be degraded separately, respectively.

[4] Zone 4: located at the bottom of the septic tank where the biological activity is minimal or no longer observed and only non-degradable or inert compounds accumulate.

Compartment 2:

[1] Zones 1 and 2: they cannot be found in the 2nd compartment since wastewater is already mixed and the dissolved oxygen concentrations are negligible.

[2] Zone 3: the anaerobic zone. Here anaerobic conversion processes of soluble and particulate organic matter (that do not settle in the 1st compartment) and residual reaction products produced in the 1st compartment take place.

[3] Zone 4: in this last zone, only non-degradable or inert compounds accumulate.

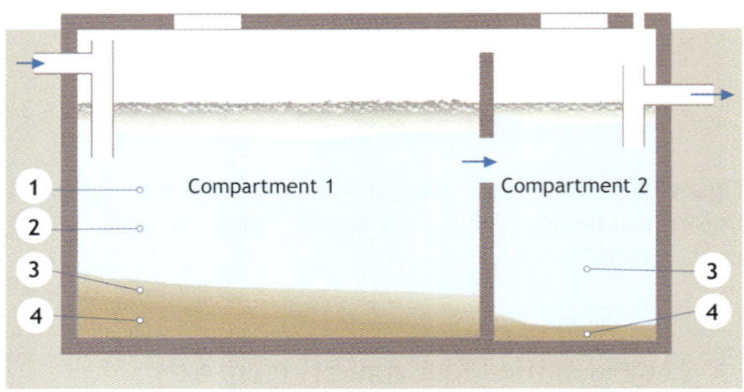

Figure 6.6 Schematic diagram illustrating the different theoretical layers within a septic tank (adapted from Tilley *et al.*, 2014).

The longer retention times of septic tanks and their configuration composed of two compartments divided by a baffle allows in the 1st compartment the development of four zones similar to those discussed previously for pit latrines. However, most of the settleable solids present in the influent wastewater settle in the 1st compartment and the rest flows to the

2nd compartment. The settleability of the solids influences the fraction retained in the 1st compartment and the fraction that flows to the 2nd compartment. Thus, in zone 3 of the 1st compartment, the fraction of the settleable solids retained in the 1st compartment degrades anaerobically and zone 4 accumulates the inert and non-degradable matter from the upper zones. It is possible that zone 3 needs to differentiate between the anaerobic degradation of soluble matter and suspended matter by splitting the zones in two. The gases generated from zone 2 and mostly from zone 3 diffuse into the adjacent zones.

The 2nd compartment is only composed of one zone 3 and one zone 4. In zone 3, the degradable matter not retained in the 1st compartment and the products and residual concentrations generated in zone 3 of the 1st compartment degrade anaerobically. Zone 4 of the 2nd compartment accumulates the remaining inert and non-degradable matter from zone 3. The wastewater flows out of the system from zone 3 of the 2nd compartment, determining the quality of the treated effluent. The gases generated in zone 3 diffuse into the headspace of the septic tank, into zone 4 and also leave through the effluent.

Being an underground system, similar to the pit latrine, septic tanks may be affected by groundwater infiltration influencing the biological and chemical conversions, as previously discussed, and also allowing the percolation of water and soluble compounds from each zone into the ground.

Overall, the portable toilet, the pit latrine and the septic tank have different conversion processes influenced by their configurations, use of water and type of service provision, location, and operation and maintenance. Tables 6.3a and 6.3b aim to provide a general overview of (*i*) the main conversion processes and (*ii*) the main transport processes that take place in these systems.

Table 6.3a General overview of the main conversion processes in portable toilets, pit latrines and septic tanks.

	Portable toilet	Pit latrine	Septic tank	
			1st compartment	2nd compartment
Retention time	Short - usually less than a few days (*e.g.* 7 days).	Long - varying from a few to several years (1-20 years).	Long - varying from a few to several years (1-20 years).	
Main aerobic conversion processes (zone 2).	(Micro-) aerobic zone of a few mm defined by the penetration of dissolved oxygen. Aerobic hydrolysis takes place but full aerobic conversion of organics is not expected.	Aerobic hydrolysis and (marginal) heterotrophic organic matter removal.	Aerobic hydrolysis and (marginal) heterotrophic removal of soluble organic matter (limited by oxygen diffusion and availability).	None. Absent. Full anaerobic compartment.
Main anaerobic conversion processes (zone 3).	Hydrolysis and fermentation.	Hydrolysis, fermentation, acetogenesis and methanogenesis	Hydrolysis, fermentation, acetogenesis and methanogenesis (mostly on settled compounds).	Hydrolysis, fermentation, acetogenesis and methanogenesis (mostly on soluble compounds).
Accumulation of inert and non-degradable matter (zone 4).	No accumulation, due to short retention times (zone 4 does not exist).	Accumulation in zone 4, due to long retention times.	Accumulation in zone 4, due to long retention times. Mostly produced by the anaerobic conversions of particulate compounds retained in 1st compartment.	Accumulation in zone 4, due to long retention times.

Table 6.3b (Part 1 of 2) General overview of the main transport mechanisms in portable toilets, pit latrines and septic tanks.

		Portable toilet	Pit latrine	Septic tank: 1st compartment	Septic tank: 2nd compartment
Zone 1	*Inputs*	• Faecal sludge and urine as function of filling rates.	• Faecal sludge and urine as function of filling rates. • Groundwater infiltration.	• Faecal sludge and urine as function of filling rates. • Diffusion of dissolved oxygen from atmosphere and of gases from zone 2. • Groundwater infiltration.	
Zone 1	*Outputs*	• Percolation of faecal sludge and urine to zone 2.	• Percolation of faecal sludge and urine to zone 2.	• Soluble compounds flow to 2nd compartment and also diffuse into zone 2. • A large fraction of particulate or suspended matter settles and reaches zone 2, the remaining fraction flows to the 2nd compartment. [Zone 1 not well defined.]	[Zone 1 is absent.]
Zone 2	*Inputs*	• Soluble and particulate compounds of faecal sludge and urine from zone 1. • Diffusion of dissolved oxygen from atmosphere.	• Soluble and particulate compounds of faecal sludge and urine from zone 1. • Diffusion of dissolved oxygen from atmosphere and of gases from zone 3 (*e.g.* methane, carbon dioxide, hydrogen sulphide). • Groundwater infiltration.	• Soluble and particulate compounds of faecal sludge and urine retained in 1st compartment. • Diffusion of dissolved oxygen from atmosphere and of gases from zone 3 (*e.g.* methane, carbon dioxide, hydrogen sulphide). • Groundwater infiltration.	
Zone 2	*Outputs*	• Percolation of aerobically converted products of faecal sludge and urine to zone 3.	• Percolation of aerobically converted products of faecal sludge and urine and inert or non-degradable compounds to zone 3. • Infiltration into the ground.	• Percolation of aerobically converted products of faecal sludge and urine and inert or non-degradable compounds to zone 3. • Infiltration into the ground.	[Zone 2 is absent.] [No oxygen is available.]

Table 6.3b (Part 2 of 2) General overview of the main transport mechanisms in portable toilets, pit latrines and septic tanks.

	Portable toilet	Pit latrine	Septic tank	
			1st compartment	2nd compartment
Zone 3 – *Inputs*	• Percolation of aerobically converted products of faecal sludge and urine from zone 2.	• Products of faecal sludge and urine from zone 2. • Inert or non-degradable compounds from zone 2. • Groundwater infiltration.	• Products of faecal sludge and urine from zone 2. • Inert or non-degradable compounds from zone 2. • Groundwater infiltration.	• Mostly soluble and the fraction of the particulate compounds of faecal sludge and urine not retained in 1st compartment. • Soluble products of faecal sludge and urine from zone 3 of 1st compartment. • Diffusion of gases generated in the zone 3 of 1st compartment (*e.g.* methane, carbon dioxide, hydrogen sulphide). • Inert or non-degradable soluble compounds from zone 3 of 1st compartment. • Groundwater infiltration.
Zone 3 – *Outputs*	• No outputs.	• Percolation of anaerobically degraded inert and non-degradable matter to zone 4. • Diffusion of gases generated to zones 2 and 4 (*e.g.* methane, carbon dioxide, and hydrogen sulphide). • Infiltration into the ground.	• Percolation of anaerobically degraded inert and non-degradable matter to zone 4 of 1st compartment • Diffusion of gases generated to zones 2 and 4 (*e.g.* methane, carbon dioxide, and hydrogen sulphide). • Infiltration into the ground.	• Effluent. • Percolation of anaerobically degraded inert and non-degradable matter to zone 4 of 2nd compartment. • Diffusion of gases generated to zone 4 (*e.g.* methane, carbon dioxide, hydrogen sulphide).
Zone 4 – *Inputs*		• Accumulation of percolated anaerobically degraded compounds and/or inert and non-degradable matter from zone 3. • Groundwater infiltration.	• Accumulation of percolated anaerobically degraded compounds and/or inert and non-degradable matter from zone 3. • Groundwater infiltration.	• Accumulation of percolated anaerobically degraded compounds and/or inert and non-degradable matter from zone 3 of 2nd compartment. • Groundwater infiltration.
Zone 4 – *Outputs*	[Zone 4 is absent.]	• Desludging. • Infiltration into the ground.	• Desludging. • Infiltration into the ground.	• Desludging. • Infiltration into the ground.

6.3.3 Onsite sanitation modelling: data collection and verification

Data need to be collected for five main purposes: (*i*) to determine the volumes of faecal sludge including urine, (*ii*) to determine the characteristics of faecal sludge and urine, (*iii*) to define the length of the reaction zones in each system, (*iv*) to assess the conversion processes, and (*v*) to estimate possible infiltrations and percolation flows. Table 6.4 suggests different sampling campaigns to assess the first four of these purposes.

Most of the samples collected for the analytical determination of standard parameters can follow the corresponding recommendations for sampling, preservation, transportation and storage (Chapter 3) prior to the conduction of the analytical tests. However, for the conduction of the required (anaerobic and aerobic) biological, physical and chemical tests, it is important to collect reliable and representative samples from each layer at different depths that have not been adulterated or disturbed. Sampling procedures from soil mechanics or studies in sediments need to be applied and followed to collect the required soil and sludge samples and transport and store them prior to the conduction of the batch activity tests of interest (Strande *et al.*, 2014).

Due to public health concerns, it is essential to assess the transport and distribution of pathogens, viruses and other harmful bacteria or organisms between the different zones. In parallel, the formation and accumulation of products, compounds and elements from the biological and chemical processes may influence the viability and inactivation of pathogens, viruses and other harmful bacteria or organisms. Several studies have been carried out on the transport of pathogens in faecal sludge and porous media (Mensah *et al.*, 2013) and these could be used to execute the required tests with samples of solids collected at different depths, and linked with the assessment and effects of potential inhibitory elements or compounds from the biological and chemical conversions at the different layers.

In the case of septic tanks, it is recommended to also conduct tracer tests for a better determination of the hydrodynamic behaviour of the system and to define the hydraulic residence time (Metcalf and Eddy *et al.*, 2014).

Infiltration and percolation mechanisms affect the transport of the components and elements of interest, also influencing the conversion processes in each zone and the performance of the system as a whole. These mechanisms and their rates are not only dependent on the rheology of the solids or characteristics of the system but also on hydrological and groundwater processes (Foppen, 2002; Halalsheh *et al.*, 2011). All these processes need to be studied in a structured manner and would probably, as with many other processes, be case-specific for each location.

The quality and reliability of the measurements need to be verified through the conduction of mass balances on water, COD, nitrogen and phosphorus. However, in addition to these balances, when reviewing the performance of anaerobic systems, molar balances also need to be performed because carbon dioxide (not accounted for in the COD balance) will be generated which affects the composition of the biogas produced, the pH and even the ADM1 model stoichiometry (Klerebezeem and Van Loosdrecht, 2006a, 2006b; Rodriguez *et al.*, 2006). Another reason for performing molar balances is that they are different anaerobic processes that are pH-dependent. If there are major differences to close the mass balances (higher than 10-15%), it will be necessary to check the results of the analytical parameters and thereafter the configuration of the system to conduct another (detailed) sampling campaign.

Table 6.4 (Part 1 of 2) Suggested sampling campaigns for data collection and verification for modelling of a portable toilet, pit latrine and septic tank.

	Portable toilet	Pit latrine	Septic tank	References
Purpose	*Determination of volumes of faecal sludge, urine or wastewater*			
Duration	1-2 days	2-3 days	2-3 days	See chapters 2, 3 and 5.
Frequency	Continuous recording of no. of users during representative periods of use.	Continuous recording of no. of users during representative periods of use.	Assessment over a few hours	
Purpose	*Determination of characteristics of faecal sludge and urine*			
Duration	1-2 days	2-3 days	2-3 days	See chapters 2, 3 and 5. Mensah *et al.* (2013)
Frequency	Every 2-3 hours	Every 2-3 hours	Every 2-3 hours	
Type of samples	Composite	Composite	Composite	
Parameters	Total COD, soluble COD, TSS, VSS, TN, NH_4-N, TP, PO_4-P, pH, microbiological analyses, sludge rheology, and dewaterability.	Total COD, soluble COD, TSS, VSS, TN, NH_4-N, TP, PO_4-P, pH, microbiological analyses, sludge rheology, and dewaterability.	Total COD, soluble COD, TSS, VSS, TN, NH_4-N, TP, PO_4-P, pH, microbiological analyses, settleable matter, and floating matter.	
Purpose	*Determination of the length of reaction zones*			
Duration	1-2 days	1-2 days	1-2 days	See Chapter 3.
Frequency	Every 2-3 hours	Every 2-3 hours	Every 2-3 hours	
Type of samples	Use of portable meters. Collection of undisturbed solids samples to perform aerobic and anaerobic activity tests.	Use of portable meters. Collection of undisturbed solids samples to perform aerobic and anaerobic activity tests.	Use of portable meters. Collection of undisturbed solids samples to perform aerobic and anaerobic activity tests.	
Sampling locations	Vertical and horizontal directions across the system.	Vertical and horizontal directions across the system.	Vertical and horizontal directions across the system in each compartment.	
Parameters	• DO (if available, use a microelectrode) and redox potential. • Conduction of experimental methods to assess aerobic and anaerobic activities.	• DO (if available, use a microelectrode) and redox potential. • Conduction of experimental methods to assess aerobic and anaerobic activities.	• DO (if available, use a microelectrode) and redox potential. • Conduction of experimental methods to assess aerobic and anaerobic activities.	

Table 6.4 (Part 2 of 2) Suggested sampling campaigns for data collection and verification for modelling of a portable toilet, pit latrine and septic tank.

	Portable toilet	Pit latrine	Septic tank	References
Purpose	*Assessment of conversion processes*			
Duration	1-2 days	1-2 days	1-2 days	See chapters 2, 3 and 5. Van Loosdrecht *et al.* (2016).
Frequency	• Every 3-4 hours (grab samples) • 24 hours (composite samples)	• Every 3-4 hours (grab samples) • 24 hours (composite samples)	• Every 3-4 h (grab samples) • 24 hours (composite samples)	
Type of samples	• Grab • Composite	• Grab • Composite	• Grab • Composite	
Sampling locations	• Grab and composite samples in each reaction zone.	• Grab samples in the influent and effluent and in the interface between reaction zones. • Composite samples in each zone.	• Grab samples in the influent and effluent and in the interface between reaction zones in each compartment. • Composite samples in each zone.	
Parameters	• Grab samples: - Total COD, soluble COD, TSS, VSS, TN, NH_4-N, TP, PO_4-P, pH, microbiological analyses, sludge rheology, and dewaterability.	• Grab samples: - Total COD, soluble COD, TSS, VSS, TN, NH4-N, TP, PO_4-P, pH, microbiological analyses, sludge rheology, and dewaterability.	• Grab samples: - Total COD, soluble COD, TSS, VSS, TN, NH_4-N, TP, PO_4-P, pH, microbiological analyses, settleable matter, and floating matter.	
Tests and methods	• Composite samples: - Aerobic and anaerobic activities. - Aerobic and anaerobic fractionations and biodegradation rates.	• Grab samples: - Off-gases (methane, carbon dioxide, hydrogen sulphide). • Composite samples: - Aerobic and anaerobic activities. Aerobic and anaerobic fractionations and biodegradation rates. - Chemical precipitation tests.	• Grab samples: - Off-gases (methane, carbon dioxide, hydrogen sulphide). • Composite samples: - Aerobic and anaerobic activities. Aerobic and anaerobic fractionations and biodegradation rates. - Chemical precipitation tests.	

6.3.4 Onsite sanitation modelling: model structure

6.3.4.1 Model structure of commonly used onsite sanitation systems

In this section, three representative and commonly used faecal sludge containment and/or treatment systems are assessed in detail with the aim of defining basic structures and highlighting the required information and assumptions that need to be gathered and proposed to model these systems. The containment systems selected and subject to a deeper discussion and assessment are the portabletoilet, the pit latrine and the septic tank. This approach is also based on the consideration that more complex systems, such as the anaerobic-baffled reactors or the upflow anerobic sludge blanket (UASB) reactors, could probably be developed based on the basic structures suggested for these three more basic units but expanded (both 'physically' by considering a higher number of interconnected reactors and also with regard to the process performance by incorporating more complex models). Therefore, in the next section, after the discussion of these three basic units, some suggestions are given to model more complex onsite containment and treatment systems (Section 6.3.4.2).

Portable toilets

The first model structure suggested is for a portable toilet (Figure 6.7). As discussed in Section 6.3.2.1, these are usually a closed system with a short retention time (of maximum a few weeks), it is composed of three zones or phases: zone 1 where the sludge retains its physical properties, zone 2 where it is distributed and contains dissolved oxygen that drives certain aerobic conversions, and zone 3 where the conditions become anaerobic and anaerobic conversions take place. In this suggested model structure, it is assumed that the relatively short retention time (of a few weeks) does not allow the complete conversion and degradation of the organics. Consequently, only a marginal degradation or conversion of the degradable matter is reached. There is no gas generation (since the conversions are not complete) and zone 4 is absent. When present, the function of zone 4 is to retain and accumulate the inert and non-degradable matter present in the influent or produced from the degradation processes. The fluxes of soluble (S) and suspended (X) compounds are indicated (Q_{1_2} and Q_{2_3}, for their transport from zone 1 to zone 2, $S_{FS,1\_2}$ and $X_{FS,1\_2}$, and from zone 2 to zone 3, $S_{FS,2\_3}$ and $X_{FS,2\_3}$, respectively) including the presence and transport of pathogens between zones ($X_{pathogens,inf}$, $X_{pathogens,1\_2}$, $X_{pathogens,2\_3}$). The system is fully closed and the only input is the discharge of faecal sludge, urine and water and the only output is the periodic emptying rate ($Q_{FS,emptying}$), resembling a fill-and-draw system. This can be considered the simplest model structure for a faecal sludge system.

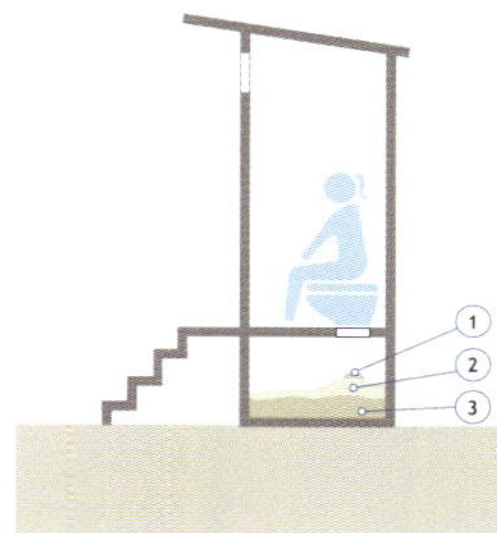

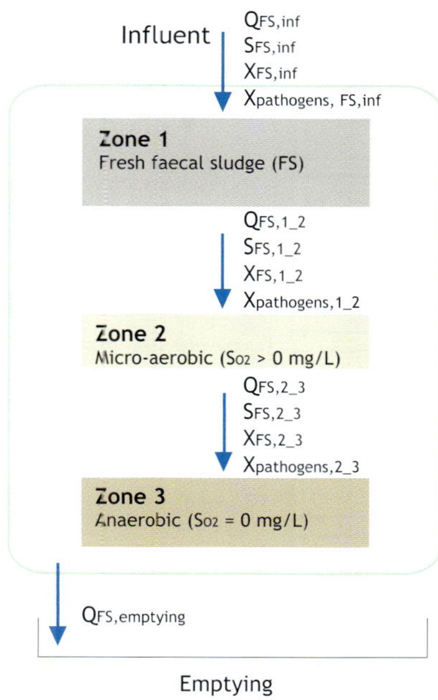

Figure 6.7 Portable toilet: suggested model structure.

Pit latrines

Pit latrines are more complex than portable toilets (Figure 6.8). Although they are subject to some similar conditions, they have longer retention times (of several months and even years) that result in the full completion of the conversion processes (mostly the anaerobic ones).

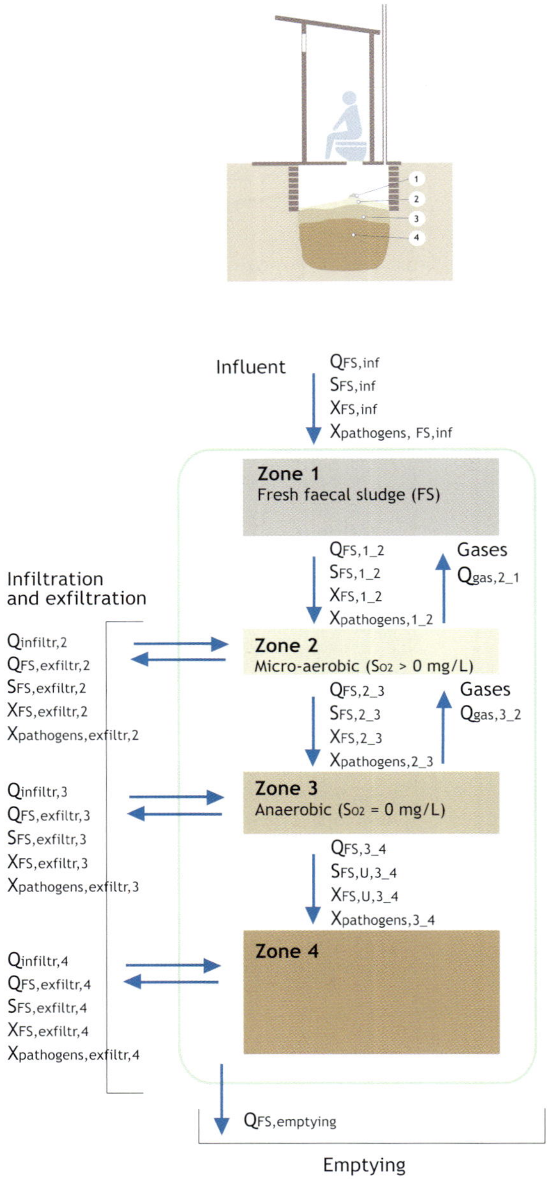

Figure 6.8 Pit latrine: suggested model structure.

This implies that the kinetics will probably not play a major role and that stoichiometric relationships can be used to describe the conversion processes. This has already been observed in studies by Brouckaert *et al.* (2013) and Todman *et al.* (2015) who were able to model the filling rates of pit latrines using basic kinetic expressions. Moreover, pit latrines are prone to infiltration and percolation. Thus, besides the effects of the transport phenomena of the sludge matrix and associated processes between zones (*e.g.* $Q_{FS,1\_2}$ that transports the soluble, $S_{FS,1\_2}$, and particulate concentrations, $X_{FS,1\_2}$ and $X_{pathogen,1\_2}$, from zone 1 to zone 2), pit latrines may also dilute their concentrations due to the infiltration of groundwater (*e.g.* $Q_{infiltr,2}$ for the infiltration in zone 2) and/or concentrate the particulate compounds because of the percolation rates (for instance, $Q_{exfiltr,2}$ to describe the exfiltration of compounds $S_{FS,exfiltr,2}$, $X_{FS,exfiltr,2}$ and $X_{pathogens,exfiltr,2}$ from zone 2). Gases and inert and non-degradable matter ($S_{FS,U}$ and $X_{FS,U}$) are usually generated, since the anaerobic conversion processes are completed. On the one hand, this leads to the transport and diffusion of gases between zones (*e.g.* $Q_{gas,2\_1}$ and $Q_{gas,3\_2}$ for the gas emissions from zone 2 to the atmosphere and from those of zone 3 to zone 2, respectively). On the other hand, due to inert and non-degradable products from the anaerobic processes remaining in zone 3, this leads to their transport from zone 3 to zone 4 ($S_{FS,U,3\_4}$, $X_{FS,U,3\_4}$) and accumulation at the bottom of the system leading to the creation of an inert zone (zone 4). Similar to the portable toilets, the model structure of the pit latrine has one major input (the sludge feed, $Q_{FS,inf}$) and one major output (the emptying rate, $Q_{FS,emptying}$), but also the infiltration ($Q_{infilt,2}$, $Q_{infilt,3}$, $Q_{infilt,4}$) and exfiltration rates ($Q_{FS,exfiltr,2}$, $Q_{FS,exfiltr,3}$, $Q_{FS,exfiltr,4}$) that may affect each zone to different degrees. These also affect the soil and groundwater quality (due to the exfiltration of the soluble and particulate compounds (*e.g.* the compounds $S_{FS,exfiltr,4}$, $X_{FS,exfiltr,4}$ and $X_{pathogen,exfiltr,4}$ flow from zone 4 into the ground).

Septic tanks

Compared to pit latrines, septic tanks usually receive a combination of faecal sludge and water (domestic wastewater) and are usually divided into two compartments (Figure 6.9). They work in a continuous mode and have long retention times (of years) that, similar to pit latrines, will result in full completion of the conversion processes (mostly the anaerobic ones). This implies that stoichiometric conversion ratios can be sufficient to provide a satisfactory description of the processes that take place in these units. Septic tanks are also prone to infiltration and percolation issues. Therefore, they have well defined inputs ($Q_{FS,inf}$, $Q_{WW,inf}$) and output (Q_{eff}) but are prone to infiltration and percolation flows. Practically all the settleable solids present in the input tend to be retained in the 1st compartment while non-settleable solids flow to the 2nd compartment ($S_{FS,inf}$, $X_{FS,1.1\_2.1}$ and $X_{pathogens,1.1\_2.1}$).

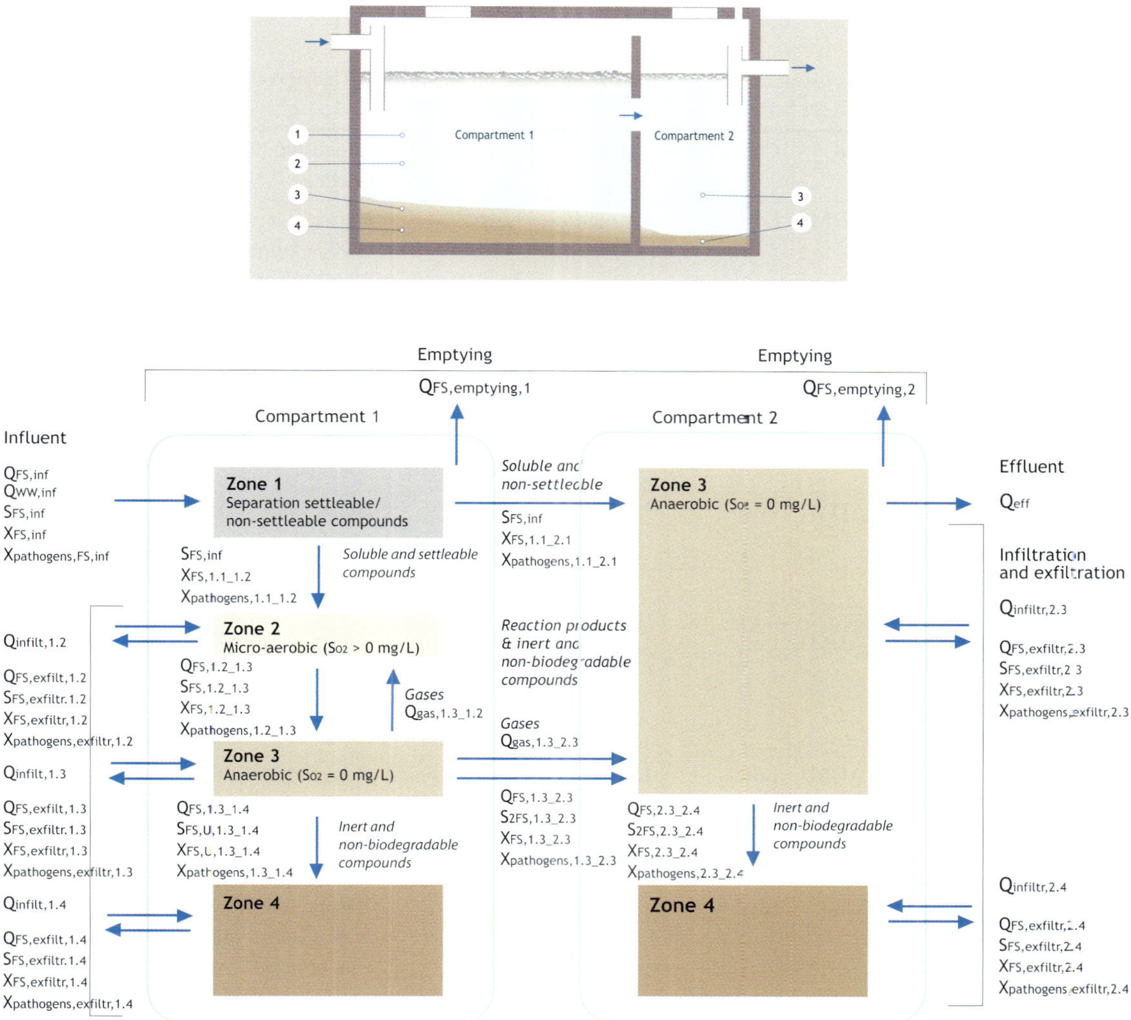

Figure 6.9 Septic tank: suggested model structure.

The settleable solids need to be measured to split the flows between the two compartments. The expected low oxygen diffusion in the 2nd compartment and the split in the flow lead to the existence of four zones in the 1st compartment (similar to those proposed for pit latrines) but only two in the 2nd compartment. In the 1st compartment, most of the processes take place in the settleable solids and soluble components and, in the 2nd compartment, in the non-settleable solids and soluble components. In addition, the 2nd compartment receives the reaction products from zone 3 of the first compartment. Consequently, a higher accumulation of solids can be expected in the 1st compartment ($S_{FS,U,1.3\_1.4}$ and $X_{FS,U,1.3\_1.4}$) than in the 2nd compartment ($S_{FS,U,2.3\_2.4}$ and $X_{FS,U,2.3\_2.4}$).

6.3.4.2 Model structures of other sanitation systems

There are several onsite and sewered sanitation technologies found in sanitation practice to which a similar approach and structure as proposed above can be applied (Table 6.5).

Table 6.5 (Part 1 of 2) Suggestions for faecal sludge modelling of sanitation technologies (list of technologies adopted from Tilley *et al.*, 2014).

Technology	Suggested models	Confidence	Track record	Applicability	Suggested literature for further reading
Portable toilets	ASM+ADM1	Low	Limited	Low	Henze *et al.* (2000), Batstone *et al.* (2000), Lopez-Zavala *et al.* (2004a, 2004b), Elmitwalli *et al.* (2006, 2011, 2013).
Pit latrines (single ventilated improved pit, double ventilated improved pit)	ASM+ADM1	Low	Medium	Low	Henze *et al.* (2000), Batstone *et al.* (2000, 2015), Brouckaert *et al.* (2013), Lopez-Zavala *et al.* (2004a, 2004b), Elmitwalli *et al.* (2006, 2011, 2013).
Septic tank with multiple units	ASM+ADM1	Low	Medium	Low	Henze *et al.* (2000), Batstone *et al.* (2000, 2015), Lopez-Zavala *et al.* (2004a, 2004b), Elmitwalli *et al.* (2006, 2011, 2013).
Fossa alterna	ASM+ADM1	Low	Limited	Low	Henze *et al.* (2000), Batstone *et al.* (2000, 2015), Mata-Alvarez *et al.* (2011), Girault *et al.* (2012),
Twin pits for pour flush	ASM+ADM1	Low	Limited	Low	Henze *et al.* (2000), Batstone *et al.* (2000, 2015), Lopez-Zavala *et al.* (2004a, 2004b), Elmitwalli *et al.* (2006, 2011, 2013).
Anaerobic baffled reactor	ADM1	Medium	Limited	Low	Barber and Stuckey (1999), Batstone *et al.* (2000), Skiadas *et al* (2000), Zhu *et al.* (2015).
Anaerobic filter	ADM1+biofilm model	Medium	Limited	Medium	Batstone *et al.* (2000, 2015), Saravanan and Sreekrishnan (2006), Rittman *et al.* (2018)

Table 6.5 (Part 2 of 2) Suggestions for faecal sludge modelling of sanitation technologies.

Technology	Suggested models	Confidence	Track record	Applicability	Suggested literature for further reading
Imhoff tank	ADM1	Medium	Medium	Medium	Batstone *et al.* (2000, 2015), Donoso-Bravo *et al.* (2011,), Mata-Alvarez *et al.* (2011), Eltmitawili *et al.* (2001, 2011, 2013), Wendland (2009).
Waste stabilisation ponds	ADM1 + ASM + hydraulic models	Medium	Medium	High	Henze *et al.* (2000), Batstone *et al.* (2000), Shilton and Harrison (2003), Alvarado *et al.* (2012), Sah *et al.* (2012).
Aerated pond	ASM	High	Medium	High	Henze *et al.* (2000), Houweling *et al.* (2005, 2008), Alvarado *et al.* (2012), Sah *et al.* (2012).
Wetlands	ADM1	High	Extensive	High	Henze *et al.* (2000), Batstone *et al.* (2000), Langergraber *et al.* (2009), Bridgham *et al.* (2013).
Trickling filter	ASM+biofilm model	Medium	Medium	Medium	Henze *et al.* (2000), Wik (2003), Rittman *et al.* (2018).
Upflow anaerobic sludge blanket system	ADM1	High	Extensive	High	Batstone *et al.* (2000, 2005), Eltmitwalli *et al.* (2001, 2011, 2013), Saravanan and Sreekrishnan (2006), Wendland (2009), Donoso-Bravo *et al.* (2013).
Activated sludge	ASM + ADM1	High	Extensive	High	Henze *et al.* (2000), Lopez-Zavala *et al.* (2004a, 2004b), Lopez-Vazquez *et al.* (2013), Brdjanovic *et al.* (2015).

Providing a thorough modelling approach for each faecal sludge collection and treatment technology falls outside the scope of this chapter. However, because limited experience in faecal sludge modelling at this relatively early stage means that additional studies and data are required, a brief overview is provided that presents different potential models that can be used to model faecal sludge containment/treatment technologies and suggested literature for further reading that can be useful to (start to) develop the required models for these systems. In Table 6.5, the levels of confidence, track record and applicability refer to the reliability of the modelling experiences, the availability of studies and papers in the literature, and the number of case studies and full-scale applications of such models, respectively.

6.3.5 Onsite sanitation modelling: characterisation of flows

Prior to characterising the flows, it is important to define the sizes of the reaction zones. However, this is not a straightforward task because very often they do not have defined physical boundaries. After reviewing the process designs and model structure of the portable toilet, the pit latrine and the septic tank (figures 6.4 to 6.9), it is likely that most of the systems will be anaerobic since the diffusion of dissolved oxygen into the contents of these units will be very low. In soils, wetlands and in particular in peat soils (which may to some extent resemble faecal sludge sanitation systems), oxygen penetration is limited to the first ten centimeters (Ball *et al.*, 1997; Armstrong *et al.*, 2000) and sometimes to even the first centimeter (Sexstone *et al.*, 1985). Thus, methane production and consumption is observed within the first 10-20 cm just below the surface (Dunfield *et al.,* 1993). Moreover, the high organic concentrations observed in wastewater and faecal sludge (higher than 500-1,000 mgCOD/L) (Lopez-Vazquez *et al*., 2013) have a high oxygen demand. Also, the methane generated in the anaerobic zones of the sanitation systems (*e.g.* zone 3 in figures 6.4 to 6.6) may intrude into the aerobic zones and consume oxygen. As a consequence, if oxygen diffusion is not enhanced (*e.g.* by mixing or external aeration) (Stenstroom and Rosso, 2010), it is highly likely that the aerobic zone proposed for the previous sanitation systems (zone 2 in figures 6.4 to 6.6) will be very small (with a thickness of just a few millimetres) or even absent. To define the size of the reaction zones, it is proposed to conduct different measurements of dissolved oxygen and redox potential profiles both vertically and horizontally within the systems. If possible and since the aerobic zone may be very small, the use of microlectrodes (Revsbech and Jørgensen, 1986) is recommended to determine the size or, more specifically, the thickness of the aerobic zone, if any. To determine the size and volumes of the anaerobic and inert zones (zones 3 and 4 in figures 6.7 to 6.9), the collection of undisturbed samples or sludge cores at different heights can be helpful to conduct anaerobic batch activity tests (as well as to assess the microbial population dynamics and sludge characteristics). For this purpose, experience gathered in other fields (*e.g.* groundwater or paleolimnology) (Glew *et al*., 2002) can be very useful to guide the collection of undisturbed and representative sludge core samples at the required heights to carry out the required activity tests and analysis (see Chapter 3). The results of the execution of aerobic and anaerobic activity tests (Van Loosdrecht *et al*., 2016), combined with microbial identification studies (McIlroy *et al*., 2015) and the characterisation of the sludge, will provide valuable information to define the size and volume of each phase and reaction zone, whereas the inert zone will start at the height where activity is minimal or even ceases.

If most of the faecal sludge process conversions are aerobic, efforts can be made to describe the aerobic activity with the application of aerobic models (Lopez-Zavala *et al*., 2004a, 2004b). The rest of the conversion processes will be anaerobic (zone 3 in figures 6.4 to 6.6) or even the whole system will be anaerobic if there are no aerobic zones (as discussed previously). To model the anaerobic conversion processes, the most suitable model is IWA ADM1 (Batstone *et al*., 2002). Since it was launched, this anaerobic model has remained state-of-the-art and, with different extensions and modifications, been successfully applied to several anaerobic conditions and systems (Batstone *et al*., 2006, 2015; Donoso-Bravo *et al*., 2011; Kythreotou *et al*., 2014). Furthermore, ADM1 has already been adapted and applied to model the anaerobic treatment and degradation of faecal sludge, black water and household solid waste in onsite sanitation systems (Wendland, 2008; Elmitwalli *et al*., 2006, 2011, 2013). However, these models have not been calibrated or validated with actual measurements from real sanitation systems. They have been used as tools to foresee and explore potential process performance and process improvements for system selection either deriving input and operational data from previous studies or from lab-scale systems. This indicates that information and experience available to model real faecal sludge systems are still limited. Furthermore, there are key structural bottlenecks related to the required ADM1 fractionation and the fractionation of faecal sludge that need to be carefully addressed, as will be discussed in later sections of this chapter.

Once the zones are known, the flows between each zone can be characterised following the recommendations given in Section 6.3.3 on data collection and verification. However, in addition to the well-known solid-liquid and gas-liquid transport mechanisms, in faecal sludge systems it will also be necessary to assess the transport of pathogens and gases in porous media. While the solid-liquid and gas-liquid transport and diffusion phenomena can be assumed to be well understood and defined based on the knowledge gathered from wastewater treatment systems (Brdjanovic *et al*., 2015), the transport of pathogens through the different zones of faecal sludge systems needs to be well-defined in order to understand and be able to describe their potential spatial distribution in faecal sludge systems. Previous reports describing the spatial distribution of pathogens in sanitation systems and past studies conducted on the transport of pathogens through porous media can be useful in this regard (Foppen *et al.*, 2007a, 2007b, 2010). Once again, an appropriate collection of sludge cores (Glew *et al.*, 2002) and the use of advanced molecular identification methods (Karst *et al.*, 2016 can provide a useful overview to understand the physical distribution of viruses, pathogens, and other organisms of relevance in onsite and sewered sanitation systems.

The transport of solids and of the products of the reactions, such as inert compounds also need to be defined as a function of the rheology of faecal sludge and the process conversion processes such as solids degradation and the generation of inert and non-degradable products. Studies on soil mechanics and peat soils can be used for this purpose. Equally important is to study the transport and/or diffusion of gases (*e.g.* methane, carbon dioxide, hydrogen sulphide) into the different zones and layers (a solid-gas transport phenomena). This is mostly because the presence or accumulation of some of these gases (*e.g.* carbon dioxide and hydrogen sulphide) will affect the potential inactivation of pathogens in a direct or indirect manner (*e.g.* carbon dioxide by affecting the pH and hydrogen sulphide through a direct inhibition or toxic effect). Research already conducted on the transport and diffusion of gases in soils, peat soils and wetlands would support this future research (Armstrong *et al.*, 2000; Aachib *et al.*, 2002, 2004; Allaire *et al.*, 2008). Understanding the transport and spatial distribution of pathogens and the generation and transport of key gases through the layers and zones of faecal sludge systems can contribute to studying potential strategies to enhance the inactivation of pathogens.

6.3.6 Onsite sanitation modelling: calibration and validation

For calibration and validation purposes, the same recommendations that apply to ASM can be followed. If the description of the performance shows that a major adjustment is needed (*e.g.* major adjustments of the kinetic parameters), the model structure and also the mass balances and data collection probably need to be revised. Sludge accumulation is the first aspect to be calibrated, followed by the most kinetically sensitive process (possibly hydrolysis or fermentation) and the rest of the kinetic processes. If the process performance and the quality of the generated flows have not been well predicted, uncertainty and sensitivity analysis can be conducted to assess which parameters have the strongest impact. Following an iterative step-wise process, the model is calibrated by adjusting the least possible number of kinetic parameters until it provides a satisfactory description of the performance of the system.

The model can be validated by assessing its capacity to predict the performance of the system using operational and environmental data from a different period than that used for the model calibration. It will need to be re-calibrated iteratively if it fails the validation step until a satisfactory validation is reached.

6.3.7 Onsite sanitation modelling: detailed characterisation

6.3.7.1 Faecal sludge characterisation and fractionation

As previously discussed, the use of dissolved oxygen meters, redox probes and microelectrodes (Sexstone *et al.*, 1985) in vertical and horizontal directions, in combination with the conduction of aerobic and anaerobic experimental methods (Van Loosdrecht *et al.*, 2016) using undisturbed core samples from sanitation systems and the characterisation at different heights of relevance, will be necessary in order to determine the extension and size of the aerobic and anaerobic zones. Once they are known, the faecal sludge needs be characterised and, more importantly for modelling purposes, it needs to be fractionated into the COD fractions of relevance. The fractionations required for aerobic models (Henze *et al.*, 2000) and anaerobic models (Batstone *et al.*, 2002) are different, yet to a certain extent similar from a biodegradability perspective (Ekama *et al.*, 2007).

In view of the limited experience and information available concerning the fractionation of faecal sludge, further research needs to focus on the determination of the required fractions through the execution of experimental methods and, whenever possible, supported by elemental composition analysis following a structured and common protocol. Furthermore, it will be very important to carry out a characterisation and fractionation campaign in different countries and regions to reach a consensus regarding the most suitable and practical steps. This will be extremely useful to develop a suitable protocol for faecal sludge characterisation and fractionation similar to those developed in the past decades for activated sludge modelling (*e.g.* the BIOMATH, HSG, WERF, and STOWA calibration protocols) (Vanrolleghem *et al.*, 2003; Langergraber *et al.*, 2004; Melcer *et al.*, 2003; Hulsbeek *et al.*, 2002; Roeleveld *et al.*, 2002).

To model the aerobic degradation of faecal sludge, a COD fractionation similar to that carried out by Lopez-Zavala *et al.* (2002, 2004a, 2004b) can be conducted using real faecal sludge. To determine the required aerobic kinetic parameters, a combination of respirometric tests (Ekama *et al.*, 1986; Kappeler and Gujer, 1992; Spanjers and Vanrolleghem, 1995; Vanrolleghem *et al.*, 1999) and activity tests can be executed (Van Loosdrecht *et al.*, 2016). The information provided by these studies will contribute to obtaining a better estimation of the aerobic COD fractionation of faecal sludge and of the hydrolysis and degradation of faecal sludge under aerobic conditions.

Since most faecal sludge collection and treatment systems are anaerobic, the determination of the faecal sludge anaerobic fractions deserves special attention in order to apply ADM1. However, although ADM1 can be recommended as the most suitable model for faecal sludge modelling, there are two major interrelated challenges for its application in this field. First is the thorough fractionation of the feeding components required by ADM1, and second, as expected, is the rather limited research and information regarding the anaerobic fractionation of faecal sludge. Thorough ADM1 fractionation necessitates the determination of the (individual) compound concentrations (using specific analytical techniques) of soluble (S) components such as sugars, aminoacids, long-chain and fatty acids, as well as those of particulate (X) components such as composites, carbohydrates, proteins and lipids. Most of these parameters can be determined following the analytical methods described in Chapter 8. For modelling implementation, determination of large numbers of individual compounds is a serious disadvantage (Kleerebezem and Van Loosdrecht, 2006a, 2006b). It is a major structural bottleneck that has been observed in reviews of the implementation of the ADM1 model (Batstone *et al.*, 2015). To overcome this bottleneck, certain approaches have been proposed: (*i*) to lump together the elemental composition of organic substrates using a limited number of widely available analyses (Kleerebezem and Van Loosdrecht, 2006b), (*ii*) to perform experimental methods to determine the anaerobic degradation kinetics needed to split the COD of a substrate into the input variables required by ADM1 (Girault *et al.*, 2012; Poggio *et al.*, 2016); and, when coupling aerobic models (*e.g.* ASM) with ADM1 for plant-wide modelling, (*iii*) to use interfaces to convert the aerobic fractionation of ASM models into the anaerobic fractionation of anaerobic models (Volcke *et al.*, 2006; Nopens *et al.*, 2009; Flores-Alsina *et al.*, 2016). In previous efforts regarding faecal sludge modelling when ADM1 was applied, Wendland (2008) carried out a direct fractionation using specific analytical techniques. However, Elmitwalli *et al.* (2011) derived the required faecal sludge fractions from previous characterisation studies where the fractions were not directly determined (Elmitwalli *et al.*, 2001; Kujawa-Roeleveld *et al.*, 2003). To overcome these gaps, a suggestion is to carry out 'anaerobic' respirometric tests (Holliger *et al.*, 2016) following a similar procedure such as that conducted by Girault *et al.* (2012) but using fresh faecal sludge. This approach will allow the faecal sludge anaerobic fractions and the hydrolysis kinetic rates required for the implementation of ADM1 to be determined. For this purpose, anaerobic respirometric tests need to be executed at different faecal sludge to anaerobic inoculum ratios. Ideally, anaerobic inoculum from real faecal sludge systems can be used but it could also be tested from different anaerobic sludge digesters (in particular, from anaerobic digesters treating primary sludge which tends to resemble faecal sludge). In parallel, the determination of proteins, lipids and carbohydrates in faecal sludge based on standard analytical techniques (Rice *et al.*, 2017) and basic procedures (Kleerebezem and Van Loosdrecht, 2006b; Girault *et al.*, 2012) can be used to support and validate the outcomes of the fractionation results. The results of the faecal sludge fractionation and its impact on faecal sludge systems modelling can be assessed by applying it to a real case or performing long-term SMA and BMP tests (Van Loosdrecht *et al.*, 2016). The conduction of SMA and BMP tests (Holliger *et al.*, 2016) will be useful to estimate the kinetic parameters of interest (hydrolysis, fermentation or acidification, acetogenesis and methanogenesis). However, the prediction of the anaerobic conversion processes will determine whether the conduction of continuous experiments is preferable, in particular to determine the faecal sludge hydrolysis kinetic rates (Batstone *et al.*, 2009; Garcia-Gen *et al.*, 2015).

According to Belia *et al.* (2009) and Nopens *et al.* (2014), there are four major locations of uncertainty that can severely affect the satisfactory calibration and validation of a model. They can be grouped as: (*i*) the inputs, (*ii*) the model, (*iii*) the model parameters and, (*iv*) technical or software aspects affecting the model. With regard to the inputs, it is important to characterise and fractionate the faecal sludge characteristics as accurately as possible and to provide a satisfactory description of the tanks and volumes. However, a major source of uncertainty is the variable generation of faecal sludge volumes, as pointed out by Brouckaert *et al.* (2013). The structure of the model and potential interfaces are another important source of uncertainty. The third group of uncertanties includes the feed model and hydraulics, and determining where the different aerobic or anaerobic zones exist, as they influence the need to use either an aerobic or an anaerobic model (ASM vs ADM1, respectively) and the interfaces required to couple the models. The last source of uncertainty is the one driven by software limitations (such as solver or numerical problems that interfere with a correct execution of the simulations). Overall, the first three sources of uncertainty can start to be analysed following the framework described in section 6.3.4.1, whereas the last one depends on the simulator or software used. In order to evaluate the uncertainty, different methods can be applied (*i*) to characterise and prioritise uncertainty by evaluating the quality of the data collected, expert elicitation, parameter estimation and sensitivity analysis, (*ii*) to increase the quality of the information by quality assurance, extended peer review and also involving the stakeholders and direct users, and (*iii*) to quantify and propagate uncertainty in the outcomes of a model (*e.g.* through the application of Gaussian error propagation, Monte Carlo simulation, among others). A detailed discussion of these methods and approaches goes beyond the scope of this chapter. Nevertheless, specialised publications on these topics can provide enough information and knowledge for their implementation to model onsite and sewered sanitation systems (*e.g.* Von Sperling *et al.*, 2020).

6.3.7.2 Inhibition and toxicity

Due to the stratification and predominance of certain processes over others (such as hydrolysis and fermentation over methanogenesis due to the differences in the growth of the microbial groups) (Van Lier *et al.*, 2008; Pratt *et al.*, 2012), the potential accumulation of ammonium and of (volatile) fatty acids with its associated drop in pH will probably lead to the inhibition of methanogenesis (Colon *et al.*, 2015). ADM1 has inhibition functions to describe the potential inhibition caused by these compounds (Batstone *et al.*, 2002). They will need to be assessed, validated and, if required, adjusted when treating and dealing with faecal sludge.

During the anaerobic degradation of organics, there is a potential risk that sulphate-reduction processes take place as a consequence of the human diet (Florin *et al.*, 1993) or intrusion of water rich in sulphates (such as seawater in faecal sludge units located close to the coastline) (Van den Brand, 2015). Consequently, anaerobic sulphate conversion processes may lead to the generation of hydrogen sulphide (H_2S) which can inhibit methanogenesis both directly (since H_2S can be toxic to methanogens and other organisms) and indirectly (due to the consumption of organics outcompeting methanogens) (Van Lier *et al.*, 2008). Sulphate-reduction processes were not included in the original ADM1 (Batstone *et al.*, 2002), but different extensions have since been developed and included (Kalyuzhnyi *et al.*, 1998; Fedorovich *et al.*, 2001; Barrera *et al.*, 2013, 2015). A similar approach can be adopted to assess and describe the potential occurrence of sulphate-reduction processes in faecal sludge systems.

Moreover, the potential toxicity caused by cleaning and sanitising solutions used in toilets, external additives or other toxic compounds (such as motor oil, batteries or solvents) also needs to be taken into consideration. For this purpose, the protocol developed by Astals *et al.* (2015) to rapidly assess any potential inhibition or toxicity effect could be adapted and tested on faecal sludge.

6.3.7.3 Pathogen inactivation

The main objective of sanitation is the assurance of basic and safe public health. As such, the safe disposal of faecal sludge and the potential inactivation of pathogens is of major importance and deserves special attention. Different authors have studied and developed expressions to describe the inactivation of pathogens in different systems (see Table 6.6). However, to date, such expressions have been only marginally incorporated into mathematical models to describe the inactivation of pathogens in faecal sludge collection and treatment systems.

pH

pH has a major influence on the inactivation of pathogens. Extreme pH levels, either low (<4.0) or high (>9.0), result in satisfactory pathogen inactivation rates (Anderson *et al.*, 2015). Mendonca *et al.* (1994) described how the pathogen inactivation observed at higher pH levels may be associated with the lysis of cells due to the disruption of the cytoplasmic membrane. Meanwhile, Russell (1992) proposes that, if the organisms cannot adjust their intracellular pH (which usually lies between a pH range of 6.0 to 8.0), at lower pH levels the accumulation of anions is responsible for the toxic effect of fermentation acids (*e.g.* acetic, propionic or butyric acids).

In treatment systems, pH is severely affected by the presence of acid-based systems and strong ions (Fairlamb *et al.*, 2003). As such, the biological and physicochemical processes occurring in faecal sludge collection systems (or promoted by external factors such as co-digestion) (Riungu *et al.*, 2018a, 2018b) or the addition of additives (Anderson *et al.*, 2015; Riungu *et al.*, 2018b) may lead to extreme pH levels that can enhance pathogen inactivation. For instance, the accumulation of acids (often interlinked to or influenced by a higher temperature) also led to a drop in pH during the (co-)treatment of faecal sludge in the studies carried out by Riungu *et al.* (2018a, 2018b). Overall, the decay rate of *E. Coli* reached up to 1.6 1/d with an accumulation of up to 16.3 g VFA/L at a pH of 4.9, whereas in a similar study (Riungu *et al.*, 2018b), concentrations of non-dissociated VFA of up to 6500 mg/L led to a full inactivation of *E. Coli* and *Ascaris Lumbricoides*. Bina *et al.* (2004) investigated the removal of faecal coliforms, Salmonella and helminth eggs using lime treatment at pH 11 and pH 12. In the Philippines (Strande *et al.*, 2014), disinfection was achieved after 30 min at pH 12, after 60 min at pH 11.5 and after 120 min at pH 11.

Magri *et al.* (2015) assessed the effects of pH in combination with concentrations of ammonia on the inactivation of adenovirus, reovirus and bacteriophagues in faecal sludge. They observed that bacteriophagues were more resistant than viruses. If the pH was higher than 8.9 and the concentrations of NH_3 reached 35 and 55 mM, the maximum time for a 3-log reduction was 35 days and 21 days at 23 °C and 28 °C, respectively. The expressions used to describe the inactivation processes were obtained by fitting the inactivation data to either an exponential decay or a lag-phase decay equation, respectively, as follows:

$$N = N_o \cdot 10^{-k \cdot t} \tag{6.1}$$

$$N = N_o \left[1 - (1 - 10^{-k \cdot t})\right]^{10n} \tag{6.2}$$

Where:

N is the final counting of bacteriophagues or viruses,

N_o is the initial counting of bacteriophagues or viruses,

k is the inactivation rate ($k = 1/t_{90}$),
t is the period of time,
t_{90} is the decimal reduction time, and
n is a parameter fitted in the regression that determines the lag phase.

It is important to underline that Magri *et al.* (2015) observed that if the biodegradable organics present in faecal sludge were hydrolysed and fermented to VFA, the pH decreased from 8.7 to 7.7. This affected the nitrogen speciation, reducing the concentration of NH_3 and consequently decreasing the inactivation effect of this compound. This indicates that if the inactivation effect of either high pH and ammonia or low pH and VFA is desirable, the hydrolysis and the fermentation processes need to be uncoupled otherwise they may counteract the inactivation effect between each other. As such, pH is a key factor that can be used and potentially enhanced (by exploring alternatives to adjusting the operating and environmental conditions through mathematical modelling) to maximise pathogen inactivation in faecal sludge systems. Interestingly, ADM1 (Batstone *et al.*, 2002) has the required expressions to estimate the pH under anaerobic conditions and also has different expressions to take into account the inhibition of methanogens at different pH levels and with different VFA concentrations. Such expressions can be expanded to consider the inactivation effect of other parameters (such as ammonia) and also the addition of external additives (such as other acids, urea or lime) to provide a better pH estimation. After the addition of a state variable to describe the outcome of certain defined pathogens, together with their required pH inactivation rates, the estimation of the pH can then be used to assess the inactivation of pathogens.

Temperature

Temperature has been reported to be an important factor for pathogen inactivation (Watanabe *et al.*, 1997). However, a thermophilic temperature range is needed (55-65 °C) for an effective inactivation (Polprasert *et al.*, 1983; Mills *et al.*, 1992a; Watanabe *et al.*, 1997). Koottatep *et al.* (2014) observed, in septic tanks operated at higher temperatures, a 3-log reduction in E. *Coli* at 50 °C and even a 6-log reduction to a level of about 10 most probable number (MPN)/100 mL at 60 °C. Their results were described with the modified Weibull expression:

$$\log \frac{N_t}{N_o} = -b_T \cdot t^n \qquad (6.3)$$

Where:
N_t is the number of microbial populations at any time,
N_o is the number of microbial populations at the initial time,
t is the contact time and b_T is a temperature coefficient, and
n is the Weibull coefficient.

In Equation 6.3, the b_T values of 1.36 and 1.71, and n of 0.26 and 0.41 were used to describe the inactivation rates at 50 and 60 °C, respectively.

In another study, Lübken *et al.* (2007) described the inactivation of pathogens with a multiple regression expression in an onsite anaerobic system used for faecal sludge treatment. For intestinal enterococci removal, the following multiple regression term was proposed:

$$n_{IE} = 98.29 - 2.2\left(\frac{1}{HRT}\right)^2 + (0.031 \cdot T) \qquad (6.4)$$

Whereas to describe the inactivation of faecal coliforms the following expression was drawn:

$$n_{IE} = 98.29 - \left(\frac{1}{HRT}\right)^2 + (0.031 \cdot T) \qquad (6.5)$$

In equations 6.4 and 6.5:
HRT corresponds to the hydraulic retention time (in days) and T to temperature, °C.

Similar to the study of Koottatep *et al.* (2014) who performed different studies in septic tanks at diverse temperatures, this study showed then considerable inactivation rates were only observed at a thermophilic temperature (55 °C) and HRT longer than approximately 5 days. However, such a high temperature range cannot be easily generated in, or provided to, most sanitation systems. It is generally those systems that enhance the composting process (such as Fossa Septica), that are directly exposed to sunlight (such as WSP) or engineered systems (such as digesters) are able to reach the required thermophilic temperature range that can lead to pathogen inactivation.

Fidjeland *et al.* (2015) modelled the inactivation of *Ascaris* eggs at different temperatures and high ammonia concentrations. For a given number of $\log_{10}$ reduction in viability (LRV), they estimate that the treatment time required to inactivate *Ascaris* eggs can be described with the following expression:

$$t = \frac{1.14 \cdot (3.2 + LRV)}{\left(10^{-3.7 + 0.062\,T}\right) \cdot NH_{3\ \ Pitzer}^{0.7}} \qquad (6.6)$$

In the previous expression,

T is the temperature, and

$NH_{3,Pitzer}$ is the activity of the ammonia ion following the Pitzer method which makes use of the software PHREEQ.

A simplified method to estimate $NH_{3,Pitzer}$ is presented by Fidjeland *et al.* (2015) using a simplified Emerson-Pitzer conversion. This conversion makes Eq. 6.6 valid and applicable under some typical conditions found in real conditions (e.g. 8.3-9.5 pH, dry matter content up to 20%, NH_{TOT} between 5 and 2,000 mM, and for temperatures between 5 and 45 °C). Similar to the description of pathogen inactivation by pH, certain expressions can be incorporated into ADM1 to describe the fate of certain defined pathogens at different temperatures.

Ammonia

Other studies have also focused on the inactivation possibilities of ammonia either present in the faecal sludge itself or after the addition of urea. Ammonia efficiently inactivates bacteria at pH levels between 9.0 and 9.5. It enters the cell membrane, increasing the internal ammonia concentration and causing the bacterial cell to pump out protons to maintain its optimum cellular pH, eventually resulting in cell death (Bujozek, 2001; Hill *et al.*, 2013). Previous studies report a reduction in numbers of organisms, including non-spore forming bacteria, viruses and parasites through urea additions to manure and faecal sludge (Nordin *et al.*, 2009; Magri *et al.*, 2015). Fidjeland *et al.* (2013) hypothesises that the intrinsic ammonia present in urine has the potential to sanitise faecal sludge if the urine is concentrated and not lost by ventilation. They observed the inactivation of *Enterococcus faecalis*, *Salmonella typhimurium* and *Ascaris suum* eggs by ammonia between 5 and 28 °C at ammonia concentrations ranging from 40 to 400 mM. *Salmonella* was fully inactivated after 2 days whereas *Enterococcus* reached a 5-log reduction between 13 and 110 days as the ammonia concentration increased from 19 to 243 mg NH_3/L. At 23-28 °C, a 3-log reduction in *Ascaris* eggs was observed within 1 to 6 months depending on the ammonia concentration as described by the Eq. 6.6 (Fidjeland *et al.*, 2015).

When ammonia is limited, the addition of urea and its subsequent hydrolysis to ammonia can lead to extreme pH levels and create a sanitising effect in combination with cell alkalisation by the ammonia released from the hydrolysis process (Fitzmorris *et al.*, 2007; Anderson *et al.*, 2015). Vinnerås *et al.* (2013) observed that, after the addition of 3% urea, *Salmonella spp.* and faecal coliforms were not detected after 5 days, *Enterococcus spp.* after 20 days, and viruses as well as viable *Ascaris* eggs were not detected after 50 days. ADM1 contains different expressions that describe the ammonia concentrations released from the hydrolysis processes of organics (Batstone *et al.*, 2002). Moreover, by making use of the pH, the species of ammonium and ammonia can be calculated and with the help of inhibition expressions their effect on the anaerobic digestion process is taken into consideration due to their damaging effect on methanogenesis. Bearing this approach in mind, the ammonia concentrations can be estimated with the use of existing ADM1 expressions and they can be coupled to the inactivation expressions previously presented to describe the inactivation of different pathogens present in faecal sludge systems.

Lactic acid

Lactic acid bacteria (LAB) have the ability to convert carbohydrates to lactic acid (Gujer *et al.*, 1986; Anderson *et al.*, 2015). Lactic acid can penetrate the cytoplasmic membrane in the associated form, resulting in a reduced intracellular pH and disruption of the trans-membrane proton motive force (Herrero *et al.*, 1985). Also, lactic acid reduces the bulk pH of the surrounding medium, influencing the activity of exo-enzymes and membrane-bound enzymes. Ligocka *et al.* (2005) observed that *Salmonella spp.* and *E. coli* in sewage sludge were inhibited under both anaerobic and aerobic conditions with lactic acid. Soewondo *et al.* (2014), conducting laboratory experiments on faeces, observed a log reduction in total coliforms of log 4 to 7.5 after enhancing the lacto-fermentation process. Zhu *et al.* (2006) reported that in addition to reducing the pH in the bulk liquid, the key antimicrobial property of lactic acid is its ability to reduce the intracellular pH of bacteria. Anderson *et al.* (2015) satisfactorily inactivated *E. Coli* using lactic acid after the addition of sugars and inoculums of LAB in 7 days. Although LAB need to be inoculated in faecal sludge systems, the fermentation and production of lactic acid can be relatively easily introduced to ADM1 following a similar approach to the one used for other carboxylic acids (*e.g.* VFA) (Nielsen *et al.*, 1991a, 1991b; Mercier *et al.*, 1992; Spann *et al.*, 2018) and for the description of pathogen inactivation in onsite sanitation systems.

Other pathogen inactivation equations

There is a vast amount of literature and research describing the inactivation of pathogens. However, the expressions that can be extrapolated and incorporated into mathematical models of onsite and sewered sanitation systems can be narrowed down to only those that contain, or are a function of, environmental and operating conditions that can be found or developed in these systems. As such, only those expressions that are a function of or dependent on the pH, temperature, dissolved oxygen concentration, and organic load are worth testing to describe the inactivation of pathogens. Practically all these equations are empirical and drawn based on laboratory, pilot or full-scale studies. Furthermore, some of the expressions are dependent and functions of different parameters depending upon the regression method or approach followed. Consequently, most of them have been developed following a 'black-box' approach without considering the actual (biochemical and physiological) inactivation mechanisms. Table 6.6 provides an overview of such pathogen inactivation expressions that could be incorporated into ADM1. For the description of the parameters in the equations, the reader is referred to the original source.

Table 6.6 Possible pathogen inactivation expressions developed for different wastewater treatment systems that could be incorporated into ADM1 to describe the inactivation of pathogens.

Pathogen removal expressions	Comment/remark	Reference
$e^{k_b} = 0.6351 \cdot (1.0281)^{T_w} \cdot (1.0016)^{C_a} \cdot (0.9994)^{BOD}$	Modified dispersion model expression applied to full-scale municipal WWTP in Brazil	Polprasert *et al.* (1983)
$K_b = K_{b,T} + K_{b,pH} + K_{b,BOD} + K_I$	Dispersion model equation applied to a pilot-scale municipal WWTP in Austria	Qin *et al.* (1991)
$K_b = 0.712 \cdot 1.166^{(T-20)}$	Completely mixed model equation applied to municipal plants in Kenya	Mills *et al.* (1992
$K_b = 0.5(1.02)^{Tw-20} \cdot 1.15^{(pH-6)^2} \cdot (0.9978)^{(BOD_5-100)}$	Plug-flow model expression applied to aerobic ponds in Jordan	Saqqar and Pescod (1992)
For coliforms: $K_b = 1.359 \cdot (1.087)^{(Tw-20)}$ For coliphages: $K_b = 0.439 \cdot (1.044)^{(Tw-20)}$	Plug-flow model equation applied to facultative ponds in Chile	Herrera and Castillo (2000)
$K_b = 0.019 \cdot (0.915)^{(Tw-20)} e^{0.171_m}$	Dispersion model applied to municipal WWTPs in France	Xu *et al.* (2002)
$K_b = K_{b,20} + K_{b,pH} \cdot pH + K_{b,DO} \cdot DO + K_{b,I} \cdot I) \cdot \theta^{(T-20)}$	Plug flow model equation applied to a laboratory-scale system in Belgium	Ouali *et al.* (2014)

6.3.8 Modelling applications, benefits and challenges

Depending upon the purpose, the application of models is meaningful during the entire lifecycle of the sanitation technology, including the design, construction, operation, and evaluation stages. Similar to wastewater treatment practice, there is a spectrum of possible use of models during the lifecycle of the onsite sanitation technology as shown in Figure 6.10. The wastewater treatment practice revealed that the most cost saving is possible when models are used in the early stage of the WWTP lifecycle, and similar expectations could be applicable to onsite sanitation technology as well. As discussed earlier in this chapter, the modelling goal determines the type and complexity of the model to be applied.

Figure 6.10 also depicts how the modelling complexity increases as the lifecycle of the sanitation technology progresses. The biggest savings are possible at the technology design phase because modelling helps to quantify scenarios at an early design stage. The quantification helps to speed up the decision-making process. High levels of uncertainty in the early design phases (*e.g.* due to faecal sludge composition) implies that large safety margins are needed (usual 150 to 200%), as such models can be simple (no calibration needed) and, during the design phase to invest in models and modelling work regularly pays back. Furthermore, practice shows that the highest financial risks are at the operational stage and modelling helps to reduce these operational risks (operational problems are often complex and more accurate models are required, *e.g.* ASM models in the case of WWTPs).

Reasons to introduce models in faecal sludge management at institutional level are: (*i*) to standardise the operation and management, control and quality assurance, (*ii*) to improve efficiency and reduce costs, (*iii*) to generate a knowledge base (organise process documentation), (*iv*) to improve internal and external communication (standardisation of information), and (*v*) to facilitate planning and decision making, etc.

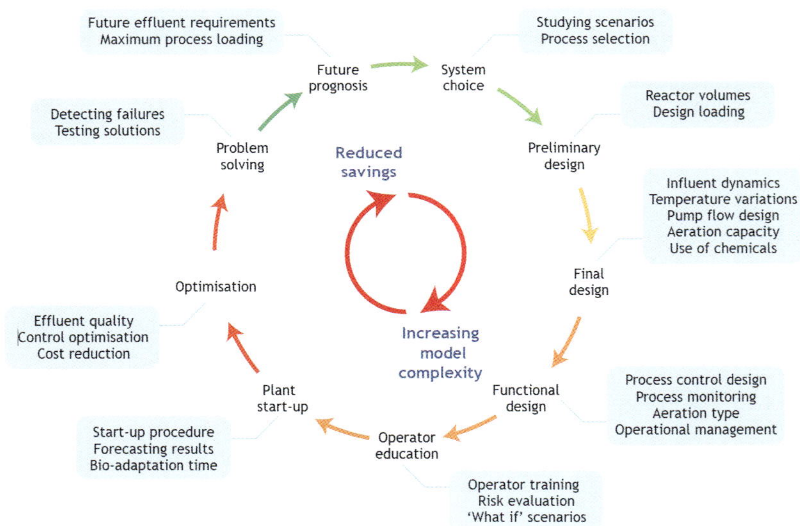

Figure 6.10 Modelling application at different stages of the sanitation technology lifecycle (adopted from Meijer and Brdjanovic, 2012).

Reasons for sanitation professionals to use models are: (*i*) to improve their work by better understanding the design and the process, (*ii*) to undertake regular training to update their skills and knowledge and introduce new and state-of-the-art technologies and approaches, (*iii*) to create a low-cost and safe platform for testing new ideas for improved operations and design, and (*iv*) to provide more efficient and improved decision-making and communication tools.

Success factors for using modelling in design can be summarised as: (*i*) following a protocol, providing realistic project planning and a practical approach, (*ii*) giving a 'bird's eye' view of the modelling project, (*iii*) defining clear modelling goals, (*iv*) keeping the model as simple as possible, and (*v*) using a standard calibration method.

However, one should be aware that by modelling, several bottlenecks may be identified such as: (*i*) choice of methods and software is important - a standardised approach is required, (*ii*) a different approach towards sanitation information systems is often needed, (*iii*) there is a continuous need to invest in education (life-long learning), and (*iv*) sharing of knowledge through a modelling platform, meetings, internet fora, and specialist groups. However, modelling practice from sewered sanitation shows that in general the use of models saves money, improves the quality of investments, is effective for management and decision making, and is an important asset for sanitation practice. Finally, the use of models in faecal sludge management is expected to have several main advantages such as: (*i*) cost reduction (especially at the design phase), (*ii*) improved management and quality control, (*iii*) optimal technological/process design using modern tools, (*iv*) the application of innovative approaches, and (*v*) the development of designs at low cost, rapidly and with confidence (Meijer and Brdjanovic, 2012).

6.4 OUTLOOK

Overall, at a micro-scale level (individual units), modelling of onsite sanitation systems can help to increase understanding about the conversions that take place in these units, contributing to improved design and operation of the onsite (and also indirectly, sewered) sanitation systems. This can be achieved by, firstly being able to describe the performance of the sanitation units to satisfactorily predict the quantity and quality of the faecal sludge generated and, secondly, based on these aspects, estimate adequate emptying and disposal practices as a function of the faecal sludge volumes and their characteristics. At a micro-scale level in onsite sanitation systems, modelling can also help to improve the design of the systems as well as their operation, enhancing, for example, the inactivation of pathogens (due to public health concerns) and increasing the generation of desired by-products (such as biogas or nutrient recovery as fertilisers).

With an increasing interest in the recovery of resources, mathematical modelling of faecal sludge might also be used as a tool to assess and develop innovative (biotechnological) practices and applications for the recovery of valuable or revalorised resources (*e.g.* methane, biodiesel, bioplastics, and nutrient-rich products) in a similar way to how it is being done in the wastewater treatment sector (Van Loosdrecht and Brdjanovic, 2014). This may be possible because the original 'raw material' (*i.e.* human excreta) is practically the same.

Moreover, by mapping and determining the type and number of sanitation systems that prevail in a region or area (in addition to the expected volumes and characteristics of the faecal sludge generated in each onsite sanitation system in accordance with the modelling studies), it is possible to estimate the overall and average faecal sludge characteristics and volumes generated in that specific region or area and to define and suggest the most appropriate practices and technologies for emptying, transporting, (co-)treating and disposing of faecal sludge. Better emptying practices and improved faecal sludge transportation to centralised plants can contribute to improving the handling of faecal sludge volumes and ultimately to achieving the goal of a CWIS approach. With a better knowledge regarding the number and types of faecal sludge systems available in a given location and considering their typical or average operating and environmental conditions, the most appropriate faecal sludge treatment technologies or practices can be selected. For instance, faecal sludge with a high biodegradable organic content can be further treated under anaerobic conditions for biogas production whereas septic sludge with a low biodegradable organic content may only need to be dewatered or dehydrated prior to safe disposal. This also requires the development of mathematical models to describe the dewatering and dehydration of faecal sludge. Also, the faecal sludge modelling aspects and considerations described in this chapter can also be applied (see Table 6.5) to improve the required and selected faecal sludge (co-)treatment process.

This chapter primarily addresses approaches to modelling of onsite faecal and septic sludge containment and treatment technologies by making maximum use of the extensive knowledge gained during more than a century of research on wastewater/sewage treatment and more than three decades of experience of using biological wastewater and sludge treatment modelling. This analogy is possible and logical because of the fact that in both cases urine and faeces are the main raw materials that enter into the sanitation system, be it sewered or onsite, and that the combination of physical, chemical and biological processes is an essential component in the treatment in both cases. As the two sectors are presently rather polarised, such an extension enables further integration of sewered and onsite sanitation technologies at a system level, which is an essential step towards a city-wide inclusive sanitation approach. Therefore, this chapter focuses on the development of approaches on how to model the selection of the most common sanitation technologies for faecal sludge containment and onsite treatment, recognising the fact that this area of interest has the most complexity yet the least understanding of all the components of the urban sanitation chain.

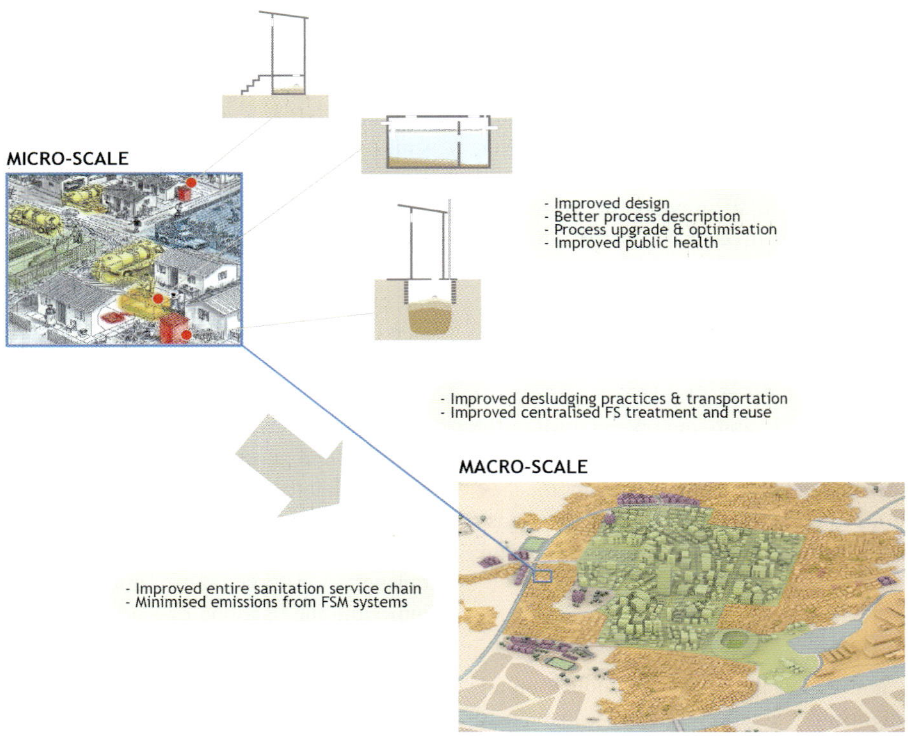

Figure 6.11 The micro- and macro-scale impact of modelling onsite sanitation systems (images adopted from Eawag).

Needless to say, this is just the tip of the iceberg concerning the modelling of FSM, whereas on the wastewater side the modelling of urban drainage and sewerage (Hvitved-Jacobsen *et al.*, 1998) and urban flood modelling (Price 2011) have advanced to the stage that can be combined with WWTP modelling and modelling of receiving waters (Hodzic *et al.*, 2011, Brdjanovic *et al.*, 2015) into an integrated urban sanitation model. Such an integrated model can be further extended towards a true CWIS model by the inclusion of the Q&Q model (Chapter 5) and combination faecal/septic sludge containment and treatment model as proposed in this chapter. Furthermore the onsite part of the model can be extended by collection and transport models (Anh *et al.*, 2018) and models for onsite centralised treatment technologies (Strande *et al.*, 2014). These models can be further integrated into a single holistic model at a city level where the challenge will be how to make all the necessary interfaces between different models so that models can properly communicate with each other. It is expected that such an integrated model will become available in the coming decade and that the first models will represent a steady state situation (*e.g.* seasonal or yearly average at the city level) and with further applications and developments, especially on the onsite sanitation side, a new generation of dynamic, real-time models will appear. However, even at this stage such an integrated dynamic model will not fully represent a CWIS model. For that it is necessary to include various business models (Strande *et al*, 2014) as well as the knowledge and application of behaviour models and citizen observatory approaches (Dreibelbis *et al.*, 2013; Fritz *et al.*, 2019) which can be seen as an attempt to extend CWIS modelling to Community and City Wide Inclusive Sanitation (CCWIS) modelling. These inclusions will increase the complexity of a CCWIS model; perhaps it will be necessary to create a simpler user interface that integrates more complex models working in the background with only some of the most essential features available to a regular (non-professional modeller) user. As the fundamental knowledge and number of models will continue to expand in the future it is to be expected that a new market for specialised modelling 'vendors' will be created and more complex modelling tasks will be outsourced to CCWIS modelling specialists.

REFERENCES

Aachib M., Aubertin M. and Mbonimpa M. (2002) Laboratory measurements and predictive equations for gas diffusion coefficient of unsaturated soils. 55th *Canadian Geotechnical Conference and 3rd joint IAH-CNC and CGS Groundwater Specialty Conference*. Niagara Falls, Ontario. 2002, 163–172.

Aachib M., Mbonimpa M. and Aubertin M. (2004). Measurement and prediction of the oxygen diffusion coefficient in unsaturated media, with applications to soil covers. *Water, Air and Soil Pollution*, 156(1–4), 163–193.

Alex J., Benedetti L., Copp K., Gernaey U., Jeppsson U., Nopens I., Pons M., Steyer J. and Vanrolleghem P. (2008). Benchmark Simulation Model No.1 (BSM1) Lund University.

Allaire S.E., Lafond J.A., Cabral A.R. and Lange S.F. (2008). Measurement of gas diffusion through soils: comparison of laboratory methods. *Journal of Environmental Monitoring*, 10(11), 1326.

Alvarado A., Sanchez E., Durazno, G., Vesvikar M. and Nopens I. (2012). CFD analysis of sludge accumulation and hydraulic performance of a waste stabilisation pond. *Water Science and Technology*, 66(11), 2370-2377.

Anderson C., Malambo D.H., Perez M.E.G., Nobela H.N., De Pooter L., Spit J., Hooijmans C.M., Van de Vossenberg J., Greya W., Thole B., Van Lier J.B. and Brdjanovic D. (2015). Lactic acid fermentation, urea and lime addition: Promising faecal sludge sanitising methods for emergency sanitation. *International Journal of Environmental Research and Public Health*, 12(11), 13871–13885.

Anh T.H., Koottatep T. and Polprasert C. (2018). Business model analysis for faecal sludge collection and transport services in Thailand and Vietnam. *Journal of Water, Sanitation and Hygiene for Development*, 8(3), 556-567

Armstrong W., Cousins D., Armstrong J., Turner D.W. and Beckett P.M. (2000) Oxygen distribution in wetland plant roots and permeability barriers to gas-exchange with the rhizosphere: A microelectrode and modelling study with Phragmites australis. *Annals of Botany*, 86(3), 687–703.

Astals S., Batstone D.J., Tait S. and Jensen P.D. (2015). Development and validation of a rapid test for anaerobic inhibition and toxicity. *Water Research*, 81, 208-215.

Barker P. and Dold P. (1997). General Model for Biological Nutrient Removal Activated-Sludge Systems: Model Application. *Water Environment Research*, 69(5), 985-991.

Bakare B.F., Foxon K.M., Brouckaert C.J. and Buckley C.A. (2012). Variation in VIP latrine sludge contents. *Water SA*, 38(4), 479–486.

Bakare, B.F., Foxon, K.M., Brouckaert, C.J. and Buckley, C. A. (2012). Variation in VIP latrine sludge contents. *Water*

SA, 38(4), 479–486.

Ball B.C., Doobie K.E., Parker J.P. and Smith K.A. (1997). The inlfuence of gas transport and porosity on methane oxidation in soils. *Journal of Geophysical Research*.

Barrera E.L., Spanjers H., Dewulf J., Romero O. and Rosa, E. (2013). The sulfur chain in biogas production from sulfate-rich liquid substrates: a review on dynamic modeling with vinasse as model substrate. *Journal of Chemical Technology & Biotechnology*, 88(8), 1405-1420.

Barrera E.L., Spanjers H., Solon K., Amerlinck Y., Nopens I. and Dewulf J. (2015). Modeling the anaerobic digestion of cane-molasses vinasse: extension of the Anaerobic Digestion Model No. 1 (ADM1) with sulfate reduction for a very high strength and sulfate rich wastewater. *Water Research*, 71, 42-54.

Batstone D.J., Keller J., Angelidaku I., Kalyuzhnyio S.V., Pavlostathis S.G., Rozzi A. and Vavilin V.A. (2002). Anaerobic Digestion Model No. 1. Scientific and Technical Report No. 13. IWA Task Group for Mathematical Modelling of Anaerobic Digestion Processes, (1), 89.

Batstone D.J., Tait S. and Starrenburg D. (2009). Estimation of hydrolysis parameters in full-scale anerobic digesters. *Biotechnology and Bioengineering*, 102(5), 1513-1520

Batstone D.J. and Virdis B. (2014). The role of anaerobic digestion in the emerging energy economy. *Current Opinion in Biotechnology*, 27, 142–149.

Batstone D.J., Hulsen T., Mehta C.M. and Keller J. (2015). Platforms for energy and nutrient recovery from domestic wastewater: A review. *Chemosphere*, 140, 2–11.

Belia E., Amerlinck Y., Benedetti L., Johnson B., Sin G., Vanrolleghem P.A., Gernaey K.V., Gillot S., Neumann M.B., Rieger L., Shaw A. and Villez K. (2009). Wastewater treatment modelling: dealing with uncertainties. *Water Science and Technology*, 60(8), 1929-1941.

Benedetti L., De Baets B., Nopens I. and Vanrolleghem P.A. (2010). Multi-criteria analysis of wastewater treatment plant design and control scenarios under uncertainty. *Environmental Modelling & Software*, 25(5), 616-621.

Bhagwan J.N., Still D., Buckley C.A. and Foxon K. (2008). Challenges with up-scaling dry sanitation technologies. *Water Science and Technology*, 58(1), 21–27.

Bina B., Movahedia H. and Kord I. (2004). The effect of lime stabilisation on the microbiological quality of sewage sludge. *Journal of Environmental Health Science & Engineering*, 1(1), 38-42.

Brdjanovic D., Van Loosdrecht M.C.M., Versteeg P., Hooijmans C.M., Alaerts G.J. and Heijnen J.J. (2000). Modeling COD, N and P removal in a full-scale WWTP Haarlem Waarderpolder. *Water Research*, 34(3), 846-858.

Brdjanovic D., Zakaria F., Mawioo P.M., Garcia H.A., Hooijmans C.M., Curko, J. and Setiadi, T. (2014). eSOS®– emergency sanitation operation system. *Journal of Water, Sanitation and Hygiene for Development*, 5(1), 156-164.

Brdjanovic D., Meijer S., Lopez-Vazquez C., Hooijmans C.M. and Van Loosdrecht M. C. M. (2015). Applications of Activated Sludge Models, IWA Publishing, London, UK.

Bridgham S.D., Cadillo-Quiroz H., Keller J.K. and Zhuang Q. (2013). Methane emissions from wetlands: biogeochemical, microbial, and modeling perspectives from local to global scales. *Global change biology*, 19(5), 1325-1346

Brouckaert C.J., Foxon K.M. and Wood K. (2013). Modelling the filling rate of pit latrines. *Water SA*, 39(4), 555-562.

Bujozek G. (2001). Influence of ammonia and other abiotic factors on microbial activity and pathogen inactivation of high-solid residues. University of Manitoba.

Chen G.H., Van Loosdrecht M.C.M., Ekama G.A. and Brdjanovic D. (2020). Biologial Wastewater Treatment: 2nd Edition. IWA Publishing, London, UK.

Dold P.L., Ekama G.A. and Marais G.v.R. (1981). A General Model for the Activated Sludge Process. *In Water Pollution Research and Development*.

Dold P.L., Jones R.M., Takacs I. and Melcer H. (2003). Practical guidance for WWTP model calibration and associated data gathering requirements. Proceedings of the *Water Environment Federation*, 2003(12), 202-224.

Donoso-Bravo A., Mailier J., Martin C., Rodríguez J., Aceves-Lara C.A. and Wouwer A.V. (2011). Model selection, identification and validation in anaerobic digestion: a review. *Water Research*, 45(17), 5347-5364

Dreibelbis R., Winch P. J., Leontsini E., Hulland K.R., Ram P. K., Unicomb L. and Luby S.P. (2013). The integrated behavioural model for water, sanitation and hygiene: a systematic review of behavioural models and a framework for designing and evaluating behaviour change interventions in infrastructure-restricted settings. *BMC Public Health*, 13(1), 1015.

Dunfield P., Dumont R. and Moore T.R. (1993). Methane production and consumption in temperate and subarctic peat soils: response to temperature and pH. *Soil Biology and Biochemistry*, 25(3), 321-326.

Ekama G.A., Dold P.L. and Marais G.v.R. (1986). Procedures for determining influent COD fractions and the maximum specific growth rate of heterotrophs in activated sludge systems. *Water Science and Technology*, 18(6), 91-114.

Ekama G.A., Sötemann S.W. and Wentzel M.C. (2007). Biodegradability of activated sludge organics under anaerobic conditions. *Water Research*, 41(1), 244-252.

Elmitwalli T.A., Van Leeuwen M., Kujawa-Roeleveld K., Sanders W. and Zeeman G. (2006). Anaerobic biodegradability and digestion in accumulation systems for concentrated black water and kitchen organic-wastes. *Water Science and Technology*, 53(8), 167–176.

Elmitwalli T., Zeeman G. and Otterpohl R. (2011). Modelling anaerobic digestion of concentrated black water and faecal matter in accumulation system. *Water Science and Technology*, 63(9), 2039–2045.

Elmitwalli T. (2013). Sludge accumulation and conversion to methane in a septic tank treating domestic wastewater or black water. *Water Science and Technology*, 68(4), 956–964.

Fairlamb M., Jones R., Takács I. and Bye C. (2003). Formulation of a general model for simulation of pH in wastewater treatment processes. Proceedings of the *Water Environment Federation*, 2003(7), 511-528

Fedorovich V., Lens P. and Kalyuzhnyi S. (2003). Extension of Enaerobic Digestion Model No. 1 with processes of sulfate reduction. *Applied biochemistry and biotechnology*, 109(1-3), 33-45.

Fidjeland J., Magri M.E., Jönsson H., Albihn A. and Vinnerås B. (2013). The potential for self-sanitisation of faecal sludge by intrinsic ammonia. *Water Research*, 47(16), 6014–6023.

Fidjeland J., Nordin A., Pecson B.M., Nelson K.L. and Vinnerås B. (2015). Modeling the inactivation of Ascaris eggs as a function of ammonia concentration and temperature. *Water Research*, 83, 153-160.

Fitzmorris K.B., Reimers R.S., Oleszkiewicz J.A. and Little M.D. (2007). Decrease of time for pathogen inactivation in alkaline disinfection systems using pressure. *Water Environment Research*, 79(4), 388-395

Flores-Alsina X., Corominas L., Neumann M.B. and Vanrolleghem P. A. (2012). Assessing the use of activated sludge process design guidelines in wastewater treatment plant projects: A methodology based on global sensitivity analysis. *Environmental Modelling & Software*, 38, 50-58.

Flores-Alsina X., Solon K., Mbamba C.K., Tait S., Gernaey K. V., Jeppsson U. and Batstone D.J. (2016). Modelling phosphorus (P), sulfur (S) and iron (Fe) interactions for dynamic simulations of anaerobic digestion processes. *Water Research*, 95, 370-382.

Florin T.H., Neale G., Goretski S. and Cummings J.H. (1993). The sulfate content of foods and beverages. *Journal of Food Composition and Analysis*, 6(2), 140-151.

Foppen J.W. (2002). Impact of high-strength wastewater infiltration on groundwater quality and drinking water supply: The case of Sana'a, Yemen. *Journal of Hydrology*, 263(1–4), 198–216.

Foppen J.W., Van Herwerden M. and Schijven J. (2007a). Measuring and modelling straining of Escherichia coli in saturated porous media. *Journal of Contaminant Hydrology*, 93(1-4), 236-254.

Foppen J.W., Van Herwerden M. and Schijven J. (2007b). Transport of Escherichia coli in saturated porous media: Dual mode deposition and intra-population heterogeneity. *Water Research*, 41(8), 1743-1753

Foppen J.W., Lutterodt G., Röling W.F. and Uhlenbrook S. (2010). Towards understanding inter-strain attachment variations of Escherichia coli during transport in saturated quartz sand. *Water Research*, 44(4), 1202-1212.

Forster C.F. (2002). The rheological and physico-chemical characteristics of sewage sludges. *Enzyme and Microbial Technology*, 30(3), 340–345.

Fritz S., See L., Carlson T., Haklay M.M., Oliver J.L., Fraisl D. and Wehn U. (2019). Citizen science and the United Nations Sustainable Development Goals. *Nature Sustainability*, 2(10), 922-930.

García-Gen S., Sousbie P., Rangaraj G., Lema J.M., Rodríguez J., Steyer J.P. and Torrijos M. (2015). Kinetic modelling of anaerobic hydrolysis of solid wastes, including disintegration processes. *Waste Management*, 35, 96-104.

Gernaey K.V, Van Loosdrecht M.C.M., Henze M., Lind M. and Jørgensen S.B. (2004). Activated sludge wastewater treatment plant modelling and simulation: state of the art. *Environmental Modelling and Software*, 19(9), 763–783.

Girault R., Bridoux G., Nauleau F., Poullain C., Buffet J., Steyer J.P. and Béline F. (2012). A waste characterisation procedure for ADM1 implementation based on degradation kinetics. *Water Research*, 46(13), 4099-4110.

Glew J.R., Smol J.P. and Last W.M. (2002). Sediment core collection and extrusion. In Tracking environmental change using lake sediments (pp. 73-105) *Springer*, Dordrecht

Gujer W. and Zehnder A.J.B. (1986). Conversion processes in anaerobic digestion. *Water Science and Technology*, 15(8/9), 127–167.

Gujer W., Henze M., Mino T., Matsuo T., Wentzel M.C. and Marais, G.v.R. (1995). The activated sludge model No. 2: biological phosphorus removal. *Water Science and Technology*, 31(2), 1-11.

Gujer W., Henze M., Mino T. and Van Loosdrecht M.C.M. (1999). Activated sludge model no. 3. *Water Science and Technology*, 39 (1), 183-193.

Halalsheh M.M., Noaimat H., Yazajeen H., Cuello J., Freitas B. and Fayyad M. (2011). Biodegradation and seasonal variations in septage characteristics. *Environmental Monitoring and Assessment*, 172(1–4), 419–426.

Henze M., Grady Jr C.L., Gujer W., Marais G.v.R. and Matsuo T. (1987). A general model for single-sludge wastewater treatment systems. *Water Research*, 21(5), 505-515.

Henze M., Gujer, W., Mino T., Matsuo T., Wentzel M.C., Marais G.v.R. and Van Loosdrecht M.C.M. (1999). Activated sludge model no. 2d, ASM2d. *Water Science and Technology*, 39(1), 165-182.

Henze M., Gujer W., Mino T. and Van Loosdrecht M.C.M. (2000). Activated sludge models ASM1, ASM2, ASM2d and ASM3. Padstow, Cornwall UK: IWA publishing.

Henze M., Van Loosdrecht M.C.M., Ekama G.A. and Brdjanovic D. (2008). Biological Wastewater Treatment- Principles, Modelling and Design, 1st ed. IWA Publishing, London, UK.

Herrera L. and Castillo G. (2000). Dynamics of a waste stabilisation pond in Chile. *Water Science and Technology*, 42, 83-90.

Herrero A., Gomez R., Snedecor B., Tolman C. and Roberts M. (1985). Growth inhibition of Clostridium thermocellum by carboxylic acids: A mechanism based on uncoupling by weak acids. *Applied Microbiology and Biotechnology*, 22(1), 53–62.

Hill G.B., Baldwin S.A. and Vinnerås B. (2013). Composting toilets a misnomer: Excessive ammonia from urine inhibits microbial activity yet is insufficient in sanitising the end-product. *Journal of Environmental Management*, 119, 29–35.

Hodzic A. Vojinovic Z. Seyoum S.D., Pathirana A., Drnda N., Meijer S.C.F. and Brdjanovic D. (2015). Chapter 10: Use of modeling for cost-effective reconstruction of urban wastewater infrastructure. Applications of Activated Sludge Models. IWA publishing.

Holliger C., Alves M., Andrade D., Angelidaki I., Astals S., Baier U. and Ebertseder F. (2016). Towards a standardisation of biomethane potential tests. *Water Science and Technology*, 74(11), 2515-2522.

Houweling C.D., Kharoune L., Escalas A. and Comeau Y. (2005). Modeling ammonia removal in aerated facultative lagoons. *Water Science and Technology*, 51(12), 139-142.

Houweling C.D., Kharoune L., Escalas A. and Comeau Y. (2008). Dynamic modelling of nitrification in an aerated facultative lagoon. *Water Research*, 42(1-2), 424-432.

Hulsbeek J.J.W., Kruit J., Roeleveld P.J. and Van Loosdrecht M.C.M. (2002). A practical protocol for dynamic modelling of activated sludge systems. *Water Science and Technology*, 45(6), 127–136.

Hvitved-Jacobsen T., Vollertsen J. and Nielsen P.H. (1998). A process and model concept for microbial wastewater transformations in gravity sewers. *Water Science and Technology*, 37(1), 233–241.

Jabłoński S., Rodowicz P. and Łukaszewicz M. (2015). Methanogenic archaea database containing physiological and biochemical characteristics. *International Journal of Systematic and Evolutionary Microbiology,* 65, 1360–1368.

Jenkins D. and Wanner J. (2014). Activated Sludge – 100 Years and Counting, IWA Publishing, London, UK.

Kalyuzhnyi, S., Fedorovich, V., Lens, P., Pol, L.H. and Lettinga, G. (1998). Mathematical modelling as a tool to study population dynamics between sulfate reducing and methanogenic bacteria. *Biodegradation*, 9(3-4), 187-199.

Kappeler J. and Gujer W. (1992). Estimation of kinetic parameters of heterotrophic biomass under aerobic conditions and characterisation of wastewater for activated sludge modelling. *Water Science and Technology*, 25(6), 125-139

Karst S.M., Albertse M., Kirkegaard R.H., Dueholm M.S. and Nielsen P.H. (2016). Chapter 8: Molecular Methods in:Experimental Methods In Wastewater Treatment. Edited by M.C.M. Van Loosdrecht, P.H. Nielsen, C.M. Lopez-Vazquez and D. Brdjanovic. ISBN: 9781780404745 (hardback), ISBN: 9781780404752 (eBook). Published by IWA Publishing, London, UK.

Kleerebezem R. and Van Loosdrecht M.C.M. (2006a). Critical analysis of some concepts proposed in ADM1. *Water Science and Technology*, 54(4), 51–57.

Kleerebezem R. and Van Loosdrecht M.C.M. (2006b). Waste characterisation for implementation in ADM1. *Water Science and Technology*, 54(4), 167–174.

Koottatep T., Phuphisith S., Pussayanavin T., Panuvatvanich A. and Polprasert C. (2014). Modeling of pathogen inactivation in thermal septic tanks. Journal of *Water, Sanitation and Hygiene for Development*, 4(1), 81.

Kuba T., Murnleitner E., Van Loosdrecht M.C.M. and Heijnen J.J. (1996). A metabolic model for biological phosphorus removal by denitrifying organisms. *Biotechnology and Bioengineering*, 52(6), 685–695.

Kujawa-Roeleveld K., Elmitwalli T., Gaillard A., Van Leeuwen M. and Zeeman G. (2003). Co-digestion of concentrated black water and kitchen refuse in an accumulation system within the DESAR (decentralised sanitation and reuse) concept. *Water Science and Technology*, 48(4), 121-128.

Kythreotou N., Florides G. and Tassou S.A. (2014). A review of simple to scientific models for anaerobic digestion. *Renewable Energy*, 71, 701-714.

Langergraber G., Rousseau D.P., García, J. and Mena J. (2009). CWM1: a general model to describe biokinetic processes in subsurface flow constructed wetlands. *Water Science and Technology*, 59(9), 1687-1697.

Ligocka A. and Paluszak Z. (2005). Capability of lactic acid bacteria to inhibit pathogens in sewage sludge subjected to biotechnological processes. *Bulletin of the Veterinary Institute of Pulawy,* 49, 23–27.

Liu J., Yu D., Zhang J., Yang M., Wang Y., Wei Y. and Tong J. (2016). Rheological properties of sewage sludge during enhanced anaerobic digestion with microwave-H2O2 pretreatment. *Water Research,* 98, 98–108.

Lopez-Vazquez C.M., Dangol B., Hoijmans C.M. and Brdjanovic D., (2013). Co-treatment of Faecal Sludge in Municipal Wastewater Treatment plants. Faecal Sludge Management , Chapter 9. IWA publishing.

Lübken M., Wichern M., Bischof F., Prechtl S. and Horn H. (2007a). Development of an empirical mathematical model for describing and optimising the hygiene potential of a thermophilic anaerobic bioreactor treating faeces. *Water Science and Technology*, 55(7), 95–102.

Lübken M., Wichern M., Letsiou I., Kehl O., Bischof F. and Horn H. (2007b). Thermophilic anaerobic digestion in compact systems: investigations by modern microbiological techniques and mathematical simulation. *Water Science and Technology*, 56(10), 19-28.

Lugali Y., Zziwa A., Banadda N., Wanyama J., Kabenge I., Kambugu R. and Tumutegyereize P. (2016). Modeling sludge accumulation rates in lined pit latrines in slum areas of Kampala City, Uganda. African Journal of *Environmental Science and Technology*, 10(8), 253-262.

Magri M.E., Fidjeland J., Jönsson H., Albihn A. and Vinnerås B. (2015). Inactivation of adenovirus, reovirus and bacteriophages in fecal sludge by pH and ammonia. *Science of the Total Environment,* 520, 213–221.

Mata-Alvarez J., Dosta J., Macé S. and Astals S. (2011). Codigestion of solid wastes: a review of its uses and perspectives including modeling. *Critical reviews in biotechnology*, 31(2), 99-111.

McIlroy S.J., Saunders A.M., Albertsen M., Nierychlo M., McIlroy B., Hansen A.A., Karst S.M., Nielsen J.L. and Nielsen P.H. (2015). MiDAS: The field guide to the microbes of activated sludge. Database 2015 (2), 1–8.

Meijer S.C.F. (2004). Theoretical and practical aspects of modelling activated sludge processes. PhD dissertation. Delft, University of Technology

Meijer S.C.F. and Brdjanovic D. (2012). A practical guide to activated sludge modeling. UNESCO-IHE Institute for Water Education.

Mendonca A.F., Amoroso T.L. and Knabel S.J. (1994). Destruction of gram-negative food-borne pathogens by high pH involves disruption of the cytoplasmic membrane. *Applied Enviromental Microbiology* , 60(11), 4009-4014.

Mensah P.Y., Kuffour R.A., Baidoo P.K. and Awuah E. (2013). The effect of different percentages of bulking agent (sawdust) on microbial quality of faecal sludge. *Water Science and Technology,* 67(8), 1728–1733.

Mercier P., Yerushalmi L., Rouleau D. and Dochain D. (1992). Kinetics of lactic acid fermentation on glucose and corn by Lactobacillus amylophilus. *Journal of Chemical Technology & Biotechnology*, 55(2), 111-121.

Metcalf and Eddy, Tchobanoglous G., Stensel H., Tsuchihashi R. and Burton F. (2014). Wastewater Engineering: Treatment and Resource Recovery, 5th ed. McGraw-Hill Professional.

Mills S.W., Alabaster G.P., Mara D.D., Pearson H.W. and Thitai W.N. (1992). Efficiency of Fecal Bacterial Removal in Waste Stabilisation Ponds in Kenya. *Water Science and Technology*, 26, 1739-1748.

Mithaiwala M. (2005). Application of Activated Sludge Model (ASM3) and Anaerobic Digestion Model (ADM1) on Wastewater Treatment Plant Anjana In Surat-India.MSc Thesis SE.05.01 UNESCO-IHE.

Murnleitner E., Kuba T., Van Loosdrecht M.C.M. and Heijnen J.J. (1997). An integrated metabolic model for the aerobic and denitrifying biological phosphorus removal. *Biotechnology and Bioengineering*, 54(5),434-450

Narayan A.S., Fischer M. and Lüthi C. (2019). Social Network Analysis for Water, Sanitation and Hygiene (WASH): Application in governance of decentralised wastewater treatment in India using a novel validation methodology. *Frontiers in Environmental Science*, 7, 198.

Nielsen J., Nikolajsen K. and Villadsen J. (1991a). Structured modeling of a microbial system: I. A theoretical study of lactic acid fermentation. *Biotechnology and bioengineering*, 38(1), 1-10.

Nielsen J., Nikolajsen K. and Villadsen J. (1991b). Structured modeling of a microbial system: II. Experimental verification of a structured lactic acid fermentation model. *Biotechnology and Bioengineering*, 38(1), 11-23.

Nopens I., Batstone D.J., Copp J.B., Jeppsson U., Volcke E., Alex J. and Vanrolleghem P.A. (2009). An ASM/ADM model interface for dynamic plant-wide simulation. *Water Research*, 43(7), 1913-1923.

Nopens I., Torfs E., Ducoste J., Vanrolleghem P.A. and Gernaey K.V. (2014). Population balance models: a useful complementary modelling framework for future WWTP modelling. *Water Science and Technology*, 71(2), 159-167.

Nordin A., Ottoson J.R. and Vinnerås B. (2009). Sanitation of faeces from source-separating dry toilets using urea. *Journal of Applied Microbiology*, 107(5), 1579–1587.

Nwaneri C., Foxon K., Bakare B. and Buckley C.A. (2008). Biological degradation processes within a pit latrine. WISA 2008 Conference.

Ouali A., Jupsin H., Ghrabi A. and Vasel J.L. (2014). Removal kinetic of Escherichia coli and enterococci in a laboratory pilot scale wastewater maturation pond. *Water Science and Technology*, 69, 755-759.

Peal A.J., Evans B.E., Sangaralingam A. Ban R., Blackett I., Hawkins I.B., Schoebitz L., Scott L., Sleigh A., Strande L. and Roda O.V., (2020). Estimating safely managed sanitation in urban areas; lessons learned from a global implementation of excreta-flow diagrams. *Frontier Enviromental Science.- Water and Wastewater management*

Peterson E., Prentice H. and Cliffs E. (1965). Chemical reaction analysis, *AIChE Journal*.

Poggio D., Walker M., Nimmo W., Ma L. and Pourkashanian M. (2016). Modelling the anaerobic digestion of solid organic waste–Substrate characterisation method for ADM1 using a combined biochemical and kinetic parameter estimation approach. *Waste Management*, 53, 40-54.

Polprasert C., Dissanayake M.G. and Thanh N.C. (1983). Bacterial die-off kinetics in waste stabilisation ponds. *Water Pollution Control Federation*, 55, 285-296.

Pratt S., Liew D., Batstone D.J., Werker A. G., Morgan-Sagastume F. and Lant P.A. (2012). Inhibition by fatty acids during fermentation of pre-treated waste activated sludge. *Journal of Biotechnology*, 159(1-2), 38-43.

Price R.K. (2011). Urban hydroinformatics: data, models and decision support for integrated urban water management. IWA Publishing, London. UK.

Qin D., Bliss P.J., Barnes D. and FitzGerald P.A. (1991). Bacterial (total coliform) die-off in maturation ponds. *Water Science and Technology*, 23(7-9), 1525-1534.

Revsbech N.P. and Jørgensen B.B. (1986). Microelectrodes: their use in microbial ecology. In *Advances in Microbial Ecology*, Springer, Boston, MA, US.

Rice E.W., Baird R.B., Eaton A.D. and Clesceri L.S. (eds.) (2017). Standard methods for the examination of water and wastewater, 23rd edition. *APHA, AWWA, and WPCR*.

Rieger L., Koch G., Kühni M., Gujer W. and Siegrist H. (2001). The EAWAG Bio-P module for activated sludge model No. 3 *Water Research*, 35(16), 3887–3903.

Rieger L., Gillot S., Langergraber G., Ohtsuki T., Shaw A., Takcs I. and Winkler S. (2012). Guidelines for Using Activated Sludge Models, IWA Publishing, London, UK.

Rittmann B.E. (2018). Biofilms, active substrata and me. *Water Research*, 132, 135-145.

Riungu J., Ronteltap M. and Van Lier J.B. (2018a). Build-up and impact of volatile fatty acids on *E. coli* and *A. lumbricoides* during co-digestion of urine diverting dehydrating toilet (UDDT-F) faeces. *Journal of Environmental Management*, 215, 22-31.

Riungu J., Ronteltap M. and Van Lier J. B. (2018b). Volatile fatty acids build-up and its effect on *E. coli* inactivation during excreta stabilisation in single-stage and two-stage systems. *Journal of Water, Sanitation and Hygiene for Development*, 8(2), 257-267

Rodríguez J., Kleerebezem R., Lema J.M. and Van Loosdrecht M.C.M. (2006). Modelling product formation in anaerobic mixed culture fermentations. *Biotechnology and Bioengineering*, 93(3), 592–606.

Roeleveld P.J. and Van Loosdrecht M.C.M. (2002). Experience with guidelines for wastewater characterisation in The Netherlands *Water Science and Technology*, 45(6), 77–87.

Rose C., Parker A., Jefferson B. and Cartmell E. (2015). The Characterisation of Faeces and Urine: A Review of the Literature to Inform Advanced Treatment Technology. *Critical Reviews in Environmental Science and Technology*, 45 (17), 1827–1879.

Rosen C., Ingildsen P., Guildal T., Nielsen T.M., Nielsen M.K., Jacobsen B. N. and Thomsen H.A. (2006). Introducing biological phosphorus removal in an alternating plant by means of control: a full scale study. *Water Science and Technology*, 53(4–5), 133.

Rubio-Rincón F.J., Lopez-Vazquez C.M., Ronteltap M. and Brdjanovic D. (2014). Seawater for phosphorus recovery from urine. *Desalination*, 348, 49–56.

Russell J.B. (1992). Another explanation for the toxicity of fermentation acids at low pH: anion accumulation versus uncoupling. Journal of *applied bacteriology*, 73(5), 363-370.

Sah L., Rousseau D.P. and Hooijmans C.M. (2012). Numerical modelling of waste stabilisation ponds: where do we stand?. *Water, Air, & Soil Pollution*, 223(6), 3155-3171.

Saravanan V. and Sreekrishnan T.R. (2006). Modelling anaerobic biofilm reactors- A review. *Journal of Environmental Management*, 81(1), 1-18.

Saqqar M.M. and Pescod M.B. (1992). Modeling Coliform Reduction in Wastewater Stabilisation Ponds. *Water Science and Technology*, 26, 1667-1677.

Sexstone A.J., Revsbech N.P., Parkin T.B. and Tiedje J.M. (1985). Direct measurement of oxygen profiles and denitrification rates in soil aggregates. *Soil Science Society of America Journal*, 49, 645-651.

Shilton A. and Harrison J. (2003). Development of guidelines for improved hydraulic design of waste stabilisation ponds. *Water Science and Technology*, 48(2), 173-180.

Sin G., Van Hulle S., De Pauw D., Van Griensven A. and Vanrolleghem P. (2005). A critical comparison of systematic calibration protocols for activated sludge models: *A SWOT analysis. Water Research*, 39(12), 2459–2474.

Sin G, Gernaey K.V., Neumann M.B., Van Loosdrecht M.C.M. and Gujer W. (2011). Global sensitivity analysis in wastewater treatment plant model applications: prioritising sources of uncertainty. *Water Research*, 45(2), 639-651.

Skiadas I., Gavala H.N. and Lyberatos G. (2000). Modelling of the periodic anaerobic baffled reactor (PABR) based on the retaining factor concept. *Water Research*, 34(15), 3725-3736.

Smolders G.J.F., Van Loosdrecht M.C.M. and Heijnen J.J. (1995). A metabolic model for the biological phosphorus removal process. *Water Science and Technology*, 31(2), 79–93.

Soewondo P., Febriana A., Handajani M. and Firdayati M. (2014). Faeces Treatment by lactic fermentation process and future perspectives of Terra Preta sanitation concept in Indonesia. *Terra Preta Sanitation*. Deutsche Bundesstiftung Umwelt, Hamburg, Germany.

Spanjers H. and Vanrolleghem P. (1995). Respirometry as a tool for rapid characterisation of wastewater and activated sludge. *Water Science and Technology*, 31(2), 105-114.

Spann R., Lantz A.E., Roca C., Gernaey K.V. and Sin G. (2018). Model-based process development for a continuous lactic acid bacteria fermentation. *Computer Aided Chemical Engineering,* 43, 1601-1606.

Stenstrom M.K. and Rosso D. (2017). Aireación y Mezclado. Tratamiento biológico de aguas residuales, 273

Strande L., Ronteltap M. and Brdjanovic D. (2014). Faecal Sludge Management – Systems Approach for Implementation and Operation. IWA Publishing, London, UK.

Strande L., Schoebitz L., Bischoff F., Ddiba D., Okello F., Englund M. and Niwagaba C.B. (2018). Methods to reliably estimate faecal sludge quantities and qualities for the design of treatment technologies and management solutions. *Journal of Environmental Management*, 223, 898-907.

Tilley E., Ulrich L., Luthi C., Reymond P. and Zurbrügg C. (2014). Compendium of Sanitation Systems and Technologies, IWA Publishing, London, UK.

Todman L.C., Van Eekert M.H.A., Templeton M.R., Hardy (née Kelly) M., Gibson W.T., Torondel B., Abdelahi F. and Ensink J.H.J. (2015). Modelling the fill rate of pit latrines in Ifakara, Tanzania. *Journal of Water, Sanitation and Hygiene for Development,* 5(1), 100–106.

Ubisi M. F., Jood T. W., Wentzel M.C., Ekama G.A. (1997). Activated sludge mixed liquor heterotrophic active biomass. *Water SA*, 23, 239-248.

Udert K.M., Larsen T.A. and Gujer W. (2003). Estimating the precipitation potential in urine-collecting systems. *Water Research,* 37(11), 71–78.

United Nations (2015). Transforming our world: the 2030 Agenda for Sustainable Development. General Assembly. Seventieth session, agenda items 15 and 116. 21 October 2015.

Vanrolleghem P.A., Spanjers H., Petersen B., Ginestet P. and Takacs I. (1999). Estimating (combinations of) Activated Sludge Model No. 1 parameters and components by respirometry. *Water Science and Technology*, 39(1), 195-214.

Vanrolleghem P., Insel G., Petersen B., Pauw D., Nopens I., Dovermann H., Weijers S. and Gernaey K. (2003). A Comprehensive Model Calibration Procedure for Activated Sludge Models. *Proceedings of the Water Environment Federation*, WEFTEC 2003 210–237.

Van den Brand T.P.H., Roest K., Chen G.H., Brdjanovic D. and Van Loosdrecht M.C.M. (2015). Occurrence and activity of sulphate reducing bacteria in aerobic activated sludge systems. *World Journal of Microbiology and Biotechnology*, 31(3), 507–516.

Van Lier J.B, Mahmoud N. and Zeeman G. (2008). Chapter 16: Anaerobic Wastewater Treatment. Biological Wastewater Treatment-Principles, Modelling and Design, 1st ed. IWA publishing.

Van Loosdrecht M.C.M. and Brdjanovic D. (2014). Anticipating the next century of wastewater treatment. *Science*, 344(6191), 1452-1453.

Van Loosdrecht M.C.M., Nielsen P.H., Lopez-Vazquez C.M. and Brdjanovic D. (eds.) (2016). Experimental methods in wastewater treatment. IWA Publishing, London, UK.

Van Veldhuizen H., Van Loosdrecht M.C.M. and Heijnen, J.J. (1999). Modelling biological phosphorus and nitrogen removal in a full scale activated sludge process. *Water Research*, 33(16), 3459-3468.

Vinnerås B., Holmqvist A., Bagge E., Albihn A. and Jönsson H. (2003). The potential for disinfection of separated faecal matter by urea and by peracetic acid for hygienic nutrient recycling. *Bioresource Technology*, 89,155–161.

Volcke E.I.P., Van Loosdrecht M.C.M. and Vanrolleghem P.A. (2006). Continuity-based model interfacing for plant-wide simulation: A general approach. *Water Research,* 40(15), 2817–2828.

Von Speriling M., Verbyla M.E. and Oliveira S.M.A.C. (2020). Assessment of Treatment Plant Performance and Water Quality Data: A Guide for Students, Researchers and Practitioners. IWA Publishing, London, UK. p. 640, ISBN 9781780409313.

Watanabe H., Kitamura T., Ochi S. and Ozaki M. (1997). Inactivation of pathogenic bacteria under mesophilic and thermophilic conditions. *Water Science and Technology*, 36(6-7), 25.

Wendland C. (2008). Anaerobic Digestion of Blackwater and Kitchen Refuse. *Journal of Biosciences*, 12.

Wentzel M.C., Loewenthal R.E. and Ekama G.A. (1988). Enhanced polyphosphate organism cultures in activated sludge systems. Part I: Enhanced culture development. *Water SA*, 14(2), 81-92.

Wentzel M.C., Ekama G.A., Loewenthal R.E. and Dold P.L. (1989a). Enhanced polyphosphate organism cultures in activated sludge systems. Part II: experimental behaviour. *Water SA*, 15(2), 71-88.

Wentzel M.C., Dold P.L. and Ekama G.A. (1989b). Enhanced polyphosphate organism cultures in activated sludge systems. Part III: Kinetic model. *Water SA*, 15(2), 89-102.

Wik T. (2003). Trickling filters and biofilm reactor modelling. Reviews in *Environmental Science and Biotechnology*, 2(2-4), 193-212.

Woolley S.M., Buckley C.A., Pocock J. and Foutch G.L. (2014a). Rheological modelling of fresh human faeces. *Journal of Water, Sanitation and Hygiene for Development,* 4(3), 484.

Woolley S.M., Cottingham R.S., Pocock J. and Buckley C. (2014b). Shear rheological properties of fresh human faeces with different moisture content, 40(2), 273–276.

Xu P., Brissaud F. and Fazio A. (2002). Non-steady-state modelling of faecal coliform removal in deep tertiary lagoons. *Water Research*, 36, 3074-3082.

Zakaria F., Thye Y.P., Hooijmans C.M., Garcia H.A., Spiegel A.D. and Brdjanovic D. (2017). User acceptance of the eSOS® Smart Toilet in a temporary settlement in the Philippines. *Water Practice and Technology*, 12(4), 832-847.

Zakaria F., Ćurko J., Muratbegovic A., Garcia H.A., Hooijmans C.M. and Brdjanovic D. (2018). Evaluation of a smart toilet in an emergency camp. *International Journal of Disaster Risk Reduction*, 27, 512-523.

Zavala M.A.L., Funamizu N. and Takakuwa T. (2004a). Modeling of aerobic biodegradation of feces using sawdust as a matrix. *Water Research*, 38(5), 1327-1339.

Zavala M.A.L., Funamizu N. and Takakuwa T. (2004b). Temperature effect on aerobic biodegradation of feces using sawdust as a matrix. *Water Research*, 38(9), 2406-2416.

Zhang L., De Schryve P., De Gusseme, B., De Muynck W., Boon N. and Verstraete W. (2008). Chemical and biological technologies for hydrogen sulfide emission control in sewer systems: A review. *Water Research*, 42, 1–12.

Zhu H., Hart C.A., Sales D. and Roberts N.B. (2006). Bacterial killing in gastric juice - Effect of pH and pepsin on Escherichia coli and Helicobacter pylori. *Journal of Medical Microbiology*, 55(9), 1265–1270.

Zhu G., Zou R., Jha A.K., Huang X., Liu L. and Liu C. (2015). Recent developments and future perspectives of anaerobic baffled bioreactor for wastewater treatment and energy recovery. Critical reviews in *Environmental Science and Technology*, 45(12), 1243-1276.

Zziwa A., Nabulime M., Kiggundu N., Kambugu R., Katimbo A. and Komakech A. (2016). A critical analysis of physiochemical properties influencing pit latrine emptying and feacal sludge disposal in Kampala Slums, Uganda. *African Journal of Environmental Science and Technology*, 10(10), 316–328.

Figure 6.12 Data obtained from the field are essential for modelling of faecal sludge containment/treatment processes (photo; UKZN PRG).

Figure 6.13 The modelling skills are becoming increasingly important for the next generations of sanitation professionals. A modern approach of modelling and simulation of urban sanitation systems can be achieved through advanced level of competence and educational training, resulting in greater confidence, deeper insight and advanced knowledge. Modelling courses nowadays are offering to young sanitation professionals a chance to comprehend the scientific, technological and engineering principles of faecal sludge treatment and modelling (photo; IHE Delft).

7

Faecal sludge simulants: review of synthetic human faeces and faecal sludge for sanitation and wastewater research

Roni Penn
Barbara J. Ward
Linda Strande
Max Maurer

OBJECTIVES

The objectives of this chapter are to:

- Introduce the concept of simulants and their applications
- Present current state of the art in simulants for faecal sludge, faeces and urine
- Compare properties between simulants and typical values observed in the field
- Introduce customisation of simulants, including advantages and constraints.

This chapter is by large part based on: Penn R., Ward B.J., Strande L. and Maurer M. (2018). Review of synthetic human faeces and faecal sludge for sanitation and wastewater research. *Water Research*, 132, 222-240.

 Methods for Faecal Sludge Analysis. Edited by Velkushanova K., Strande L., Ronteltap M., Koottatep T., Brdjanovic D. and Buckley C. ISBN: 9781780409115. Published by IWA Publishing, London, UK.

7.1 INTRODUCTION

Presented in this chapter is a critical literature review of two categories of simulants: synthetic faeces and faecal sludge (FS), together with how to select and further customise simulants for experimentation depending on the specific properties of interest. Simulants play an important role in research, and have a long history in wastewater research and other fields. The high variability of faeces (Rose *et al.,* 2015) and faecal sludge collected from onsite systems (Strande *et al.,* 2014) makes it difficult to obtain consistent samples and therefore execute repeatable experiments. Moreover, due to the potentially pathogenic content of human excreta, working with real faecal matter involves special safety precautions. Working with synthetic faecal matter can alleviate these challenges. Replicable experiments are important in order to gain an understanding of the specific role of different mechanisms. For example, faeces and faecal sludge simulants could be used to investigate rheological properties (*e.g.* pumpability), energy content, or anaerobic digestion. In addition, simulants could be used to investigate how flows in the sewer network affect operations and maintenance (*e.g.* clogging with solids). The applications are diverse, and demonstrate the range of applicability of simulants in faecal sludge and wastewater research. Although there are no 'perfect' simulants, they can be adapted depending on the specific physical, chemical, or thermal properties that are of interest, and allows for research of properties, when faecal sludge is not available.

Investigations involving human faeces are of great importance in many fields of research, such as medicine (Lewis and Heaton, 1997; Bekkali *et al.,* 2009), sanitary product development (such as diapers, toilets, etc.) (Stern and Holtman, 1987; Palumbo and D'acchioli, 2001), operation and maintenance of sewer systems (Butler *et al.,* 2003; Penn *et al.,* 2017), and implementation of faecal sludge collection and treatment for onsite sanitation systems (Wignarajah *et al.,* 2006; Bassan *et al.,* 2014; Colón *et al.,* 2015). Development of synthetic faeces and faecal sludge is a challenging task due to their high variability depending on diet, lifestyle and geographical location (Rose *et al.,* 2015). In this chapter, we focus on synthetic faeces and faecal sludge developed for sanitation research, hence resembling human faeces and faecal sludge in specific physical and chemical properties.

When discussing simulants, it is important to understand the difference between faeces and faecal sludge. Faecal sludge is the faecal waste stored within onsite sanitation technologies. In addition to faeces it includes everything that goes into the toilet, for example, urine, flush water, greywater, anal cleansing materials and municipal solid waste (Strande *et al.,* 2014). Faecal sludge differs significantly from fresh faeces alone; it is typically much more dilute due to the addition of liquids. Additionally, its characteristics are highly variable due to differences in storage duration, storage temperature and storage technology, and can range from fresh, to partially degraded, to completely stabilised (Strande *et al.,* 2014).

Three distinct recipes for synthetic faecal sludge have been reported in the literature. Their intended purposes include research into anaerobic digestion (Zuma, 2013; Colón *et al.,* 2015) and pit latrine emptying (Radford *et al.,* 2015). Together with the synthetic faeces recipes presented in this chapter, they could also be used as a basis for the development of improved faecal sludge simulants in the future.

Synthetic faeces have been developed to address many sanitation-related research questions. Most of the developed simulants mimic specific physical, chemical or thermal characteristics of human faeces important to the research objectives for which they are developed. Physical properties such as shape, size, density and rheology are of importance for simulating phenomena such as faeces settling, transport in sewer pipes, dewatering, viscous heating for pathogen destruction, and physical disintegration (*e.g.* Butler *et al.,* 2003; Veritec Consulting Inc. & Koeller and Company, 2010; Podichetty *et al.,* 2014). Chemical properties including chemical and biological oxygen demand, nutrient concentration, pH and conductivity are of importance for simulating biological disintegration, treatment of faeces and biogas production (*e.g.* Kaba *et al.,* 1989; Wignarajah *et al.,* 2006; Miller *et al.*, 2015). Elemental composition (C, H, N, O) and heating properties are of importance for analysing energy recovery and for using the faeces for soil amendment *e.g.* biochar or compost production (*e.g.* Ward *et al.*, 2014; Colón *et al.*, 2015; Onabanjo *et al.*, 2016a). Studies on the fate of faeces in sewers and in onsite sanitation systems include their movement, settling and physical disintegration together with biochemical disintegration. For these kinds of investigations a simulant is required that represents closely a combination of chemical, biological and physical properties of faeces and faecal sludge. Such a simulant is still missing in the literature.

This chapter provides a critical literature review of synthetic faeces and faecal sludge used for human waste-related research. Based on this overview a modified simulant recipe is presented that is applicable for studying the fate of faeces in sewers and in onsite sanitation systems. A series of experimental results show how these properties can be selectively manipulated by making changes in the recipe and an explicit preparation procedure can be found in the appendix of this chapter.

7.2 CHARACTERISTICS OF FAECES AND FAECAL SLUDGE

7.2.1 Faeces

Faecal solids are composed of proteins, fats, fibre, bacterial biomass, inorganic materials and carbohydrates. Their chemical and physical characteristics vary widely depending on health and diet, as presented in Table 7.1.

Table 7.1 Chemical and physical properties of faeces identified in the literature.

Property	Parameter	Range (amount/cap.d)	Range (other units)	Median
Chemical	Wet mass	35-796 g[1,6]		128 g/cap.d[6]
	Water content		63-86 wt %[6]	75 wt %[6]
	Protein		2-25 wt % of solids weight (+50% of bacterial biomass)[6]	
	Fibre	0.5-24.8 g[6]		6 g/cap.d[6]
	Carbohydrates	4-24 g[6]	25 wt % of solids weight[6]	9 g/cap.d[6]
	Fats	1.9-6.4 g[6]	8.7-16 wt % of solids weight[6]	4.1 g/cap.d[6]
	Bacteria content		25-54 wt % of solids weight[6] $100\text{-}2{,}200 \cdot 10^{12}$ cells/kg[6]	
	BOD	14-33.5 g[6]		
	COD	46-96 g[6]		
	TN	0.9-4 g[6]	5-7% wt % of solids weight[6]	1.8 g/cap.d[6]
	VS		92 wt % of TS[6]	
	pH		5.3-7.5[6]	6.6[f], 7.15 (avg)[2]
	Calorific value	0.21-1.45 MJ[6]		0.55 MJ/cap.d[6]
Physical	Shape		Type 1 (hard lumps) - type 7 (watery diarrhoea)[5]	3.6 (avg.) [6]
	Viscosity		3,500-5,500 cPs at 50 rpm[3]	
	Density		<1 g/mL for 10-15% of healthy humans[1]	1.06-1.09 (avg.)[2,4]

[1]Levitt and Duane, 1972; [2]Ciba-Geigy, 1977; [3]Yeo and Welchel, 1994; [4]Brown *et al.*, 1996; [5]Lewis and Heaton, 1997; [6]Rose *et al.*, 2015

The average number of stools produced by adults is one per day (Ciba-Geigy, 1977). The median daily wet mass of faeces produced per person is 128 g (Rose *et al.,* 2015), which falls within the reported full range of 35-796 g reported by Ciba-Geigy (1977) and Rose *et al.* (2015). Wyman *et al.* (1978) compared average stool sizes of 20 people (average of 10 samples from each individual). They identified that 250 g/stool and 111.3 g/stool were the maximum averaged weights for the male and female participants, respectively, in the study. In their review of faeces characteristics Rose *et al.* (2015) further report that live and dead bacteria comprise between 25 and 54% of the dry weight of faeces. The median water content in faeces is 75%, with a range of 63–86% across the mean values of the studies. Variations in water content and faecal mass are attributed to differences in fibre intake, as non-degradable fibre absorbs more water in the colon and degradable fibre stimulates growth of bacterial biomass (Eastwood, 1973; Garrow *et al.,* 1993; Reddy *et al.,* 1998). Rose *et al.* (2015) report that volatile solids comprise 92% of the total solids (TS) fraction of faeces. Faeces pH ranges between 5.3-7.5, with biological oxygen demand (BOD) between 14.0 and 33.5 g/cap.day and chemical oxygen demand (COD) between 46 and 96 g/cap.day (Rose *et al.,* 2015).

Faeces are also highly variable in their physical structure. This variability can be characterised by the Bristol Stool Form Scale introduced by Lewis and Heaton (1997) for assessing intestinal transit rate. The scale categorises stools into one of seven types, ranging from type 1 (hard lumps) to type 7 (watery diarrhoea). Types 3 and 4 ('hard, lumpy sausage' and 'loose, smooth snake') are classified as normal stool forms. Onabanjo *et al.* (2016 a) identified the moisture content of each stool classification ranging from ~50% (type 1) to >80% (type 7). The Bristol Scale has been used to assess stool form in the study of gastrointestinal disorders (*e.g.* Garsed *et al.,* 2014, Nolan *et al.,* 2015). Woolley *et al.* (2014) measured the rheological properties of fresh human faeces. They showed that with increasing shear rate the apparent viscosity measurements of the samples decreased. For any given shear rate, higher apparent viscosities were associated with lower moisture contents. Viscosity measurements of runny to solid faeces were found to be in the ranges of 3,500-5,500 cP (Yeo and Welchel, 1994). According to the US National Bureau of Standards (NBS) faeces are characterised by density of 1.06 g/mL (Brown *et al.,* 1996). 10-15% of healthy humans produce stools that float (have a density less than 1.0 g/mL) due to trapped gas in the faeces (Levitt and Duane, 1972).

7.2.2 Faecal sludge

Faecal sludge originates from onsite sanitation technologies, and has not been transported through a sewer. It is raw or partially digested, a slurry or semisolid, and results from the collection, storage or treatment of combinations of excreta and blackwater, with or without greywater (Strande *et al.,* 2014). Blackwater is defined as wastewater generated by the toilet, and includes excreta as well as flush water, anal cleansing water and/or dry anal cleansing materials (Tilley *et al.,* 2014). Greywater contains all other domestic wastewater flows including bathing, washing, laundry and kitchen (Gross *et al.,* 2015).

Typical quantities and qualities of faecal sludge are difficult to determine due to the variety of onsite sanitation technologies in use, such as pit latrines, septic tanks, aqua privies, and dry toilets. They further depend on the design and construction of the sanitation technology, how the technology is used, how the faecal sludge is collected, and the frequency of collection (Strande *et al.,* 2014). Recent findings have indicated that faecal sludge characteristics are correlated to the containment technology, but that there is not always a discernible difference between faecal sludge from public toilets and households (Strande *et al.,* 2018). The lack of standardised methods for the characterisation of faecal sludge further contributes to the variability in the measured parameters.

The important parameters to be considered for faecal sludge characteristics are similar to those of faeces and are presented in Table 7.2.

Figure 7.0 Prof. Chris Buckley of UKZN PRG presents the faeces simulant at the Reinvent the Toilet Fair in 2012, Bill & Melinda Gates Foundation, Seattle, U.S., (photo: D. Brdjanovic).

Table 7.2 Physical and chemical properties of human faeces and faecal sludge compared with simulants.

Properties	Faeces	Synthetic faeces	Faecal sludge	Synthetic faecal sludge
Shape	From 'hard lump' to 'watery diarrhoea'; 'hard lump sausage' and 'loose smooth snake' are normal forms[c]	Cylinder[9]		
Length (cm)		8-10[9,21]		
Diameter (cm)		2.5-3.4[9,21]		
Volume (ml)	90-169 (for women)[3] 82-196 (for men)[3]			
Density (kg/L)	1.06-1.09[3,8]	1.02-1.06[9,21]	1.0-2.2[33]	0.8-2.2[33]
Viscosity (cP)	3,500-5,500[8]	1,000-40,000[8,28]	$8.9 \cdot 10^{-1}$-$6 \cdot 10^{9}$[29]	
Dewatering rate (g/m^2.min)	350-400 (for regular stool, very high for runny faeces)[6]	50-400[8]	11 (% of TS in the dewatered cake)[41,42]	4.5 (% of TS in the dewatered cake)[41,42]
Shear strength (Pa)			<1,760[33]	9-10,000[33]
COD_{total}	0.6-1.5 % of TS[17,32,34]	1.3 % of TS[30,32]	7,000-106,000 mg/L[24,26]	73[30] 12,500-72,800 mg/L[25,30]
$COD_{soluble}$	0.38 % of TS[30]			1,000-48,300 mg/L[25,30]
BOD	14-33.5 g/cap.d[34]		600-40,000· mg/L[13,26]	
TN	2-7 % of TS[17,20,32,34]	2.8 % of TS[30,32]	50-1,500 mg/L[21,28]	880-7,200 mg/L[25,30]
$N\text{-}NH_3$ (% of N_{total})	<7[32]	<3.02[32]		
C/N	5-16[32]	17.3[32]		
pH	4.6-8.4[5,30,32,34]	5.3[30,32]	6.7-8.5[19,26]	5.5-7.73[25,30]
EC (mS/cm)		5.7[30]	2.2-14.6[39]	14.4[30]
TS (%)	14-37[18,32,34]	18.4[16,32]	0.5-40[39,23]	1.7-85.0[25,30,33]
VS	80-92 % of TS[1,4,5,32,34,37]	86.8-88.5 % of TS[30,32,37]	7,000-52,000 mg/L[24,26]	78.9-79.9 % of TS[30] 1,600 -1,800 mg/L[25]
C (% of TS)	44-55[16,31,36,38]	43.4-47.3[11,32,36,37,38]	27.8-28.8[40]	
H (% of TS)	7.0-7.6[31,37,38]	6.2-7.2[11,32,36,37,38]	4.2[40]	
N (% of TS)	1.1-18[5,16,31,37,38]	2.1-7.2[11,36,37,32,38]	3.0-3.2[40]	
O (% of TS)	21-32[31,37,38]	30-42[11,32,36,37,38]		

Fe (µg/kgTS)	72,381-1,230,769[5]	59,950[30]		
Zn	64,762-660,256 µg/kgTS[5]	46,210 µg/kgTS[30]	646-918 ppm[40]	
Ni	1,016-34,615 µg/kgTS[5]	1,289[30]	24-30 ppm[40]	
Co (µg/kgTS)	254-3,846[5]	642[30]		
Mn (µg/kgTS)	46,857-236,539[5]	6,25[30]		
Mo (µg/kgTS)	1,148-12,180[5]	1,555[30]		
Cu (µg/kgTS)	24,889-125,641[5]	5,654[30]	114-216 ppm[40]	
B (µg/kgTS)		3,524[30]		
S	0.5-1.6 % of TS[5,31]	0.06-0.19 % of TS [11,22,32]		388-1,300 mg/L[25]
P (% of TS)	0.39-4.93[5]	0.28[11]	1.5-0.95[40]	
Calorific value (MJ/kg)	17.2-25.1[2,5,10,12,31,34,37]	17.5-22.36[35,37]	11-19[27,39]	
Ash (% of TS)	9.7-14.6[31,37]	13.15[37]	47-59[40]	
Biogas yield	0.16-0.53 NLbiogas/gCOD[1,5,15] 53.1 $mLCH_4/gVS$[43]	0.44 NLbiogas/gCOD[30]	45-50 $mLCH_4/gVS$[43,44]	0.24 $NLCH_4/gVS$[25] 0.12-0.37 NLbiogas/gCOD[30]
Average methane (% vol)		63[30]		38-60[30]
$Sulphate_{soluble}$ (mg/L)				88-392[25]
Total protein (mg/L)	3.2-16.2 g/cap.d[34]			2,874-8,835[25]
$Protein_{soluble}$ (mg/L)				497-1,723[25]
Total carbohydrates (mg/L)	4-24 g/cap.d[34]			660-3,812[25]
Lipids (g/gTS)	0.09-0.16[34], 4.2 g/d[5]			0.03-0.30[25]
Total fiber (g/gTS)	0.25[34]			0.33-0.79[25]
Hemicellulose (g/gTS)				0.15-0.31[25]
Cellulose (g/gTS)				0.03-0.34[25]
Lignin (g/gTS)				0.03-0.16[25]

[1]Snell, 1943; [2]Lovelady, 1970; [3]Levitt and Duane, 1972; [4]Fry, 1973; [5]Ciba-Geigy, 1977; [6]Wyman *et al.*, 1978 ; [7]Meher *et al.*, 1994; [8]Yeo and Welchel, 1994; [9]Brown *et al.*, 1996; [10]Girovich, 1996; [11]Tennakoon *et al.*, 1996; [12]Spellman, 1997; [13]Heinss *et al.*, 1999; [14]Koottatep *et al.*, 2001; [15]Park *et al.*, 2001; [16]Eawag, 2002; [17]Jönsson *et al.*, 2005; [18]Wignarajah *et al.*, 2006; [19]Henze *et al.*, 2008; [20]Barman *et al.*, 2009; [21]Veritec Consulting Inc. & Koeller and Company, 2010; [22]Serio *et al.*, 2012; [23]Still and Foxon, 2012; [24]Bassan *et al.*, 2013; [25]Zuma, 2013; [26]Appiah-Effah *et al.*, 2014; [27]Muspratt *et al.*, 2014; [28]Podichetty *et al.*, 2014; [29]Strande *et al.*, 2014; [30]Colón *et al.*, 2015; [31]Monhol and Martins, 2015; [32]Miller *et al.*, 2015; [33]Radford *et al.*, 2015; [34]Rose *et al.*, 2015; [35]Yerman *et al.*, 2015; [36]Ilango and Lefebvre, 2016; [37]Onabanjo *et al.*, 2016a; [38]Onabanjo *et al.*, 2016b; [39]Gold *et al.*, 2017a; [40]Gold *et al.*, 2017b, [41]Ward *et al.*, 2017a; [42]Ward *et al.*, 2017b; [43]Bourgault, 2019; [44]Rose *et al.*, 2014

Table 7.3 Recipes for faeces and faeces sludge simulants.

Component	Composition of solid content (wt %)																		
	Faeces simulant												Faecal sludge simulants						
Simulant number (#)	#4	#5	#6	#7	#8	#9	#10	#11	#12[E]	#13[F]	#14a	#14b	#15	#16	#17	#18	#19	#20	#21
Source	(2)	(2)	(1,3)	(4)	(5)	(13)	(11)	9,10)	(12)	(8)	(14)		(6)	(9)			(7)		
Cellulose	65.1[A]		33	15	37.5	12.4	15	10			10[G]	10[G]							
Wheat	11																		
Psyllium	6.6[B]	25[B]																	
Poly(oxyethylene)	11																		
Polyvinyl pyrrolidone		75																	
Potassium sorbate	0.7																		
Burnt sienna[C]	2.8																		
Yellow ochre[C]	1.4																		
Raw umber[D]	1.4																		
Torpulina			25																
E.coli			7	30															
Baker's yeast					37.5	32.8	10	30			0	3							
Yeast extract											30	27							
Casein			10																
Oleic acid			20					20			20	20							
KCl			2		4			2			2	2							
NaCl			2					2			2	2							
$CaCl_2$			1			11.3		1			1	1							
Polyethylene glycol				20															
Psyllium husk				5		24.3	15	17.5			17.5	17.5							
Peanut oil				20	20	17.5	5												
Miso paste				5		10.95	30	17.5			17.5	17.5							
Inorganics				5															
Dried coarsely ground vegetable matter (mg)				50															
$Ca_3(PO_4)_2$							5												

CaH_2PO_4					1														
Propylene glycol						10.95	20												
Soybean paste									62.6	52.2									
Rice									34.4	28.5									
Salt									0	19.3									
Ethanol									3										
Walnuts																20.23	39.08	62.83	77.6
Hay flour															79.4	60.56	39.08	20.94	
$Na_2HPO_4{\cdot}12H_2O$															6.35	6.71	6.14	7.41	6.29
NH_4HCO_3															14.25	12.49	15.71	8.82	16.11
Kaolin clay (ultra-fine particle size)													67						
Compost (by dry mass)													33						
1 ml synthetic urine (Table 7.5) + 0.4 g wet simulant #11														100					

[1]Kaba *et al.*, 1989; [2]Yeo and Welchel, 1994; [3]Tennakoon *et al.*, 1996; [4]Wignarajah *et al.*, 2006; [5]Danso-Boateng *et al.*, 2012; [6]Radford and Fenner, 2013; [7]Zuma, 2013; [8]Podichetty *et al.* 2014; [9]Colón *et al.*, 2015; [10]Miller *et al.*, 2015; [11]Yermán *et al.*, 2015; [12]Ilango and Lefebvre, 2016; [13]Onabanjo *et al.*, 2016a; [14]Penn *et al.*, 2019; [A]Powdered cellulose; [B]Fibrall® psyllium hydrophilic mucilloid; [C]Reddish brown and yellowish pigments; [D]Hydrous silicates and oxides of iron and manganese; [E]Water was added as 39.8% of total ingredients; [F]Water was added as 35.5% of total ingredients; [G]Microcrystalline cellulose. The components' composition is made up of dry solids. Water can be added in different amounts to adjust to various TS concentrations.

7.3 SYNTHETIC FAECES AND FAECAL SLUDGE FOUND IN THE LITERATURE

Appropriate simulants for faeces and faecal sludge should be able to reflect the range of physical, chemical, biological and thermal characteristics relevant for the research objective. This specifically includes:

- Physical characteristics *e.g.* represented by the Bristol Stool Form Scale (for faeces simulants).
- Shapability into the characteristic faeces cylinder, and can be made to float or sink (for faeces simulants).
- Viscosity and dewatering properties.
- Chemical and biological properties including COD, BOD, TN, pH, EC, TS, VS, elemental composition, biogas potential.
- Thermal properties, such as calorific value and ash content.
- Ability to physicality disintegrate with a resulting aqueous suspension having similar chemical properties to real disintegrated faeces (for faeces simulants) and biologically degrade in a typical way (for faecal sludge simulants).

This wide variety of faecal and faecal sludge properties poses a substantial challenge for creating a universal synthetic replacement and such an optimal simulant has not yet been developed. Simulants found in the literature were developed to reproduce specific characteristics of human faeces or faecal sludge, depending on the research objectives, with varying degrees of success. All the developments were successful in producing a simulant that is safe to use and does not represent any biohazard.

7.3.1 Physical parameters

The simulants discussed in the following sections are designed to reflect specific physical properties of human faeces and faecal sludge such as shape, rheology or density. As faeces are distinct from faecal sludge each type of simulant is discussed separately.

7.3.1.1 Faeces simulants

Butler *et al.* (2003) prepared artificial faeces for laboratory investigation of gross solids movement in sewers (referred to here as simulant #1). Solids were represented with plastic cylinders with a diameter of 3.4 cm, length of 8 cm and density of 1.06 g/ml, following the US NBS solid (Brown *et al.*, 1996). Penn *et al.* (2018) implemented similar solids for examining their movement in real sewers. Two techniques for tracking the gross solids were developed; using light sticks tracked by computerised light detector and RFID (radio frequency identification)-based tracking. They further analysed the effect of reduced sewer flows on the movement of the solids (Penn *et al.*, 2017).

Maximum Performance (Map) in the USA (Veritec Consulting Inc. & Koeller and Company, 2010) developed a media for testing toilet flush performance (simulant #2). In a Toilet Fixture Performance Testing Protocol, they define a test media (*i.e.* synthetic faeces) to comprise the following: 'one or more 50±4 g test specimen consisting of one of the following (*i*) soybean paste contained in latex casing (cased media), tied at each end forming a sausage shape or (*ii*) the same quantity consisting of extruded soybean paste (uncased raw media), and four loosely crumpled balls of toilet paper. Each test specimen will be approximately 100±13 mm in length and 25±6 mm in diameter.' A similar media was developed by DIN (German Industrial Norm/European Norm, 2006). The US Environmental Protection Agency's (EPA's) WaterSense program (EPA WaterSense, 2014) adopted Map's protocol and indicated that a 'high efficiency' toilet should successfully and completely clear 350 g of the test specimen from the fixture in a single flush in at least four out of five attempts.

All the above inert simulants were developed to reflect the shape, size and density of real faeces. These simulants were mainly used for investigating solids movement in sewers and in drainage pipes of buildings and for investigating the flushing performance of toilet user interfaces. Simulant faeces with varying densities and shapes as described in the Bristol Stool Form Chart (Lewis and Heaton, 1997) can be produced by modification of these physical simulants. Simulants can be further modified to represent other type of solids found in sewers such as FOGs (fats, oils and greases) by producing them from materials with various densities. With the increasing number of in-sink food waste disposals, the discharge of FOGs to sewers is widely increasing (Thyberg *et al.,* 2015) and hence investigating their transport in sewers is of significance. These simulants do not disintegrate and therefore are not impacted by the shear stress present in the system and their chemical properties are not reflected. It is also important to realise that the rheological properties of these simulants differ significantly from the real material.

Podichetty *et al.* (2014) evaluated the application of viscous heating for the destruction of pathogens in faeces. Heat was generated within faecal simulants by applying shear stress with an extruder. They found, based on a literature review, several alternative materials displaying the same shear thinning behaviour as human stools, and demonstrating similar viscosity profiles with changing shear rate as reported by Woolley *et al.* (2013). The alternatives included contents from pig caecum (a section of the pig lower intestine) (Takahashi and Sakata, 2002), content from chicken caecum (Takahashi *et al.,* 2004), wheat flour (Podichetty *et al.,* 2014), different types of mashed potatoes (Podichetty *et al.,* 2014) and simulant stool (Susana.org, 2008), simulant #13 presented in Table 7.3. While wheat flour had the closest match to the rheological behaviour of human faeces, they selected red potato mash since it had a higher resemblance in terms of moisture content (simulant #3). Their choice of red potato mash as a faecal simulant was confirmed by its structural, thermal and viscoelastic properties (Singh *et al.,* 2008). Simulant #13 (Table 7.3) showed poor rheological resemblance to human faeces. Rheological behaviour of the various simulants is presented in Figure 7.1.

Viscous heating of the red potato mash (simulant #3) was not compared to viscous heating performance of real human faeces. Further, this simulant was not tested for its density or whether it could be representative of faeces shape. It can reasonably be assumed that this simulant will poorly represent the chemical characteristics of human faeces, as it lacks important components such as bacterial content, fibre, proteins and inorganic matter.

Yeo and Welchel (1994) patented a synthetic faeces for simulating the dewatering rate of human stools. It was developed to be used as a substitute for real faeces in the testing and development of diapers. They examined 32 formulations using different components. Many of their attempts were based on a commercially available synthetic faeces, FECLONE®BFPS-4 powder (simulant #4, Table 7.3) from Silicone Studio of Vallez Forge, Pa. FECLONE®BFPS-4 was reported to have a viscosity of 2,276-4,032 cP which is comparable to human stools, but a substantially higher dewatering rate of 524-535 $g_{water}/m^2_{simulant}$.min. In comparison, viscosity and dewatering values of human faeces were reported as 3,500-5,500 cp and 350-400 g_{water}/m^2_{faeces}.min for regular faeces (Table 7.1 and Table 7.2).

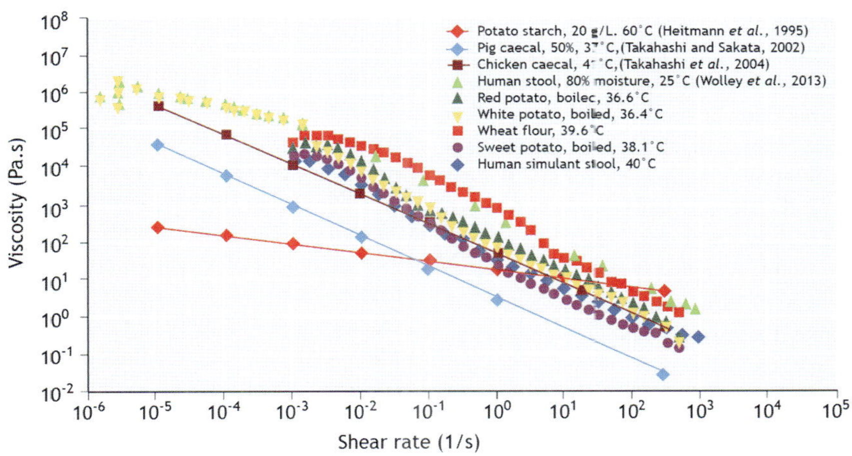

Figure 7.1 Rheological behaviour of the various simulants (Podichetty *et al.*, 2014).

The units of the dewatering rate (g/m^2.min) include m^2 of material, which is determined according to the measurement procedure reported in Yeo and Welchel (1994). Since such a unit is not applicable to be used easily for other research purposes, the authors of this paper converted the unit to $g_{water}/L_{material\ (simulant\ or\ real\ faeces)}$.min according to the methods described in Yeo and Welchel (1994). The converted results were found to be 110.1-125.9 g_{water}/L_{faeces}.min and 164.9-168.0 $g_{water}/L_{simulant}$.min for regular faecess and the simulant, respectively. The viscosity was measured in centipoise at 50 revolutions per minute using a model RVT viscometer manufactured by Brookfield Engineering Laboratories, Inc., Stoughton, Mass. The shear rate at which the viscosity was measured was not given.

Yeo and Welchel (1994)'s best-performing simulant (simulant #5, Table 7.3) was composed of 15% Polyvinylpyrrolidone (PVP), 5% psyllium mucilloid and 80% water. By varying the weight percentage of the soluble to insoluble components, the molecular weight of the soluble component (PVP) and the water content, the viscosity of the simulant could be varied along the Bristol Stool Scale. Therefore, the viscosity can be adjusted to between 1,000 to 40,000 cP covering the range of real human stools (Table 7.4 and Table 7.5). When the simulant was adjusted to a viscosity range of 3,500-5,500 cp (similar to that of human stools), a dewatering rate of 50-400 g/m^2.min (15.73-125.9 g/L.min after conversion) was reported. The simulant was found to bind water to a better extent than other commercially used alternatives. The alternatives included mashed potatoes, brownie mix, peanut butter and pumpkin filling, and were reported to have a dewatering rate of over 500 g/m^2.min (157.3 g/L.min after conversion), much higher than human faeces. A proper mixture of both water soluble (84 wt % of total solids) and water insoluble (16 wt % of total solids) components was necessary to achieve low dewatering rates while keeping the water content relatively constant at 70-90% of the total weight. The authors also found that water-soluble components which had an average molecular weight of over 10,000 g appeared to provide lower dewatering rates. They further reported that adding saturated fat to the solids portion at less than 2 wt % of total simulant weight resulted in reduction of both the surface tension and dewatering rate of the compound.

According to Wignarajah *et al.* (2006), the drawback of the simulant developed by Yeo and Welchel (1994) is its inability to act as a faeces-like substrate for microbial activity. The addition of PVP resulted in much higher nitrogen levels than are typically found in faeces.

7.3.1.2 Faecal sludge simulants

The physical properties of faecal sludge are different from faeces. Hence, investigations making use of faecal sludge require different simulants from those used for faeces. However, as faeces are an essential ingredient in faecal sludge, some of the simulants described in the previous section can be a base for the development of faecal sludge simulants.

Radford and Fenner (2013) developed a synthetic faecal sludge to represent the physical characteristics of pit latrine sludge (simulant #15, Table 7.3). It was developed for studying pit-emptying performance by vacuum trucks, specifically for systems in southern Africa. It was composed of a mixture of compost, kaolin clay, and water. The authors calculated the shear strength of faecal sludge as <400 Pa from a previous study of sludge densities in pit latrines (Boesch and Schertenleib, 1985). The simulant could be modified to have a shear strength from 60 to 900 Pa, which replicated and exceeded the full range of shear strengths found in faecal sludge. The simulant densities were in the range of some faecal sludges (800-1,200 kg/m^3) but were not representative of sludge with elevated sand content, which has a much higher density (up to 2,200 kg/m^3).

Radford *et al.* (2015) expanded the recipe developed by Radford and Fenner (2013) by proposing two simulants to cover the entire range of faecal sludge densities and shear strengths. The simulants were further developed to be used for research on emptying various types of containment systems (*e.g.* septic tanks, pour-flush systems, pit latrines, and urine-diverting dry toilets). While detailed recipes for these simulants were not described in the literature, their components were provided. Simulant 'a' replaced the compost in simulant #15 with topsoil, it further included (like simulant #15) kaolin clay and a range of water contents. Their second simulant (simulant 'b') contained milorganite organic fertiliser derived from sewage sludge, as well as salt, vinegar and a range of water contents. Both simulants were found to represent the full range of shear strengths reported for faecal sludge, but had different densities of 1,400 kg/m^3 and 980 kg/m^3 for simulants 'a' and 'b', respectively. Simulant 'a' was used for a Water for People-led research project in Uganda as those materials were locally available. Milorganite was recommended for faecal sludge processing technology development testing in the USA because it is easy to obtain in that country and can be standardised. Thorough validation of the faecal sludge simulants was impossible because there have been limited characterisation studies of the rheological properties and 'pumpability' of actual faecal sludge.

7.3.2 Chemical, biological, and thermal parameters

The previously discussed simulants were developed to mimic specific physical properties of human faeces and faecal sludge, and are unlikely to reflect their chemical properties. Various simulants reflecting specific chemical, biological, and thermal properties of human faeces and faecal sludge have also been developed. These chemical and biological properties are mostly defined as chemical oxygen demand (COD), total nitrogen (TN), pH, electrical conductivity (EC), total solids (TS), volatile solids (VS), elemental composition, and biogas potential. Important thermal properties are calorific value and ash content. Some of these simulants provide very high chemical, biological, and/or thermal resemblance to human stools and faecal sludge. However, many lack a physical resemblance to faeces and faecal sludge.

7.3.2.1 Faeces simulants

The first attempts to simulate the chemical composition of faeces were made by Kaba *et al.* (1989) for investigating faeces treatment by onsite oxidation (simulant #6, Table 7.3). The treatment was carried out by electrochemical incineration of waste. Bhardwaj *et al.* (1990) reported that oxidation of real faeces and oxidation of this simulant, with urine serving as an electrolyte, occurred at the same potential. Their simulant was developed based on the assumption that faeces solids are made up of one-third microorganisms from the intestinal flora, one-third undigested fibre and the rest is lipids and inorganic material. Tennakoon *et al.* (1996) made use of this simulant for investigating electrochemical treatment of human wastes in a packed bed reactor.

Based on the simulants developed by Kaba *et al.* (1989) (simulant #6 Table 7.3) and Yeo and Welchel (1994) (simulant #5 Table 7.3), Wignarajah *et al.* (2006) developed synthetic faeces formulations for NASA's development of onsite waste processing for its space missions (simulant #7, Table 7.3). These recipes focus primarily on representing the water-holding capacity, rheology and the chemical composition of real faeces. They replaced the oleic acid suggested by Kaba *et al.* (1989) with peanut oil due to its high fraction of oleic acid (50-80%). Additionally, they replaced the casein (protein) in the original recipe with miso paste, composed of 38% protein, 21% fat, 20% fibre and 4% minerals. *E.coli* was the only organism used. In their simulant, Wignarajah *et al.* (2006) opted to use the nitrogen-free polyethylene glycol to represent the water-holding capacity instead of PVP based on lessons learned from the high nitrogen content of simulant #5 (Table 7.3). The resulting product was reported to be more chemically similar to faeces than the previously developed simulants #5 and #6 (Table 7.3). Wignarajah *et al.* (2006) produced five different versions to represent different aspect of faeces: water-holding capacity, rheology and chemical composition. They indicated that each version may be best used for different studies. Table 7.4 presents the function of the different components in the basic recipe proposed by Wignarajah *et al.* (2006).

Table 7.4 Functions of the components in the synthetic faeces #7 (Wignarajah *et al.*, 2006).

Component	Function
E.coli	Bacteria debris
Cellulose	Fibre/carbohydrate
Polyethylene glycol	Water retention
Psyllium husk	Dietary fibre/carbohydrate
Peanut oil	Fat
Miso paste	Proteins/ fats/ fibre/minerals
Inorganics	Minerals
Dried coarsely ground vegetable matter	Undigested vegetable matter

It should be noted that even though some of the ingredients presented in Table 7.4 contain water (*e.g.* miso paste), 'solids content' refers to all of the recipe's ingredients excluding deionised water.

Simulant #7 (see Table 7.3) is the basis of the synthetic faeces used by a number of research groups focusing on the energy recovery from faeces and its treatment in onsite sanitation systems. Ward *et al.* (2014) and Danso-Boateng *et al.* (2012) used it to simulate the energy content of carbonised faeces. Danso-Boateng *et al.* (2012) modified this simulant for investigating the conversion of biomass within faeces into char using hydrothermal carbonisation (HTC). Their modified recipe is presented as simulant #8 in Table 7.3. No information on the purpose of their modification or the simulant's resemblance to faeces was reported. Ward *et al.* (2014) evaluated solid fuel char briquettes produced from faeces. They found that although both the faeces and the simulant (simulant #7 as in Table 7.3) had similar calorific values, the char produced from synthetic faeces had a higher calorific value compared to char produced from real faeces. They attributed this difference to the low inorganic content of the simulant in comparison with real faeces. They further showed that the faecal char had a comparable calorific value to wood char. The energy content was reported as 25.57 MJ/kg and 29.53 MJ/kg for chars produced from faeces and synthetic faeces, respectively at a pyrolysis temperature of 300 °C. Increasing the pyrolysis temperature to 750 °C decreased the energy content of the chars to 13.83 MJ/kg and 18.92 MJ/kg for faeces and synthetic faeces, respectively. Onabanjo *et al.* (2016a) and Yermàn *et al.* (2015) adapted simulant #7 (Table 7.3) to investigate the combustion performances of faeces. Their modifications can be found as simulants #9 and #10 in Table 7.3. The result presented by Onabanjo *et al.* (2016a) showed good representation of human faeces regarding parameters effecting combustion including calorific value, VS, ash content and element chemical composition (as shown in Table 7.2). Yermàn *et al.* (2015) validated the combustion performance of the simulants with the performance of dog faeces.

Colón *et al.* (2015) modified simulant #7 to investigate anaerobic digestion of undiluted synthetic faeces and urine, and Miller *et al.* (2015) looked at supercritical oxidation of a similar simulant to treat faecal sludge. This simulant (simulant #11), shows high chemical and biological resemblance to human faeces (Table 7.2). Colón *et al.* (2015) further adjusted the simulant for trace metal contents since trace metals play an important role in the growth of methanogens and methane formation. The adjustment was made by adding a trace element solution with the following composition: $FeCl_2 \cdot 4H_2O$, 28.6 mg/kgTS; H_3BO_3, 1.14 mg/kgTS; $MnCl_2 \cdot 4H_2O$, 1.91 mg/kgTS; $CoCl_2 \cdot 6H_2O$, 2.29 mg/kgTS; $ZnCl_2$, 1.34 mg/kgTS; $NiCl_2 \cdot 6H_2O$, 0.48 mg/kgTS; $CuCl_2 \cdot 2H_2O$, 0.29 mg/kgTS; $NaMoO_4 \cdot 2H_2O$, 0.48 mg/kgTS $FeCl_2 \cdot 4H_2O$, 28.6 mg/kgTS; H_3BO_3, 1.14 mg/kgTS. The results shown by Colón *et al.* (2015) demonstrated that anaerobic digestion of undiluted human simulant excreta in simple unmixed digesters is feasible and yields biogas, which is a valuable by-product of the treatment. As it was not relevant to their studies, no attempt was made to match the physical properties of their simulant to that of real human stools.

Of the previously addressed modifications to simulant #7, four of them use active baker's yeast instead of *E.coli* to represent microbial material (see Table 7.3). The inorganic fraction was supplied by various salts including calcium phosphate (Ward *et al.,* 2014), a mixture of calcium phosphate and potassium chloride (simulant #8, Table 7.3), or a mixture of calcium chloride, sodium chloride, and potassium chloride (simulant #11, Table 7.3). The quantities of the other components of simulant #7 were only slightly modified (Table 7.3) and no further information was given for those modifications.

Simulant #11, developed by Colón *et al.* (2015), was the only one thoroughly analysed for chemical properties important for wastewater treatment (including COD_{total}, $COD_{soluble}$ TN, pH, EC, TS VS and elemental composition). It showed a high chemical resemblance to human faeces (Table 7.2). It further showed adequate potential for production of biogas. However, based on the personal experience of the authors of this chapter, the large amount of baker's yeast included in this recipe makes it physically very different from real human faeces as it inflates like bread dough, and yields a sticky, unshapable slime.

Ilango and Lefebvre (2016) used miso paste (a mixture of soybean paste, rice, salt, ethanol and water) as a chemical approximation of faeces for a study of biochar production from faeces (simulant #12, Table 7.3). This simulant was found to have a similar elemental composition to faeces along with comparable moisture content and calorific value (as shown in Table 7.2). While this recipe produced a successful simulant for pyrolysis studies, a similar simulant was also evaluated by Podichetty *et al.* (2014) (simulant #13, Table 7.3) in the previously discussed rheology studies and deemed to be a poor physical representation of human faeces. Both studies provide similar compositions for a miso paste-based simulant.

Simulant #11 and simulant #12 (Table 7.3) appear to provide good approximations of faeces in terms of the chemical properties (as indicated in Table 7.2), while Simulant #11 (Table 7.3) showed good resemblance to the chemical properties important for wastewater treatment. It further showed high compatibility in terms of its elemental content important for energy and nutrient recovery and similar biogas production to faeces. Simulant #12 had similar elemental composition and heating properties to faeces, both important factors for energy recovery from faeces. However, they both proved to poorly resemble the physical properties of faeces.

7.3.2.2 Faecal sludge simulants

Fresh faecal sludge can be represented as a combination of faeces and urine with the option to include flush water, greywater, anal cleansing material, municipal solid waste, or other constituents depending on the system. Faecal sludge emptied from onsite containment or arriving at a treatment facility undergoes biological degradation, contributing to the various chemical and physical characteristics that a simulant will need to address. Two simulants were found in the literature intended to represent the chemical and biological properties of faecal sludge for anaerobic digestion research (Zuma, 2013; Colón *et al.*, 2015). In addition, a recipe for synthetic urine (Colón *et al.*, 2015) and a few recipes for synthetic greywater were developed (Gross *et al.*, 2015). These can be combined with synthetic faeces for the preparation of synthetic faecal sludge. Examples for these simulants are presented in tables 7.5 and 7.6.

Table 7.5 Synthetic urine (Colón *et al.*, 2015).

Component	Amount (g/L)
Urea	14.2
Creatinine	3
Ammonium citrate	2
NaCl	8
KCl	1.65
$KHSO_4$	0.5
$MgSO_4$	0.2
KH_2PO_4	1.75
$KHCO_3$	0.5

Colón *et al.* (2015) mixed 300 ml of a modified urine simulant (Table 7.5) developed by Putnam (1971) with 120 g of wet simulant #11 in their studies of onsite anaerobic digestion of undiluted fresh faecal sludge (to produce simulant #16, Table 7.13). Their simulant was required to have chemical similarity to facilitate growth of anaerobic bacteria (specifically, COD_{total}, $COD_{soluble}$, N, N-NH_3, C:N, pH, EC, P, Fe, Zn, Ni, Co, Mn, Mo, B, Cu). For adjusting the simulant to contain missing trace elements (important for methanogen growth), the same trace element solution described with the discussion of their faeces simulant (simulant #11, Table 7.3), was added. The simulant had specific gas production of 0.12-0.37 NL biogas/gCOD (gas volume at 237 K and 1 atm). Since the time that this study was originally published, biomethane potential values of 47.3 $mLCH_4$/gVS for faecal sludge, and 53.1 $mLCH_4$/gVS for faeces have been published, suggesting that further characterisation of may be necessary to establish realistic biomethane potential targets for faeces and faecal sludge simulants (Bourgault, 2019).

Zuma (2013) developed synthetic faecal sludge for representing the chemical and biological properties of faecal sludge for anaerobic digestion testing. Five different recipes were developed by varying the proportions of hayflour, ground walnuts, sodium phosphate ($Na_2HPO_4 \cdot 12H_2O$), and ammonium bicarbonate (NH_4HCO_3) (simulants #17-21, Table 7.3). This simulant was found to have a comparable biomethane potential to dairy manure, with 0.237 $NLCH_4$/gVS after 24 days and 0.24 NL CH_4/gVS after 40 days at 37 °C for the simulant and the dairy manure, respectively. Sludge parameters

TS, VS, TSS, and VSS were easily adjusted for the entire ranges present in faecal sludge by varying the ingredient ratios. COD could be varied with hay flour content. Nutrients could be adjusted with sodium phosphate and ammonium carbonate, and sulphate content was adjustable by varying the walnut content. Recipes with more hay flour had higher lignin and cellulose, and recipes with more walnut had higher lipid levels. The range of values achievable for these simulants is presented in Table 7.2. This simulant needs further development to be able to model a broader range of faecal sludge characteristics. The authors found that they were unable to replicate sludge with a VS/TS ratio lower than 0.85, which seriously limits applicability in the case of more stabilised faecal sludge. VS/TS ratios for faecal sludge samples collected during discharge at treatment facilities typically range between 0.43 and 0.73 (Gold *et al.*, 2017a. The physical properties of this simulant were not reported.

Table 7.6 Recipes for synthetic greywater (adopted from Gross *et al.*, 2015)

Constituent	Greywater[(1)] (mg/L)	Greywater[(2)] (g/100 L)	Bath[(3)] (g/100 L)	Laundry[(3)] (amount/100 L)	Laundry and bath[(4)] (amount/100 L)
Ammonium chlorine	75				
Soluble starch	55				
Potassium sulphate	4.5				
Sodium sulphate Na_2SO_4		3.5		4 g	
Na_2PO_4				4 g	
Sodium dihydrogen phosphate	11.4	3.9			
Sodium bicarbonate $NaHCO_3$		2.5		2 g	
Boric acid		0.14			
Lactic acid		2.8	3		
Synthetic soap					
Body wash with moisturiser			30		
Conditioner			21		
Shampoo	0.022	72	19		86 mL
Liquid hand soap			23		
Bath cleaner			10		
Liquid laundry fabric softener				21 mL	
Liquid laundry detergent				40 mL	
Laundry		15			At recommended concentrations for hard water
Kaolin	25				
Clay		5			
Test dust			10	10 g	
Sunscreen/moisturiser		1/1.5			
Toothpaste		3.25	3		
Deodorant		1	2		
Vegetable oil		0.7			1 mL
Secondary effluent		2 L	2 L	2 L	To give final concentration of 10^5-10^6 CFU of total coliforms

[1]Friedler *et al.*, 2008; [2]Diaper *et al.*, 2008; [3]NSF, 2011; [4]BSI, 2010; CFU: colony-forming unit.

In addition to faeces and urine, greywater is an important component of some faecal sludge, especially within higher economic brackets that are likely to have piped water and septic tanks (Strande *et al.*, 2014; Schoebitz *et al.*, 2016). Recipes for synthetic greywater contain ingredients typically found in real greywater such as a variety of personal hygiene products, chemicals used in the home, and bacteria. The mixture of these substances typically yield similar levels of pH, COD, BOD_5, TSS and surfactants usually found in greywater. Greywater characteristics are influenced by the type of flows contained within the greywater (*e.g.* kitchens, showers, sinks, laundry etc.), cultural and socioeconomic variables, climate and geographical variables and quality of the source water (Gross *et al.*, 2015). Recipes for synthetic greywater found in the literature and in government standards are presented in Table 7.6.

7.4 DISCUSSION

Based on the literature review, there have been successful simulants mimicking specific physical and chemical properties of human faeces and faecal sludge. A summary of the reviewed simulants and their similarity to human faeces and faecal sludge is presented in Table 7.7 and 7.8. The differences in the simulant properties are readily apparent in Table 7.8, since each was developed to mimic specific faeces and faecal sludge characteristics applicable to the study undertaken, but ignore most others. A clear distinction can be made between the physical (simulants #1 to#5 and #15) and chemical, biological, and thermal simulants (#6 to #14, #16 to #21). Almost none of the simulants adequately represent both chemical and physical properties. The information provided in Table 7.8 can support the selection of the simulant to be used or to be further developed for any intended research. For example, in wastewater research of sewer systems and onsite sanitation systems a combination of some of these properties is of importance. Such investigations include faeces movement and faeces and faecal sludge settling, dewatering, and physical and biochemical disintegration. A first attempt to combine these properties in one faecal simulant is described in detail in Section 7.4.1.

To date, constituents of interest, such as odour, pathogens, pharmaceuticals, hormones and comprehensive COD fractionation, have not been included in faecal sludge simulants. The development of simulants including COD fractionation (*e.g* inert and slowly and readily biodegradable fractions of COD is important for the study of biochemical properties of faecal sludge during onsite storage and treatment. Odours can be simulated by real or synthetic components, such as hydrogen sulphides, methyl sulphides and benzopyrrole derivatives (Moore *et al.*, 1987). Sato *et al.* (2001a,b) found that sulphur-containing components were 2.2% of the total gaseous fraction, while the nitrogenous benzopyrrole compounds were only about 0.3%. Ammonia occurred at 6.3%.

It is important to note that faecal sludge is highly variable and it differs significantly from fresh faeces. As seen in this review, the development of faecal sludge simulants is in its preliminary stages. The importance of FSM has only been acknowledged relatively recently (Moe *et al.*, 2006; WHO, 2017), which is a possible reason for the comparative lag in simulant development. One reason for the complexity of developing representative simulants is due to the lack of comprehensive characterisation data for faecal sludge, although, with the increasing awareness of the importance of FSM, this data is becoming more readily available (Gold *et al.*, 2017a). The lack of available information on faecal sludge characteristics makes it difficult to validate simulant performance. Faeces is an important constituent of faecal sludge, which typically also includes additional components such as urine, greywater, flush water, and/or solid waste, and with varying levels of biological and physical degradation (Chapter 2). The comprehensive review of faeces, urine, and greywater simulants presented in this paper will support the further development of faecal sludge simulants. This will be valuable for conducting research to understand what is occurring during onsite storage of faecal sludge, to develop treatment technologies, and to enhance potential for resource recovery (Diener *et al.*, 2014; Muspratt *et al.*, 2014; Gold *et al.*, 2017a,b).

In the discussion on faeces and faecal sludge simulant development it is also important that average values are targeted for desired simulant characteristics. However, in reality, the characteristics of faeces and faecal sludge vary widely depending on health and diet, storage time, containment technology, and usage patterns (Chapter 2). Further research is necessary prior to the development of simulants that reflect regional and dietary dependent variations. To achieve this, it will be important to identify which parameters are most sensitive to such effects and how much impact they have on the purpose of the simulant.

Table 7.7 Summary description of all the simulants.

	Sim. #	Reference	Description	Investigation
Faeces simulants	1	(4,17)	Plastic cylinders with detecting device	Investigating gross solids movement in sewers
	2	(5, 7, 11)	Soybean paste in a latex casing	Testing toilet performance (connected to sewers and off-grid)
	3	(12)	Red potato mash	Viscous heating of faeces for pathogen destruction
	4	(2)	Water soluble polymer (for water-holding capacity), fibre and water	For testing diapers
	5	(2)		
	6	(1,3)	Variations on a recipe containing bacteria, water, retention component, fibre, fat, proteins and minerals	Electrochemical oxidation for treatment of faeces
	7	(6,13)		Wastewater treatment in space vehicles (6) Production of char briquettes from faeces (13)
	8	(8)		Production of char briquettes from faeces through hydrothermal carbonisation
	9	(19)		Combustion performance of human faeces
	10	(19)		
	11	(14,15)		Anaerobic digestion of faeces and urine (14) Supercritical oxidation to treat FS (15)
	14	(20)		Physical disintegration of faeces under sewer flow conditions, biological disintegration of faeces in onsite systems and optimisation of FS treatment
	12	(18)	Mixture of soybean paste, rice, salt, ethanol and water	Biochar production from faeces
	13	(11)		
Faecal sludge simulants	15	(9,16)	Mixture of compost, kaolin clay and water	For studying pit-emptying procedure
	16	(14)	Same as simulant #11 + addition of synthetic urine	Anaerobic digestion of FS
	17-21	(10)	Mixture of hay flour, ground walnuts, sodium phosphate and ammonium carbonate	
	22	(20)	Same as simulant #14 + addition of synthetic urine	Dewatering studies of FS

[1]Kaba *et al.*, 1989; [2]Yeo and Welchel, 1994; [3]Tennakoon *et al.*, 1996; [4]Butler *et al.*, 2003; [5]German Industrial Norm/European Norm, 2006; [6]Wignarajah *et al.*, 2006; [7]Veritec Consulting Inc. & Koeller and Company, 2010; [8]Danso-Boateng *et al.*, 2012; [9]Radford and Fenner, 2013; [10]Zuma, 2013; [11]EPA WaterSense, 2014; [12]Podichetty *et al.*, 2014; [13]Ward *et al.*, 2014; [14]Colon *et al.*, 2015; [15]Miller *et al.*, 2015; [16]Radford *et al.*, 2015; [17] Yermán *et al.*, 2015; [18]Ilango and Lefebvre 2016; [19]Onabanjo *et al.*, 2016a; [20]Penn *et al.*, 2018.

Table 7.8 Summary comparison of human faeces simulants.

Simulant #	1	2	3	4	5	6	7	8	9	10	11	12	13	14	15	16	17-21	22
Reference	(4,20)	(5,7,11)	(5)	(2)	(2)	(1,3)	(6)	(8)	(19)	(17)	(14,15)	(18)	(12)	(21)	(9,16)	(14)	(10)	(21)
Shape	+	+									–			+				
Density	+	+												+	+			
Physical disintegration	–	–												+				
Viscosity	–	–	+	+	+		○					–	–	+	+			+
Dewatering	–	–		–	+		○											–
Water content	–	–	+	+	+	+	○	+	+	+	+	+	+	+	+	+	+	+
COD_{total}	–	–					○				+			+		+	+	+
$COD_{soluble}$	–	–									□			□		□	□	□
TN	–	–					○				+			+		+	+	+
NH_3-N and NH_4-N	–	–					○				+			+		+	+	+
C/N	–	–									+					+		
BOD	–	–																
PH	–	–					○				+			+		+	+	+
EC	–	–					○				□			□		+		+
TS	–	–	+	+	+	+	○	+	+	+	+			+		+	+	+
VS	–	–					○		+		+			+		+	–	+
Elemental composition	–	–				+	+		+		+	+	+					
S	–	–				+					+							
P	–	–															+	
Fe	–	–									+					+		
Zn	–	–									+					+		
Ni	–	–									+					+		
Co	–	–									+					+		
Mn	–	–									+					+		
Mo	–	–									+					+		
Cu	–	–									+					+		
B	–	–									+					+		
Calorific value	–	–					+	+	+			+	+					
Ash content	–	–							+				+					
Biogas yield											+					+		
Odour	–	–	–	–		–	–	–	–	–	–	–	–	–	–	–	–	–
Pathogens	–	–	–	–		+	+	–	–	–	–	–	–	–	–	–	–	–

[1] Kaba *et al.*, 1989; [2] Yeo and Welchel, 1994; [3] Tennakoon *et al.*, 1996; [4] Butler *et al.*, 2003; [5] German Industrial Norm/European Norm, 2006; [6] Wignarajah *et al.*, 2006; [7] Veritec Consulting Inc. & Koeller and Company, 2010; [8] Danso-Boateng *et al.*, 2013; [9] Radford and Fenner, 2013; [10] Zuma, 2013; [11] EPA WaterSense, 2014; [12] Podichetty *et al.* 2014; [13] Ward *et al.*, 2014; [14] Colón *et al.*, 2015; [15] Miller *et al.*, 2015; [16] Radford *et al.*, 2015; [17] Yermán *et al.*, 2015; [18] Ilango and Lefebvre, 2016; [19] Onabanjo *et al.*, 2016a; [20] Penn *et al.*, 2017; [21] Penn *et al.*, 2018.

+ Validated with real faeces or faecal sludge;

□ Reported value for synthetic, but no available data to compare to real faeces or faecal sludge;

– Not expected to be comparable to real faeces or faecal sludge (based on reported literature, other literature values, and experiences of authors; blank box: not enough data to make a conclusion;

○ Reported to be comparable to faeces or faecal sludge but no results provided (simulants #7-10 and simulant #13 are based on this recipe);

Blank box: not enough data to make a conclusion.

7.4.1 Development of a new simulant

For research into the fate of excreta in urban sewers and in onsite sanitation systems, both the chemical/biological aspects of faeces and faecal sludge and their physical properties are important. Investigations of their fate include their physical motion (movement, settling, sedimentation, and dewatering) and their physical and biochemical disintegration in sewer pipes and in onsite sanitation systems. Based on the information gained in the literature review, it is possible to create an adapted simulant with mixed physical and chemical properties that can be used for such investigations. Detailed instructions on the simulant preparation and recommended storage practices can be found in Example 7.1 in the appendix of this chapter.

7.4.1.1 Synthetic faeces

It is required that the new simulant represents a range of physical characteristics based on the Bristol Stool Form Scale. It should be able to be modified from soft to hard by adding different amounts of water, should be shaped into the characteristic faeces cylinder, sausage, or snake, and be able to be controlled as to whether it floats or sinks. The desired simulant should also possess a similar viscosity and dewatering rate to real faeces. Additionally, it should have similar chemical composition to faeces including COD, TN, pH, EC, TS, VS and elemental composition. It should be able to disintegrate in water and the resulting aqueous suspension should have similar chemical properties to disintegrated faeces. By looking at Table 7.8, one can see that both simulant #11 and simulant #12 (from Table 7.3) showed high chemical and biological resemblance in their elemental composition but poor physical resemblance in their shape and rheological properties. Indeed, none of the simulants with proper physical parameters has a representative chemical composition. The modification of the physically-related simulants to represent additional chemical properties was found to be impracticable. Simulants #11 and #12 were therefore the best candidates for further development. Simulant #11 shows high compatibility in its chemical properties important for wastewater-related research, including COD, TN, TS, and VS. Baker's yeast is used to represent microbial biomass and to produce floating stools (due to gas produced by the yeast). However, the quantity included in this recipe creates an unfavourable physical structure. It produced a gassy and sticky material that floated when added to water, but was too sticky to be shaped into a cylinder. Figure 7.2A illustrates the high gas production, shown by the many bubbles in the beaker. The stickiness of the material is shown in Figure 7.2B. Use of active yeast also contributes to quick biological changes within the synthetic material, which is undesirable if reduced sample variability is a priority. An ideal simulant would be storable and resistant to physical or biological change over a span of at least several days in order to maximise reproducibility of experiments. We hypothesised that by adapting the baker's yeast content of simulant #11 (Table 7.3), a physically representative simulant could be produced, while still maintaining its chemical and biological resemblance.

Although Simulant #12 also looks like a good candidate for further development efforts were focussed on simulant #11 at this stage.

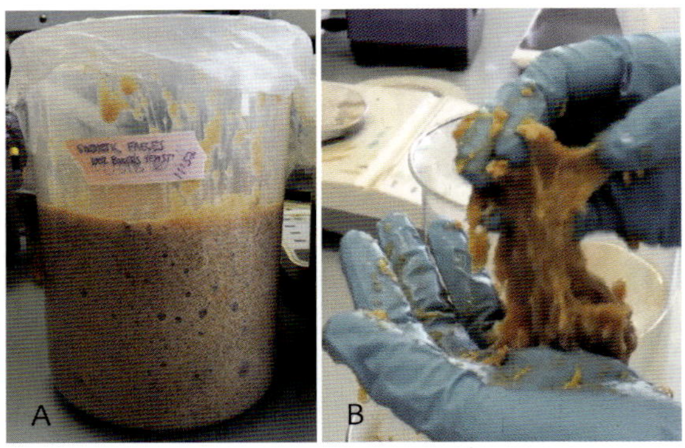

Figure 7.2 Synthetic faeces containing 30 wt % of the solids content baker's yeast. Photo A shows the mixture after standing at room temperature for 1.5 hours and photo B shows a sample of the mixture.

Two substitutions for the bacterial content (*i.e.* baker's yeast) of the adapted simulant were evaluated for shape formation (*i.e.* whether it could be shaped into a cylinder) and density. These substitutions include yeast extract and baking soda. The resulting optimal recipe was then analysed for its chemical and physical properties.

Replacing baker's yeast with yeast extract resulted in a simulant with representative physical properties (shape formation, viscosity, and density) and chemical properties (COD, TN, ES, pH, TS and VS). These results are shown in Table 7.9. Compared to simulant #11 the physical properties of the modified simulant were improved while the well-represented chemical properties were not affected. In addition, the disintegration of the modified simulant in turbulent flow revealed a disintegration mode similar to that of human faeces, with a similar time span (Penn *et al.*, 2019).

The density of this modified simulant was found to be 1.07 g/mL. Since faeces densities can be <1 g/mL (Table 7.2), two rising agents were tested as to whether they could be used to manipulate the density without losing the shaping capabilities. The two rising agents used were: (*i*) baker's yeast, which generates gas through fermentation, and (*ii*) sodium bicarbonate, which produces gas through a chemical reaction with acids in the mixture. The optimum quantity of baker's yeast was identified as 3 wt % of solids content. This amount of baker's yeast produced faeces with roughly the same buoyancy as water, with an average density of 0.99 g/mL. A range of water contents can be added to represent the span present in human faeces: from 65-80% moisture. When lower than 80 wt % water content is required, the portion of baker's yeast can be increased to a maximum of 5 wt % of solids content (in the case of a solid containing 65 wt % water) in order to facilitate quicker gas production. The density and viscosity of the modified simulant could be altered with varied yeast extract and water content fractions, respectively.

Replacing baker's yeast with sodium bicarbonate did not provide satisfying results. The minimum quantity of bicarbonate required for sufficient gas production to yield floating was 3 wt % of solids content in the recipe. However, the resulting product had an undesirable fluffy, sticky structure, and did not pass the shape formatting test.

Table 7.9 Comparison of chemical and physical properties of synthetic faeces from this study simulants #14a and #14b (Table 7.3) with real and artificial faeces from the literature.

Properties	Parameters	This chapter		Literature	
		Simulant #14a	Simulant #14b	Human faeces	Simulant faeces
Chemical	COD_{total} (gCOD/gTS)	1.117±0.056	1.194±0.162	0.567-1.450[10]	
	$COD_{soluble}$ (gCOD/gTS)	0.624±0.017	0.551±0.048	1.24[6]	1.33[9]
	TN (% of TS)	3.56±0.13	4.05±0.22		
	pH	5.4	5.2	5-7[10]	
	EC	6.06±0.17	6.40±0.25	2-3[6,8]	2.75[9]
	TS (%)	20.65±0.29	20.79±0.30	5.0-8.0 (avg. 6)[10]	
	VS (% of TS)	87.61±0.13	87.93±0.07	4.6-8.4[9]	5.3[9]
Physical	Viscosity (cPs at 50 rpm)	6,360	4,640		5.7[9]
	Density (g/ml)	1.07±0.02	0.98±0.05	14-37[10]	

[1]Snell, 1943; [2]Fry, 1973; [3]Meher *et al.*, 1994; [4]Yeo and Welchel, 1994; [5]Brown *et al.*, 1996; [6]Jönsson *et al.*, 2005; [7]Wignarajah *et al.*, 2006; [8]Barman *et al.*, 2009; [9]Colón *et al.*, 2015; [10]Rose *et al.*, 2015;
Note: Average ± standard deviation were calculated from three replicates;
Note: Results are for synthetic faeces containing 80 wt % water.

7.4.1.2 Synthetic faecal sludge

The synthetic faeces developed by the authors as described in Section 7.4.1.1 (simulant #14(a), Table 7.3) was combined with synthetic urine (Table 7.5) and water to produce a synthetic faecal sludge for dewatering studies. The simulant was chemically very similar to simulant #16 (Table 7.3) and to fresh faecal sludge, however it displayed a 60% reduced dewaterability compared to real fresh faecal sludge. In this case, dewaterability is defined as the percentage of dry solids in the dewatered cake after centrifugation, which was 11% and 4.5% for fresh faecal sludge and synthetic fresh faecal sludge, respectively (Ward *et al.*, 2017a, b). This is likely due to the high water-binding affinity of the psyllium husk included in the simulant. For further dewatering experiments, a faeces simulant with a reduced proportion of water-binding components could be evaluated.

7.5 CONCLUSIONS

The use of synthetic faeces and synthetic faecal sludge enables replicable experimentation, while simultaneously reducing health risks. There are a multitude of simulants for faeces in the literature, however, they are still relatively scarce for faecal sludge. At this stage, simulants have for the most part been developed for specific purposes, and simulants that are mutually representative of physical, chemical, biological and thermal properties are still lacking. It is important to develop recipes including COD fractionations for detailed biochemical process, and potentially other properties such as pharmaceuticals and hormones, pathogens and odours. The compilation of existing simulants in this chapter has been valuable for the identification of strengths and weaknesses of simulants, and areas for further research.

A critical analysis of the literature yields the following conclusions:

- Synthetic faeces and faecal sludge are very useful for conducting research, but cannot entirely replace research with real faeces and faecal sluge.
- As with any surrogate, the results have to be validated with real faeces and faecal sludge.
- Standardisation and validation of other results can be significantly increased through the use of standard methods for the characterisation of faeces and faecal sludge.

APPENDIX 7.1 EXAMPLES FOR THE DEVELOPMENT OF FAECES AND FAECAL SLUDGE SIMULANTS

Example 7.1 Development of a new faeces simulant by Eawag, Switzerland

For research into the fate of excreta in urban sewers and in onsite sanitation systems, both the chemical/biological aspects of faeces and its physical properties are important. Investigations of its fate include its physical motion (movement, settling and sedimentation) and its physical and biochemical disintegration in sewer pipes and in onsite sanitation systems. Both types of chemically related simulants (simulant #11 (Table 7.3) and simulant #12 (Table 7.3) showed poor physical resemblance, as discussed above (Table 7.8). Similarly, none of the simulants with proper physical parameters have an adequate chemical composition. In the following experimental sections, substitutions for the bacterial content (*i.e.* baker's yeast) of the adapted simulant were evaluated for shaping capability and density. These substitutions include yeast extract and baking soda. The resulting optimal recipe was then analysed for its chemical and physical properties.

Material and methods

Chemicals and materials used

Table 7.10 lists the materials and chemicals used for preparation of the simulant.

Table 7.10 Chemicals and materials used for the simulant

Component	Chemical/material	CAS number
Yeast extract	Sigma Aldrich	8013-01-2
Cellulose	Sigma Aldrich	9004-34-6
Oleic acid	MP Biomedicals LLC	112-80-1
NaCl	Merck KGaA	7647-14-5
KCl	Fluka Chemika GmbH	7447-40-7
$CaCl_2 \cdot 2H_2O$	E. Merck	10035-04-8
Baker's yeast	Dry, Betty Bossi, COOP, Switzerland	
Psyllium husk	Govinda Nature GmbH	
Miso paste	Seasoned soybean paste HACCP, TS content ~48%	

Measurement methods

Total chemical oxygen demand (COD_{total}), soluble COD ($COD_{soluble}$), total nitrogen (TN), ammonium nitrogen (NH4-N), total solids (TS), volatile solids (VS), pH, and electrical conductivity (EC) were determined based on standard methods (Rice *et al.*, 2017). Hach LCK test kits were used to measure COD_{total} and $COD_{soluble}$, TN, and NH_4-N with a Hach DR 6000 spectrophotometer. EC and pH were measured using a WTW Multi 3320 following the procedure described in Colón *et al.* (2015), by diluting synthetic faeces in DI water at a 1:5 w:v ratio. Viscosity was measured with a Brookfield DVII-LV viscometer using a #64 spindle at 50 rpm with a 30-second measurement time.

Physical structure of the synthetic faeces, *i.e.* its shaping capabilities, was evaluated by attempting to shape it into a cylinder, following the normal stool form according to the Bristol Stool Form Chart (Lewis and Heaton, 1997). Approximately 100 g of synthetic faeces was handled and rolled gently into a cylinder, while wearing wetted nitrile gloves. If the material was too sticky, gooey, or liquid to form a cylinder, it failed the shape test.

Buoyancy of the synthetic faeces was evaluated by placing a piece of prepared substance in a beaker filled with water. Floating or sinking performance of the faeces was recorded.

The estimated density was measured by weighing a 40 g portion of simulant and placing it in a 1,000 mL graduated cylinder filled with 600 mL of deionised water. The increase in volume was measured, and the density was calculated. In order to reduce the uncertainty in this measurement, a pycnometer could be used in future experiments. An average and standard deviation from three repetitions was calculated.

Base synthetic faeces recipe

The range of recipes for preparation of 1 kg of synthetic faeces is presented in Table 7.11.

Table 7.11 Ingredients for basic recipe of the simulants S80 and S65, all quantities are in grams.

Water content (% TS)(A)	80% (S80)		65% (S65)	
	SB80(B)	SE80(C)	SB65(B)	SE65(C)
Yeast extract	65.06	72.29	105.42	126.51
Baker's yeast	7.23	0.00	21.08	0.00
Microcrystalline cellulose	24.10	24.10	42.17	42.17
Psyllium	42.17	42.17	73.80	73.80
Miso paste	42.17	42.17	73.80	73.80
Oleic acid	48.19	48.19	84.34	84.34
NaCl	4.82	4.82	8.43	8.43
KCl	4.82	4.82	8.43	8.43
$CaCl_2 \cdot H_2O$	2.75	2.41	4.81	4.81
DI Water	758.7	758.7	577.72	577.72
Final mass 'faeces'	1,000.00	1,000.00	1,000.00	1,000.00

[A]The water content was determined by TS measurements; [B]Simulants starting with SB contain baker's yeast and yeast extract; [C]Simulants starting with SE contain only yeast extract.

Table 7.12 Results of physical testing for synthetic faeces S80 with different quantities of rising agents (baker's yeast and sodium bicarbonate).

Rising agent	Amount of rising agent added (wt % of solids content in recipe[A])	Shapable?[B]	Floats?[B]	Waiting time (h)[C]
None	0.0	yes	no	1.5
Baker's yeast	30.0*	no	yes	1.5
	15.0	no	yes	1.5
	10.0	no	yes	1.5
	5.0	no	yes	1.5
	3.0	yes	yes	1.5
	1.4	yes	yes	3.0
	0.9	yes	yes	48.0
Baking soda	15.0	no	yes	1.5
	3.0	no	yes	1.5
	1.0	no	no	1.5
	0.4	yes	no	1.5

[A]Original recipe from Colón *et al.* (2015); [B]Results from synthetic faeces made with 80% water (actual water content obtained from TS measurements of the simulant). In each recipe, wt % rising agent + wt % yeast extract = 30 wt % of solids content; [C]Time needed for the mixture to stand at room temperature

Experiments

Identification of the base recipe

With the goal of producing a simulant to be used for investigating the fate of faeces in sewer systems and in onsite sanitation systems, which will resemble human faeces in both its physical and chemical properties, an adaption of one of the reviewed simulants was undertaken. Modifying the physical simulants also represent also the chemical properties of human faeces was found to be impracticable. Both simulant #11 (Table 7.3) and simulant #12 (Table 7.3) showed high compatibility in terms of elemental content but poor physical resemblance in terms of shapable capabilities and rheology (Table 7.8).

Simulant #11 shows high compatibility in its chemical properties important for wastewater related research, including COD, TN, TS, and VS. Baker's yeast is used to represent microbial biomass and to produce floating stool (due to gas produced by the yeast). However, the quantity included in this recipe creates an unfavourable physical structure, as explained above and later demonstrated in the results. It was hypothesised that by adapting the baker's yeast content of simulant #11 (Table 7.3), a physical representative simulant can be produced, while still maintaining its chemical resemblance.

In order to consider simulant #12 (Table 7.3) as a good base for further development, its additional wastewater-related chemical properties (*i.e.* COD, TN, TS, VS) would first need to be analysed. If these results showed a close resemblance to human faeces, its shaping capabilities and density would then need to be further adjusted to replicate those of human faeces. However, since the results from adapting simulant #11 showed good chemical and physical resemblance, modification of simulant #12 was not investigated further.

Density adjustments

After identifying the base recipe, a series of experiments were performed to adjust the density of the simulants. For each formulation of yeast and baking soda, shapable capability and floating tests were conducted. The time required for the simulant to float was recorded.

Quantities of two rising agents, baker's yeast, which generates gas through fermentation, and sodium bicarbonate, which produces gas through a chemical reaction with acids in the mixture, were tested to determine whether they could be used to manipulate the density without losing the shapable capabilities. These tests were conducted on simulants containing 80% and 65% water, S80 and S65 respectively, corresponding to the reported maximum and minimum water content expected in human faeces. The corresponding ingredients are listed in Table 7.11. The water content was determined by TS measurements and not only by the volume of water added, since miso paste also contains water.

Different formulations of baker's yeast and yeast extract were tested. The total yeast content was held constant at 30% (dry weight by dry weight), but the ratio of these two forms of yeast were varied. Reduced quantities of baker's yeast were replaced by respective quantities of yeast extract. Quantities of baker's yeast examined were 0, 0.9, 1.4, 3, 5, 10, 15, and 30 wt % of the recipe's solids content. The activity of the yeast depends on the temperature, amount of yeast added and substrate availability. The optimal quantity of baker's yeast was determined when a simulant obtained the desired cylinder shape and buoyancy properties after approximately 1.5 h at room temperature (23 °C). 1.5 hours is the minimum time required for the psyllium husks to gel. It further should enable a relatively 'comfortable' time range (not less than an hour with preference for longer) in which the simulant maintains its physical structure.

Replacing baker's yeast with sodium bicarbonate as an alternative to the biological gas production was further examined. Quantities of bicarbonate examined were 0.4, 1, 3, 5 and 15 wt % of the recipe's solids content.

Physical and chemical properties

Once the optimum formulation was obtained, chemical properties and viscosity of two types of simulant S80 were evaluated. These simulants include SB80, made with baker's yeast and yeast extract, and SE80, made with only yeast extract. As addition of bicarbonate showed poor results, simulants containing bicarbonate were not analysed further for their chemical properties and viscosity. Density was evaluated for these two types and for SB65 and SE65, *i.e.* simulant S65 made with baker's yeast and yeast extract, and only yeast extract, respectively.

Results and discussion

Physical structure

Synthetic faeces SE80 and SE65, *i.e.* both simulants not containing baker's yeast, immediately sank when added to water, with an average density of 1.07 g/ml and standard deviation (SD) of 0.02 for SE80 and a density of 1.12 g/ml with SD of 0.05 for SE65. Densities resemble the density of an NBS solid

(Brown *et al.*, 1996). These simulants were easily shaped (Figure 7.3) and sank when placed in standing water.

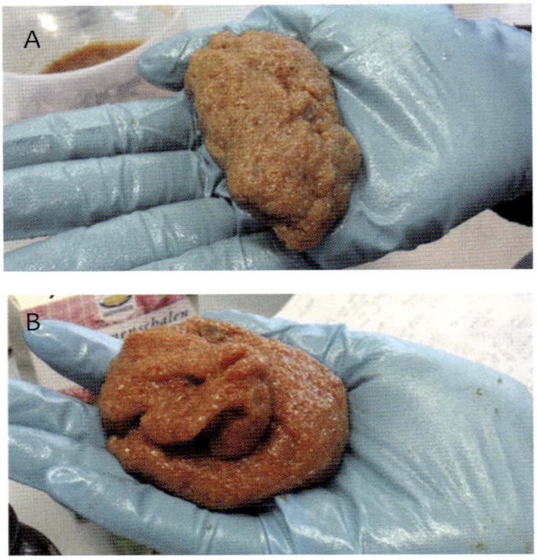

Figure 7.3 Simulants with 30 wt % of solids content yeast extract and no baker's yeast: A) SE65; B) SE80.

A summary of the physical characteristics of synthetic faeces made with the different amounts of rising agents (baker's yeast and sodium bicarbonate) is shown in Table 7.12. The results presented are for simulant S80. Adding baker's yeast contents of more than 3 wt % created a gassy and sticky material that floated when added to water, but was too sticky to be shaped into a cylinder. The resultant simulant did not represent the physical structure of human faeces. An extreme example can be depicted in Figure 7.2 where one can observe high gas quantities, shown by the many bubbles in the beaker (Figure 7.2A) and a very sticky material that could not be shaped into a cylinder (Figure 7.2B). Addition of smaller quantities of baker's yeast (1.4 wt % of solids content or lower) resulted in a long delay in yeast activation. These simulants eventually floated in water, but only standing at room temperature for between 3 hours to 2 days.

The optimum quantity of baker's yeast was therefore identified as 3 wt % of solids content, *i.e.* simulant SB80. This amount of baker's yeast produced faeces with roughly the same buoyancy as water, with an average density of 0.99 g/mL and SD of 0.05 (Figure 7.4). Simulants SB65 required a longer period of 4 hours (compared to the 1.5 hours mentioned above) for the yeast to produce sufficient gas to enable floating of the stool. Increasing baker's yeast quantity to 5 wt % of solids content enabled floating of the simulant, while maintaining its physical properties, in a shorter period of 2 hours. The average density of this simulant was found to be 0.96 g/ml with SD of 0.005. Replacing baker's yeast with sodium bicarbonate did not provide satisfying results. For simulant S80 the minimum quantity of bicarbonate required for sufficient gas production to yield floating was 3 wt % of solids content in the recipe. However, the resulting product had an undesirable fluffy, sticky structure, and did not pass the shapable capability test.

Figure 7.4 Density tests for synthetic faeces (A) S65 and (B) S80.

Table 7.13 Properties of the two simulants identified as most closely representing the range of human faeces.

Simulant		Density (g/mL)	
		Average	SD
Sinking stool (baker's yeast is not added)	SE80	1.07	0.020
	SE65	1.12	0.050
Floating stool (baker's yeast is added)	SB80	0.99	0.050
	SB65	0.96	0.005

As a result of these physical tests, two recipes were identified to most closely represent the range of human faeces, according to Table 7.13.

The addition of baker's yeast resulted in a simulant with a weaker structure, corresponding to the lower viscosity measured. Simulant made with baker's yeast was less robust to handle, and disintegrated more rapidly upon immersion in water than simulant made without baker's yeast. Higher water content also resulted in a simulant with decreased structural strength. Ongoing research conducted by the authors of this paper includes examination of physical disintegration of faeces in turbulent flow conditions. The experiments are being conducted on the reported simulant and verified by real human faeces.

Chemical composition

The modified simulant was analysed for chemical properties of interest to wastewater treatment and compared to properties found in the literature as presented in Table 7.14 (Snell, 1943; Fry, 1973; Meher *et al.*, 1994; Jönsson *et al.*, 2005; Wignarajah *et al.*, 2006; Barman *et al.*, 2009; Colón *et al.*, 2015; Rose *et al.*, 2015). Results are presented only for simulants S80. The synthetic faeces developed in this study provide compatible chemical and physical properties resembling real human faeces. The simulants are appropriate candidates for replacing human faeces in investigations into faeces physical and biochemical disintegration in sewer systems and in onsite sanitation systems.

Table 7.14 Comparison of chemical and physical properties of synthetic faeces SE80 and SB80 with real and simulant faeces from the literature.

Properties	Parameters	This chapter		Literature	
		SE80 (Table 7.10)	SB80 (Table 7.10)	Human faeces	Simulant faeces
Chemical	COD_{total} (gCOD/gTS)	1.117±0.056	1.194±0.162	0.567-1.450[(10)], 1.24[(6)]	1.33[(9)]
	$COD_{soluble}$ (gCOD/gTS)	0.624±0.017	0.551±0.048		
	TN (% of TS)	3.56±0.13	4.05±0.22	5-7[(10)], 2-3[(6,8)]	2.75[(9)]
	pH	5.4	5.2	5.0-8.0[(10)], 4.6-8.4[(9)]	5.3[(9)]
	EC	6.06±0.17	6.40±0.25		5.7[(9)]
	TS (%)	20.65±0.29	20.79±0.30	14-37[(10)], 15-35[(7)]	18.4[(9)]
	VS (% of TS)	87.61±0.13	87.93±0.07	92[(10)], 80-92[(1,2,3)]	88.5[(9)]
Physical	Viscosity (cP)	6,360	4,640	3,500-5,500[(4)]	
	Density (g/mL)	1.07±0.02	0.98±0.05		1.06[(5)]

[1]Snell, 1943; [2]Fry, 1973; [3]Meher *et al.*, 1994; [4]Yeo and Welchel, 1994; [5]Brown *et al.*, 1996; [6]Jönsson *et al.*, 2005; [7]Wignarajah *et al.*, 2006; [8]Barman *et al.*, 2009; [9]Colón *et al.*, 2015; [10]Rose *et al.*, 2015; Note: Average ± standard deviation calculated from three replicates; Note: Results are for synthetic faeces S80.

Recommendation for recipes

Based on detailed chemical and physical characterisation, the two most suitable recipes have been selected for providing chemical and physical properties similar to those of human faeces. Recommended recipes are presented in Table 7.15.

Table 7.15 Recommended recipes for synthetic faeces solids.

Component	Composition of solids content (wt %)	
	Yeast extract	Baker's yeast + yeast extract
Baker's yeast	0.0	3.0
Yeast extract	30.0	27.0
Microcrystalline cellulose	10.0	10.0
Psyllium husk	17.5	17.5
Miso paste	17.5	17.5
Oleic acid	20.0	20.0
NaCl	2.0	2.0
KCl	2.0	2.0
$CaCl_2$	1.0	1.0

A range of water contents can be added to represent the range present in human faeces: from 65 to 80% moisture. Baker's yeast should be added if floating faeces are desired. When lower than 80 wt % water content is required, the portion of baker's yeast can be increased to a maximum of 5 wt % of solids content in order to facilitate quicker gas production.

Figure 7.5 presents the steps for making synthetic faeces as outlined in this case study.

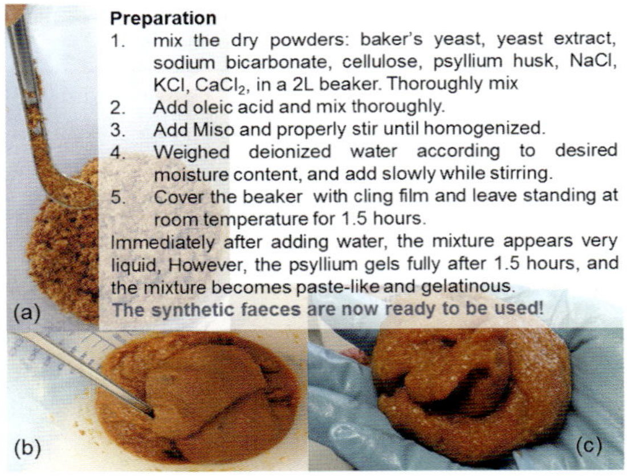

Figure 7.5 Procedure for preparation of synthetic faeces; (a) mixture of synthetic faeces prior to addition of water; (b and c) mixture of prepared synthetic faeces containing ~80% water (b) prepared mixture after standing for 1.5 hours; (c) structured faeces. Tip: Use wet hands, preferable gloves, because the synthetic faeces contain substantial amount of oil.

Recommended storage practices

Baker's yeast produces gas via a biological process which is time and temperature sensitive. It was observed that the mixture should be held at room temperature for at least 1.5 hours but not more than 4 hours in order to produce the required amount of gas for floating synthetic faeces. Results were obtained at room temperature of ~23 °C; higher temperatures will shorten the time interval, and lower temperatures will lengthen it. The synthetic faeces can be refrigerated for a period of not more than 24 hours if they contain baker's yeast or one week if they do not contain baker's yeast. Additionally, both mixtures can be held in the freezer for a longer period of time (not evaluated for more than one month). The frozen synthetic faeces containing baker's yeast should be allowed to reach room temperature, until the point at which the yeast will again become active. Activity can be confirmed by examining the floating of the simulant. Further investigations are needed to verify that chemical and physical properties of the simulant will not change due to freezing as freezing and thawing may change the properties of the recipe material. ■

Example 7.2 *Development of new simulants by the Pollution Research Group of UKZN, S. Africa*

The Pollution Research Group based at the University of KwaZulu-Natal (UKZN PRG) in Durban, South Africa, has developed, tested and characterised synthetic simulants for fresh faeces and faecal sludge for use in laboratory trials to test treatment methods and processes. In addition to making reproducible substrates for laboratory experimentation, another need for a faecal sludge simulant that arose was health and safety in the handling of faecal matter, for example testing and demonstrating new toilet technologies during the Reinvent the Toilet Fair in India in 2014. Presented in this example are the evolution of the developed synthetic simulants and their comparison to faecal sludge, faeces and other synthetic simulants reported in the literature. The process of simulant development is presented in two stages - the development of a synthetic simulant for the Reinvent the Toilet Fair, and the further development of simulants for laboratory testing.

Development of a synthetic simulant for prototype testing of innovative toilet technologies

The purpose of this study was to develop a uniform simulant that matched as closely as possible with the properties of faecal sludge, to be used for the demonstration of innovative sanitation treatment systems during the Reinvent the Toilet Fair. The properties are presented in Table 7.16, and the appearance of the final simulant is shown in Figure 7.6.

During this study, experiments were carried out with various recipes for faecal simulants found in the literature, mainly from the University of Colorado Boulder, Duke University and Wignarajah *et al.* (2006). All the simulants were tested for the following properties: moisture content; total, fixed, volatile and suspended solids; sludge volume index; chemical oxygen demand; pH; density; thermal conductivity; heat capacity; calorific value; rheology and particle size distribution, and then compared to the same properties for faecal sludge and fresh faeces. The faecal sludge and faeces samples were obtained from onsite sanitation facilities (dry and wet ventilated improved pit latrines, community ablution blocks and urine diversion toilets) in the eThekwini Municipal area around Durban, South Africa (Velkushanova *et al.*, 2019, Zuma *et al*, 2015). Standard operational procedures were followed for all the analysed properties and repeated for all samples in order to ensure compatibility.

Figure 7.6 Faecal sludge simulant developed for prototype technologies at the Reinvent the Toilet Fair in India, 2014 (www.mentalfloss.com/article/56003/recipe-fake-poop).

Thirteen trial recipes, named Simulant Trials (ST) 1 to 13 (Table 7.16) were prepared, and modified in order to match more closely the properties of faecal sludge and faeces. The recipe for each simulant was prepared by adding the ingredients following the sequence presented in Figure 7.7.

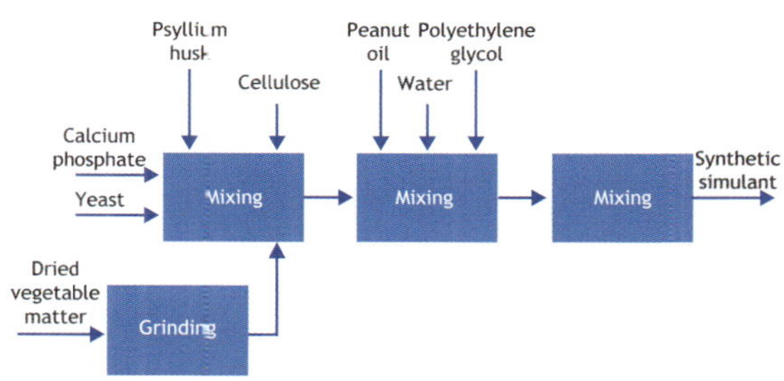

Figure 7.7 Process flow diagram showing the procedure followed in the preparation of synthetic faecal simulants at UKZN PRG.

Table 7.16 A list of ingredients and the respective quantities used in the trials /development of a simulant for the RTT Fair in India (2014).

Component used	Simulant trials												
	ST1	ST2	ST3	ST4	ST5	ST6	ST7	ST8	ST9	ST10	ST11	ST12	ST13
Instant yeast (g)			50			50				18.8	18.8	37.5	37.5
Nutritional yeast (g)	75			75.0			75	75	18.8				
Fresh yeast (g)		30			30								
Baker's yeast (g)													
Cotton balls & paper towels (g)	37.5												
Sawdust (g)								37.5					
Shredded tissue paper (g)		10	75	37.5	10		37.5						
Cotton linters & tissue paper (g)											37.5	6.25	37.5
Sawdust & tissue paper (g)									37.5				
Paper towels (g)						75							
Cotton linters (g)										37.5			
PEG-400 (g)	50		100	50		100	50	50	12.5	14	7	12.5	14
Psyllium husk (g)	12.5	17.5	75	12.5	17.5	75	12.5	12.5	12.5	12.5	12.5	12.5	12.5
Peanut oil (g)	50		25	12.5		25	12.5	25	25	25	25	20	20
Oleic acid/olive oil (g)		20			5								
Miso paste (g)	12.5	17.5	150	12.5	17.5	150	12.5	12.5	12.5	12.5	12.5	12.5	12.5
Calcium phosphate (g)	12.5		25	12.5		25	12.5	12.5	12.5	6.25	6.25	12.5	12.5
NaCl (g)		2			2								
KCl (g)		2			2								
$CaCl_2$ (g)		1			1								
Water (mL)	400	400	400	400	400	400	500	500	450	900	900	400	360

Table 7.17 Comparison of properties of faecal sludge simulants with real faecal sludge and faeces.

Type of onsite sanitation facility	Total solids	Moisture content	Suspended solids	Volatile solids	Ash	Sludge Volume Index	pH	COD	Heat capacity	Thermal conductivity	Calorific value	Density
	%	%	(mg/L)	(g/g)[A]	(g/g)[A]	(mL/mg)		(g/g)[A]	(J/kg.K)	(W/m.K)	(MJ/kg)	(kg/m³)
Dry VIP	21	79	577	0.58	0.42	0.04	7.60	0.69	2,530	0.54	14.06	1,379
Wet VIP	21	79	402	0.54	0.46	0.06	7.59	0.69	2,422	0.55	13.08	1,447
UDT	40	60	245	0.45	0.27	0.23	7.54	0.49	2,150	0.38	12.93	1,450
CAB	23	77	139	0.49	0.51	0.51	7.44	0.65	3,268	0.60	14.31	1,350
Fresh faeces	24	76		0.87	0.13			0.66			22.64	
ST1	35	65	109	0.96	0.04	0.09	6.37	0.97	2,337	0.45	24.22	1,384
ST2	11	89	121	0.87	0.13	0.82	5.08	0.65	2,878	0.52	24.38	1,272
ST3	42	58	396	0.90	0.10	0.01	4.96	0.83	2,573	0.44	19.37	1,232
ST4	34	66	105	0.91	0.09	0.19	5.54	2.28	2,281	0.42	18.94	1,268
ST5	11	89	88	0.86	0.14	0.91	5.03	0.73	2,920	0.53	17.48	936
ST6	42	58	121	0.86	0.14	0.41	4.10	2.15	3,001	0.38	20.30	1,340
ST7	14	86	78	0.93	0.07		6.37		2,181	0.56		1,756
ST8	29	71	27	0.92	0.08	4.36	5.52	14.51	2,691	0.49	24.14	1,756
ST9	24	76	20	0.89	0.11	2.42	5.64	2.16	3,312	0.50	21.89	1,308
ST10	13	87	144	0.88	0.11	1.73	5.95	2.70	2,868	0.55	21.95	1,068
ST11	15	85	113	0.88	0.12	1.32	5.97	1.80	3,199	0.52	24.18	1,316
ST12	**19**	**81**	**141**	**0.75**	**0.12**	**1.76**	**5.92**	**1.59**	**2,700**	**0.50**	**22.17**	**1,300**
ST13	23	77	177	0.85	0.25	1.52	5.91	1.04	3,040	0.50	22.17	1,156

[A](g/g dry sample)

The comparative results between faecal sludge and faeces, and the simulant faecal sludge are presented in Table 7.17 and Figure 7.8. On the basis of these results, the recommended simulant for the Reinvent the Toilet Fair was ST12 (Table 7.16). It was recommended that the recipe should be prepared by adding the ingredients in the indicated sequence with constant stirring until a smooth and homogeneous texture is achieved.

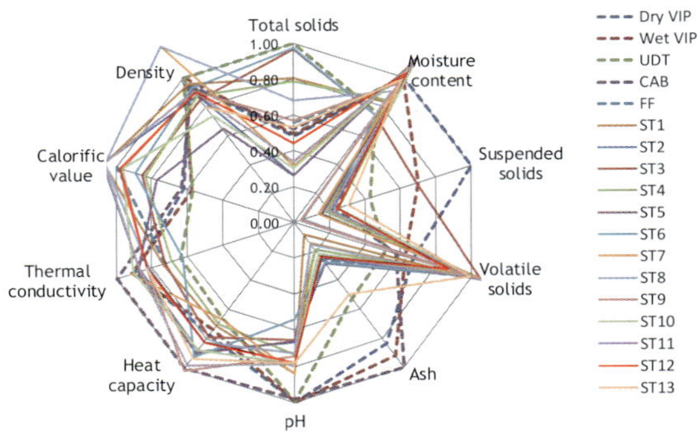

Figure 7.8 Comparison between different properties of faecal sludge with the synthetic simulants to establish which of the developed simulants had the best match with faecal sludge from different onsite sanitation facilities.

Further development and improvement of synthetic simulant for laboratory experiments and testing of pilot technologies

Following the Reinvent the Toilet Fair, the UKZN PRG continued to conduct experiments to improve the simulant ST12. The modifications were based on feedback for improvement by users of the simulant, and also to simplify the preparation process. For the sake of simplicity, in the following text simulant ST12 is hereafter referred to as S1, and the subsequent modified simulant as S2 (Table 7.18). The goal was that S1 would resemble more closely properties of faecal sludge from onsite containments, and S2 would resemble more closely properties of fresh faeces with a smoother consistency. S1 (Figure 7.10) was consecutively modified by the substitution of ingredients and adjustment of ratios of ingredients to create S2. In addition, five other synthetic simulants based on recipes in Wignarajah *et al.* (2006) were produced (Figure 7.9) in order to compare their properties with the developed S2, and to verify the properties of S2. These simulants were selected as they were used as a base for the development of S1. They represented different simulants of fresh faeces, but actual characteristics/properties were not reported in Wignarajah *et al.* (2006). Based on the characterisation carried out by UKZN PRG, the two simulant recipes S6 and S7 (Wignarajah *et al.*, 2006) were selected for comparison with the rest of the simulants in this study as they had the closest match to properties of fresh faeces, and they are presented in this case study.

Table 7.18 A list of recipes used in the formulation of synthetic faecal simulants S1 and S2.

Ingredient represents	Component used (in g)	S1	S2
Micro-organisms	Instant yeast	18.20	18.21
Cellulose	Cellulose (powder)		2.13
	Cotton linters (50%) and shredded paper (50%)	3.10	
Water retention	PEG (polyethene glycol) 400	6.08	6.08
Fibre/carbohydrate	Psyllium husk	6.08	6.09
Fat	Peanut oil	9.70	9.71
Fibre/protein/fats	Miso paste	6.08	6.08
Minerals	Calcium phosphate	6.08	6.08
Vegetable matter	Vegetable matter		1.04
Water	Water	194.03	113.76

Figure 7.9 Mixed simulants for analysis at the UKZN PRG laboratory: A) PRG (S2), B) NASA2 (S6), C) NASA1, D) NASA3, E) NASA5 and F) NASA4 (S7).

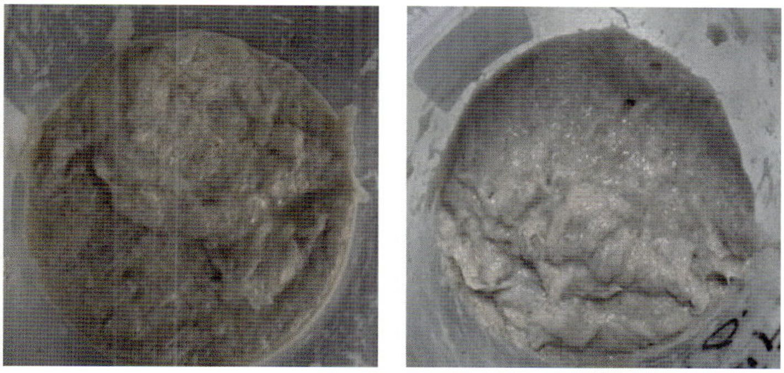

Figure 7.10 Faecal sludge simulant S1.

A description of all the synthetic simulants that were compared with faeces and faecal sludge in this case study are included in Table 7.19.

Table 7.19 A description of the labels of simulants, real fresh faeces and faecal sludge.

Label	Description
S1	Simulant developed by UKZN PRG in 2014 for the RTT Fair in India (2014).
S2	Simulant developed by UKZN PRG in 2015 based on modifying the simulant, S1.
S3	Faecal sludge simulant developed by Deering et al. (2018), based on modifying the simulant, S1.
S4	Fresh faeces simulant #14 (a) developed by Penn *et al.*, 2017.
S5	Fresh faeces simulant developed by Colón *et al.*, 2015.
S6	Fresh faeces simulant developed by Wignarajah *et al.*, 2005 (combination 2).
S7	Fresh faeces simulant developed by Wignarajah *et al.*, 2005 (combination 4).
FF1	Fresh faeces properties reported by Nwaneri, 2009.
FF2	Fresh faeces properties reported by Jönsson *et al.*, 2005.
FF3	A set of eight fresh faeces samples (a-h) on rheological properties reported by Woolley *et al.*, 2014.
FS1	Faecal sludge properties from dry VIPs reported by UKZN PRG, 2014.
FS2	Faecal sludge properties from CABs reported by UKZN PRG, 2014.

Samples of the selected simulants were analysed to provide a characterisation of the following properties: pH, COD_{total}, density, total solids, moisture content, volatile solids, ash, viscosity, thermal conductivity, heat capacity, calorific value and rheology. These properties were compared against properties of faecal sludge collected from ventilated improved pit (VIPs) latrines in Durban. All the samples were analysed using the standard operational procedures (SOPs) that are presented in this book. The source of fresh faeces (FF) samples used for comparison with simulants is presented in Table 7.19. Table 7.20 summarises some of the data available in the literature on variations of properties. A comparison of the properties of simulants S1 and S2 to other simulants and faeces and faecal sludge is presented in Table 7.21.

Table 7.20 Physical, chemical and thermal properties of fresh human faeces and faecal sludge found in the literature.

Faecal matter type	Parameter / property	
	Fresh faeces	Faecal sludge
Total solids (%)	14-37 [1,2,3]	4-91 [9,10]
Moisture content (%)	63-86 [1,2,3]	9-96 [9,10]
Volatile solids (g/g dry sample)	0.80-0.92 [1]	0.01-0.84 [9,10]
Ash (g/g dry sample)	0.08-0.20 [1]	0.16-0.99 [9,10]
Density (kg/L)	1.06-1.09 [4]	0.54-2.34 [2,9,10]
COD (gCOD/gTS)	0.6-1.5 [1,2]	0.01-5.01 [9,10]
pH	4.6-8.4 [2,5]	4.5-9.1 [9,10,11]
Heat capacity (J/kg.K)	3,200-4,200 [6]	707-4,773 [9,10]
Thermal conductivity (W/m.K)	0.35-0.6 [7]	0.09-0.79 [9,10]
Calorific value (MJ/kg)	20-25 [7], 15.1-25.1 [8]	2-25 [9,10]

[1]Rose *et al*., 2015; [2]Penn *et al*., 2017; [3]Wignarajah *et al*., 2006; [4]Levitt and Duane, 1972; [5]Colón *et al*. 2015; [6]Makununika, 2016; [7]Chikava and Velkushanova, 2014; [8]Wierdsma *et al*. 2014; [9]Zuma *et al*., 2015; [10]Velkushanova, 2014; [11]Afolabi and Sohail, 2017.

Table 7.21 Comparison of properties of faecal matter simulants developed by the UKZN PRG (S1 and S2), simulants presented in the literature and described in Table 7.18, and samples of faecal sludge and faeces described in Table 7.19.

Label	Chemical properties		Physical properties					Thermal properties		
	COD total (gCOD/gTS)	pH	Density (kg/m^3)	Total solids (%)	Moisture content (%)	Volatile solids (%)	Ash	Thermal conductivity (W/m.K)	Heat capacity (J/kg.K)	Calorific value (MJ/kg)
S1	1.59	5.9	1,300	19	81	70	30	0.50	2,700	22
S2	1.19	5.3	1,081	29	71	83	17	0.45	3,476	
S3			1,365	19	81	76	24	0.32	2,609	
S4	1.12	5.4	1,070	21	79	88	12			
S5	1.12	5.2	980	21	79	87	12			
S6	1.33	5.3	1,060	18	82	88	12			
FF1	1.24	5.1		24	76	79	21			23
FF2	1.13	5.3		22	78	84	16			20
FF3	1.45	7.0		14	86	89	11			21
FS1	0.69	7.6	1,379	21	79	58	42	0.54	2,530	14
FS2	0.65	7.4	1,350	23	77	49	51	0.60	3,268	14

Physical properties

Overall, based on a comparison of physical properties, simulant S2 demonstrated the closest match to fresh faeces and therefore can be used as a substitute for fresh faeces in applications targeting these physical properties. Simulant S1 was more suited as a substitute for faecal sludge in applications targeting specifically the total solids, moisture content and density. This is outlined in more detail in the following sections and in Table 7.20.

Chemical properties

A comparison of the chemical properties of faecal simulants (S1 and S2) was carried out relative to fresh faecal samples, faecal sludge (from household VIP latrines and CABs) and other simulants found in literature. The pH of simulant S1 (5.92) was higher than, but comparable to that of S2 (5.29). The pH values of both synthetic simulants were comparable to those of fresh faeces (FF1 and FF2) and other synthetic faecal simulants (S4, S5, S6 and S7), but lower than faecal sludge (FS1 and FS2). Nonetheless, the pH of both simulants (S1 and S2) falls within the range for both fresh faeces and faecal sludge as indicated in the literature (Table 7.20). The COD of simulant S1 was higher compared to faecal sludge (FS1 and FS2) and fresh faeces. In contrast, the COD of simulant S2 was comparable to that of fresh faeces (FF1 and FF2) and synthetic simulants (S4 and S6), but it was also higher compared to faecal sludge (FS1 and FS2). It is however, important to note that the COD of both simulants (S1 and S2) fall within the range of fresh faeces and faecal sludge as indicated in literature (Table 7.20). In overall, simulants S1 and S2 demonstrated properties similar to fresh faeces and faecal sludge for applications targeting chemical properties of faecal matter such as COD and pH.

Thermal properties

The thermal conductivity of simulant S1 (0.5 W/m.K) was similar to simulant S2 (0.45 W/m.K and for both synthetic simulants it was comparable to that of faecal sludge samples from dry VIP toilets (FS1) and other faecal simulants S6 and S7. Both simulants (S1 and S2) indicated thermal conductivity properties that fall within the range for both faecal sludge and fresh faeces (Table 7.21). It was also observed that the thermal conductivity of simulant S3 is considerably lower compared to that of S1 and S2; this was attributed to the use of brewer's yeast instead of instant yeast (Deering *et al.*, 2018) though no further tests or analysis were presented by the authors to validate this argument. The heat capacity of simulant S1 (2,700 J/kg.K) is lower as compared to that of S2 (3,476 J/kg.K). However, it can be observed that the heat capacity of simulant S1 is similar to that of faecal sludge from VIP toilets (FS1) whereas that of S2 compares well with that of faecal sludge from community ablution blocks (FS2) which was more diluted. In general, the heat capacity of simulants S1 and S2 fall within the range indicated for both fresh faeces and faecal sludge (Table 7.21). The calorific value of simulant S1 (22 MJ/kg) was comparable to that of fresh faeces (FF1 and FF2), but was considerably higher relative to that of faecal sludge from household VIP latrines and CABs. The calorific value of simulant S2 was not analysed. Overall, simulants S1 and S2 demonstrated properties similar to fresh faeces and faecal sludge for applications targeting thermal properties of faecal matter, namely thermal conductivity and heat capacity.

Mechanical properties

The set of mechanical properties analysed were rheological properties and particle size distribution. The simulants demonstrated shear thinning (and viscosity reduction) with higher shear rate: behaviour similar to faecal sludge and fresh faeces (Figure 7.11).

A comparison of the viscosity with shear rate of faecal simulants (S1 and S2) relative to fresh faeces samples (FF_3a - FF_3h), faecal sludge from VIP toilets (dFS_3a - dFS_3e) is shown in Figure 7.11. It can be observed that both simulants S1 and S2 demonstrated shear thinning (and viscosity reduction) with higher shear rate: behaviour similar to faecal sludge and fresh faeces, although the results for both S1 and S2 showed behaviour more similar to fresh faeces than faecal sludge. A comparison of the particle size distribution of faecal simulant S1 relative to fresh faeces samples (FF) is shown in Figure 7.12. The size classes for simulant S1 and fresh faeces were similar, but there was a difference in the volume density as indicated by the position of the peaks. More investigations are required to improve on the particle size distribution of simulant S1.

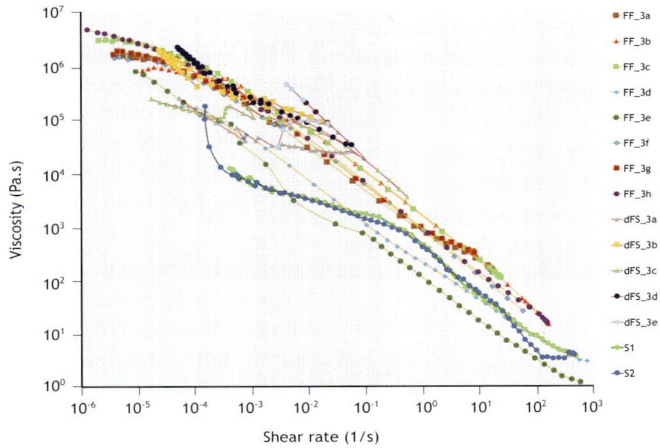

Figure 7.11 A comparison of the relationship between viscosity and shear rate for simulants S1 and S2, fresh faeces (FF) samples and dry faecal sludge (dFS) samples.

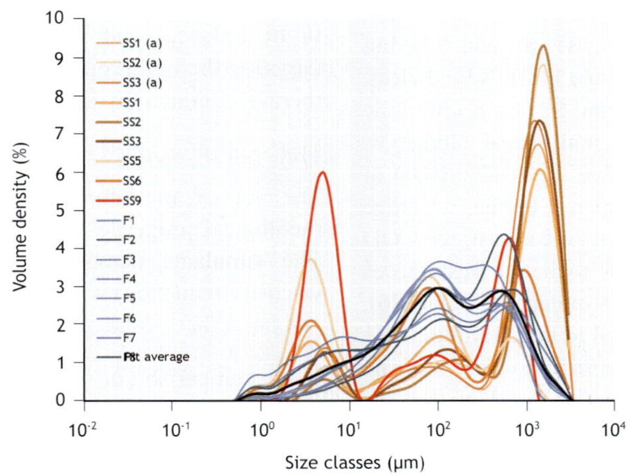

Figure 7.12 A comparison of the particle size distribution of simulant S1 and fresh faeces (FF) samples.

Conclusions and recommendations

Based on the results of the characterisation presented here, it was concluded that simulant S1 most closely mimicked the properties of: (*i*) faecal sludge moisture content, total solids content, density, thermal conductivity and heat capacity; and (*ii*) fresh faeces pH, calorific value, COD_{total} and rheological properties. S2 closely resembled fresh faeces for all the measured properties in this study. In addition, S2 was easier to mix and handle logistically during analysis. There is some degree of overlap with S1 and S2 in their comparison to fresh faeces and faecal sludge in pH, total solids, moisture content and thermal conductivity. Therefore, either S1 or S2 are recommended for usage in applications where the specific parameters are most closely replicated (Table 7.22).

Table 7.22 Simulants S1 and S2's resemblance to fresh faeces and faecal sludge.

Parameter/property	Fresh faeces	Faecal sludge
Total solids	S1, S2	S1
Moisture content	S1, S2	S1
Volatile solids	S2	
Ash	S2	
Density	S2	S1
Calorific value*	S1	
pH	S1, S2	
COD_{total}	S2	
Thermal conductivity	S1, S2	S1, S2
Heat capacity	S2	S1, S2
Viscosity vs shear rate	S1, S2	
Shear stress vs shear rate	S1, S2	

REFERENCES

Afolabi O.O. and Sohail M. (2017). Microwaving human faecal sludge as a viable sanitation technology option for treatment and value recovery–A critical review. *Journal of Environmental Management*, 187, 401-415.

Appiah-Effah E., Nyarko K.B., Gyasi S.F. and Awuah E. (2014). Faecal sludge management in low-income areas: a case study of three districts in the Ashanti region of Ghana. *Journal of Water Sanitation and Hygiene for Development*, 4, 189-199.

Barman S.M., Barrett K.E., Boitano S. and Brooks H. (2009). Ganong's review of medical physiology, 23rd edition, McGraw-Hill, USA.

Bassan M., Nguyen V.A., Holliger C. and Strande L. (2014). Optimisation of the Sampling and Laboratory Methods for Faecal Sludge. Sandec News. Sandec/Eawag. http://www.eawag.ch/fileadmin/Domain1/Abteilungen/sandec/publikationen/EWM/Laboratory_Methods/optimization_sampling.pdf

Bassan M., Tchonda T., Yiougo L., Zoellig H., Mahamane I., Mbéguéré M. and Strande L. (2013). Characterisation of faecal sludge during dry and rainy seasons in Ouagadougou, Burkina Faso. Proceedings of the 36th WEDC International Conference, Nakuru, Kenya, 1-5.

Bekkali N., Hamers S.L., Reitsma J.B., Van Toledo L. and Benninga M.A. (2009). Infant Stool Form Scale: Development and Results. *The Journal of Pediatrics*, 154(4), 521-526.

Bhardwaj R.C., Sharma D.K. and Bockris J. O'M. (1990). Report to NASA on Grant NAGW 1779, Aug.

Bourgault C. (2019). Characterisation and quantification of faecal sludge from pit latrines (Doctoral dissertation). Université Laval, Quebec, Canada.

Bösch, A. and Schertenleib, R. (1985) Emptying on-site excreta disposal systems. Report 03/85. International Reference Centre for Waste Disposal, Eawag, Dübendorf, Switzerland.

Brown D.M., Butler D., Orman N.R. and Davies J.W. (1996). Gross solids transport in small diameter sewers. *Water Science and Technology*, 33(9), 25-30.

BSI (2010). Greywater Systems BS 8525-1&2:2010. British Standards Institute.

Butler D., Davies J.W., Jefferies C. and Schütze M. (2003). Gross solid transport in sewers. Proceedings of the Institution of Civil Engineers: Water and Maritime Engineering 156(2), 175-183.

Ciba-Geigy AG (1977). Wissenschaftliche Tabellen Geigy. Ciba-Geigy Limited.

Chikava F. K. and Velkushanova K. (2014). Properties of fresh human faeces. University of KwaZulu-Natal, Durban.

Colón J., Forbis-Stokes A.A. and Deshusses M.A. (2015). Anaerobic digestion of undiluted simulant human excreta for sanitation and energy recovery in less-developed countries. *Energy for Sustainable Development*, 29, 57-64.

Danso-Boateng E., Holdich R., Wheatley A.D., Martin S. and Gyi D.E. (2012). A toilet system based on hydrothermal carbonisation. https://dspace.lboro.ac.uk/2134/11289

Deering N., Briggs K., Bryant L. and Way C. (2018). Characterisation of faecal sludge simulants for drying experiments. 41st WEDC International Conference: Transformation towards sustainable and resilient WASH services, 2018.

Diaper C., Tofl M. and Storey M. (2008). Greywater Technology Testing Protocol. CSIRO: Water for a Healthy Country National Research Flagship, Vol. 1835-0.95X. Victoria, Australia.

Diener S., Semiyaga S., Niwagaba C.B., Muspratt A.M., Gning J.B., Mbéguéré M., Ennin J.E., Zurbrugg C. and Strande L. (2014). A value proposition: Resource recovery from faecal sludge—Can it be the driver for improved sanitation?. *Resources, Conservation and Recycling*, 88, 32-38.

Dinauer R.C. (1982). Methods of soil analysis. ASA, SSSA. Madison, Wisconsin, USA.

Eastwood M. (1973). Vegetable fibre: Its physical properties. Proceedings of the Nutrition Society, 32(03), 137–143.

EAWAG (2002). Faecal sludge treatment. Eidgenössische Anstalt für Wasserversorgung, Abwasserreinigung und Gewässerschutz (Swiss Federal Institute for Environmental Science and Technology/ETH), Dübendorf, Switzerland.

EPA WaterSense (2014). The US Environmental Protection Agency's (EPA's) WaterSense program https://www3.epa.gov/watersense/partners/toilets_final.html

Friedler E., Katz I. and Dosoretz C.G. (2008). Chlorination and coagulation as pretreatment for greywater desalination. *Desalination,* 222(1-3), 38-49.

Fry L.J. (1973). Methane digesters for fuel gas and fertiliser. Woods Hole, Massachusetts: The New Alchemy Institute [http://journeytoforever.org/biofuel_library/MethaneDigesters/MDToC.html (accessed May 2017)].

Garrow J.S., James W.P.T. and Ralph A. (1993). Human nutrition and dietetics (9th ed.). Edinburgh: Longman Group.

Garsed K., Chernova J., Hastings M., Lam C., Marciani L., Singh G., Henry A. and Hall I. (2014). A randomised trial of ondansetron for the treatment of irritable bowel syndrome with diarrhoea. *Gut*, 63, 1617–25.

German Industrial Norm/European Norm, WC-Becken und WC-Anlagen mit angeformtem Geruchsverschluss; Deutsche Fassung EN 997:2003+A1:2006.

Girovich M.J. (1996). Biosolids treatment and management: processes for beneficial use. Taylor & Francis.

Gold M., Ddiba D.I.W., Seck A., Sekigongo P., Diene A., Diaw S., Niang S., Niwagaba C. and Strande L. (2017b). Faecal sludge as a solid industrial fuel: a pilot-scale study. *Journal of Water Sanitation and Hygiene for Development,* 7(2), 243-251.

Gold M., Harada H., Therrien J.D., Nishida T., Cunningham M., Semiyaga S., Fujii S., Dorea C., Nguyen V.A. and Strande L. (2017a). Cross-country analysis of faecal sludge dewatering. Environmental Technology.

Gross A., Maimon A., Alfiya Y and Friedler E. (2015). Greywater reuse. Taylor and Francis Group, CRC Press.

Henze M., Van Loosdrecht M.C., Ekama G.A. and Brdjanovic D. (2008). Biological wastewater treatment: Principles, Modelling and Design, IWA Publishing.

Heinss U., Larmie S.A. and Strauss M. (1999). Characteristics of faecal sludges and their solids-liquid separation. Eawag/Sandec.

Ilango A. and Lefebvre O. (2016). Characterizing Properties of Biochar Produced from Simulated Human Faeces and Its Potential Applications. *Journal of Environmental Quality*, 45, 734-742.

Jönsson H., Daky A., Jeppsson U., Hellström D. and Kärman E. (2005). Composition of urine, faeces, greywater and biowaste for utilisation in the URWARE model. Gothenburg, Sweden: Chalmers University of Technology Publisher, [http://www.iea.lth.se/publications/Reports/LTH-IEA-7222.pdf (accessed May 2017)].

Kaba L., Hitchens G.D. and Bockris J.O'M. (1989). Electrochemical Incineration of Wastes. Tech. Paper #891510

Koottatep T., Polprasert C., Oanh N.T.K., Montangero A. and Strauss M. (2001). Sludges from onsite sanitation systems-low-cost treatment alternatives. In IWA Conference on Water and Wastewater Management for Developing Countries.

Levit M.D. and Duane W.C. (1972). Floating stools—flatus versus fat. *New England Journal of Medicine*, 286, 973-975.

Lewis S. and Heaton K. 1997. Stool form scale as a useful guide to intestinal transit time. *Scandinavian journal of gastroenterology*, 32, 920-924.

Lovelady H.G. and Stork E.J. (1970). An improved method for preparation of faeces for bomb calorimetry. *Clinical Chemistry,* 16, 253–254.

Makununika B.S.N. (2016). Thermal Drying of Faecal Sludge from VIP latrines and Characterisation of Dried Faecal Material. MSc Eng. Chemical Engineering Dissertation, University of KwaZulu-Natal, Durban.

Meher K., Murthy M. and Gollakota K. (1994). Psychrophilic anaerobic digestion of human waste. *Bioresource Technology,* 50, 103–106.

Miller A., Espanani R., Junker A., Hendr, D., Wilkinson N., Bollinger D., Abelleira-Pereira J., Deshusses M., Inniss E. and Jacoby W. (2015) Supercritical water oxidation of a model faecal sludge without the use of a co-fuel. *Chemosphere,* 141, 189-196.

Moe C.L. and Rheingans R.D. (2006). Global challenges in water, sanitation and health. *Journal of Water and Health,* 4(S1) 41-57.

Monhol F.A.F. and Martins M.F. (2015). Cocurrent combustion of human faeces and polyethylene waste. *Waste and Biomass Valorization*, 6(3), 425-432. doi:10.1007/s12649-015-9359-2

Moore J.G., Jessop L.D. and Osborne D.N. (1987). Gaschomatographic and mass spectrometric analysis of the odor of human feces. *Gastroenterology*, 93, 1321-1329.

Muspratt A.M., Nakato T., Niwagaba C., Dione H., Kang J., Stupin L., Regulinski J., Mbéguéré M. and Strande L. (2014). Fuel potential of faecal sludge: calorific value results from Uganda, Ghana and Senegal. *Journal of Water Sanitation and Hygiene for Development*, 4(2), 223-230.

Nolan J.D., Johnston I.M., Pattni S.S., Dew T., Orchard T. and Walters J.R. (2015). Diarrhea in Crohn's Disease: investigating the role of the ileal hormone Fibroblast Growth Factor 19. *Journal Crohn's Colitis,* 9, 125–131.

NSF (2001). NSF/ANSI 350-2011, Onsite Residential and Commercial Water Reuse Treatment Systems.

Nwaneri C. Foxon K. Bakare B. Buckley C. WISA 2008 conference. South Africa: WISA; 2008. Biological degradation processes within a pit latrine. 1, 18–22 May 2008, Sun City (p. 1)

Onabanjo T., Kolios A. J., Patchigolla K., Wagland S.T., Fidalgo B., Jurado N. and Cartmell E. (2016a). An experimental investigation of the combustion performance of human faeces. *Fuel,* 184, 780-791. doi:10.1016/j.fuel.2016.07.077

Onabanjo T., Patchigolla K., Wagland S.T., Fidalgo B., Kolios A., McAdam E. and Cartmell E. (2016b). Energy recovery from human faeces via gasification: A thermodynamic equilibrium modelling approach. *Energy Conversion and Management*, 118, 364-376.

Palumbo G. and D'Acchioli V. (2001). Disposable absorbent article having improved in-use storage capacity for low and medium low viscosity faeces. Google Patents, US6190369 B1.

Park, J.A., Hur, J.M., Son, B.S., Lee, J.H., (2001). Effective treatment of night soil using anaerobic sequencing batch reactor (ASBR). *Korean J. Chem. Eng.* 18 (4), 486-492.

Penn R., Schuetze M., Jens A. and Friedler E. (2018). Tracking and simulation of gross solids transport in sewers. *Urban Water Journal*, 15(6), 584-591.

Penn R., Schütze M., Alex J. and Friedler E. (submitted) Tracking and simulation of gross solids transport in sewers.

Penn R., Ward B.J., Strande L. and Maurer M. (2018). Review of synthetic human faeces and faecal sludge for sanitation and wastewater research. *Water Research*, 132, 222-240.

Penn R., Maurer M., Michalec F.G., Scheidegger A., Zhou J., and Holzner M. (2019). Quantifying physical disintegration of faeces in sewers: Stochastic model and flow reactor experiments. *Water Research*, 152, 159-170.

Podichetty J.T., Islam M.W., Van D., Foutch G.L. and Johannes A. (2014). Viscous heating analysis of simulant faeces by computational fluid dynamics and experimentation. *Journal of Water Sanitation and Hygiene for Development*, 4, 62-71.

Putnam D.F. (1971). Composition and concentrative properties of human urine.

Radford J. and Fenner R. (2013). Characterisation and fluidisation of synthetic pit latrine sludge. *Journal of Water Sanitation and Hygiene for Development*, 3, 375-382.

Radford J., Malinga S., Drummond G., Atayo H., Whitesell A. and Sugden S. (2015). Latrine desludging pump development using a simple test for simulant strength: A case study from Uganda. *Journal of Water Sanitation and Hygiene for Development*, 5, 620-624.

Rice E.W., Baird R.B., Eaton A.D. and Clesceri L.S. (eds.) (2017). Standard methods for the examination of water and wastewater, 23rd edition. *APHA, AWWA, and WPCR*.

Rose C., Parker A., Jefferson B. and Cartmell E. (2015). The characterisation of feces and urine: a review of the literature to inform advanced treatment technology. *Critical reviews in environmental science and technology*, 45, 1827-1879.

Rose C.M., Parker A.H. and Cartmell E. (2014). The biochemical methane potential of faecal sludge from on-site sanitation instalments in low income countries, *12th Specialized Conference on Small Water and Wastewater Systems and 4th Specialized Conference on Resource Oriented Sanitation*, Muscat, Sultanate of Oman.

Reddy S., Sanders T., Owen R. and Thompson M. (1998). Faecal pH, bile acid and sterol concentrations in premenopausal Indian and white vegetarians compared with white omnivores. *British Journal of Nutrition*, 79(06), 495–500.

Sato H., Hirose T., Kimura T., Moriyama Y., Yamashita T and Nakashima Y. (2001a). Analysis of the tablemalodours volatile substances of human wastes: feces and urine. *Journal of Health Science*, 47(5), 483-490.

Sato H., Morimatsu H., Klmura T., Moriyama Y., Yamashita T and Nakashima Y. (2001b). Analysis of the malodours volatile substances of human feces. *Journal of Health Science*, 48(2), 179-185.

Schoebitz L., Niwagaba C.B. and Strande L. (2016). SFD Promotion Initiative, Kampala, Uganda. Sandec/Eawag.

Serio M.A., Cosgrove J.E., Wójtowicz M.A., Wignarajah, K. and Fisher J. (2012). Pyrolysis Yields from Microwave-Assisted Heating of Solid Wastes, 42nd International Conference on Environmental Systems, San Diego, CA.

Singh N., Isono N., Srichuwong S., Noda T. and Nishinari K. (2008). Structural, thermal and viscoelastic properties of potato starches. *Food Hydrocolloids*, 22(6), 979–988.

Snell J.R. (1943). Anaerobic digestion: III. Anaerobic digestion of undiluted human excreta. *Sewage Works Journal*, 15(4), 679–701.

Spellman F.R. (1997). Incinerating biosolids. Taylor & Francis.

Stern B.A. and Holtman D.C. (1987). Disposable urinary and fecal waste containment product. Google Patents, US 5476459.

Still, D.A. and Foxon, K. (2012). Tackling the challenges of full pit latrines. Water Research Commission.

Strande L., Schoebitz L., Bischoff F., Ddiba D. and Niwagaba C. (2018). Methods to reliably estimate faecal sludge quantities and qualities for the design of treatment technologies and management solutions. *Journal of Environmental Management*, 223(1): 898-907.

Strande L., Ronteltap M. and Brdjanovic D. (2014) Faecal sludge management: Systems approach for implementation and operation. IWA Publishing, London.

SuSanA – Sustainable Sanitation Alliance (2008) MISO (Soybean Paste) 14 March 2008. http://forum.susana.org/media/kunena/attachments/52/certificate.pdf (Accessed: 03 June 2017).

Takahashi T., Goto M. and Sakata T. (2004). Viscoelastic properties of the small intestinal and caecal contents of the chicken. *British Journal of Nutrition*, 91(6), 867-872.

Takahashi T. and Sakata T. (2002). Large particles increase viscosity and yield stress of pig cecal contents without changing basic viscoelastic properties. *Journal of Nutrition*, 132(5), 1026-1030.

Tennakoon C.L.K., Bhardwaj R.C. and Bockris J.O. (1996). Electrochemical treatment of human wastes in a packed bed reactor. *Journal of Applied Electrochemistry*, 26(1), 18-29.

Thyberg K. L., Tonjes D. J. and Gurevitch J. (2015). Quantification of food waste disposal in the United States: A meta-analysis. *Environmental Science and Technology*, 49(24), 13946-13953

Tilley E., Ulrich L., Lüthi C., Reymond P. and Zurbrügg C. (2014). Compendium of Sanitation Systems and Technologies. 2nd Revised Edition. Swiss Federal Institute of Aquatic Science and Technology (Eawag). Dübendorf, Switzerland. ISBN: 978-3-906484-57-0

Veritec Consulting Inc. & Koeller and Company (2010). Maximum Performance (MaP) Testing Toilet Fixture Performance Testing Protocol, Version 4.

Velkushanova K. (2014). Mechanical properties of faecal sludge Pollution Research Group, South Africa: University of KwaZulu-Natal, South Africa.

Velkushanova K.V., Zuma L. and Buckley C. (2019). Mechanical properties of faecal sludge. *Gates Open Research*, 3(1582), p.1582.

Ward B.J., Morgenroth E. and Strande L. (2017a). Dewatering Characterisation of Synthetic Faecal Sludge Project. Sandec news (18), 8. http://www.eawag.ch/fileadmin/Domain1/Abteilungen/sandec/publikationen/EWM/Treatment_Technologies/Dewatering_characterisation_BJ_Ward.pdf

Ward B.J., Morgenroth E. and Strande L. (2017b). Water Hub @ NEST: Faecal Sludge Dewatering for Resource Recovery [Poster]. Exhibited at 2017 Eawag Symposium, Duebendorf, Switzerland. https://www.researchgate.net/publication/320489990_Water_Hub_NEST_Faecal_Sludge_Dewatering_for_Resource_Recovery

Ward B.J., Yacob T.W. and Montoya L.D. (2014). Evaluation of solid fuel char briquettes from human waste. *Environmental science & technology*, 48, 9852-9858.

Wignarajah K., Litwiller E., Fisher J.W. and Hogan J. (2006). Simulated human faeces for testing human waste processing technologies in space systems. SAE Technical Paper.

Woolley S.M., Buckley C.A., Pocock J. and Foutch G.L. (2014). Rheological modelling of fresh human faeces. *Journal of Water Sanitation and Hygiene for Development,* 4(3), 484-489.

World Health Organisation and UNICEF (2017). Progress on drinking water, sanitation and hygiene: 2017 update and SDG baselines.

Wyman J.B., Heaton K.W., Manning A.P. and Wicks A.C.B. (1978). Variability of colonic function in healthy subjects. *Gut,* 19(2), 146-150.

Yeo R.S. and Welchel D.N. (1994). United State Patent Number 5356626, October 18.

Yermán L., Hadden R.M., Carrascal J., Fabris I., Cormier D., Torero J.L. and Cheng Y. (2015). Smouldering combustion as a treatment technology for faeces: Exploring the parameter space. *Fuel*, 147, 108-116.

Zuma L. (2013). Development of a synthetic faecal sludge. EPP Fellowship – Eawag & EPFL, Eawag Partnership Program for Developing Countries.

Zuma L., Velkushanova K. and Buckley C. (2015). Chemical and thermal properties of VIP latrine sludge. *Water SA*, 41, 534-540.

Wignarajah K., Litwiller E., Fisher J.W. and Hogan J. (2006). Simulated human feces for testing human waste processing technologies in space systems. *SAE Transactions*, 424-430.

Woolley S., Cottingham R., Pocock J. and Buckley C. (2014). Shear rheological properties of fresh human faeces with different moisture content. *Water SA*, 40, 273-276.

http://www.eawag.ch/fileadmin/Domain1/Abteilungen/sandec/publikationen/EWM/Laboratory_Methods/optimization_sampling.pdf

http://www.gatesfoundation.org/What-We-Do/Global-Development/Water-Sanitation-and-Hygiene.

http://www.npr.org/sections/health-shots/2012/08/08/158447235/why-is-the-worlds-largest-foundation-buying-fake-poop

Ward B.J., Yacob T.W. and Montoya L.D. (2014). Evaluation of solid fuel char briquettes from human waste. *Environmental science & technology*, 48, 9852-9858.

Wignarajah K., Litwiller E., Fisher J.W. and Hogan J. (2006). Simulated human faeces for testing human waste processing technologies in space systems. SAE Technical Paper.

Woolley S.M., Buckley C.A., Pocock J. and Foutch G.L. (2014). Rheological modelling of fresh human faeces. *Journal of Water Sanitation and Hygiene for Development,* 4(3), 484-489.

World Health Organisation and UNICEF (2017). Progress on drinking water, sanitation and hygiene: 2017 update and SDG baselines.

Wyman J.B., Heaton K.W., Manning A.P. and Wicks A.C.B. (1978). Variability of colonic function in healthy subjects. *Gut,* 19(2), 146-150.

Yeo R.S. and Welchel D.N. (1994). United State Patent Number 5356626, October 18.

Yermán L., Hadden R.M., Carrascal J., Fabris I., Cormier D., Torero J.L. and Cheng Y. (2015). Smouldering combustion as a treatment technology for faeces: Exploring the parameter space. *Fuel*, 147, 108-116.

Zuma L. (2013). Development of a synthetic faecal sludge. EPP Fellowship – Eawag & EPFL, Eawag Partnership Program for Developing Countries.

Zuma L., Velkushanova K. and Buckley C. (2015). Chemical and thermal properties of VIP latrine sludge. *Water SA*, 41, 534-540.

Wignarajah K., Litwiller E., Fisher J.W. and Hogan J. (2006). Simulated human feces for testing human waste processing technologies in space systems. *SAE Transactions*, 424-430.

Woolley S., Cottingham R., Pocock J. and Buckley C. (2014). Shear rheological properties of fresh human faeces with different moisture content. *Water SA*, 40, 273-276.

http://www.eawag.ch/fileadmin/Domain1/Abteilungen/sandec/publikationen/EWM/Laboratory_Methods/optimization_sampling.pdf

http://www.gatesfoundation.org/What-We-Do/Global-Development/Water-Sanitation-and-Hygiene.

http://www.npr.org/sections/health-shots/2012/08/08/158447235/why-is-the-worlds-largest-foundation-buying-fake-poop

8

Laboratory procedures and methods for characterisation of faecal sludge

Konstantina Velkushanova
Merlien Reddy
Thabiso Zikalala
Bhekumuzi Gumbi
Colleen Archer
Barbara J. Ward
Nienke Andriessen
Stanley Sam
Linda Strande

OBJECTIVES

The objectives of this chapter are to:

- To provide methods for protecting health and safety during collection, handling, transportation, storage, and disposal of faecal sludge.
- To provide information required to adapt and develop standard methods for faecal sludge characterisation, including quality assurance and quality control (QA/QC) strategies and selection of appropriate methods.
- To provide an overview of existing methods for faecal sludge analysis being used in partner laboratories.

DISCLAIMER: In this book, brand names, suppliers, and manufacturer's information are for illustration purposes only, and no endorsement is implied. Equivalent results can be achieved with apparatus and materials other than those presented here. Meeting the performance requirements of the particular method is the responsibility of the sampling team and laboratory. Such examples in this chapter are noted with the symbol [D].

 Methods for Faecal Sludge Analysis. Edited by Velkushanova K., Strande L., Ronteltap M., Koottatep T., Brdjanovic D. and Buckley C. ISBN: 9781780409115. Published by IWA Publishing, London, UK.

8.1 INTRODUCTION

Faecal sludge management is a relatively new and rapidly developing field. There is a need for increased scientific knowledge and understanding of faecal sludge characteristics. To help fill this gap, standard methods that can be used in scientific research and monitoring of treatment plants are needed. However, the characterisation of faecal sludge is challenging due to the high variability and wide range of characteristics, as presented in Chapter 1, which requires rigorous sampling methods, quality assurance and quality control (QA/QC) measures, and well-defined procedures to reduce the uncertainties in data generation and analysis. When using existing standard methods developed for other sample matrices, these methods need to be validated and adapted for the specific type of faecal sludge being analysed. Presented in this chapter are examples of how laboratories are adapting existing standard methods and manufacturers' test kits for this difficult matrix, and also developing new methods.

When adapting existing methods for faecal sludge, appropriate methods for more liquid matrices (*e.g.* from wastewater) or more solid matrices (*e.g.* from fresh faeces, soil and food science) need to be selected. As explained in Chapter 2, four types of faecal sludge are defined in this book as liquid (< 5% TS), slurry (5-15% TS), semi-solid (15-25% TS), and solid (> 25% TS). Due to the range of TS concentrations, mass concentrations in this book are expressed as both weight/volume and weight/weight. Unfortunately, it is not possible to make a hard and fast rule about when to analyse volumetrically (weight/volume) or gravimetrically (weight/weight), and in practice this has to be determined by how accurately volumes of faecal sludge can be measured. In general, it is practical to analyse liquid and slurry samples volumetrically and semi-solid and solid samples gravimetrically, but in reality, analysis needs to be done consistently, and samples can span the entire range of concentrations. Whenever possible, it is therefore recommended to report the density of samples, as this can be used to convert between concentrations, and TS analysis can be done in parallel to report the results in weight/weight of a dry sample mass. Due to the wide range of TS concentrations, faecal sludge samples may require different preparation steps before characterisation, compared with the sample preparation described in existing standard methods for other matrices. Sludge with higher TS concentrations may require additional preparation techniques such as mixing, blending, diluting, centrifuging and filtering. For example, the method for characterising the pH of faecal sludge involves sample preparation techniques from the soil standard methods for high TS samples, techniques from food science for semi-solid samples, and techniques from wastewater methods for slurry and liquid samples. For liquid sample matrices such as treatment plant effluent, established standard methods for wastewater analysis should be applicable.

Common concerns when adapting existing methods for faecal sludge include homogenisation, dilutions, sample size, and filtration. Homogenisation needs special attention in order to obtain representative samples, and as discussed in Chapter 3, sampling plans need to be carefully designed and executed to produce repeatable results. Filter pore size needs to be adjusted to address issues with clogging. Special pre-treatment steps such as treatment with activated carbon may be required when employing colorimetric methods to ensure the inherent colour of the sample matrix does not influence the measurement. Faecal sludge in general has higher organic loads than wastewater or digested sludge, hence more dilutions will be required in the sample preparation stage. In other cases, the analytical measurements or the size of the sample are modified in order to adapt the method to faecal sludge. Settleability and dewaterability methods are also modified to reflect operational differences at faecal sludge treatment plants compared with wastewater treatment plants, for example, allowing for different settling or mixing times to match actual process conditions. For analysis of helminths, the UKZN PRG laboratory found that the USEPA Ascaris method was not giving adequate quantitative results for faecal sludge, as it is often mixed with sand, soil and other materials, and the silica particles can be confused as helminth eggs under the microscope. To address this, the UKZN PRG developed a method that includes dissolving the silica particles and additional sieving - UKZN PRG helminth method (Method 8.8.1.4).

Methods for fresh faeces also need adaptation for use with faecal sludge, as illustrated by the example of faecal sludge simulants in Chapter 7.

The methods presented in this chapter are the first step towards the development of standard methods for analysis of faecal sludge. Additional method development and adaptation is still required by laboratories around the world. The TS method is the most complete method that is presented, as TS is one of the most basic and fundamental characteristics to report for faecal sludge and is often the source of many errors. The chapter also includes a method for sand content, which is important to consider with faecal sludge, for example when interpreting VS to TS ratios. Other categories of methods are not comprehensive, and are based on the experience of the partner laboratories. For example, COD is the only method presented for the analysis of organic content, as experience indicates that it is a representative measurement and it is robust to the variability and levels of organic loads in faecal sludge. Other methods to consider include biochemical oxygen demand (BOD), volatile fatty acids (VFA), total organic carbon (TOC), and biomethane potential (BMP), which will be included in future editions of the book. A general overview of metal analysis is provided, but only one example of a digestion. For nutrient analysis, options of manufacturers' kit-based methods, spectrophotometers, and titrations are presented but not for all parameters. Methods for characterising crude fat and fibre are included, but further methods for measuring other fibre fractions and crude protein are not yet included in this book. Throughout this chapter, information is provided about the standard methods being adapted, the extent of adaptation and validation that has taken place for faecal sludge, examples of implementation, and links to further resources.

Method development and adaption for new sample matrices is a standardised experimental procedure. For example, in the Standard Method for Method Development and Evaluation (1040), Rice *et al.* (2017) present the steps of single operator characteristics, analysis of unknown samples, and method ruggedness. The first step is to evaluate the systematic error or bias and the precision of the method for the specific sample matrix. The systematic error is the inherent fluctuation in a method. For example, if the same person was analysing replicates of the same sample, the variability within the results would be used to calculate the systematic error. To evaluate this, faecal sludge samples need to be prepared across the range of TS that will be evaluated, and then spiked with known concentrations of a standard of the constituent to be measured, across the concentration range of the method. Then, each of the samples is measured 10 times to calculate the systematic error and precision. The second step requires analysing blind samples that were prepared independently from the laboratory conducting the analysis. This means that the laboratory doing the analysis does not know the concentration in advance, but the known concentration is used after analysis to evaluate the accuracy of the method. The third step is making minor changes to the method, such as mixing time, sample size, temperature, or pH, and based on the results, the ruggedness of the method is calculated. Important quality assurance and quality control (QA/QC) measures to have in place for method development are covered in Section 8.3. In addition, important components of method development and experimental procedure are covered throughout this book, including setting up a laboratory (Chapter 2), how to develop and execute a sampling plan (chapters 3 and 5), and experimental design (Chapter 4).

For the future, as more laboratories develop new methods and adapt existing methods for the analysis of faecal sludge, collaborative testing can take place, where blind samples are analysed in parallel in multiple laboratories, as a step further to the establishment of standard methods for faecal sludge analysis. This can also help to develop and standardise new lower-cost methods developed specifically for faecal sludge, such as measurement of sludge colour using image analysis (Method 8.6.8, Ward *et al.*, 2021), *in situ* characterisation of rheological properties in pit latrines using a portable penetrometer (Chapter 3), and development of field laboratory methods for characterisation of faecal sludge in resource-limited settings (Chapter 2).

Presented in this chapter is a general overview of factors to consider when conducting the laboratory analysis. The reader is referred to additional references throughout the text for more in-depth knowledge. Section 8.2 provides an overview of the health and safety and occupational safety measures for risk prevention, Section 8.3 summarises quality assurance measures, Section 8.4 provides an overview of the included methods, and guidelines for how to select methods for implementation based on criteria such as available budget, required accuracy, sample preparation and testing time, and laboratory capacity, Section 8.5 provides information about packaging and shipping faecal sludge samples, Section 8.6 contains the methods for chemical and physicochemical characteristics, Section 8.7 contains the methods for physical and mechanical characteristics, and Section 8.8 contains the methods for biological characteristics.

8.2 HEALTH AND SAFETY (H&S)

Specific concerns with setting up laboratories for faecal sludge analysis are covered in Section 2.5. For a further discussion of EHS (environment, health, and safety) management systems, including laboratory safety, chemical handling and management, emergency planning, evaluating hazards and risks, working with equipment, and management of waste, please refer to the free online guidelines, Prudent Practices in the Laboratory: Handling and Management of Chemical Hazards (National Research Council, 2011). Care must be taken when handling faecal sludge, and it should always be considered a hazardous biological agent (HBA) due the content of pathogens. Pathogens pose risks to human health that require strict health and safety procedures when working with faecal sludge in the laboratory and the field. People involved at every step of the sampling and analytical process need to take precautions. These precautions include required vaccinations that will depend on regionally prevalent diseases and the extent of exposure to faecal sludge (*e.g.* cholera, tetanus, polio, typhoid fever, hepatitis). As discussed in Section 2.5.1, the laboratory workflow includes the transmission chain, and preventative and protective measures. Pathways for human contamination need to be carefully considered in advance, and appropriate procedures put in place to manage them. Pathways of contamination include the following (European Agency for Safety and Health at Work (EU-OSHA):

- Reservoir: the source of the infective agent (*e.g.* faecal sludge sample)
- Portal of exit: the biological agent leaves the reservoir and/or is transmitted to another reservoir
- Mode of transmission: direct (*e.g.* inhalation), semi-direct (*e.g.* transmitted through dirty hands); or indirect (*e.g.* transmitted on a contaminated surface)
- Portal of entry: respiratory tract (*e.g.* inhalation), digestive tract (*e.g.* ingestion), skin (*e.g.* existing or new injury), or mucus (*e.g.* splashing into eyes)
- Potential hosts: employees or workers who have been exposed via contamination pathways.

In addition to pathogens, general occupational health and safety are an essential part of faecal sludge laboratory practices. The right to health and safety at work is stipulated in the constitution of the World Health Organisation (WHO) and is supported by a number of international and local organisations and regulations. 'Occupational risk factors' include chemical, physical, biological or other agents that may cause harm to an exposed person in the workplace. Examples of categories of occupational risk factors are carcinogens (*e.g.* 150 known probable human carcinogens (IARC, 2012)), airborne particulates, noise, ergonomic stressors, risk factors for injuries, and exposure to hazardous biological agents. For more information, the reader is referred to international organisations and guidelines for occupational health and safety listed below[1].

8.2.1 Monitoring and responsibilities

Setting up the work flow in a laboratory conducting faecal sludge analysis is covered in Section 2.5.1,

[1] a) Wolf J., Prüss-Ustün A., Ivanov I., Mugdal S., Corvalán C., Bos R., Neira M. and World Health Organization (2018). Preventing disease through a healthier and safer workplace.
b) World Bank (1984). Occupational health and safety guidelines. Office of Environmental Affairs, Washington DC.
c) Occupational Health and Safety Administration (OSHA), https://www.osha.gov.
d) European Agency for Safety and Health at Work (EU-OSHA), http://osha.europa.eu
e) World Bank (2019). Health, Safety and Dignity of Sanitation Workers an Initial Assessment.

including receiving samples. In order to achieve an environment where accidents and hazards are minimised, there needs to be a clear chain of command and defined responsibilities. This includes laboratory management (Section 2.5.3) and health and safety practices (Section 2.5.2). The laboratory manager has the overall responsibility for health and safety and delegates specific health and safety responsibilities to all persons working in the laboratory. The laboratory manager also has the responsibilities to issue annual and *ad hoc* reports on the laboratory activities and procedures, raise concerns relating to health and safety, and promote and ensure compliance with all health and safety procedures. Line managers are accountable for workplace health and safety and the welfare of employees in their chain of command while working in the laboratory. They are responsible for establishing safety goals and objectives under their responsibility, consulting and motivating their staff to adopt good health and safety practices, and providing the necessary risk controls and maintenance procedures to ensure a safe and healthy workplace. Everyone working in the laboratory is required to make every effort to ensure the health and safety of themselves and their co-workers, and that their acts or omissions do not affect the health and safety of other workers. They must also have undergone the appropriate safety training, and have an intimate understanding of their own workplace in order to be in a position to identify, report, and minimise risks. They should always be prepared to communicate with their line managers as the first point of contact in the workplace, and with community representatives when sampling in the field. It is especially important that everyone working in a laboratory immediately reports any health and safety issues to management as they arise in the workplace. To ensure a safe environment and compliance, there should be no repercussions for employees who identify risks or hazards, and this should in fact be rewarded. In addition, a health and safety committee should be formed that is responsible for organising general health and safety committee meetings and keeping up to date with the newest regulations and guidelines. Members of the committee should be clearly informed regarding their responsibilities, including investigation of incidents, monthly laboratory inspections and reports that are communicated with the laboratory manager, line managers and/or all persons working in the laboratory.

To mitigate the risks of working with faecal sludge, a number of steps for assessing the degree of risk associated with biological and other types of hazards should be followed, such as: identifying the hazards, identifying who may be affected and how they could be harmed, evaluating the risks and taking precautions, and documenting, implementing and regularly updating these steps. Based on this process, standard operating procedures and risk assessments must be developed to minimise the risks and mitigate the hazards.

8.2.2 Standard operating procedures (SOPs)

All laboratory operations should follow standard operating procedures (SOPs), including what to do in the case of accident, exposure, and spills of faecal sludge, for example information on flushing with eye baths, emergency showers, and how to properly contain spills. SOPs are written instructions describing in detail the steps to be performed, and also include procedures for sampling, transportation, analysis, use of equipment, quality assurance and quality control (QA/QC), calibration, and documentation of the entire sample chain of command. SOPs should be written by laboratory personnel who are the most knowledgeable on experimental processes, and should be regularly reviewed by the laboratory manager and the health and safety committee. The SOPs should cross-reference all other related SOPs and expand upon them where necessary. When developing a SOP, the following should be considered:

- Type, quantity, and nature of the hazardous materials used
- Location of use, including fume hood or other containment devices
- Process details
- Available safety equipment, including personal protective equipment (PPE)
- Waste collection, storage, and disposal requirements
- Decontamination procedures.

Table 8.1 sets out examples of procedures where health and safety SOPs should be developed to reduce potential risks through the entire chain of command - from faecal sludge sampling, through sample collection and transportation, to sample storage and disposal.

Table 8.1 Examples of SOPs for health and safety in activities dealing with faecal sludge.

No.	Activity	Potential risks	Potential controls	Related section
1	Travel to field location	Defective roads causing vehicle accidents	- Ensure vehicles are roadworthy and insured; ensure driver has appropriate licence. - Ensure mobile phones are fully charged. - Carry emergency numbers and First Aid kit to field.	Section 8.2.3.4
2	Sampling in community	Bullying and intimidation by members of community	- Municipality consent and site induction is essential. Communicate with municipality and selected liaison officers from the community. - Arrange introductions to caretaker and/or householder in charge of facility before sampling begins, keep updated on activities.	Section 8.2.3.4
3	Sample collection and transportation	Illnesses due to contact with faecal sludge	- Wear PPE, including safety boots, overalls, elbow length and sharps-resistant gloves, dust masks and goggles. Vaccinations against Hepatitis A and B, tetanus, and typhoid must be current. Use anthelmintic medicines (if necessary) and have stool samples tested 3 times a year for helminths.	Section 8.2.3.1 Section 8.2.3.4
		Sample spill	- Ensure sample containers are securely closed before removing from the facility; secure container or box with a secondary containment at the back of the vehicle. - In the case of spillages or splashes: - in environment: clean with disinfectant and paper towels, - on person: rinse affected area with water, dry with paper towel, sanitise affected area with sanitiser, *e.g.* 70% ethanol.	
4	Storage of samples at laboratory	Biological samples stored in an unsafe manner resulting in contamination risk to environment and/or personnel	- Label samples appropriately and record sample number and other details on data sheet. - Store samples in a designated location, under appropriate storage conditions at 4 °C, and maintain a database of all the samples.	Section 8.4.1
6	Cleaning/ washing of PPE	PPE stored or disposed of in an unsafe manner resulting in contamination risk to environment and/or personnel	- Wash or spray sharps-resistant gloves used in the field with ethanol, place in a plastic bag for transportation, and wash well using anti-bacterial detergent. - Before leaving the field, place overalls and safety boots into plastic bags, then wash on the same day. Never clean PPE at home. - Place gloves and dust masks into a separate plastic bag and dispose of in allocated areas for contaminated waste in the laboratory. - Wash and disinfect goggles in the laboratory. - All PPE used during laboratory work with faecal sludge must be either safely disposed of, or disinfected and stored in designated areas.	Section 8.2.3.1
7	Disposal of samples	Health and environmental risk	- The correct PPE must be worn. - Discard small amounts of sample in a drain connected to a sewer.	Section 8.4.3

8.2.3 Handling of faecal sludge

This section gives the health and safety guidelines to take into consideration during field sampling and laboratory analysis, including preparation, sample collection and transportation. The provided guidelines can vary depending on the specific local context and regulations and each laboratory may adopt and modify them accordingly. Chapter 3 contains a detailed description of different techniques and tools for faecal sludge sampling, how to calculate the needed volume and number of samples, and the importance of PPE and health and safety during sampling.

8.2.3.1 Personal protective equipment

To ensure safe laboratory practices are followed, there are many references that are freely available to develop SOPs and laboratory training, including 'Promoting Chemical Laboratory Safety and Security in Developing Countries and Security' (National Research Council, 2020). The hierarchy of health and safety levels of protection are discussed in Section 2.5.2 and include elimination, substitution, engineering controls, administrative controls, and personal protective equipment (PPE). In general, good PPE practices to prevent infection or contamination include:

- *Good housekeeping*
  Any equipment taken out of the laboratory into a 'clean' environment should be handled only with clean gloves and disinfected with 70% ethanol. Clean all contaminated equipment and surfaces thoroughly with soap and water, and then disinfect with 70% ethanol for 30 sec. Dispose of faecal sludge samples and chemicals properly. If possible, autoclave contaminated materials prior to disposal or reuse (*e.g.* samples, glassware, tools). Dispose of materials in properly allocated and labelled waste bins with lids (*e.g.* biohazard). Adequately wash hands with soap and water for 20-30 sec, followed by hand sanitiser. Small cuts or wounds on the skin must be adequately covered with a protective barrier prior to handling of faecal sludge.
- *Laboratory coat and/or safety overalls*
  Laboratory coats and overalls provide a basic level of protection from accidental chemical and sample spills, and from spreading contamination in 'clean' environments. Coats and overalls need to be fully buttoned at all times. Laboratory coats should not be worn outside of the laboratory, and sampling PPE (*e.g.* used for sampling in the field) should be removed upon return from the field. Jackets and overalls must never be worn in clean areas outside the laboratory (*e.g.* office space). It is also recommended to cover exposed skin with long trousers.
- *Safety shoes*
  Closed, toed shoes and/or rubber boots are recommended when working in a faecal sludge laboratory. Shoes should be non-slip, non-porous and preferably have an impact-resistant front (*e.g.* toes and insoles protected with metal).
- *Eye protection*
  Eye protection should be worn at all times while working with faecal sludge, especially during activities involving risk of splashing such as sample collection, preparation and analysis. Safety goggles protect the eyes and areas immediately surrounding the eyes, while face shields protect more of the face. Safety goggles and face shields can be worn over prescription glasses.
- *Gloves*
  Disposable gloves must be worn at all times when handling faecal sludge. Depending on the type of work, select gloves that provide resistance to abrasion, tearing, punctures, heat and chemicals. In a faecal sludge laboratory, all surfaces, equipment and consumables should be regarded as contaminated. Do not pick up the telephone or touch the door knob, or other common surfaces, while wearing gloves. Do not handle faecal sludge samples directly with gloved hands, as faecal sludge can contain sharp items that could tear or rip gloves. Use appropriate tools such as a spatula or forceps.
- *Masks*
  Respiratory protection minimises the risk of infection through aerosols, inhalation, ingestion and splashing of faecal sludge. FFP2 disposable particulate/filtering masks provide the minimum level of protection from sample splashes, aerosol inhalation and odours. If working with large samples or sampling in the field, FFP3 masks or

half respirator masks are recommended. If respirators are provided in the laboratory, disinfect and inspect them before use and check for proper fit. FFP2 disposable particulate/filtering masks or FFP3 half respirator masks are recommended for sampling faecal sludge in the field (see Chapter 3).

8.2.3.2 Vaccinations / inoculations

Inoculations provide protection against infectious diseases that are associated with the types of samples received and analysed in the laboratory. For work related with any faecal sludge activities, examples of required vaccines are: Tetanus, Polio, Typhoid fever, Hepatitis A and B. Additional inoculations may be required depending on the prevalence of local diseases and the extent of exposure.

8.2.3.3 'Clean' and 'dirty' work areas

It is important to keep clean and dirty work areas throughout sample collection and processing to ensure protection from pathogens. This includes designated areas for putting on PPE before and after field sampling, and also areas of laboratories where samples are processed during sample preparation and disposal (see Chapter 2). Any activities that are not related to handling of faecal sludge should not be carried out in the 'dirty' areas to prevent exposure to pathogens. This includes normal office work and data analysis. During sampling and analysis, the containers, boxes, tools and equipment used for faecal sludge handling should be placed on washable surfaces. Any 'clean' items that are frequently removed from the laboratory, *e.g.* cameras and data sheets, should be stored in allocated areas on a workbench, and be handled with clean gloves.

8.2.3.4 Sampling

Protocols for the collection and transport of samples are covered in detail in Chapter 3 including the necessary equipment, proper recording of processes, chain of custody, transport, and a field sampling kit. Specific SOPs for sampling are included in Table 8.1. Additional concerns for health and safety include:

- Ensure the sampling vehicle has a First Aid kit and all the passengers have valid identification, permission, and documentation of medical insurance.
- Ensure mobile phones are fully charged and check network coverage.
- Be sure to inform the line manager of the planned itinerary and estimated time of return to the laboratory.
- Prepare sufficient drinking water and weather protection (*e.g.* hats, sunscreen, and waterproof clothing).
- Avoid working alone in the field.
- After sampling, contain the tools tightly in a container or bag to prevent contamination.
- Tightly seal the sampling containers and store them in a large, solid cool box during transportation to avoid spills.
- Brush shoes and spray shoes (including soles) and protective clothing with 70% ethanol after sample handling is complete.
- Dispose of used gloves in an allocated waste bag after sample handling and cleaning of equipment is complete. Reusable gloves (*e.g.* sharp-resistant) must be placed in a separate bag for disinfection in the laboratory.
- Wash hands for 20-30 sec with soap and water, followed by hand sanitiser.
- Upon arrival at the laboratory, clean the sampling tools and PPE in the designated area.
- Leave the washed and disinfected tools and sampling containers in the wash area until dry, then store safely.
- Send PPE clothing to the laundry facility.

8.3 QUALITY ASSURANCE AND QUALITY CONTROL (QA/QC)

Once samples have been collected and transferred to the laboratory, protocols need to be in place to ensure representative, comparative, and repeatable results when carrying out analytical methods. The goal of quality assurance and quality control (QA/QC) is to ensure precise and accurate test results within acceptable limits. The purpose of quality assurance specifically is to give relevant, reliable and timely test results that are correctly interpreted. Quality assurance includes the laboratory procedures that are carried out to produce defensible results with reliable precision and accuracy, and to ensure that the laboratory functions efficiently and effectively. Each laboratory should have an operation manual, with a defined set of procedures that includes all the necessary aspects to demonstrate the laboratory's competence and to ensure and document the quality of its analytical data (Rice *et al.*, 2017). Quality assurance is a total process whereby the overall quality of laboratory results can be guaranteed (WHO, 2011). This includes, but is not limited to, sample collection and processing, good laboratory practice and management skills (*e.g.* roles and responsibilities of staff, training of staff, tracking of samples, SOPs, structures for clear reporting and documenting, maintenance of equipment, calibration curves, protocols for determining method detection limits (MDLs), and sample disposal). Quality assurance also includes quality control. Quality control is primarily concerned with the control of errors in the performance of sampling and analysis, and verification of the accuracy and precision of test results (*e.g.* standards, blanks, duplicates, and standard reference materials). Quality control should be practical, achievable and affordable. The broad aim of quality control is that results are both accurate and precise, which ensures that results from one laboratory are comparable with the results from any other laboratory in the world, if the same method is followed. This can even then be optionally verified through sample exchange. For further information on implementing a quality assurance programme, and how to apply for accreditation, see ISO 9001 and ISO/IEC 17025:2017.

8.3.1 Training

It is important that all staff are adequately trained for the tasks they have to perform. Training must be documented so that management and other personnel can verify that staff are competent to conduct the duties required of them. Laboratory training, a competency assessment, SOPs and a hazard identification risk assessment (HIRA) are required before any person is permitted to access the laboratory or use laboratory equipment for the analysis of faecal sludge.

8.3.2 Standard operating procedures

Standard operating procedures (SOPs) were defined in Section 8.2.2. They provide the core of most of the day-to-day operations of any quality assurance programme. SOPs are based on internal experience, manuals and published references. In practice, SOPs should clearly present the procedural steps in a way that avoids potential differences in interpretation, thereby avoiding subtle changes in the way methods are performed or equipment is used. Such differences can impact the overall quality of the results. An SOP should be clear, concise and contain all the relevant information to perform the procedure it describes. In addition, it should include the methods and frequency of calibration, MDLs, maintenance and quality control, remedial action to be taken in the event of malfunction or loss of control, and information on how to properly dispose of samples and required laboratory consumables (*e.g.* including biowaste, and handling of toxic chemicals). SOPs must be regularly revisited and updated by experienced laboratory technicians.

8.3.3 Laboratory facilities

Resources are required for regular laboratory work as well as for the additional workload associated with quality assurance. It is essential resources of space, staff, equipment and supplies are sufficient for the volume of work. The environment in which the work is conducted must be controlled to eliminate interferences. In practice, anything that restricts the efficient running of the laboratory would be a cause for concern. Further details on setting up a faecal sludge laboratory are provided in Section 2.5.

Basic laboratory equipment and glassware as defined by the American Chemical Society (ACS) (2020) include a balance, beakers, beaker tongs, Bunsen burner, burette, clay triangle, crucible, crucible tongs, Erlenmeyer flask, evaporating dish, forceps, funnel, graduated cylinders, mortar and pestle, pipette bulb, ring stand, scoopula, striker, stirring rod, test tube, test tube clamp, test tube rack, thermometer, utility clamp, volumetric flasks, volumetric pipettes, wash bottle, watch glass, and wire gauze. Their website presents pictures of the equipment, and a brief description. Further required equipment and analytical machines will depend on the specific procedures being conducted in the laboratory.

8.3.4 Sample chain of command

Protocols for the collection and transport of samples are covered in detail in Chapter 3 including the necessary equipment, proper recording of the process, chain of custody, transport, and a field sampling kit. Upon delivery of faecal sludge samples to the laboratory, another important activity is how the samples are received, organised and stored for analysis. These activities are an essential part of the chain of command and need to be done in a systematic manner. Improper sample storage can compromise the analytical results and incomplete labelling can lead to unidentifiable sources of the samples, incomplete data sets, and increase the potential need to repeat the sampling and analysis. All incoming samples should be systematically labelled according to a standard system that has the sample number, project name, date, and name of the sampler clearly labelled on the sample container. Assign different IDs (a coding system) to the samples and write on the container with permanent marker. If undertaking a large sampling campaign, assign a number to each container and keep a separate ID (coding key). After labelling, place the samples in storage (see Section 8.4.1). If storing the samples in a cold room with many other samples, consider developing a sample register for improved sample management. This information must also be stored electronically (a database) using a sample coding key (ID) system that enables the samples to be tracked to their source. Data from paper copies should be transferred to the electronic sample database as soon as possible. The database must be updated regularly and stored in a specifically allocated directory. Basic characteristics may be recorded at this time, for example a photograph, and the mass and type of the faecal sludge. If dealing with samples of individual fresh faeces, use the Bristol Stool Chart for form (Lewis and Heaton, 1997), and for more details refer to Chapter 7. The procedures for receiving and logging faecal sludge samples can vary depending on the specific local context and regulations, and each laboratory should adopt and modify them accordingly.

8.3.4.1 Laboratory photographs and notebooks

Photographs are useful to capture at this stage, and can be a part of the labelling and sample management system. This is useful to understand the texture and appearance of the sample during data analysis and discussions, or to help recover the sample's ID at a later stage in cases when they have not been recorded properly. Two people are recommended for this task – one to handle the samples and the other to handle the camera, in order to prevent contamination of the equipment. The sample handler should open the lids of the sample containers and make sure that the sample inside and all the details on the containers, such as sample name, dates and times are visible for the photos. The person holding the camera should take the photos making sure that all the details are captured. If working with a large number of samples, photographing can be done in a manageable batch order - *e.g.* five to ten samples at a time. A data sheet can be used to record details such as amount of sample, type and condition of sample (*e.g.* visibly degraded, covered with maggots, liquid or solid, and specific odours). After completing the photography, the sample handler should close all the sample containers and return them to storage. Save the updated database (with photos) in a designated electronic directory. Use 70% ethanol spray to disinfect the camera and surfaces as a preventative measure. A sample collection also frequently includes questionnaire data. It is important that this information is all stored electronically in the same format for data analysis. For a detailed description of sample collection with questionnaires and photographs, and possible methods for data analysis, see Ward *et al.* (2021).

It is very important to keep a detailed notebook documenting every step taken while carrying out laboratory methods. Further information on keeping laboratory notebooks ('lab books'), and a general reference for other ethical guidelines when conducting research, can be found in Marcrina (2005) and Barker (2005). Lab books are important for transparency and ethical reporting of results, and also for troubleshooting laboratory procedures. For example, if there are two different balances in the laboratory, when weighing samples, it should always be recorded which balance is used. If one is discovered to be malfunctioning, then it is clear which samples need to be re-evaluated. A general rule is that it is best to err on the side of too much detail. Other general rules include: lab books must be bound (pages are not removable), pages should be numbered and dated in chronological order, pages cannot be removed, information must be recorded in ink, if an error is made a line should be drawn through the text followed by the correct information (absolutely no white-out or correction fluid), blank pages or spaces must not be left (if left intentionally, draw a line through the space), entries cannot be modified (additional information can be added as a new entry in chronological order), and any attachments should be permanently placed in the lab book (*e.g.* a print-out of a calibration curve attached with staples). In addition, lab books are the property of the laboratory and should be tracked, recorded, and archived. Laboratory books cannot be removed from the laboratory, although photocopies or carbon copies can be made for this purpose.

8.3.4.2 Equipment maintenance and calibration

The laboratory must regularly service and maintain all the equipment, which is monitored by a competent and trained laboratory manager or experienced laboratory technicians. Frequent checks on the reliability of equipment must also be performed. To reduce the use of malfunctioning equipment, calibration and maintenance records of all instruments must be stored in a filing system, allowing for the operational status of all the apparatus to be monitored. Analytical equipment should be serviced at regular intervals, according to the manufacturer's instructions and recommendations in standard methods. The frequency and complexity of routine servicing is dependent on the instrument and use. Along with this, the laboratory should perform daily inspections and weekly maintenance of all the equipment and develop a monthly maintenance plan. Each laboratory should develop their own equipment maintenance plan based on frequency of use, age of equipment, user experience, service plans, maintenance checks, and availability of funds. Further guidelines on equipment maintenance and calibration are provided in ISO/IEC 17025:2017, particularly for accredited laboratory systems (IEC, 2017).

As described in Prichard and Barwick (2003), instrument calibration is a part of the majority of laboratory analysis. As it is such an important step in an analytical method, it is essential to have a good understanding of how to set up a calibration experiment and evaluate the results. Calibration entails making a set of standards of known concentration, measuring the instrument response to the standards, and then establishing the relationship between the instrument response and concentrations of the analyte. Detailed information is provided by the LGC group, and the report can be downloaded free of charge (Prichard and Barwick, 2003). Included is information on how to determine the number of standards and range of concentrations that should be used, including a blank. Whether or not the standards are made up in the sample matrix or a solvent will depend on the type of analysis and the analytical machine. The accuracy of some methods can be improved by using an internal standard in standards and samples. In general, seven standards should be used, and they should be evenly spaced across the range of concentrations to be measured with the majority of concentrations falling in the middle range. The standards should be analysed in random order, and the results must be plotted. The statistical analysis of the results is critical, and can be a common source of inaccurate results. It is important to know when to fit the curve through zero, and how to check for bias or leverage due to outliers. The residual standard deviation is used as an estimate of the uncertainty in the predicted concentration values. It is also important to reduce the uncertainty of concentrations through quality control measures such as the use of proper glassware, and an adequate grade and purity of chemicals (Prichard and Barwick, 2003).

The instrument lower-level detection limit (IDL) needs to be quantified by using replicate measurements of aliquots from one standard. The lower limit of detection (LOD) is the lowest concentration that can be statistically detected from a blank. However, the method detection limit (MDL) also needs to be calculated, and to take into account uncertainties that can be introduced during analysis at each step of the method (*e.g.* dilutions, digestions, sub-sampling, sample matrix, type of instrument, and laboratory skill). The MDL is the minimum value that can be reported with 95% confidence that the measured value is above zero. The MDL should be calculated with replicate measurements of aliquots from the actual sample matrix, with low level spikes

2-10 times the expected MDL, together with blanks (minimum 7 of each). This analysis is performed in triplicate on three separate days, and should be repeated annually. The lowest level of MDL that can be reported is calculated as the mean determined concentration plus three times the standard deviation of a set of method blanks.

Any value less than the MDL is reported as non-detectable (ND). There is more variability in the MDL, so it is normally expected to be higher than the instrument detection limit. For more information on calculating the MDL, see USEPA 40 CFR Part 136 Appendix B, and tutorials that can be found by searching for 'method detection limit' at the USEPA website[2] (USEPA, 2017).

For most environmental samples, the actual level of quantification (LOQ) will be 5-10 times the MDL (Rice *et al.,* 2017). However, this is for environmental samples that can be considered to be more 'clean' than faecal sludge, so it is very important to calculate the method LOQ for each specific method and sample matrix. The LOQ will depend on defined levels of precision and accuracy, and can be reliably achieved during routine operations. Calibration curves must include a standard with the LOQ reported by laboratories. A common definition is 10 times the reagent water blank signal (Rice *et al.,* 2017). The LOD, LOQ and MDL are all important parameters to consider when comparing results between laboratories.

8.3.4.3 Reporting of results

At the end of the analysis, the final product of the laboratory is the reported data to be used for data analysis, either by clients, other institutional departments, regulators, or for research publications. Quality assurance ensures that the data is suitable for use in an assessment. This includes the final stages of reporting and interpreting the results. The data should be examined at many stages in the quality assurance system and no data should be reported if it is out of range of the methods. Reports must be prepared according to an agreed procedure, and they must accurately reflect the findings of the study. They should include reference to all the calibration and quality control data, any problems or limitations encountered during the study, and all the calculations or correction factors. Data should be reported in standard units, as described in the SOP. Results should also include the method uncertainty, and correct representation of significant digits (Method 1050B). Whenever possible, the open sharing of raw data is strongly encouraged, to ensure transparency of the results and to increase the overall knowledge of characteristics of faecal sludge.

8.3.4.4 Checking compliance

In order to maintain the quality assurance system, it is necessary to periodically check each area of the laboratory for compliance with the quality assurance system. The audit must be independent, hence the need for a quality assurance officer who reports directly to the highest level of management, or an external auditor. Laboratories should routinely monitor and assess the quality of the testing process in the pre-analytical, analytical and post-analytical phases from sampling and transporting to reporting of the obtained faecal sludge data.

The pre-analytical phase encompasses the following procedures:

- Laboratory training, laboratory safety, number of trained personnel available
- Sample collection, labelling, transport, processing before testing, and storage
- Number, types and sources of samples tested
- Chemical reagent storage conditions, selection of test kits and regular monitoring of the expiry dates
- Regular recording of all information, proper data storage systems.

The analytical phase encompasses the laboratory analysis using laboratory methods:

- Written standard operating procedure manual
- Testing performance, and performance and preventive maintenance of equipment
- Reagent preparation and correct use

[2] https://www.epa.gov/cwa-methods

- Inclusion of internal and/or external quality control
- Quality control monitoring procedure.

The post-analytical phase encompasses all the steps that occur following the analysis:

- Maintaining records for traceability of the documented results, calibrations, standards, calculations and feedback
- Data entry and storage (computer or hard copy)
- Data interpretation and reporting
- Reviewing and addressing queries.

8.3.5 Quality control

Quality control consists of the operational techniques (internal and external) used by the laboratory staff for continuous assessment of accuracy and precision. Internal quality control focuses on the individual method and tests its performance against the precision and accuracy of the given SOPs. Quality control external to the laboratory is a way of establishing the accuracy of analytical methods and procedures and the representativeness and repeatability of sampling by comparing the results obtained in one laboratory with the results obtained by others conducting the same analysis on the same material. In the future, for methods of faecal sludge analysis to become standardised, reference laboratories will have to send out sets of samples with known concentrations of variables to a group of participating laboratories. Each participating laboratory will then analyse the samples for the specified variables and report the results to the reference laboratory. Examples of quality control measures in general to ensure accuracy and precision are provided in Table 8.2, and the required minimum quality control steps are included in each of the methods.

Precision and accuracy play an important role in quality control of experimental analysis and measuring errors. Although they are often used interchangeably, they have different meanings and are independent from each other. Accuracy refers to how close a measurement is to the true value, while precision refers to how close measurements using the same method and equipment are to each other. Two important elements of precision are repeatability and reproducibility. Repeatability is the variation observed when the same person repeats the same method using the same equipment. Reproducibility is the variation observed when different people repeat the same method using the same equipment. A measurement can be very accurate but not precise, or very precise but not accurate. The best quality control is achieved when an analysis is both accurate and precise. This is further illustrated in Figure 8.1, with the example of a dartboard, where the bull's eye is the true value and darts hitting the dartboard close to it are accurate. A illustrates neither accuracy nor precision; B illustrates precision (close repetitions) but not accuracy; C illustrates accuracy but not precision, and D illustrates both accuracy and precision (Byron Inouye, n.d.).

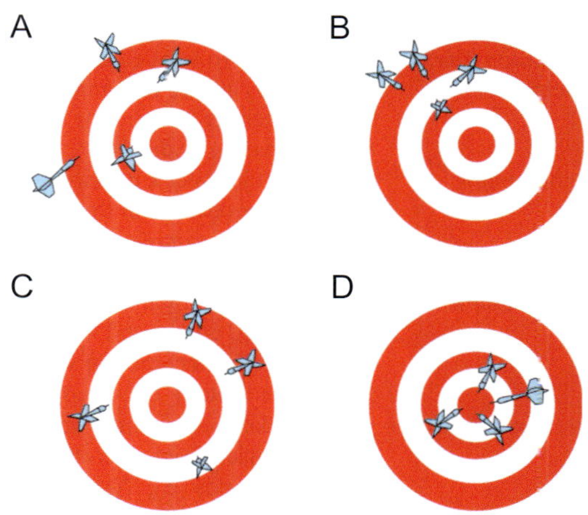

Figure 8.1. Dartboards showing different accuracy and precision scenarios. A: neither accuracy, nor precision; B: precision (close repetitions) but not accuracy; C: accuracy but not precision; and D: both accuracy and precision (required result of QC measures). Source: Byron Inouye, 2020.

Different measures for quality control are presented in Table 8.2. In practice, different methods to measure parameters will be selected based on the required precision, sensitivities and cost. Once a method has been validated and embedded into routine use in the laboratory with an SOP, it is necessary to ensure that it continues to produce satisfactory results.

Table 8.2 Example of quality control steps and measures for faecal sludge analysis to ensure precision and accuracy.

Quality control measure	Description	Main purpose	Minimum frequency
Field replicates	Multiple samples that are collected at the same time and place, and analysed separately.	Identify whether the sample collection is reproducible.	Depending on analytical goal.
Laboratory replicates	One sample that is split into subsamples in the laboratory, and each one is then analysed separately.	Identify whether the laboratory analysis is reproducible.	After every 10 samples, should be within 10% of each other.
Pipette volume	Weighing a known volume of pipetted water.	Identify the accuracy of the pipette and technician pipetting.	Before starting a new experiment and after every 100 samples.
Standard curve	Standard concentrations purchased from high quality laboratories, and/or self-prepared, used for device calibration and for evaluating accuracy of analytical devices.	Determine accuracy of instrument.	Frequency is dependent on analytical method, but recommended to include at least one standard or set of standards in every series of analyses.
Continuing calibration verification	An individual standard that is analysed between every batch of 10-15 samples.	Identify whether the analysis is accurate and ensure instrument is not drifting.	After every 10-15 sample analyses.
Calibration blank (machine blank)	A blank (often deionised water) is used to ensure that analytical devices are not over or under-estimating or drifting.	Identify background contamination and drifts of the analytical device.	Before every series of analyses.
Method blank	A method blank is deionised water processed in the same fashion as samples and used to 'zero' the instrument.	Detect any contamination during the course of analysis and sample preparation.	After every 15-20 samples.
Spiked samples	Adding a known amount of the analysed compound to a sample to test the accuracy of sample concentration measurements.	Determine the accuracy of the method for different sample matrices.	Depends on objective of study.
Reference standards	Standards in a sample matrix purchased from reference laboratories (*e.g.* National Institute of Standards and Technology (NIST), UK Bioresearch Centre (UKBRC)) to evaluate the percentage recovery during sample preparation and analysis (*e.g.* acid digestion for metals).	Identify the percentage recovery of the specific compound that is tested for throughout the entire method (sample preparation and analysis).	Depends on objective of study.
Internal standard	An analyte that is similar to the compound of interest spiked into standards (*e.g.* for use with GC-MS, LC-MS, HPLC).	Monitor retention times and calculate concentrations.	Depends on type of analysis.
External analysis	Sending out a sample for external analysis to verify the accuracy of laboratory results.	Estimate analysis precision and accuracy.	Depends on objective of study.

A validity check standard sample should be included in a batch of samples to monitor precision, standard deviation and accuracy. Precision and accuracy checks allow for quality control of data quality, including calculations and records, standard solutions, reagents, equipment and quality control materials. Some controls such as duplicates ensure precision (reproducibility), whereas other controls ensure accuracy (*e.g.* standards and blanks). If any of the quality control procedures indicates that a problem exists, corrective action must be taken immediately. Quality control measures should also be tracked over time, and reported with the data when published. A discussion of how to statistically interpret results of blanks can be found in Statistics in Analytical Chemistry: Part 40 - Blanks (Coleman and Vanatta, 2010).

8.4 SELECTION OF THE APPROPRIATE METHOD FOR THE PURPOSE OF CHARACTERISATION

In this book, faecal sludge is classified into four categories based on the total solids (TS) content: liquid (TS < 5%), slurry (TS 5-15%), semi-solid (TS 16-25%), and solid (TS > 25%), as defined in Chapter 2. This is important, as the type of faecal sludge determines the selection of appropriate sample preparation, analytical methods and/or steps that are taken within the methods. For example, semi-solid and solid types of sludge may require greater dilution and/or centrifugation in addition to filtration for particular methods, while the liquid sludge may not require dilution and may require only filtration to remove suspended particles. Specific concerns for sample preparation are detailed in each of the methods. The category of sludge is also relevant in regard to how results are reported, either as a mass/volume concentration (mg/L) or as a mass/mass concentration (g/g). In general, semi-solid to solid samples are reported as mass concentration, and liquid and slurry samples as concentration. However, it is not possible to make a hard and fast rule, and this will also depend on the range of samples being analysed. Sample density and TS concentrations can be used to convert results between concentrations and mass fractions (Section 2.2).

Selection of the most suitable method for characterisation depends on the purpose, required level of accuracy, cost of analysis (often a limiting factor), access to analytical machines, and the laboratory capacity required to undertake specific analyses, as discussed in Chapter 2. In general, the level of accuracy, cost of analysis, and access to analytical machines are related to each other, with more expensive laboratory equipment and analytical machines leading to increased accuracy. However, the highest level of accuracy is not always necessary. For example, nitrate test strips might be appropriate to get a rough idea of concentrations when working in the field, or in the laboratory when determining a required dilution to fall within the spectrophotometer method limits. If analysing ammonia, the choice of using a test kit based on the phenate method *versus* a titration will depend on factors such as the available budget for purchasing test kits, the costs of different manufacturers' test kits *versus* the cost of chemicals, access to a spectrophotometer, and access to titration equipment or an automated titration system. For analysis of heavy metals, the appropriate method will depend on whether total metals or available metals (*e.g.* plant available) are of interest, or for dewatering if settling or filtration (*e.g.* drying beds) are of interest. Therefore, an assessment needs to be undertaken before the analysis is conducted, to weigh these factors when selecting the most appropriate method for the specific context.

As discussed in chapters 2, 3 and 5, the number of samples is also a factor contributing to cost. Frequently, budgets limit the possibility of collecting a number of samples required for 95-99% confidence, and decisions on how to ensure representativeness need to be taken carefully. Another way to reduce costs is to carefully define the objectives and then only analyse metrics that answer the specific objectives. This might sound obvious, but *E. coli* should not necessarily be selected as a parameter of interest based solely on having the laboratory capacity. For example, when setting up experiments to scale up inline dosing of conditioners for enhanced dewatering performance as described in Chapter 4, TSS and CST are the metrics of interest, and the additional cost of *E. coli* analysis would not be justified. In addition, the total number of parameters that are analysed contributes to increasing levels of data analysis.

The methods provided in this chapter are summarised in Table 8.3, including estimates of time needed for sample preparation and analysis.

Table 8.3 Summary of methods presented in this chapter, by the properties being analysed: chemical and physico-chemical (Section 8.6), physical (Section 8.7) and biological (Section 8.8).

No.	Method	Preparation time	Analysis time
8.6	CHEMICIAL AND PHYSICO-CHEMICAL PROPERTIES		
8.6.1	*Solids and moisture content*		
8.6.1.1	Total solids and moisture content - volumetric and gravimetric method by oven drying	< 1 hr	> 24 hr
8.6.1.2	Volatile and fixed solids - ignition method	< 5 min	< 2 hr
8.6.1.3	Total suspended solids and total dissolved solids - oven-drying method	< 1 hr	> 24 hr
8.6.1.4	Volatile suspended solids - ignition method	< 5 min	< 2 hr
8.6.1.5	Total solids and moisture content - thermal balance (moisture analyser) method	< 30 min	< 30 min
8.6.1.6	Sand content	< 30 min	< 3 hr
8.6.2	*Chemical oxygen demand (COD)*		
8.6.2.1	Chemical oxygen demand - closed reflux spectrophotometric method	< 30 min	< 30 min
8.6.2.2	Chemical oxygen demand - closed reflux titrimetric method	> 24 hr	< 3 hr
8.6.3	*Fat and fibre*		
8.6.3.1	Crude fat - Soxhlet extraction method	> 24 hr	> 24 hr
8.6.3.2	Crude fibre - filtration method	> 24 hr	> 6 hr
8.6.4	*Nitrogen*		
8.6.4.1	Total nitrogen - spectrophotometric method	< 30 min	< 30 min
8.6.4.2	Ammonium - colorimetric (test strip method)	< 15 min	< 30 min
8.6.4.3	Ammonium - phenate spectrophotometric method	< 30 min	< 30 min
8.6.4.4	Ammonia - distillation and titration method	< 1 hr	< 1 hr
8.6.4.5	Nitrite - colorimetric (test strip method)	< 15 min	< 30 min
8.6.4.6	Nitrite - spectrophotometric method	< 30 min	< 30 min
8.6.4.7	Nitrate - colorimetric (test strip) method	< 15 min	< 30 min
8.6.4.8	Nitrate - cadmium reduction spectrophotometric method	< 30 min	< 30 min
8.6.4.9	Total Kjeldahl nitrogen - distillation and titration method	> 30 min	> 3 hr
8.6.5	*Phosphorus*		
8.6.5.1	Total phosphorus and orthophosphate - spectrophotometric method	< 30 min	< 30 min
8.6.5.2	Orthophosphate - colorimetric (test strip) method	< 15 min	< 30 min
8.6.6	*pH and electrical conductivity*		
8.6.6.1	pH - electrode method	> 10 min	< 15 min
8.6.6.2	Electrical conductivity - electrode method	> 10 min	< 15 min
8.6.7	*Elemental analysis*		
8.6.7.1	Metals - overview	Dependent on selected method	Dependent on selected method
8.6.7.2	Metals - acid digestion for environmentally available metals	< 2 hrs	< 30 min
8.6.7.3	Ultimate analysis - total carbon, hydrogen, nitrogen, oxygen, and sulphur	Dependent on selected method	Dependent on selected method
8.6.7.4	Chlorine - colorimetric (test strip) method	< 15 min	< 30 min
8.6.7.5	Chlorine - spectrophotometric method	< 30 min	< 30 min

8.6.7.6	Chloride - colorimetric (test strip) method	< 15 min	< 30 min
8.6.7.7	Chloride - spectrophotometric method	< 30 min	< 30 min
8.6.8	*Colour and turbidity*		
8.6.8.1	Colour - visual comparison method	< 30 min	< 15 min
8.6.8.2	Turbidity - nephelometric method	5 min	5 min
8.6.9	*Settleability and dewaterability*		
8.6.9.1	Jar test	30 min	< 60 min
8.6.9.2	Capillary suction time	5 min	< 15 min
8.6.9.3	Water activity	< 10 min	< 15 min
8.6.9.4	Sludge volume index	< 15 min	< 3 hr
8.7	PHYSICAL PROPERTIES		
8.7.1	*Physical and mechanical*		
8.7.1.1	Density - mass and volume measurement method	< 15 min	< 15 min
8.7.1.2	Density - volume displacement method	< 15 min	< 30 min
8.7.1.3	Particle size - laser light scattering method	5 min	< 15 min
8.7.1.4	Rheological properties - rheometer method	< 10 min	< 30 min
8.7.1.5	Liquid limits - cone penetrometer method	10 min	< 15 min
8.7.1.6	Plastic limits - thread-rolling method	30 min	< 60 min
8.7.1.7	Compressibility and stickiness - texture analyser method	< 15 min	< 15 min
8.7.2	*Physical and thermal*		
8.7.2.1	Thermal conductivity - thermal conductivity analyser method	< 10 min	< 15 min
8.7.2.2	Calorific value - bomb calorimeter method	> 24 hr	< 15 min
8.8	BIOLOGICAL PROPERTIES		
8.8.1	*Pathogens*		
8.8.1.1	*E. coli* and total coliforms - colony forming unit method	> 30 min	< 60 min
8.8.1.2	*E. coli*, faecal coliforms, and total coliforms - most probable number method	< 30 min	< 30 min
8.8.1.3	Bacteriophage - plaque assay method	< 3 hr	< 2 hr
8.8.1.4	Helminths - microscopy method	< 30 min	< 30 min

8.4.1 Faecal sludge storage and preservation

As discussed in Section 3.12, proper preservation helps ensure that no significant changes in composition occur before the analyses are made. It is best to analyse samples immediately upon arrival at the laboratory. However, time and capacity constraints do not always allow this. Therefore, short-term storage is frequently required (*e.g.* up to a few days at 4 °C), and sometimes longer-term storage is required. Some properties are more affected by sample storage conditions and durations than others. Whenever possible, longer-term storage with the addition of preservatives should be avoided, as adding preservatives can also change the composition of the sample and can affect the properties. In this case, it is recommended to only use preservative in a sub-sample of the original sample. Examples of preservation methods depending on the intended analysis include pH control, chemical addition, the use of amber and opaque bottles, filtration prior to storage, drying, dry-freezing, refrigeration (4 °C), and freezing (-20 °C). However, all methods of preservation may be inadequate when applied to suspended matter. Preservative should not be added if volatile, semi-volatile or microbial contaminants are to be analysed, unless specified in the standard operational procedure. For the analysis of stable compounds, such as the total metal analysis of properly stored and dried samples, longer-term storage is allowable. Information on specific preservation techniques is included in each method, and more information can be found in Method 1060C (Rice *et al.,* 2017).

8.4.2 Faecal sludge sample preparation for analysis

Before analysis, samples are usually homogenised, diluted and, if necessary, filtered or centrifuged. The purpose of homogenisation is to ensure that the sample is thoroughly-mixed and that the analysis is representative of the sample. Further processing may be necessary in order to ensure that the sample is able to be analysed using a specific method or instrument. This can include drying, grinding, sieving, diluting, filtering, and/or centrifuging. The sample preparation methods vary depending on the type of faecal sludge (solid, semi-solid, slurry or liquid), and also on the type of analysis. For example, blending and dilution are appropriate for preparation for chemical analysis such as COD and ammonia but not for other methods. For rheological and viscosity analysis, the samples must be analysed as received because blending or mixing will change the structure of the sample and affect the results of the analysis. Similarly, dilution should not be applied to samples that are going to be analysed for total or volatile solids content. Information on specific preparation techniques is included in each method. It is important to consider the different processing requirements for all of the intended methods, as this will dictate the required sample volume and laboratory preparation pathways. An example of this is provided in Figure 3.23, for a series of preparations with the same sample that includes blending/not blending, centrifugation, and further dilutions.

8.2.4.1 Homogenisation of samples

Homogenisation is especially important for faecal sludge, as it is typically a very heterogeneous and complex matrix, and inadequate homogenisation can be a significant source of error in the final results. Homogenisation techniques include blending, shaking, vortexing, mixing, stirring, and grinding. Prior to homogenisation, remove any large non-faecal materials, such as stones, plastics, textiles, hair, maggots, and rubbish.

For liquid and slurry sludge (TS < 5%, TS 5-15%, respectively), mix the sample by gently inverting the container until the settled particles are in suspension, or apply rapid stirring with a ladle so the sample can be distributed prior to any settling. If slurry samples are too thick to homogenise in this fashion, use the method for semi-solid and solid sludge. Whether or not samples should be blended is specified in each method, and will depend on whether or not destroying the sludge structure would interfere with the subsequent analysis. For semi-solid and solid sludge (TS > 15%), mix the entire sample using a stainless steel rod (or another appropriate tool), until visibly homogenised. Blending before dilution may not be suitable for these types of samples because of their high TS content. For dried resource recovery products, such as dried sludge, samples can be air-dried and then ground using a mortar and pestle or mechanical grinder to homogenise.

8.2.4.2 Dilution of samples

Depending on the type of analysis and the type of faecal sludge, dilution may be necessary so that the sample concentration is within the quantification range for the specific method. For example, the COD concentration in undiluted faecal sludge is frequently relatively high and requires a series of dilutions. Care needs to be taken with dilution, as it can be a significant source of error in the final results, especially with highly heterogeneous faecal sludge. The level of uncertainty of volumetric measurements from glassware and pipettes needs to be taken into account to know what can be accurately measured. For example, the level of uncertainty for a 1 mL pipette is ± 0.01 mL, for a 10 mL volumetric flask ± 0.01 mL, and for a 10 mL graduated cylinder ± 0.1 mL. Pipetting small volumes into large volumes for dilution needs to be avoided, and in general, a series of dilutions should be used for more than two orders of magnitude. For example, for a 1:1,000 dilution, pipette 1 mL into 9 mLs for a total of 10 mL diluted sample volume, and repeat this three times; and never pipette 1 mL into 999 mL. Serial dilutions are defined as repeating the same dilution step over and over, and hence represent a geometric series (*e.g.* 1/10, 1/100, 1/1,000, or 1/3, 1/9, 1/27). It is of utmost importance that at each step in the dilution, the sample is thoroughly mixed before making the next dilution. Depending on the method, this could be stirring, shaking, vortexing, or even mixing in a blender; exact protocols need to be developed for each method and local context.

For liquid and slurry types of faecal sludge (TS < 15%), sub-samples for dilution are measured out volumetrically. Mix the sample as described in the 'homogenisation' section and measure out an appropriate volume for dilution into a volumetric flask, and then add distilled water to a specific volume. A serial dilution will be necessary depending on the required concentration range, and could be followed by filtration or centrifugation to separate the suspended solids depending on the specific method. For semi-solid and solid sludge (TS > 15%), tare a beaker on an analytical balance and then weigh out the sample into the beaker. Calculate the volume of required dilution water, and weigh it out in a separate tared container. Add a volume of the dilution water to the sample, and mix it well, then transfer to the blender. Reserve some of the volume of dilution water to rinse the beaker that held the sample and also transfer the wash water to the blender. For example, this could mean weighing out between 1.8 g and 2.0 g into a 50 mL beaker, and then rinsing it in in series to obtain the required volume of 200 mL. Blend the diluted sample at the highest setting until homogenised, then pour the diluted sample into a bottle for analysis or storage. It is not necessary to remove all of the solids from the blender as the sample has been thoroughly mixed.

The results should be represented as either a volumetric concentration (mg/L) or as a mass fraction by total dry solids (g/g TS). In general, it is not recommended to report results as a wet basis mass fraction (g/g wet sample), due to the high variability in percentage solids or moisture in faecal sludge samples. This means that density (mass per volume) and TS (mass of TS per volume) should also be measured for each sample for accurate reporting and to allow for comparison between results. The definition of the dilution factor that is used in the methods is:

$$\text{Dilution factor} = \frac{\text{Final adjusted volume of dilution}}{\text{Original volume of sample aliquot to be diluted}}$$

For example, if 10 mL is the final dilution volume and 1 mL was the original sample volume, then the dilution factor would be 10.

8.2.4.3 Filtration

Filtration of samples could include sieving of dried ground samples prior to metals digestion, or filtration of liquid samples to remove suspended solids. For example, as described in Method 8.6.1.3 for total suspended solids, acceptable glass fibre filters will range in pore size from 0.45 μm to 2.0 μm depending on the thickness of sludge and clogging of the filters. In addition to pore size, clogging is addressed through the process of sample dilution for semi-solid to solid samples, as explained in the method description. The type of filter material will also depend on the intended method, and is explained in each section; for example, glass fibre filters are specific to methods that use a 550 °C muffle furnace, and paper filters can be used in nutrient methods.

8.2.4.4 Centrifugation

Centrifugation can be used instead of, or in addition to, filtration to separate suspended particles from the sample. For liquid faecal sludge (TS < 5%), mix and blend the sample as described in the 'homogenisation' section and transfer an aliquot of the sample into a clean glass beaker. Withdraw sample aliquots of equal volume into one or more centrifuge tubes. Place an even number of centrifuge tubes diagonally opposite each other to balance the centrifuge. If there are an odd number of centrifuge tubes filled with sample, fill an additional tube with the same volume of water to balance the centrifuge. Centrifuge the tubes at a specific g-force and time. Decant the filtrate (supernatant) into a clean glass beaker and discard the pellet. Centrifuge the samples until the required volume for sample analysis is obtained. Centrifugation can also be used for separation of the suspended solids in slurry, semi-solid, and solid samples (TS > 5%); however, centrifugation will be more effective if the samples are diluted beforehand. In cases where the sludge is very difficult to filter, centrifugation can be used prior to filtration to remove most of the suspended material.

8.4.3 Sample and chemical disposal

Waste minimisation and pollution prevention in the laboratory is the preferred approach to managing laboratory waste. However, when the remaining material is disposed of, it is important to ensure the protection of public and environmental health, and disposal needs to be included in each SOP. For more detailed information on methods of disposal, see National Research Council (2011).

Faecal sludge samples and laboratory equipment should be contained and sterilised to eliminate all pathogens. All contaminated equipment should be disinfected before washing, storage, or disposal. If possible, it is recommended to autoclave materials contaminated with faecal sludge before discarding them. After sterilisation, waste can be handled safely and disposed of in the local waste collection. Incinerated or pasteurised samples can also be contained and disposed of in the general waste. If autoclaving is not possible, small volumes of faecal sludge samples (*e.g.* 1-5 L) should be flushed down a drain that is connected to the sewer, and the whole area then properly sterilised. If this is not possible, store the waste samples in a sealed container in a cold room or refrigerator at 4 °C. The person responsible for the samples (*e.g.* the student, researcher, or laboratory personnel) then needs to arrange for the waste samples to be discarded at a local wastewater or faecal sludge treatment plant. If a faecal sludge treatment plant is not available, the samples may be discarded in the laboratory toilet facilities. This, however, should be done cautiously due to potential blockages and contamination. The toilet facilities should be well washed and disinfected after the sample disposal.

It is important to document in SOPs which acids, bases, salts, and solutions can be poured down the drain with dilution, and which chemicals may not be disposed of in this fashion under absolutely any conditions (*e.g.* heavy metals, pesticides, oils, nitrite). A management plan needs to be in place for disposal of toxic compounds.

Containers need to be labelled for separate collection of waste that is safe to be managed with the local waste collection, *e.g.* biohazard and sharps, and waste chemicals. Material such as broken glassware, sharp objects and fine powders that could harm workers during collection also needs to be separately collected. For more information on developing waste handling procedures, refer to Prudent Practices in the Laboratory: Handling and Management of Chemical Hazards (National Research Council, 2011).

8.5 SHIPPING AND RECEIVING OF FAECAL SLUDGE SAMPLES AND EQUIPMENT

In some cases, faecal sludge samples may be shipped locally or internationally; for example, when the required analysis is not locally available, when samples are collected from remote areas for comparison among multiple laboratories of the same samples, or for research studies. In this case, the sending and receiving parties need to fulfil specific regulatory requirements and documentation based on the local context and requirements of the shipping company. These may include import-export permits for biological samples issued by the relevant authorities (*e.g.* see the Department of Health example in Figure 8.2), and a material transfer agreement (MTA) signed by the sender and receiver organisations. Further information and details on specific local regulations can be obtained by local courier companies. Sufficient time between application for all the documents and the shipping date must be planned, including customs regulations. For example, in some countries this process may take up to 3 months.

The sample packaging method depends on the type of faecal sludge samples to be shipped. For example, oven- dried and/or autoclaved samples are relatively easy to ship due to their reduced risk of leakage and contamination whereas more liquid samples need special care. Liquid samples need to be contained in order to minimise the risk of leakage. Samples should be packed in sealed, airtight containers with headspace, wrapped in double containment (*e.g.* watertight plastic bags) along with absorbent material in case of leakage, and shipped with ice packs (for very short transport times) or dry ice. Below in Figure 8.3 are examples of packaging, but specific guidelines must be obtained from each courier company, and further information on packaging instructions can be found through the International Air Transport Association (IATA). Samples should be shipped as soon as possible after collection to ensure that they are well preserved. In addition, the shipping time should be as short as possible to avoid biodegradation processes in the samples and accumulation of biogas in the shipping containment.

DEPARTMENT OF HEALTH
Private Bag X828 PRETORIA 0001
Inquiries: Mr JR Mokonoto Tel.: (012) 312-0395 Fax: (012) 312-0287

APPLICATION FOR AN IMPORT PERMIT FOR BIOLOGICAL SUBSTANCES

Person applying for an import permit:	
NAME	
RANK/POSITION	

Organisation:			
NAME			
ADDRESS			
TEL. NO.		FAX. NO.	

Specific substance(s) for which an import permit is required:	
SUBSTANCE	QUANTITY
Period during which import will take place	

Contact person and organisation supplying the substance(s):			
NAME: PERSON			
NAME: ORGANISATION			
ADDRESS			
TEL. NO.		FAX. NO.	

Purpose(s) for which substance(s) is(are) to be used. Although detail is not required, the specific purpose(s) must be clearly stated:

SIGNATURE OF APPLICANT:.......................................DATE:...............................

Figure 8.2 Example of an export permit application in South Africa (source: Department of Health, Pretoria, South Africa).

For primary containment, samples should be shipped in rigid plastic bottles or containers sealed with a watertight, screw-on cap (see Figure 8.3 A). Samples should be packed directly from the refrigerator to maintain temperature. Absorbent material should be packed around the primary containers before sealing the plastic bags. For secondary containment, primary containment containers are packed into watertight, sealed, strong plastic bag/s (see Figure 8.3 B and C). Absorbent material must be packed on the outside of the secondary containment. There must be a sufficient amount of absorbent material inside of the package to ensure absorbance of the sample volume in the event of leakage. Examples of absorbent materials are sponges, vermiculite, paper towels, or wooden chips. Sealed plastic bags with the sample containers must be packed inside a rigid container to ensure that samples are fixed during shipping (see Figure 8.3 E). A cardboard or a rigid outer packing (*e.g.* a hard-plastic cooler box) is recommended (see Figure 8.3 F). It may be additionally packed into a larger box if required by shipping regulations. Include paperwork inside the shipping box explaining the contents of the shipment and the regulations it falls under.

Cooling packs, ice or dry ice may be used in the shipping container (see Figure 8.3 D). Samples should be kept at 4 °C for the entire duration of shipping. General cooling rules for samples are:

- The use of ice packs is recommended, *e.g.* gel packs or hard ice packs. Water produced from melting ice and condensation should be well contained within the packaging to prevent leakages.
- Dry ice can be used; however, ensure that there is a safe escape of the carbon dioxide gas from the container There are specific limits for dry ice quantities and this needs to be confirmed with the shipping company.
- No use of liquid nitrogen.

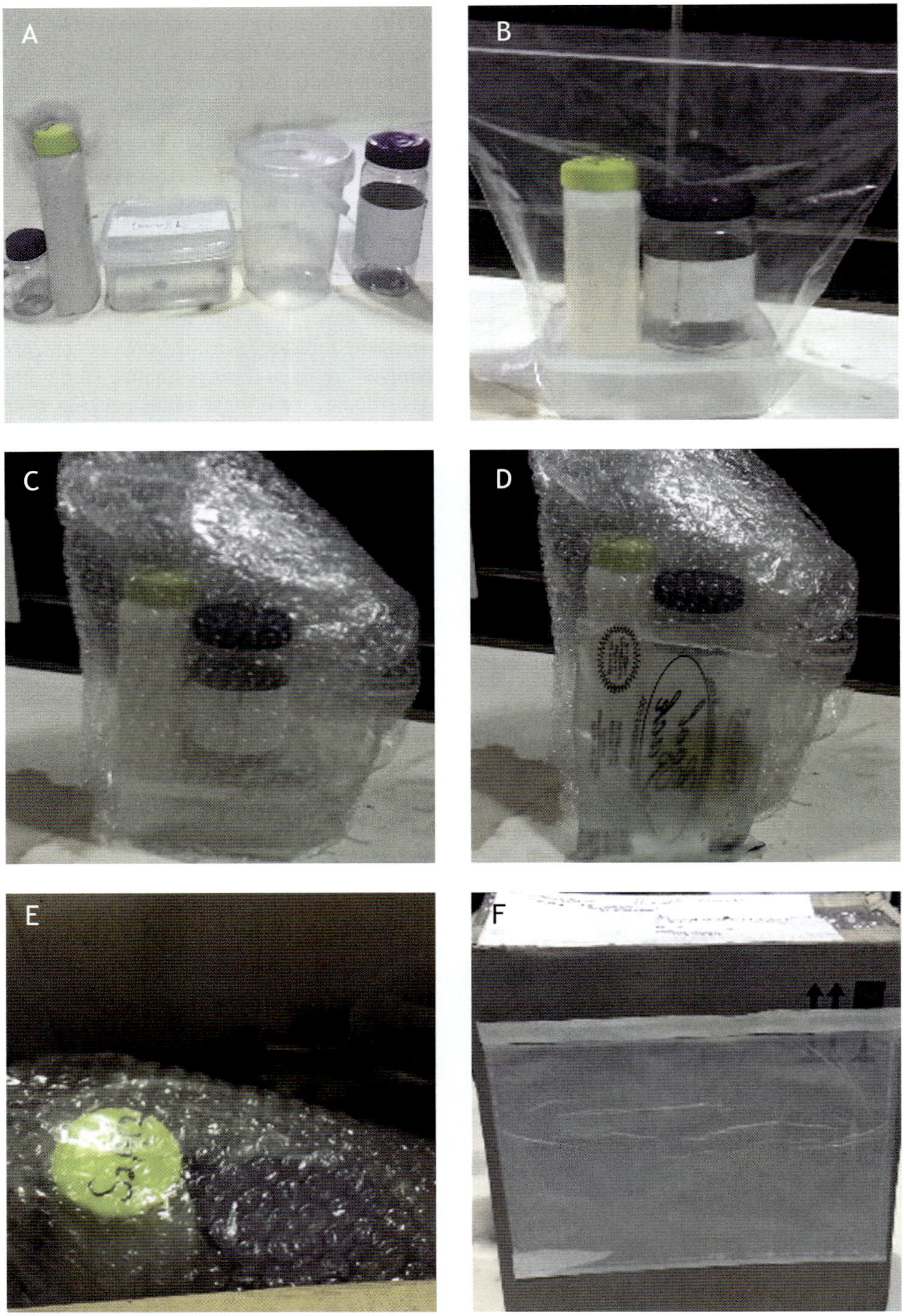

Figure 8.3 Examples of packaging in the laboratory (photos: UKZN PRG).

8.6 CHEMICAL AND PHYSICO-CHEMICAL PROPERTIES

8.6.1 Solids and moisture content

Total solids is a term applied to the material left in a vessel after evaporation of a sample and its subsequent drying in an oven at a defined temperature. Total solids are comprised of total suspended solids (TSS), total dissolved solids (TDS), fixed solids (FS) (ash) and volatile solids (VS). Total solids are determined for all types of faecal sludge – liquid, slurry, semi-solid and solid. The same methods are used to determine TS and moisture content; the total mass of a sample before the analysis is the sum of its TS and moisture content. Sand content (measured as silica as an indicator of soil content in faecal sludge) is the concentration of sand in the TS of an unfiltered faecal sludge sample. Sand can influence faecal sludge treatment processes (*e.g.* dewatering), increase abrasion of mechanical equipment, and affect the quality of faecal sludge treatment end products.

8.6.1.1 Total solids and moisture content – volumetric and gravimetric methods by oven drying[3]

8.6.1.1.1 Introduction

Total solids (and/or moisture content) is one of the most commonly used faecal sludge parameters, and is used for almost every design or management decision. For example, for making decisions on treatment design, settling, or emptying. A known volume (the volumetric method) and/or weight (the gravimetric method) of a thoroughly-mixed sample is evaporated to a constant weight in a crucible (porcelain or silica) or an aluminium weighing boat, in a drying oven at 103-105 °C; the remaining solids are cooled down to room temperature in a desiccator to avoid absorption of air moisture and then re-weighed. The residual material remaining in the crucible are TS, and can consist of organic and inorganic material, and dissolved, suspended or volatile matter.

The gravimetric method is recommended for semi-solid and solid types of sludge, as it is often difficult to measure volumes accurately for sludge with higher TS concentrations. For more liquid types of sludge, either the gravimetric or volumetric method can be used. However, these are general recommendations, and a final decision of which method to use needs to be assessed for each application individually. Conversion between volumetric and gravimetric measurements can be done if the density is known (Chapter 2). Density of faecal sludge can easily be measured by weighing a known volume of sludge (Method 8.7.1.1). When doing such conversions, it is always recommended to measure the actual density of the specific samples, and this becomes even more important with samples at the higher range of % TS.

Solid and semi-solid sludge types can form a water-trapping crust if the initial rate of drying is too high. This can be avoided by placing the samples in the drying oven at a lower temperature, and gradually increasing the temperature of the oven until the prescribed temperature of 103-105 °C is reached.

8.6.1.1.2 Safety precautions

- General health and safety (H&S) procedures specific for conducting laboratory analysis of faecal sludge are presented in Section 8.2. Before conducting this method, it is important to be familiar with Section 8.2.3 to ensure safety measures are properly carried out.
- Appropriate personal protective equipment (PPE) should be used; specific details are covered in Section 8.2.3.1.
- Always conduct the TS analysis in a room with sufficient airflow and an exhaust system.
- Wear gloves suitable for withstanding high temperatures when placing and removing crucibles from the oven.
- Use appropriate mechanical tools, such as metal tongs, to remove crucibles and trays after drying in the oven to avoid direct contact with hot surfaces.

[3] The volumetric method is based on Method 2540B of the Standard Methods for the Examination of Water and Wastewater. The gravimetric method is based on ASTM E1756–08 Method A and on Method 2540G of the Standard Methods for the Examination of Water and Wastewater. Both methods should be cited as: Rice *et al.* (2017) as described in Velkushanova *et al.* (2021).

8.6.1.1.3 Apparatus and instruments

- Porcelain crucibles or aluminium weighing boats
- Desiccator with dry desiccant
- Drying oven
- Analytical balance with four decimal places
- Spatula
- Stainless steel tray (optional, to move crucibles in and out of the oven)
- Heat-resistant gloves
- Pencil
- Thermometer (for quality control procedure)
- Set of standard calibration weights (for quality control procedure)

8.6.1.1.4 Quality control

General information on quality assurance and quality control (QA/QC) is provided in Section 8.3. Information on standards, operating conditions and interferences that are specific to this method includes:

- The analytical balance and oven must be checked and calibrated weekly.
- Check the temperature throughout the oven area by placing a calibrated thermometer on each shelf. After 30 min, check the temperature at each level against the oven setting. Using the same method, also check for temperature differences between the front and back of the oven. Adjust the oven setting if necessary. If temperatures are uneven on the shelves, check the insulation.
- To calibrate the analytical balance, place a standard calibration weight on the balance and weigh. Adjust the balance manually if necessary. Do this with the whole range of weights from the calibration set. Make sure to include a standard weight of a mass similar to the mass of the expected sample + crucible.
- Make sure the desiccant in the desiccator is not saturated, otherwise samples can absorb water while cooling down in the desiccator. Routinely dry the desiccant in the oven at 105 °C (or at a different temperature, depending on the manufacturer's instructions), prior to the colour indicating that the desiccant is nearly saturated.
- Always keep the lid of the desiccator on and use a lubricant on the rim to ensure airtight sealing. Do not overload the desiccator.
- The volume or mass of the wet sample used should be chosen so that the drying will yield a residue between 2.5 and 200 mg of the dried sample (in general approximately 30 mL for the volumetric method, or 10-20 g for the gravimetric method, but this will depend on the type of sludge).
- For solid, semi-solid and slurry samples: limit the sample to no more than 10-20 g faecal sludge, otherwise the sample will take too long to dry and can form a moisture-trapping crust on top. If crust formation is occurring, the samples should be placed in the oven at a lower temperature initially, gradually increasing the temperature until 103-105 °C is reached.
- For liquid samples, the volume of the sample can be higher because the TS content is much lower. The proportion of the weight of the sample to the weight of the porcelain or aluminium crucible should be also taken into account, so that weight differences in the sample can be measured accurately.
- Make sure that the samples are fully cooled in a desiccator to ambient temperature prior to weighing.
- Sludges that contain highly mineralised water with a significant concentration of calcium, magnesium, chloride and sulphate can be hygroscopic and require prolonged drying, complete desiccation and rapid re-weighing.
- Exclude larger, inconsistent or floating particles from the sample if it is determined that their inclusion can affect the final result (*e.g.* hair, stones, glass, and maggots).
- Disperse visible floating oil and grease with a blender or stainless steel mixing rod before withdrawing a sample portion for analysis.

8.6.1.1.5 Sample preservation

Samples should be analysed as soon as possible. If samples cannot be analysed immediately, they should be stored in a refrigerator at 4 °C for no longer than 7 days and, if TSS or VSS analysis is conducted, no longer than 48 hours. Before starting analysis, let the samples return to ambient temperature. Do not freeze the samples.

8.6.1.1.6 Sample preparation

- Uniformly mix all the samples using a stainless steel rod (or other appropriate tool) in order to have thoroughly mixed representative samples. If desired, samples can also be blended (see Section 8.4.2).

- Measure out an appropriate sample volume/mass (indicatively 30 mL for the volumetric method, or 10-20 g for the gravimetric method) which will yield a residue between 2.5 and 200 mg of dried sample, by using a volume measuring cylinder or analytical balance. With very dilute faecal sludge samples, a pipette can be used. For other sludge types, clogging of the pipette will occur, and therefore using a graduated cylinder to measure volume is recommended.

8.6.1.1.7 Analysis protocol

Preparation of equipment

- Pre-heat the oven to 103-105 °C.
- If analysing multiple samples or replicates at the same time, number the bottom of the crucible with a pencil and record in a laboratory notebook which sample and replicate is in which number crucible to distinguish between crucibles. If using aluminium weighing boats, the replicates can also be marked by scratching the number on the weighing boat with a sharp item.
- Place the clean crucible in the oven at a temperature of 103-105 °C for 1 hr prior to use (to remove any moisture). After drying, place the crucible in the desiccator and allow it to cool down to room temperature. Keep the crucible in the desiccator until the next step.
- Note: if measuring volatile solids after the TS, prepare the crucible in a furnace at 550 °C for 15 min prior to use to remove any potential residual organic material from previous measurements. Only porcelain crucibles should be used (see Method 8.6.1.2).

Procedure

- Remove the crucible from the desiccator and weigh it using the analytical balance. Record the weight of the dry, empty crucible (W_1).
    - For the gravimetric method (semi-solid to solid sludge):
        - Weigh out 10-20 g mass of the sample to the weighed crucible using a spatula.
        - Record the wet mass + mass of the crucible (W_2).
    - For the volumetric method (liquid to slurry sludge):
        - Measure 30 mL of the sample volume using a measuring cylinder and record the exact volume of the sample (V_{sample}).
        - Transfer to the weighed crucible. Rinse the cylinder with small volumes of distilled water to dislodge heavy particles. Make sure that all the particles are transferred to the crucible. Add the washings to the crucible but note, calculations must be based on the sample volume and exclude the volume of the washings.
- Oven-dry the sample at 103-105 °C for at least 24 hr or until a constant weight is achieved (which could take longer). To do this, cool and weigh the sample as described below, place the sample back in the drying oven for 1 hr and cool and weigh again. Repeat the steps of drying, cooling and weighing until a constant weight is obtained, or until the weight change is less than 0.5 mg, or 4% of the previous measurement. The length of drying time needs to be evaluated for each specific type of sample, and revisited periodically.
- Take the sample out of the oven and place it in the desiccator to reach room temperature.
- Weigh the dry mass of sample + crucible using an analytical balance and record the weight (W_3).

8.6.1.1.8 Calculation

Liquid and slurry samples (volumetric method):

Total Solids in wet sample (mg/L) =

$$\frac{(W_3\ (g)-W_1(g)) \times 1{,}000{,}000}{V_{sample}\ (mL)}$$

Total Solids in wet sample (g/L) =

$$\frac{(W_3(g)-W_1(g)) \times 1{,}000}{V_{sample}\ (mL)}$$

Semi-solid and solid samples (gravimetric method):

Total Solids in wet sample (g/g) =

$$\frac{(W_3(g)-W_1\ (g))}{(W_2(g)-W_1\ (g))}$$

Moisture content in wet sample (g/g) =

$$\frac{(W_2\ (g)\text{-}W_3(g))}{(W_2\ (g)\text{-}W_1(g))}$$

Moisture content (%) =

$$\frac{(W_2\ (g)\text{-}W_3(g))}{(W_2\ (g)\text{-}W_1(g))} \times 100(\%)$$

Where:

W_1 = Crucible mass (g)
W_2 = Wet sample mass + crucible mass before drying (g)
W_3 = Dry sample mass + crucible mass after drying (g)
V_{sample} = Volume of sample used (mL)

For an explanation of the conversion of these units into %TS, refer to Chapter 2, Section 2.2.

8.6.1.1.9 Data set example

Described in Engund *et al.* (2019) and Strande *et al.* (2018) are the collection of 60 faecal sludge samples in Hanoi, Vietnam, and 180 samples in Kampala, Uganda. Solids analysis for TS, TSS, VS, VSS, and fixed solids were carried out and reported as concentrations. The complete raw data set is available using the link below[4].

A faecal sludge sample was collected from a ventilated improved pit latrine in Durban, South Africa. It was analysed gravimetrically in six replicates using Method 8.6.1.1. The average COD (g/g wet sample) was 0.23. The results for TS and moisture content are presented in Table 8.4 (source: unpublished data UKZN PRG).

Table 8.4 Mass of samples before and after the analysis and analysis results for the gravimetric method.

Sample no.	Crucible mass (g) (W_1)	Sample mass (g)	Sample + crucible (g) (W_2)	Residue + crucible mass after drying (g) (W_3)	Moisture (g/g wet sample)	Total solids (g/g wet sample)
1-a	64.7232	19.9688	84.6920	69.4310	0.7642	0.2358
1-b	48.0356	20.0035	68.0391	52.7174	0.7660	0.2340
1-c	38.6685	20.0007	58.6692	43.2768	0.7696	0.2304
1-d	36.5180	20.0119	56.5299	41.2682	0.7626	0.2374
1-e	41.1442	20.0934	61.2376	45.8654	0.7650	0.2350
1-f	34.8260	20.0226	54.8486	39.5203	0.7655	0.2345
Average					0.7655	0.2345
SD					0.0023	0.0023

8.6.1.2 Volatile and fixed solids – ignition method[5]

8.6.1.2.1 Introduction

The dry sample residue from Method 8.6.1.1 is ignited at 550 °C for 30 min or until constant weight. The remaining ash represents the fixed (inorganic) solids, while the weight lost on ignition represents the volatile solids (organic matter) in faecal sludge. For more details, see Chapter 2.

8.6.1.2.2 Safety precautions

- General health and safety (H&S) procedures specific for conducting laboratory analysis of faecal sludge are presented in Section 8.2. Before conducting this method, it is important to be familiar with Section 8.2.3 to ensure safety measures are properly carried out.

[4] https://doi.org/10.25678/0000tt.
[5] This method follows Method 2540E of the Standard Methods for the Examination of Water and Wastewater, and should be cited as: Rice *et al.*, (2017)

- Appropriate personal protective equipment (PPE) should be used; specific details are covered in Section 8.2.3.1.
- Always conduct the volatile solids analysis in a room with sufficient airflow and preferably with an exhaust system.
- Never remove crucibles or trays by directly touching objects in the furnace, even if heat resistant gloves are worn. Use appropriate metal tools (such as stainless steel tongs) to place and remove crucibles and trays from the furnace to avoid direct contact with hot surfaces. Always wear heat-resistant gloves suitable for withstanding high temperatures.

8.6.1.2.3 Apparatus and instruments

- Porcelain crucibles
- Desiccator with dry desiccant
- Muffle furnace that can reach temperatures of 550 °C
- Analytical balance with four decimal places
- Stainless steel tray (optional, to move crucibles in and out of the furnace)
- Stainless steel tongs (to move crucibles in and out of the furnace)
- Heat-resistant gloves
- Pencil

8.6.1.2.4 Quality control

General information on quality assurance and quality control (QA/QC) is provided in Section 8.3. Information on standards, operating conditions and interferences that are specific to this method includes:

- The analytical balance and furnace must be checked and calibrated weekly.
- Check the temperature throughout the furnace area by placing a calibrated thermocouple on each shelf or reading the temperature with a laser thermometer.
- After 30 min, check the temperature at each level against the furnace setting. Using the same method, also check for temperature differences between the front and back of the furnace. Adjust the furnace setting if necessary. If temperatures are uneven on the shelves, check the insulation.
- To calibrate the analytical balance, place a standard calibration weight on the balance and weigh. Adjust the balance manually if necessary. Make sure to use a standard weight of a mass similar to the mass of the expected sample + crucible.
- Limit the sample to no more than 200 mg of residue after ignition at 550 °C (initial faecal sludge mass before TS analysis 10-20 g).
- Sludges that contain highly mineralised water with a significant concentration of calcium, magnesium, chloride and sulphate can be hygroscopic and require prolonged drying, proper desiccation and rapid re-weighing.

8.6.1.2.5 Sample preservation

Samples should be analysed as soon as possible. If samples cannot be analysed immediately, they should be stored in a refrigerator at 4 °C for no longer than 7 days and, if TSS or VSS analysis is conducted, no longer than 48 hours. Before starting the analysis, let the samples return to room temperature (20 °C). Do not freeze the samples.

8.6.1.2.6 Sample preparation

Dry the samples to constant weight to remove moisture content, following Method 8.6.1.1.

8.6.1.2.7 Analysis protocol

Preparation of equipment

- Pre-heat the furnace to 550 °C temperature before inserting the sample.
- Before conducting TS analysis (Method 8.6.1.1), position clean, dry crucibles in the furnace at 550 °C for 1 hr to remove any potential organic matter.

Procedure

- Ignite the residue from the TS in a muffle furnace at a temperature of 550 °C for 20 min. Note: for some solid and semi-solid faecal sludge samples, 20 min might not be enough, as they might need more time to combust all the volatile matter. For each type of sludge, check that constant weight is achieved after 20 min. To do this, cool and weigh the sample as described below, place the sample back in the furnace for 10 min and cool and weigh again. Repeat the steps of drying, cooling and weighing until a constant weight is obtained, or until weight change is less than 4% of the previous measurement. The length of combustion time needs to be evaluated for each specific type of sample, and revisited periodically.

- Transfer the crucibles to the stainless tray and let them cool partially until cool enough to transfer to a desiccator.
- Transfer to the desiccator for final cooling. Do not overload the desiccator.
- Weigh the crucible on the analytical balance as soon as it has cooled to ambient temperature and record the weight (W_4).

8.6.1.2.8 Calculation

Liquid and slurry samples (volumetric method):

Volatile solids in wet sample (g/L) =

$$\frac{(W_3\,(g)-W_4\,(g)) \times 1{,}000}{V_{sample}\,(mL)}$$

Fixed solids in wet sample (g/L) =

$$\frac{(W_4(g)-W_1(g)) \times 1{,}000}{V_{sample}\,(mL)}$$

Where:

W_1 = Crucible mass (g)

W_2 = Crucible mass + wet sample mass (g)

W_3 = Crucible mass + sample after drying (g)

W_4 = Crucible mass + sample after incinerating (g)

(W_3-W_1) = Sample mass after drying (g)

(W_4-W_1) = Sample mass after incinerating (g)

V_{sample} = Sample volume used (mL)

Slurry, semi-solid and solid samples (gravimetric method):

Volatile solids in wet sample (g/g) =

$$\frac{(W_3\,(g)-W_4(g))}{(W_2(g)-\,W_1\,(g))}$$

Volatile solids in dry sample (g/g) =

$$\frac{(W_3\,(g)-W_4\,(g))}{(W_3(g)-W_1(g))}$$

Volatile solids (%TS) =

$$\frac{(W_3(g)-W_4(g))}{(W_3(g)-W_1(g))} = \frac{VS\,(\frac{g}{g})}{TS\,(\frac{g}{g})} \times 100\,(\%)$$

Fixed solids in wet sample (g/g) =

$$\frac{(W_4\,(g)-W_1\,(g))}{(W_2\,(g)-W_1\,(g))}$$

Fixed solids in dry sample (g/g) =

$$\frac{(W_4(g)-W_1(g))}{(W_3(g)-W_1(g))}$$

Fixed solids (%TS) =

$$\frac{(W_4(g)-W_1(g))}{(W_3(g)-W_1(g))} = \frac{\text{Fixed solids}\,(\frac{g}{g})}{TS\,(\frac{g}{g})} \times 100\,(\%)$$

Note: for values of W_1 to W_3 and how to calculate them, see Method 8.6.1.1.

8.6.1.2.9 Data set example

A faecal sludge sample was collected from a ventilated improved pit latrine in Durban, South Africa. It was analysed in six replicates using Method 8.6.1.2. The average initial samples mass was 5 g dry weight - the same dry samples from Section 8.6.1.1.9 were used for ignition. The average VS (g/g dry sample) was 0.56. The results for VS and fixed solids are presented in Table 8.5 (source: UKZN PRG).

Table 8.5 Mass of samples before and after the analysis for the ignition method.

Sample no.	Volatile solids (g/g dry sample)	Fixed solids (g/g wet sample)	Fixed solids (g/g dry sample)
1-a	0.5574	0.1044	0.4426
1-b	0.5673	0.1013	0.4327
1-c	0.5896	0.0946	0.4104
1-d	0.5499	0.1069	0.4501
1-e	0.5571	0.1041	0.4429
1-f	0.5599	0.1032	0.4401
Average	0.5635	0.1024	0.4365
SD	0.0140	0.0042	0.0140

8.6.1.3 Total suspended solids and total dissolved solids – oven drying method[6]

8.6.1.3.1 Introduction

The TSS method is used to determine the efficiency of treatment technologies, such as settling tanks and biological filters. The measured volume of a thoroughly-mixed sample is vacuum-filtered through a dried, pre-weighed glass fibre filter. The filters and residue are then dried to a constant weight at 103-105 °C. The increase in weight of the filter represents the total suspended solids. Total dissolved solids are the TS minus the TSS.

For faecal sludge, clogging of the filters is a common problem. For this reason, this method is only suitable for liquid and slurry samples. If clogging occurs, the method can be adapted by dilution of the sample and/or choosing a larger pore size (maximum up to 2.0 µm), but needs to be carefully documented.

8.6.1.3.2 Safety precaution

- General health and safety (H&S) procedures specific for conducting laboratory analysis of faecal sludge are presented in Section 8.2. Before conducting this method, it is important to be familiar with Section 8.2.3 to ensure safety measures are properly carried out.
- Appropriate personal protective equipment (PPE) should be used; specific details are covered in Section 8.2.3.1.
- Always conduct the total suspended solids analysis in a room with sufficient airflow and an exhaust system.
- Wear gloves suitable for withstanding high temperatures when removing crucibles from the oven.
- Use appropriate mechanical tools, such as metal tongs, to remove crucibles and trays after drying in the oven to avoid direct contact with hot surfaces.

8.6.1.3.3 Apparatus and instruments

- Analytical balance with four decimal places
- Büchner funnel with a rubber bung and fitting conical filtration flask
- Vacuum pump with a rubber tubing
- Glass fiber filters (GF/C grade) ranging in size from 0.45 µm to 2.0 µm depending on the thickness of the sludge and clogging of the filters. It is important to use GF/C grade to withstand 550 °C and that the filter diameter matches the Büchner funnel diameter.
- Desiccator with dry desiccant
- Aluminium weighing boats or porcelain crucible
- Drying oven
- Graduated cylinder
- Forceps
- Pencil
- Stainless steel tray (optional, to move the crucibles in and out of the oven)
- Heat-resistant gloves
- Thermometer (for the quality control procedure)
- Set of standard calibration weights (for the quality control procedure)

8.6.1.3.4 Quality control

General information on quality assurance and quality control (QA/QC) is provided in Section 8.3. Information on standards, operating conditions and interferences that are specific to this method includes:

- The analytical balance and oven must be checked and calibrated weekly.
- Check the temperature throughout the oven area by placing a calibrated thermometer on each shelf. After 30 min, check the temperature at each level against the oven setting. Using the same method, also check for temperature differences between the front and back of the oven. Adjust the oven setting if necessary. If temperatures are uneven on the shelves, check the insulation.
- To calibrate the analytical balance, place a standard calibration weight on the balance and weigh. Adjust the balance manually if necessary. Do this with the whole range of weights from the calibration set. Make sure to at least use a standard weight of a mass similar to the mass of the expected sample + crucible.
- Make sure the desiccant in the desiccator is not saturated, otherwise samples can absorb water

[6] This method is adapted from Method 2540D of the Standard Methods for the Examination of Water and Wastewater, and should be cited as: Rice *et al.* (2017) as adapted in Velkushanova *et al.* (2021).

while cooling down in the desiccator. Routinely dry the desiccant in the oven at 105 °C (or at a different temperature, depending on the manufacturer's instructions), prior to the colour indicating that the desiccant is nearly saturated.

- Always keep the lid of the desiccator on and use a lubricant on the rim to ensure airtight sealing. Do not overload the desiccator.
- The volume or mass of the wet sample used should be chosen so that the drying will yield a residue between 2.5 and 200 mg of the dried sample (in general around 30 mL for the volumetric method, or 10-20 g for the gravimetric method, but this will depend on the type of sludge).
- For solid, semi-solid and slurry samples: limit the sample to no more than 10-20 g faecal sludge, otherwise the sample will take too long to dry and can form a moisture-trapping crust on top. If crust formation is occurring, samples should be placed in the oven at a lower temperature initially and the temperature gradually increased until 103-105 °C is reached.
- For liquid samples, the volume of the sample can be higher as the TS content is much lower. The proportion of the weight of the sample to the weight of the porcelain or aluminium crucible should also be taken into account, so that weight differences in the sample can be measured accurately.
- Make sure samples are fully cooled in a desiccator to ambient temperature prior to weighing.
- Sludges that contain highly mineralised water with a significant concentration of calcium, magnesium, chloride and/or sulphate can be hygroscopic and require prolonged drying, proper desiccation, and rapid re-weighing.
- Exclude larger, inconsistent or floating particles from the sample if it is determined that their inclusion can affect the final result (*e.g.* hair, stones, glass and maggots).
- Glass fibre filters are delicate, especially when wet, and care should be taken not to rip or damage them during filtration and handling. If a filter is damaged during filtration, particles might not be captured or pieces of the filter could be washed away, which will lead to measurement errors. Filters need to be prepared as described in Section 8.6.1.3.7.

8.6.1.3.5 Sample preservation

Samples should be analysed as soon as possible. If samples cannot be analysed immediately, they should be stored in a refrigerator at 4 °C for no longer than 48 hours. Before starting the analysis, let the samples return to ambient temperature. Do not freeze the samples.

8.6.1.3.6 Sample preparation

- Thoroughly mix all the samples using a stainless steel rod (or other appropriate tool) in order to have representative samples. For liquid samples, invert the closed sample bottle with the sample about 3 times.
- When measuring total dissolved solids, in addition to following this method, TS should be measured (following Method 8.6.1.1 or Method 8.6.1.5).

8.6.1.3.7 Analysis protocol

Equipment preparation

- Pre-heat the oven to 103-105 °C.
- Rinse the Büchner funnel with distilled water.
- Place the Büchner funnel with the rubber bung (stopper) on top of the filtration flask to seal the apparatus.
- Attach the filtration flask to a vacuum pump.
- If analysing multiple samples or replicates at the same time, mark each crucible/aluminium weighing boat with a unique identification number/letter. Number the crucible with a pencil or scratch the number into the aluminium weighing boat and note down which sample and replicate is in which number crucible to be able to distinguish between samples later.
- Pre-wash the glass fibre filter: place a filter onto the funnel (rough side up), apply the vacuum, and rinse three times with an aliquot of distilled water.
- Place the washed filter in a crucible or aluminium weighing boat and place in the oven at a temperature of 103-105 °C for 1 hr, prior to use (to remove any moisture). Afterwards, place the crucible with the filter in the desiccator and allow it to cool to room temperature. Always keep the rough side of the filter up.
- Note: if measuring volatile suspended solids after the total suspended solids, prepare the filter + crucible at 550 °C for ≥15 min in a muffle furnace instead of in the oven prior to use to remove any

potential residual organic material from previous measurements. Only porcelain crucibles should be used (see Method 8.6.1.2).

Procedure

- Weigh the filter + crucible or aluminium weighing boat on a balance and record its mass (W_1).
- Place the filter into a Büchner funnel, with the rough side up.
- Measure out a 30 mL sample volume using a graduated cylinder. Note: choose the sample volume to yield between 2.5 and 200 mg residue. For slurry sludge, measure 20 mL sample using a graduated cylinder. (Use less sample volume if the dried residue is more than 200 mg or use a smaller pore size if the dried residue is lower than 2.5 mg).
- Wet the filter with distilled water to seal the edges of the filter to the surface of the funnel.
- Turn on the vacuum pump.
- Pour the sample onto the filter, keeping the sample in the middle of the paper.
- Wash the graduated cylinder with distilled water until thoroughly rinsed (at least 2-cylinder volumes). Ensure all the particles are washed onto the filter.
- Pour rinse water onto the filter. For liquid and slurry samples >5 % TS, wash with at least two successive volumes of 10 mL distilled water and pour the rinse into the filter. Allow complete drainage between washings, and continue suction until all the traces of water are removed.
- If the sample is clogging the filter during filtration, dilute the sample using an appropriate dilution factor (*e.g.* 1:5 or 1:10) and filter the diluted sample. Note: the dilution factor needs to be reported and accounted for when calculating the total suspended solids concentration.
- If clogging still occurs even with the dilutions (*i.e.* if filtration takes >10 min to complete), then the next size larger pore size filter should be used. It is very important to document this and report it in the methods. In general, the smallest pore size possible in the range 0.45 μm to 2.0 μm should be used.
- When filtration is complete, remove the filter with forceps gently along the edge of the filter paper and then lift slowly (or first with a spatula and then forceps).
- Remove the paper with a pair of forceps, taking care not to tear the paper.
- Carefully place the filter in its marked crucible or aluminium weighing boat, rough side (containing the sample) facing up.
- Place in the oven at 103-105 °C for at least 2 hr, until constant weight is achieved. To do this, cool and weigh the sample as described below, place the sample back in the drying oven for 1 hr and then cool and weigh again. Repeat the steps of drying, cooling and weighing until a constant weight is obtained, or until weight change is less than 0.5 mg. The length of drying time needs to be evaluated for each specific type of sample, and revisited periodically.
- Remove from the oven, place in the desiccator and cool to room temperature.
- Weigh the crucible or weighing boat with the filter on the analytical balance and record the mass (W_2).

8.6.1.3.8 Calculation

Total suspended solids (g/L) =

$$\frac{(W_2(g)\text{-}W_1(g))}{V_{sample}(L)} \ (\times \text{ DF if using dilution factor})$$

W_1 = Weight of filter + crucible/aluminium weighing boat before drying (103-105 °C) (g)

W_2 = Weight of residue + filter + crucible/aluminium weighing boat after drying (103-105 °C) (g)

V_{sample} = Volume of sample used (L)

DF = Dilution factor

Total dissolved solids (g/L) =

$$\text{Total solids (g/L)} - \text{Total suspended solids (g/L)}$$

8.6.1.3.9 Data set example

A faecal sludge sample was collected from a ventilated improved pit latrine in Durban, South Africa. It was analysed in six replicates using Method 8.6.1.3. The average TSS (g/L) was 0.37. The results for TSS are presented in Table 8.6 (source: unpublished data UKZN PRG).

Table 8.6 Total suspended solids obtained by oven drying methods.

Sample no.	Filter paper mass (g) (W_1)	Residue + filter mass after drying (g) (W_2)	Sample volume (L)	Total suspended solids (g/L)
1-a	0.4146	0.4289	0.0300	0.4767
1-b	0.4186	0.4313	0.0300	0.4233
1-c	0.4289	0.4446	0.0300	0.4298
1-d	0.4287	0.4427	0.0300	0.4292
1-e	0.4137	0.4264	0.0300	0.4233
1-f	0.4268	0.4276	0.0300	0.0267
Average				0.3682
SD				0.1685

8.6.1.4 Volatile suspended solids – ignition method[7]

8.6.1.4.1 Introduction

The dry sample residue from Method 8.6.1.3 is ignited at 550 °C to constant weight. The remaining ash represents the fixed (inorganic) suspended solids, while the weight lost on ignition represents the volatile suspended solids (organic matter).

8.6.1.4.2 Safety precaution

- General health and safety (H&S) procedures specific for conducting laboratory analysis of faecal sludge are presented in Section 8.2. Before conducting this method, it is important to be familiar with Section 8.2.3 to ensure safety measures are properly carried out.
- Appropriate personal protective equipment (PPE) should be used; specific details are covered in Section 8.2.3.1.
- Always conduct the VS analysis in a room with sufficient airflow and preferably with an exhaust system.
- Never remove crucibles or trays by directly touching objects in the furnace, even if heat resistant gloves are worn. Use appropriate metal tools (such as stainless steel tongs) to place and remove crucibles and trays from the oven to avoid direct contact with hot surfaces. Always wear heat resistant gloves suitable for withstanding high temperatures.

8.6.1.4.3 Apparatus and instruments

- Muffle furnace that can reach temperatures of 550 °C
- Desiccator with dry desiccant
- Analytical balance with four decimal places
- Porcelain crucibles
- Heat resistant gloves
- Stainless steel trays (optional, to move crucibles in and out of the furnace)
- Tongs (to move crucibles in and out of the furnace)
- Pencil

8.6.1.4.4 Quality control

General information on quality assurance and quality control (QA/QC) is provided in Section 8.3. Information on standards, operating conditions and interferences that are specific to this method includes:

- The analytical balance and furnace must be checked and calibrated weekly.
- Check the temperature throughout the furnace area by placing a calibrated thermocouple on each shelf or read the temperature with a laser thermometer.
- After 30 min, check the temperature at each level against the furnace setting. Using the same method, also check for temperature differences between the front and back of the furnace. Adjust the furnace setting if necessary. If temperatures are uneven on the shelves, check the insulation.
- To calibrate the analytical balance, place a standard calibration weight on the balance and weigh. Adjust the balance manually if necessary. Make sure to use a standard weight of a mass similar to the mass of the expected sample + crucible.
- Limit the sample to no more than 200 mg of residue after ignition at 550 °C (the initial faecal sludge mass before TS analysis 10-20 g).
- Sludges that contain highly mineralised water with a significant concentration of calcium,

[7] This method follows methods 2540D and E of the Standard Methods for the Examination of Water and Wastewater, and should be cited as: Rice *et al.*, (2017).

magnesium, chloride and sulphate can be hygroscopic and require prolonged drying, proper desiccation, and rapid re-weighing.

8.6.1.4.5 Sample preservation

Samples should be analysed as soon as possible. If samples cannot be analysed immediately, they should be stored in a refrigerator at 4 °C for no longer than 48 hr. Before starting analysis, let the samples return to ambient temperature. Do not freeze the samples.

8.6.1.4.6 Sample preparation

Filter and dry the samples to constant weight, following Method 8.6.1.3.

8.6.1.4.7 Analysis protocol

Preparation of equipment

- Pre-heat the furnace to 550 °C temperature before inserting the samples.
- Before conducting analysis, position clean, dry crucibles in the furnace at 550 °C for 1 hr to remove any potential organic matter (see Section 8.6.1.3.7).

Procedure

- Ignite the crucibles containing glass fibre filters + residue from the total suspended solids measurement in a muffle furnace at a temperature of 550 °C for 20 min. Note: for some solid and semi-solid faecal sludge samples, 20 min might not be enough, as they might need more time to combust all the volatile matter. For each type of sludge, check that a constant weight is achieved after 20 min. To do this, cool and weigh the sample as described below, place the sample back in the furnace for 10 min, and cool and weigh again. Repeat the steps of drying, cooling and weighing until a constant weight is obtained, or until the weight change is less than 4 % of the previous measurement. The length of combustion time needs to be evaluated for each specific type of sample and revisited periodically.
- Transfer the crucibles to the stainless tray and let them cool partially until cool enough to transfer to the desiccator.
- Transfer to a desiccator for final cooling. Do not overload the desiccator.
- Weigh the crucible on the analytical balance as soon as it has cooled to ambient temperature and record the weight (W_3).

8.6.1.4.8 Calculation

Volatile suspended solids in wet sample (g/L) =

$$\frac{(W_2\ (g) - W_3\ (g))}{V_{sample}\ (L)} (\times \text{DF, if used})$$

W_2 = Weight of residue + filter + crucible after drying (103-105 °C) (g)

W_3 = Weight of residue + filter + crucible after ignition in furnace at 550 °C (g).

V_{sample} = Volume of sample used (L)

DF = Dilution factor. Note: if a dilution was used during the determination of total suspended solids (Method 8.6.1.3), this should be taken into account in the calculation.

8.6.1.4.9 Data set example

A faecal sludge sample was collected from a ventilated improved pit latrine in Durban, South Africa. It was analysed in six replicates using Method 8.6.1.4. The average VSS (g/L) was 0.03. The results for VSS are presented in Table 8.7 (source: unpublished data UKZN PRG).

Table 8.7 Volatile suspended solids obtained by oven drying and ignition methods.

Sample no.	Residue + filter mass after drying (g) (W_2)	Sample volume (L)	Residue + filter mass after incineration (g) (W_3)	Volatile suspended solids (g/L)
1-a	0.4289	0.0300	0.4150	0.0313
1-b	0.4313	0.0300	0.4190	0.0274
1-c	0.4446	0.0300	0.4300	0.0318
1-d	0.4427	0.0300	0.4264	0.0355
1-e	0.4264	0.0300	0.4140	0.0279
1-f	0.4276	0.0300	0.4268	0.0018
Average				0.0260
SD				0.0122

8.6.1.5 Total solids and moisture content – thermal balance (moisture analyser) method[8]

8.6.1.5.1 Introduction

A moisture analyser is designed to determine relative moisture content in small samples of various substances, by measuring the change of weight due to water evaporation during convective drying. This method is applicable for all types of sludge – liquid, slurry, semi-solid and solid; however, samples with a higher moisture content will have a longer drying and measurement time. Method 8.6.1.1 and Method 8.6.1.5 are equally suitable to determine TS and moisture content, and should be selected depending on the availability of equipment.

8.6.1.5.2 Safety precautions

- General health and safety (H&S) procedures specific for conducting laboratory analysis of faecal sludge are presented in Section 8.2. Before conducting this method, it is important to be familiar with Section 8.2.3 to ensure safety measures are properly carried out.
- Appropriate personal protective equipment (PPE) should be used; specific details are covered in Section 8.2.3.1.
- Always conduct the TS analysis in a room with sufficient airflow and an exhaust system.
- Do not place any flammable substances on or near the moisture analyser.

8.6.1.5.3 Apparatus and instruments

- Aluminium weighing boats
- Thermal balance (moisture analyser)

8.6.1.5.4 Quality control

General information on quality assurance and quality control (QA/QC) is provided in Section 8.3. Information on standards, operating conditions and interferences that are specific to this method includes:

- Before using a moisture analyser, make sure the instrument was left on for a sufficient period of time (see Section 8.6.1.5.7).
- Minimise external environmental influences such as air draught, vibrations or direct sunlight.
- Ensure the analyser is levelled. This is essential for testing liquid samples, which must be at uniform level in the sample container.
- Exclude larger, inconsistent or floating particles from the sample if it is determined that their inclusion could affect the final result (*e.g.* hair, stones, glass and maggots).
- Disperse visible floating oil and grease with a blender or stainless steel mixing rod before withdrawing a sample portion for analysis.

8.6.1.5.5 Sample preservation

Samples should be analysed as soon as possible. If samples cannot be analysed immediately, they should be stored in a refrigerator at 4 °C for no longer than 7 days. Before analysis, let the samples return to ambient temperature. Do not freeze the samples.

8.6.1.5.6 Sample preparation

Thoroughly mix all the samples using a stainless steel rod (or other appropriate tool) in order to have representative samples. If desired, samples can also be blended (see Section 8.4.2).

8.6.1.5.7 Analysis protocol

Equipment preparation

- Switch the instrument on. Wait until the analyser completes its self-examination and finishes heating up. To deliver accurate results and enable the moisture analyser to reach the required operating temperature, it must be switched on for at least 20-30 minutes every time before use. The program must be set to end when the sample mass changes less than 0.05% of mass per minute.
- Check that the temperature is 105 °C for moisture analysis.

Procedure

- Press 'Start Program' and follow prompts on the display screen; this can vary per model and brand.
- Open the lid of the moisture analyser, place the clean and empty weighing boat on the weighing cradle.
- Close the cover gently and tare the boat weight; the LCD screen should now show weight as '0'

[8] This method follows Method B of the ASTM Standard Test Method for Determination of Total Solids in Biomass (E1756-08) and should be cited as: ASTM (2015).

ready for loading the sample.

and a flashing icon to indicate that the machine is ready for loading the sample.

- Lift the lid of the moisture analyser and then evenly spread 1-3 g of the wet sample on the weighing boat.
- Close the cover gently.
- The halogen light will start to heat the sample until it reaches a steady reading. Note: this process usually takes between 2-15 min, depending on the sample weight and its moisture content.
- Record the moisture reading (before lifting the lid); this is the end of the drying procedure.
- Press 'Stop' and lift the lid to end the current testing.

8.6.1.6 Sand content[9]

8.6.1.6.1 Introduction

The residual ash from Method 8.6.1.2 is washed with 0.1 M HCl and combusted at 550 ºC. The leftover residual is silica, and is typically reported as the 'sand' content as an indicator of soil content in faecal sludge.

8.6.1.6.2 Safety precautions

- General health and safety (H&S) procedures specific for conducting laboratory analysis of faecal sludge are presented in Section 8.2. Before conducting this method, it is important to be familiar with Section 8.2.3 to ensure safety measures are properly carried out.
- Appropriate personal protective equipment (PPE) should be used; specific details are covered in Section 8.3.
- Always conduct the volatile solids analysis in a room with sufficient airflow and preferably with an exhaust system.
- Never remove crucibles or trays by directly touching objects in the furnace, even if heat-resistant gloves are worn. Use appropriate metal tools (such as stainless steel tongs) to place and remove crucibles and trays from the oven to avoid direct contact with hot surfaces. Always wear heat-resistant gloves suitable for withstanding high temperatures.

8.6.1.6.3 Required chemicals

- Distilled water
- 0.1 M HCl solution

8.6.1.6.4 Required apparatus and instruments

- Analytical balance
- Porcelain crucibles
- Ashless filter paper (*e.g.* Whatman grade 44)
- Heat-resistant gloves
- Stainless steel tray (optional, to move the crucibles in and out of the oven)
- Stainless steel tongs (to move the crucibles in and out of the furnace)
- Forceps
- Desiccator with dry desiccant
- Furnace (operating at 550 °C)
- Vacuum filtration setup
- Pencil
- Thermometer (for quality control procedure)
- Set of standard calibration weights (for the quality control procedure)

8.6.1.6.5 Quality control

General information on quality assurance and quality control (QA/QC) is provided in Section 8.3. Information on standards, operating conditions and interferences that are specific to this method includes:

- The analytical balance and oven must be checked and calibrated weekly.
- Check the temperature throughout the furnace area by placing a calibrated thermocouple on each shelf or reading the temperature with a laser thermometer.
- After 30 min, check the temperature at each level against the oven setting. Using the same method, also check for temperature differences between the front and back of the oven. Adjust the oven setting if necessary. If the temperatures are uneven on the shelves, check the insulation.
- To calibrate the analytical balance, place a standard calibration weight on the balance and weigh. Adjust the balance manually if necessary. Make sure to use a standard weight of a mass similar to the mass of the expected sample + crucible.

[9] This method is from the German Association for Water Wastewater and Waste guidelines, and should be cited as: Method DWA-M 383 for the determination of sand content (DWA, 2008).

- The filters used for sand content analysis must be ashless.
- Make sure that the desiccant in the desiccator is not saturated, otherwise the samples may absorb water while cooling down in the desiccator. Routinely dry the desiccant in the oven at 105 °C (or at a different temperature, depending on the manufacturer's instructions), prior to the colour indicating that the desiccant is nearly saturated.
- Always keep the lid of the desiccator on and use a lubricant on the rim to ensure airtight sealing. Do not overload the desiccator.
- Make sure the samples are fully cooled in a desiccator to ambient temperature prior to weighing.

8.6.1.6.6 Sample preservation

Samples have to be 100 % dry and should be stored in a desiccator after volatile solids analysis (Method 8.6.1.2) until the start of the sand content analysis.

8.6.1.6.7 Sample preparation

Sand content analysis is performed on the 100% dry ash remaining from the volatile solids analysis (Method 8.6.1.2).

8.6.1.6.8 Analysis protocol

- Preheat the furnace to 550 °C.
- Place the filter on the filter apparatus and rinse it with 3 portions of approximately 20 mL of demineralised water.
- Place one crucible with one filter paper in the furnace at 550 °C for 15 min.
- Take the porcelain crucible with the filter paper out of the oven (using appropriate heat-resistant tools) and place them in the desiccator until they reach ambient temperature.
- Take the porcelain crucible and filter paper out of the desiccator and weigh them (W_1).
- Place the weighed filter paper in the (vacuum) filtration apparatus.
- Transfer all of the residue from the crucible left after the volatile solids analysis (*i.e.* fixed solids, Method 8.6.1.2) onto the paper filter inside the filtering apparatus. Ensure that all the residue is transferred onto the filter paper. If needed, rinse the crucible with the 0.1 M HCl solution to remove the last bits of the ash and pour onto the filter paper. Wash the residue on the paper filter with the 0.1 M HCl solution. Keep rinsing until the mass of residue stays constant and the filtrate is clear. If the filtrate is clear from the beginning, rinse at least three times with an aliquot of HCl solution.
- Place the paper filter inside the weighed porcelain crucible using tweezers.
- Place the porcelain crucible into the furnace operating at 550 °C for two hours.
- Remove the porcelain crucible from the furnace. Wait for it to partially cool and place it in the desiccator. Wait until the porcelain crucible has reached ambient temperature.
- Remove the porcelain crucible from the desiccator. Place it on the analytical balance and note the weight (W_2).

8.6.1.6.9 Calculations

Liquid and slurry samples (volumetric method):

Sand content $\left(\frac{g}{L}\right)$ =

$$\frac{(W2\ (g) - W1\ (g))}{\text{Sample volume from TS analysis (mL)}} \times 1{,}000\ \left(\frac{mL}{L}\right)$$

Sand content (%TS) =

$$\frac{\text{Sand content}\ \left(\frac{g}{L}\right)}{TS\ \left(\frac{g}{L}\right)} \times 100\ (\%)$$

Sand content (% ash) =

$$\frac{\text{Sand content}\ \left(\frac{g}{L}\right)}{\text{Ash}\ \left(\frac{g}{L}\right)} \times 100\ (\%)$$

Slurry, semi-solid, and solid samples (gravimetric method):

Sand content of wet weight $\left(\frac{g}{g}\right)$ =

$$\frac{(W2\ (g) - W1\ (g))}{\text{Wet sample weight from TS analysis (g)}}$$

Sand content of dry weight $\left(\frac{g}{g}\right)$ =

$$\frac{(W2\ (g) - W1\ (g))}{\text{Weight of dry solids from TS analysis (g)}}$$

Sand content (%TS) =

$$\frac{\text{Sand content }\left(\frac{g}{g}\right)}{\text{TS }\left(\frac{g}{g}\right)} \times 100\ (\%)$$

W1 = Weight of dried filter + crucible
W2 = Weight of filter + crucible + residue

8.6.1.6.10 Data set example

In Gold *et al.* (2017), 73 faecal sludge samples from septic tanks, lined pit latrines, unlined pit latrines and johkasou[10] tanks in Uganda, Vietnam and Japan, and 18 samples from wastewater treatment plants in Switzerland, were analysed for sand content following Method 8.6.1.6. The results were used to determine influence on dewaterability. They report a 75% confidence interval (where 75% of the data is expected to lay) ranging from 7-9% sand of TS for lined johkasou septic tanks from Japan, 9-33% sand of TS for wastewater sludges from Switzerland, to a maximum of 45-69% sand of TS from unlined pit latrines in Uganda. When higher %TS of dewatered sludge are due to high sand content, it can have a negative impact on resource recovery.

8.6.2 Chemical oxygen demand (COD)

Chemical oxygen demand (COD) is defined as the amount of oxygen that is required for the chemical oxidation of organic matter using a strong chemical oxidant. For faecal sludge, it is used as a proxy for the organic matter content of a sample. For example, it is measured to characterise sludge prior to treatment, or during monitoring of treatment processes. COD is one of multiple parameters to measure organic content of a sample. Other parameters include biochemical oxygen demand (BOD), volatile fatty acids (VFA), total organic carbon (TOC), and biomethane potential (BMP), which are not yet included in this book. Common perception in the sector, is that COD is more accurate than BOD for faecal sludge analysis, due to the high variability and concentrations of organic matter. More information on these methods can be found in Rice *et al.* (2017).

8.6.2.1 Chemical oxygen demand – closed reflux spectrophotometric method[11]

8.6.2.1.1 Introduction

Chemical oxygen demand (COD) is a proxy measurement of the amount of organic matter in water or wastewater. The principle of the COD test is that organic compounds are oxidised to carbon dioxide with a strong oxidising agent such as dichromate under acidic conditions. The amount of oxidising agent required to completely oxidise the organic matter in the sample is compared to the equivalence of oxygen. COD can be determined by an open reflux method, a closed reflux titrimetric method (Method 8.6.2.2) or a closed reflux colorimetric method. The method described here is the closed reflux colorimetric method which uses a spectrophotometer to measure absorbance. The principle is that when a sample is digested, the dichromate ion oxidises COD in the sample and the chromium is reduced from Cr^{6+} to Cr^{3+}. These chromium species are coloured and absorb in the visible region of the spectrum. Dichromate ion ($Cr_2O_7^{2-}$) is extremely absorbent in the 400-nm region, whereas the chromic ion (Cr^{3+}) is extremely absorbent in the 600-nm region.

Commercial test kits based on standard methods for measuring COD are available, with pre-packaged individual aliquots of the necessary chemical in pillows (dry chemicals) and vials (liquid chemicals). Commonly used COD test kits from manufacturers such as Hach and Merck employ the closed reflux colorimetric method.

The example provided here is the Merck COD spectrophotometric test[D] (Merck, 2020d) for samples with concentrations of 15-300 mg COD/L, and it is based on the manufacturer's protocol for water and wastewater using the standard method 5220 (Rice *et al.*, 2017). For faecal sludge, samples must be diluted

[10] Johkasou is the Japanese word for on-site wastewater treatment.
[11] This method should be cited as: Method 5220 D (Rice *et al.*, 2017), and if test kits are used, also as per the manufacturer's directions, including any modifications.

to prevent false high readings associated with turbid solutions.

8.6.2.1.2 Safety precautions

- General health and safety (H&S) procedures specific for conducting the laboratory analysis of faecal sludge are presented in Section 8.2. Before conducting this method, it is important to be familiar with Section 8.2.3 to ensure safety measures are properly carried out.
- Appropriate personal protective equipment (PPE) should be used; specific details are covered in Section 8.2.3.1.

8.6.2.1.3 Required chemicals

- Distilled water

8.6.2.1.4 Required apparatus and instruments

- Spectrophotometer that operates in the region of 420-600 nm (*e.g.* Merck, Hatch, and Hanna)
- Heating block capable of heating to 150 °C, with fitting digestion tubes
- Analytical balance with weighing boats (slurry to solid samples)
- Blender (slurry to solid samples)
- Pipettes and pipette tips (2 mL or 5 mL)
- Reaction cells with reagents (supplied by the manufacturer)
- Volumetric flask (1 L)
- Glass beakers (50 or 100 mL)
- Glass storage bottle
- Laboratory tissue.

8.6.2.1.5 Quality control

General information on quality assurance and quality control (QA/QC) is provided in Section 8.3. Information on standards, operating conditions and interferences that are specific to this method includes:

- Perform quality control with COD standard solution with every test batch (or on a daily or weekly basis, depending on the testing load).
- For spectrophotometric measurements, the cuvette must be clean. Before analysis, wipe with a clean, dry paper tissue.
- Measurement of turbid solutions yields false high readings. For faecal sludge, samples should be diluted with an appropriate dilution factor through serial dilutions based on the type of sludge for accurate measurements.
- To ensure accuracy of measurement, blank samples must be prepared and analysed together with the samples.
- Chlorides react with silver ions to form a silver chloride precipitate that inhibits the catalytic activity of silver, and this must be avoided.
- To prevent chloride interference, samples suspected to have high chloride concentration must be analysed for chloride (Method 8.7.5.6) prior to COD analysis. For chloride concentration > 2,000 mgCl$^-$/L, samples must be diluted with distilled water before COD analysis.
- The measurement value remains stable for up to 60 min after the end of the reaction time; thus, the spectrophotometric measurement should be conducted within that timeframe.

8.6.2.1.6 Sample preservation

- Samples should be analysed as soon as possible. If samples cannot be analysed immediately, they should be stored in a refrigerator at 4 °C for no longer than 24 hr. If samples cannot be analysed within that time period, they should be acidified to pH ≤ 2 by adding concentrated sulphuric acid.
- For samples preserved with sulphuric acid, pH must be adjusted to 7 with sodium hydroxide solution of a known normality (5.0 N NaOH is recommended) before analysis.
- Samples must be thawed to room temperature before analysis is performed.
- Homogenise the samples prior to COD analysis.

8.6.2.1.7 Sample preparation

For liquid, slurry, semi-solid or solid samples:

- Samples containing concentrations of COD beyond the range of the test kit must be diluted appropriately with distilled water, following serial dilutions.

For slurry to solid samples with an estimated range of 15-300 mg COD/L:

- Weight out between 1.8 g and 2.0 g thoroughly-mixed faecal sludge sample into a beaker.
- Dilute the sample gravimetrically and transfer to a blender to a total of 250 g, as described in Section 8.4.2.

- If necessary, adjust the dilution based on the specific sludge characteristics.
- Blend for 30 sec.
- Transfer the contents of the blender into a 1 L volumetric flask, rinse the blender twice with 250 mL distilled water and dilute to the mark with distilled water.
- Transfer this 1 L solution into a plastic bottle and store at 4 °C until ready for testing.
- For samples with an estimated range of 500 to 10,000 mg COD/L, accurately weigh 0.1 g of homogenised faecal sludge sample into a 50 ml beaker and follow the steps described above.
- Measure the COD concentration according to the analysis protocol.
- If required, TS analysis should be performed on the samples so that the results of the COD measurement can be expressed as mg COD/gTS.

8.6.2.1.8 Analysis protocol

Calibration

Follow the spectrophotometer manufacturer's instructions for calibration, since calibration procedure differs between instruments. It may be necessary to calibrate the instrument before every reading, or it may only be necessary to perform periodic calibration checks to determine when calibration is necessary.

- Prepare a series of at least four different concentrations of a standard solution, making sure to include the lowest and highest concentrations of the kit testing range. It is typical to do serial dilutions or dilutions with a uniform interval including the lowest and highest concentrations.
- Determine the COD concentration of the standard solutions.
- Multiply the answer by the appropriate dilution factor and report the results in mg COD/L.
- Prepare a calibration curve by plotting the instrument response against the standard concentrations.

8.6.2.1.9 Procedure

This method is for a measuring range of 15-300 mg/L COD.

- Heat up the heating block or digester to 148 °C.
- Pipette 2 mL of the sample and 2 mL of the blank into separate reaction cells.
- Tightly screw the screw cap on the cells.
- Mix the content of the cells vigorously.
- Heat the reaction cells at 148 °C in the preheated heating block for 2 hr.
- Remove the hot cells from the heating block and allow to cool in a test tube rack. Do not cool with cold water.
- Wait 10 min, swirl the cells, and return to the rack for complete cooling to room temperature (cooling time at least 30 min).
- Wipe the reaction cells with a laboratory tissue to clean it (*e.g.* remove water spots, fingerprints)
- Measure the COD concentration in the spectrophotometer.

8.6.2.1.10 Calculation

Liquid and slurry samples:

Result of analysis (mg/L)

$$\text{Final concentration}\left(\frac{\text{mg}}{\text{L}}\right) = \text{A}\left(\frac{\text{mg}}{\text{L}}\right) \text{ x DF}$$

$$\text{Dry basis}\left(\frac{\text{g}}{\text{g}}\right) = \frac{\text{Final concentration}\left(\frac{\text{mg}}{\text{L}}\right)}{\text{Total solids concentration}\left(\frac{\text{mg}}{\text{L}}\right)}$$

Where:

A = Spectrophotometric measurement value (mg/L)
DF = Dilution factor (F/I)
F = Final diluted volume (L)
I = Initial sample volume (L)

Slurry, semi-solid and solid samples:

$$\text{Dry basis}\left(\frac{\text{g}}{\text{g}}\right) = \frac{\frac{\text{A}\left(\frac{\text{mg}}{\text{L}}\right)}{1{,}000} \times \frac{\text{V}}{\text{M}}\left(\frac{\text{L}}{\text{g}}\right)}{\text{Total solids content}\left(\frac{\text{g}}{\text{g}}\right)}$$

Where:

A = Spectrophotometric measurement value (mg/L)
V = Volume of dilution (L)
M = Wet weight of sludge used in sample preparation (g)

8.6.2.1.11 Data set example

COD is measured in most characterisation studies for faecal sludge and it is essential for the design of faecal sludge treatment plants. In a study to quantify and characterise faecal sludge from urban areas in Douala, Cameroon, Maffo *et al.*, (2019) used the COD spectrophotometric test kit to measure the COD of seven faecal sludge composite samples collected within a day from trucks at a dumping site in Douala. The COD concentrations measured ranged from of 28,900-73,150 mg/L with an average of 39,925 mg/L and a standard deviation of 15,034 mg/L.

Similarly, Semiyaga *et al.* (2016) used the COD spectrophotometric method in the characterisation of faecal sludge from lined and unlined pit latrines. For 22 faecal sludge samples from lined pit latrines, the average COD was 65,521 ± 43,960. The 10 unlined pit latrines sampled had an average COD of 132,326 ± 43,786 mg/L. Variability in COD concentration was evident in the 3 settlements where samples were taken, with an average COD of 107,137 ± 32,542 mg/L, 75,120 ± 28,778 mg/L, and 20,794 ± 8,456 mg/L for the Bwaise II, Kibuye and Kamwokya settlements, respectively.

Gold *et al.* (2017) also determined the COD concentration of faecal sludge and wastewater samples from different types of containments in Uganda, Vietnam, Japan and Switzerland in a cross-country analysis of FS dewatering. The results of COD for septic tanks in Uganda, septic tanks in Vietnam, johkasou systems (onsite technologies) in Japan and digested wastewater sludge in Switzerland were in the range of 7.2-21.5 g/L, 15.3-31.5 g/L, 10.1-16.5 g/L and 17.0-22.1 g/L, respectively. Using percentage TS (% TS) after centrifugation as a metric of dewaterability, it was observed that COD correlated to a decreased dewaterability (decreased final % TS), a relationship that has also been observed in wastewater.

8.6.2.2 Chemical oxygen demand – closed reflux titrimetric method[12]

8.6.2.2.1 Introduction

A sample is digested at 150 °C for 2 hours in a concentrated dichromate solution, using silver sulphate as a catalyst and mercuric sulphate as a masking agent to prevent chloride interference. The dichromate will be partially reduced by the oxidisable material present in the sample. The excess dichromate is then titrated with ammonium iron (II) sulphate and the COD value calculated from the amount of dichromate used.

The half reaction for the reduction of dichromate is:

$$Cr_2O_7^{2-} + 14\ H^+ + 6e^- \rightarrow 2\ Cr^{3-} + 7\ H_2O$$

The remaining dichromate is titrated with a standard ammonium iron (II) sulphate solution:

$$Cr_2O_7^{2-} + 6\ Fe^{2+} + 14\ H^+ \rightarrow 6\ Fe^{3+} + 7\ H_2O + 2\ Cr^{3+}$$

The equivalence point is indicated by a sharp colour change from blue-green to reddish brown as the ferroin indicator undergoes reduction from iron (III) to the iron (II) complex.

8.6.2.2.2 Safety precautions

- General health and safety (H&S) procedures specific for conducting laboratory analysis of faecal sludge are presented in Section 8.2. Before conducting this method, it is important to be familiar with Section 8.2.3 to ensure safety measures are properly carried out.
- Appropriate personal protective equipment (PPE) should be used; specific details are covered in Section 8.2.3.1.
- In general, alternatives to methods that use toxic metals should be investigated, the smallest possible volumes should be used, and extreme care needs to be taken in the safe handling and disposal of used chemicals.
- Handle concentrated acid with care.

[12] This method is adapted from Method 5220C of the Standard Methods for the Examination of Water and Wastewater and should be cited as: Rice *et al.* (2017), as described in Velkushanova *et al.* (2021).

- Use eye and hand protection when preparing acid or handling colour reagents. Ensure when handling concentrated acid that an acid-proof lab coat and acid-proof gloves are used. For more detailed information on selecting the correct type of glove, consult the glove comparison chart (Berkeley Environment, Health & Safety).
- Prepare and keep the reagents in a fume hood.
- Dispose waste containing mercury according to H&S procedures.

8.6.2.2.3 Required chemicals

Standard potassium dichromate (K_2Cr_6O) - digestion solution: 0.13 M

- Potassium dichromate ($K_2Cr_6O_7$)
- 32% concentrated sulphuric acid (H_2SO_4)
- Mercuric sulphate ($HgSO_4$)
- Distilled water.

Sulphuric acid (H_2SO_4)/silver sulphate reagent (Ag_2SO_4) - COD reagent

- Silver sulphate crystals or powder
- Concentrated sulphuric acid (H_2SO_4)

Ferroin indicator

- 1,10-phenentroline monohydrate
- Ferrous sulphate (FeSO4·7H2O)

Ferrous ammonium sulphate (FAS) ($Fe(NH_4)_2(SO_4)_2{\cdot}6H_2O$): 0.10 M

- Ferrous ammonium sulphate ($Fe(NH_4)_2(SO_4)_2{\cdot}6H_2O$)
- 98% concentrated sulphuric acid
- Distilled water

Potassium hydrogen phthalate (KHP) - standard solution

- Potassium hydrogen phthalate (KHP)
- Distilled water

8.6.2.2.4 Required apparatus and instruments

- Heating block capable of heating to 150 °C or microwave digester
- Digestion tubes (use borosilicate glass tubes if using a block digester or Teflon-coated digestion vessels if using a microwave digester) (the number depends on the number of samples, replicates and blanks)
- Titration setup with burette in a metal clamp (or automatic titration unit)
- 100 mL Erlenmeyer flasks (the number depending on the number of samples)
- 5 mL pipette and pipette tips
- Blender
- Analytical balance
- Crucible
- 2 x 1 L volumetric flasks
- 2.5 L volumetric flask
- 10 mL volumetric flask
- Drying oven
- Pestle and mortar

8.6.2.2.5 Quality control

General information on quality assurance and quality control (QA/QC) is provided in Section 8.3. Information on standards, operating conditions and interferences that are specific to this method includes:

- Check the accuracy of the measurement procedure by using a standard solution of known COD. This calibration should be done with every set of samples. Dilute and reanalyse any samples exceeding the highest standard on the calibration curve.
- Prepare and analyse blanks with every set of samples.
- In samples with a high level of chloride (>500 $mgCl^-/L$), the chloride can undergo oxidation, resulting in an incorrectly high measured value for COD. This error is overcome by the addition of mercuric sulphate to samples before digesting. The chloride ions are then eliminated from the reaction by forming a soluble mercuric chloride complex. This, however, should not be common practice because the sample might become more toxic.
- The sample volume should be based on the volume of the digestion tubes. The method described below is based on 50 mL digestion tubes, but other sizes are available (the digestion tubes must match the heating block/microwave digester used). For smaller or larger tubes, the sample and reagent volumes should be adjusted to fit the tube size appropriately, but the same ratios as presented in this method should be used. For more information on selecting the correct sample

and reagent volumes, consult Table 5220:I in Rice *et al.* (2017).

- Rinse the glassware with 20% H_2SO_4 to remove any organic residues from previous use.

8.6.2.2.6 Sample preservation

- The samples should be analysed as soon as possible. If they cannot be analysed immediately, they should be stored in a refrigerator at 4 °C for no longer than 24 hr. If the samples cannot be analysed within that time period, they should be acidified to pH ≤ 2 by adding concentrated H_2SO_4.
- The KHP standard solution is stable if stored in a refrigerator at 4 °C and can be stored for a period of 3 months; however, with any visual signs of biological growth it needs to be immediately discarded.

8.6.2.2.7 Sample preparation

This method is valid within a concentration range of 40-3,600 mgCOD/L

For liquid, slurry, semi-solid or solid samples:

- Samples containing concentrations of COD beyond the concentration range must be diluted appropriately with distilled water, following serial dilutions. The sample needs to be homogenised thoroughly (with a blender or stirring rod) before diluting.

For slurry to solid faecal sludge:

- Accurately weigh 0.1 g homogenised faecal sludge sample into a beaker with distilled water.
- Dilute the sample gravimetrically and transfer to a blender to a total of 250 g, as described in Section 8.4.2.
- If necessary, adjust the dilution based on the specific sludge characteristics.
- Blend for 30 sec.
- Transfer the contents of the blender into a 1 L volumetric flask, rinse the blender twice with 250 mL distilled water and dilute to the mark with distilled water.
- Transfer this 1 L solution into a plastic bottle and store at 4 °C until ready for testing.
- If required, TS analysis should be performed on the samples so that the results of the COD measurement can be expressed as mgCOD/gTS.

8.6.2.2.8 Analysis protocol

Preparation of reagents, chemicals and standard solutions

- Standard potassium dichromate (K_2Cr_6O)
  Digestion solution: 0.13 M for faecal sludge samples. Molarity of the digestion solution is adapted here from Rice *et al.* (2017) to fit the COD range commonly seen in faecal sludge. When analysing faecal sludge with a COD ranging between 40-400 mg/L, using a digestion solution of 0.017 M (following Rice *et al.*) is recommended.
  - Dry $K_2Cr_6O_7$ in the oven (150 °C) for 2 hr and let it cool to ambient temperature in a desiccator. Afterwards, weigh 39.93 g of dried $K_2Cr_6O_7$ and transfer into a 1 L volumetric flask, and add approximately 500 mL distilled water.
  - To this volumetric flask, add 167 mL of 32% concentrated sulphuric acid (H_2SO_4) and 13.3 g mercuric sulphate ($HgSO_4$). This reaction produces heat!
  - Completely dissolve the mercuric sulphate by swirling the volumetric flask or by using a magnetic stirrer. Allow the flask contents to cool to room temperature and then dilute the mixture with distilled water to the mark of the volumetric flask.
- Sulphuric acid $HgSO_4$/silver sulphate reagent Ag_2SO_4 (COD reagent)
  - Add 26 g silver sulphate crystals or powder to the 2.5 L volumetric flask of concentrated sulphuric acid while mixing using a magnetic stirrer.
  - Leave for 2 days to dissolve and mix well before use.
- Ferroin indicator
  - Dissolve 1.485 g 1,10-phenentroline monohydrate and 0.695 g ferrous sulphate ($FeSO_4 \cdot 7H_2O$) in distilled water and dilute to 100 mL in a 100 mL volumetric flask.
- Ferrous ammonium sulphate (FAS) ($Fe(NH_4)_2(SO_4)_2 \cdot 6H_2O$): 0.10 M
  - Dissolve 39.2 g $Fe(NH_4)_2(SO4)_2 \cdot 6H_2O$ in distilled water.
  - Add 20 mL of 98% concentrated sulphuric acid and dilute to 1 L in a volumetric flask.

Standardise daily (or before every set of analysis) by titration against the standard potassium dichromate digestion solution.

Calibration

- Prepare a potassium hydrogen phthalate (KHP) stock solution. Gently grind KHP with pestle and mortar. Dry the KHP at 110 °C until a constant weight is obtained. Weigh 340 mg of dried KHP into a 1 L volumetric flask and dissolve the content with distilled water up to the mark. This solution has a theoretical COD of 400 mg COD/L. The stock solution is diluted to make up four points of standard solutions.
- For samples with high expected COD values, dissolve 425 mg KHP into a 1 L volumetric flask. This KHP solution has a theoretical COD of 500 mg COD/L.
- Test the standard solutions following the COD procedure.
- Prepare a calibration curve by plotting instrument response against standard concentration and calculating the linear regression line. Compute the sample concentration by adjusting the sample response with the offset of the standard curve.

Standardisation

Use a standard $K_2Cr_2O_7$ digestion solution (daily, or before every analysis set) to correct any variation in the concentration of the FAS, following the COD measurement procedure below. The molarity of the FAS is calculated by the following equation:

Molarity of FAS =

$$\frac{\text{Volume } K_2Cr_2O_7 \text{ solution titrated (mL)}}{\text{Volume FAS used in titration (mL)}} \times 0.10$$

Procedure

Equipment preparation

- If using a heating block for digestion, preheat the heating block to 150 °C.

Sample digestion

- Add 5 mL prepared faecal sludge sample into each digestion tube. Label each tube with the corresponding sample identification.
- Add 5 mL distilled water into another digestion tube (blank).
- Add 3 mL potassium dichromate digestion solution into each tube.
- Add 7 mL sulphuric acid reagent (with silver sulphate) into each tube. Pour the acid down the wall of the flask while the flask is tilted. If the sample is too concentrated, it will turn green and further dilution of the sample must then be performed.
- Digest the samples at 150 °C for 2 hrs. Follow the manufacturer's instructions for the microwave digester or for the heating block.

Titration procedure

- Let the samples cool to room temperature in the digester after digestion.
- Pour the content of each digestion tube into an individual 100 mL Erlenmeyer.
- Record the starting volume of the FAS titrant standard in the burette.
- Add 2 drops of ferroin indicator to each Erlenmeyer.
- Titrate the excess dichromate in the digested mixture with standard ferrous ammonium sulphate.
- Titrate from a bright green-orange to a red-brown end-point.
- Record the final volume of the FAS titrant standard in the burette. The difference is the volume delivered.
- Repeat the above procedure with the blank sample.

8.6.2.2.9 Calculations

Molarity of FAS =

$$\frac{\text{Volume } K_2Cr_2O_7 \text{ solution titrated (mL)}}{\text{Volume FAS used in titration (mL)}} \times 0.10$$

COD (mg/L) =

$$\left(\frac{\text{Volume FAS used in blank titration (mL)}}{\text{Sample volume (mL)}}\right) \times$$

$$(\text{Molarity of FAS} \times 8{,}000 \times \text{DF}) -$$

$$\left(\frac{\text{Volume FAS used in titration (mL)}}{\text{Sample volume (mL)}}\right) \times$$

$$(\text{Molarity of FAS} \times 8{,}000 \times \text{DF})$$

COD on dry basis (g/g) =

$$\frac{\text{COD (mg/L)} \times 0.001}{\text{Total solids (g/g)}}$$

Where:
FAS = Ferrous ammonium sulphate (see the method description)
8,000 = Milliequivalent weight of oxygen × 1,000 mL/L
DF = Dilution factor (V/M), if used
M = Mass of sludge used in sample preparation
V = Total volume (L) of diluted sample

8.6.2.2.10 Data set example

In Ward *et al.* (2021), COD was measured using the closed reflux titrimetric method for 465 faecal sludge samples from 421 pit latrines and septic tanks in Lusaka, Zambia. For septic tanks, the median COD was 53.3 g/L and the average COD was 72.1 g/L, with a standard deviation of 56.9 g/L. For pit latrines, the median COD was 121.1 g/L and the average COD was 122.6 g/L, with a standard deviation of 65.5 g/L. The heterogeneity of faecal sludge is reflected in the high standard deviations. As a quality control measure, laboratory triplicates were done for every 10th sample. The relative standard error on the replicates was 8%. The entire raw data set is included with publication[13].

Two faecal sludge samples were collected in Durban, South Africa: one from a ventilated improved pit latrine and another one from a urine diversion toilet. These were analysed in four replicates using Method 8.6.2.2. The average COD (g/g dry sample) were 0.77 and 0.64 for VIP and UDDT toilets, respectively. The results for COD are presented in Table 8.8 (source: unpublished data UKZN PRG).

Table 8.8 COD values for VIP and UDDT toilets analysed by the COD closed reflux titrimetric method.

	VIP	UDDT
Samples	COD (g/g dry sample)	COD (g/g dry sample)
1	0.7769	0.6305
2	0.7967	0.6244
3	0.8099	0.6335
4	0.7879	0.6365
Average	0.7728	0.6374
SD	0.0272	0.0122

8.6.3 Fat and fibre

Crude fat and crude fibre are components of the organic matter in a sample. Characterisation of crude fat and crude fibre content is useful in assessing the resource recovery potential of faecal sludge. Crude protein is not covered in this section, but may also be of interest for characterising samples for resource recovery applications. Methods for characterisation for crude protein can be found in FAO (2011) and Regulation (EC) No 152/2009 – Determination of the Content of Crude Protein. These methods, however, could require additional method development for faecal sludge samples.

8.6.3.1 Crude fat – Soxhlet extraction method[14]

8.6.3.1.1 Introduction

In the proximate system of analysis, crude fat is the fraction of the sample that is soluble in organic solvents, such as ether. The crude fat fraction contains mostly fatty acids, but can also include waxes. Crude fat can be used to characterise resource recovery end products from faecal sludge treatment, for example black soldier fly larvae, to determine their value as animal fodder and/or use as bio-oil or a feedstock for biodiesel. Crude fat can also be measured directly on dewatered or dried faecal sludge to determine its suitability as a feedstock or co-feedstock for larvae rearing or vermiculture. Since several lipid extraction methods may be considered, the source material and lipid type present in the sample play an important role

[13] https://doi.org/10.25678/00037X
[14] This method is adapted from Crude Fat - Ether Extract Method (FAO, 2011) including recommendations from Regulation (EC) No 152/2009 Determination of Crude Oils and Fats. This method should be cited as: Crude Fat - Ether Extract Method (FAO, 2011) as adapted in Velkushanova *et al.* (2021).

in determining the appropriate extraction method. The most widely used is the Soxhlet extraction method. This method uses a selected solvent of choice, such as petroleum ether, to extract fat from a dried, ground sample. Then petroleum ether is distilled, leaving behind fat as residue. The fat is then dried and quantified by weighing, and is reported as a mass fraction (% crude fat).

8.6.3.1.2 Safety precautions

- General health and safety (H&S) procedures specific for conducting laboratory analysis of faecal sludge are presented in Section 8.2. Before conducting this method, it is important to be familiar with Section 8.2.3 to ensure safety measures are properly carried out.
- Appropriate personal protective equipment (PPE) should be used; specific details are covered in Section 8.2.3.1.
- Petroleum ether is flammable and has a very low flash point. Ensure that electrical equipment is earthed (grounded) and that sparks are not generated in the vicinity. Ensure that the ether is fully evaporated before heating flasks in the oven, to avoid fire or explosion.
- Do not inhale petroleum ether vapour. Always use a fume hood when decanting or handling open containers. Ensure that no solvent escapes out of the condenser (the water in the condenser must be cool). Check the temperature of the water bath regularly. Ensure the joints are sealed and the glassware is not cracked. Conduct the extraction in a well-ventilated area.
- Acids must be handled with care.
- Eye and hand protection must be used when preparing acids. Ensure when handling concentrated acids that an acid-proof lab coat and acid-proof gloves are used. For more detailed information on selecting the correct type of glove, consult the glove comparison chart (Berkeley Environment[D], Health & Safety).
- Never add water to acid. Add acid to water (slowly and carefully).
- Be extra cautious when heating acids and always use a fume hood during digestion.
- Inspect the glassware before heating acids and bases. Never use cracked glassware.

8.6.3.1.3 Required chemicals

- Petroleum ether (40-60 °C, evaporation residue ≤ 20 mg/L).
- Hydrochloric acid (3 M) (if hydrolysis pre-treatment is necessary).

8.6.3.1.4 Required apparatus and instruments

- Pestle and mortar
- Extraction thimbles and glass wool or fat-free cotton wool
- 250 mL round-bottomed flask
- Silicon carbide chips or glass beads
- Soxhlet extractor
- Condenser fitted to the Soxhlet
- Retort stand and clamps
- Heating block (heating mantle, hot water bath – able to maintain 40-60 °C)
- Cooling unit and water pump (circulate cold water through the condenser – approx. 15 °C)
- Rotary evaporator
- Vacuum grease
- Desiccator with dry desiccant
- Analytical balance
- Fume hood
- Drying oven or vacuum oven
- Beaker or conical flask (if hydrolysis pre-treatment is necessary)
- Hot plate (if hydrolysis pre-treatment is necessary)
- Buchner funnel filtration apparatus and vacuum pump with fat-free filter paper (if hydrolysis pre-treatment is necessary)
- Litmus paper (if hydrolysis pre-treatment is necessary)

8.6.3.1.5 Quality control

General information on quality assurance and quality control (QA/QC) is provided in Section 8.3. Information on standards, operation conditions, and interferences that are specific to this method include:

- Commercially available oils can be used to prepare quality control standards. Quality control samples should be stored in a refrigerator (2-8 °C) and changed every 6 months.
- This method is for samples with a crude fat content lower than 20%. Samples with a higher fat content should be pre-treated with a preliminary extraction (see FAO, 2011).

- Depending on the composition of the sample, oils and fats may not be fully extractable without prior hydrolysis pre-treatment. When developing this method for a new faecal sludge type or feedstock, it will be necessary to perform crude fat analysis with and without prior hydrolysis. If the amount of crude fat measured with prior hydrolysis is higher than the result achieved without hydrolysis, the amount obtained with hydrolysis should be considered the true crude fat value and prior hydrolysis should be incorporated as part of the method (EC, 2009).
- Ensure the desiccant in the desiccator is not saturated, otherwise samples may absorb water while cooling down in the desiccator. Desiccant should be routinely dried in the oven at 105 °C (or at a different temperature, depending on the manufacturer's instructions), prior to the colour indicating that the desiccant is nearly saturated.
- Always keep the lid of the desiccator on and use a lubricant on the rim to ensure airtight sealing. Do not overload the desiccator.
- Ensure the samples are fully cooled in a desiccator to ambient temperature prior to weighing.
- Follow the quality control recommendations for calibration of the analytical balance and the drying oven described in Method 8.6.1.1.

8.6.3.1.6 Sample preservation

To preserve, samples can be freeze-dried, oven-dried at 100 ± 3 °C, or dried in a vacuum oven at 80 ± 2 °C to a constant weight, following Method 8.6.1.1. Freeze-dried or oven-dried samples can be stored in a cool, dry place for an extended period of time.

8.6.3.1.7 Sample preparation

- Determine the amount of faecal sludge required for analysis - *e.g.* 5.0 g of dried, ground sludge is required per crude fat analysis. It is recommended to prepare at least 11 g of dried, ground sludge to allow for duplicate measurements and determination of the moisture content of a dried sample. Ensure that a sufficient amount of faecal sludge is dried to allow for at least 11 g of dried, ground sample, taking into account that some sample may be lost during the grinding and sieving. The amount of wet sludge required will be dependent on the TS of the sample.
- Before drying, uniformly mix all the faecal sludge using a stainless steel rod (or other appropriate tool) in order to have thoroughly-mixed representative samples. If desired, samples may also be blended (see Section 8.4.2).
- Freeze-dry or oven-dry (100 ± 3 °C) or vacuum-dry (80 ± 2 °C) samples to a constant weight, following Method 8.6.1.1.
- Grind the dried samples (*e.g.* with a pestle and mortar or coffee grinder) to pass through a 1 mm sieve.
- Measure the moisture content of the dried, ground sample using either Method 8.6.1.1 or Method 8.6.1.5. Note the moisture content of the final dried sample.

8.6.3.1.8 Analysis protocol

Calibration

Standard curve and error estimation:

- Use commercially available oil as a standard reference (a known value of the fat present).
- Add 2 g defatted sand/silica to a thimble, add 0.00625 g oil to the sand in the thimble, add more defatted sand/ silica to make the total weight of 5 g ~ 0.125% crude fat.
- Continue as follows:
  0.0125 g ~ 0.25%
  0.025 g ~ 0.5%
  0.05 g ~ 1%
  0.075 g ~ 1.5%
  0.1 g ~ 2%
  0.125 ~ 2.5%
  0.15 ~ 3%
  0.175 ~ 3.5%
  0.2 ~ 4%
- Perform Soxhlet extraction on each dilution of the standard oil.
- Plot a curve of the % crude fat determined.
- Determine if there is any error and the degree of error in the method.

Hydrolysis pre-treatment (optional, based on sample characteristics)

- Weigh out 5 g of the dried and ground sample into a beaker or conical flask. Record the exact weight of the sample to 4 decimal places (W_1).
- Add 100 mL of hydrochloric acid and silicon carbide chips or glass beads to the beaker or

conical flask. Cover the beaker with a watch glass or connect a reflux condenser to the conical flask.
- Boil the mixture gently on a hot plate for 1 hr. Swirl occasionally to keep the material from sticking to the sides of the beaker or flask.
- Cool and filter through a pre-moistened fat-free double filter paper using a Buchner funnel with a vacuum.
- Wash the residue with cold distilled water until the filtrate is neutral (check with Litmus paper).
- Carefully transfer the filter paper and residue into an extraction thimble and dry in a drying oven for 1 hr at 100 ± 3 °C or a vacuum oven at 80 ± 2 °C.
- Remove the thimble with filter paper from the drying oven, cover with glass wool or fat-free cotton wool and proceed with the extraction (see the next section).

Extraction
- Weigh out 5 g dried and ground sample into a thimble. Record the exact weight of the sample to 4 decimal places (W_1).
- Cover the thimble with a wad of glass wool or fat-free cotton wool, then place the thimble in the Soxhlet chamber.
- Add silicon carbide chips or glass beads (5) to a 250 mL round-bottomed flask.
- Weigh the 250 mL round-bottomed flask and record to 4 decimal places (W_2).
- Pour 150 mL petroleum ether into the 250 mL round-bottomed flask.
- Connect the 250 mL round-bottomed flask to the Soxhlet apparatus (see Figure 8.4).
- Ensure the 250 mL round-bottomed flask, Soxhlet and condenser are connected and the joints are greased with vacuum grease (always grease after adding the contents in the Soxhlet and the round-bottomed flask). Ensure that the 250 mL round-bottomed flask sits snugly in the heating pocket.
- Connect the water cooling system to the condenser and ensure there is a flow of cold water (approximately 15 °C) through the condenser and back into the cooling bath (the condenser: the bottom nozzle - water flows in, and the top nozzle - water flows out), see Figure 8.4.
- Extract for at least 6 hr, to get at least 60 siphons.
- Turn off the heating mantle and allow the 250 mL round-bottomed flask to cool down to room temperature.

Drying
- Remove the 250 mL round-bottomed flask and connect to the rotary evaporator.
- Evaporate the solvent until the 250 mL round-bottomed flask is almost free from solvent.
- Leave the 250 mL round-bottomed flask overnight in a fume hood, to ensure all the solvent is evaporated.
- Dry the 250 mL round-bottomed flask with residue for 1.5 hr in the drying oven set at 100 ± 3 °C.
- Cool in the desiccator until room temperature.
- Weigh the 250 mL round-bottomed flask with residue and record to 4 decimal places (W_3).

8.6.3.1.9 Calculation

$$\% \text{ crude fat} = \frac{W_3 - W_2}{W_1} \times 100\%$$

Where:

W_1 = Initial weight of sample (g)
W_2 = Weight of 250 mL round-bottomed flask (g)
W_3 = Weight of 250 mL round-bottomed flask (g) with fat residue

8.6.3.1.9 Data set example

Gold *et al.* (2020) quantified lipids in human faeces to assess their potential as a feedstock for black soldier fly larvae rearing. Lipids were extracted from freeze-dried samples using ether extraction with petroleum ether at 40-60 °C following hydrolysis with 3 M hydrochloric acid following Regulation (EC) No 152/2009 Determination of Crude Oils and Fats. Crude fat content in human faeces was measured at 20.9 % of the dry mass. Nyakeri *et al.* (2019) used ether extraction to quantify crude fat in black soldier fly larvae reared using faecal sludge as a feedstock. Larvae were dried and ground, and the residue was dried at 110 °C after extraction, instead of the 80 °C specified in this method. Analyses were performed in triplicate. The average crude fat content in larvae fed with faecal sludge was reported as 184 ± g/kg, which can be converted to 18.4% of dry mass.

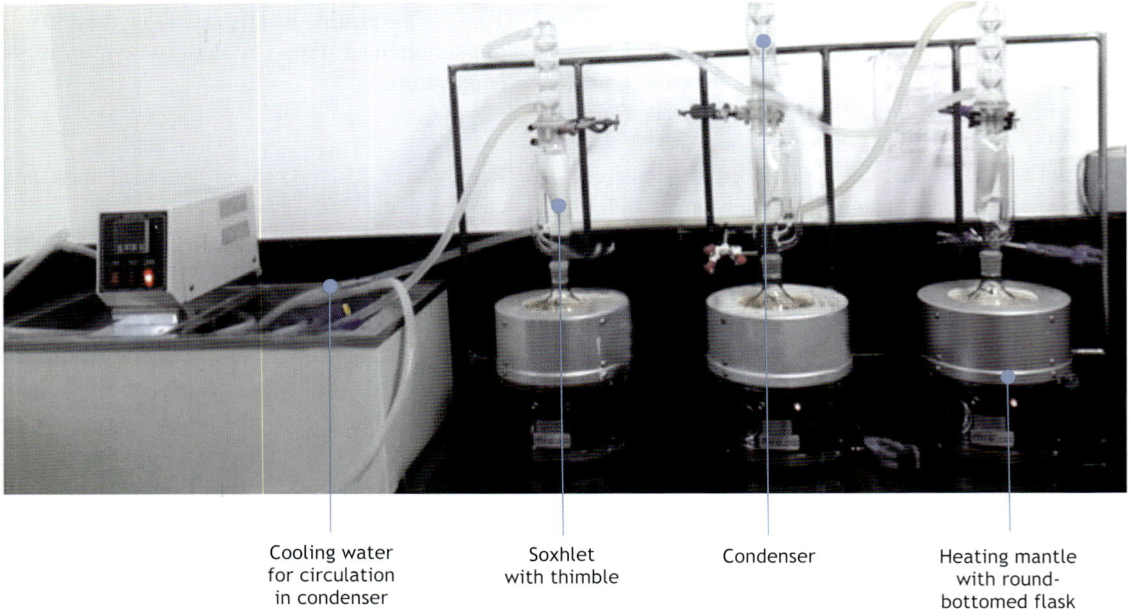

Figure 8.4 Soxlet setup (photo: UKZN PRG).

8.6.3.2 Crude fibre – filtration method[15]

8.6.3.2.1 Introduction

Determinations of crude fibre are used for quality control and specifications of animal feeds by regulatory agencies (Mertens, 2003). This method involves the extraction of fibre from a defatted sample using dilute sulphuric acid followed by potassium hydroxide solution. The mass difference after extraction, filtration, and incineration in a muffle furnace is equal to the mass of crude fibre (reported as % crude fibre). While this method is the standard for regulatory agencies, and is robust and easy to measure for many feedstocks and in many laboratories, crude fibre is not reflective of actual available dietary fibre, and does not fully capture all cellulose, pentosans, and lignins in the sample. Other methods of characterising dietary fibre may be more appropriate, depending on the purpose of characterisation, including acid detergent fibre (ADF), neutral detergent fibre (NDF), and total dietary fibre (TDF). These methods are described in detail in FAO (2011) and Novotny *et al.* (2017).

This method is applicable for determination of samples with a crude fibre content higher than 1%. If the sample contains >10% fat, the fat must first be extracted with petroleum ether prior to beginning the analysis. Fat extraction can be performed as part of the characterisation of crude fat, described in Method 8.6.3.1. The defatted sample from the crude fat analysis can be subsequently analysed for crude fibre following this method.

8.6.3.2.2 Safety precautions

- General health and safety (H&S) procedures specific for conducting laboratory analysis of faecal sludge are presented in Section 8.2. Before conducting this method, it is important to be familiar with Section 8.2.3 to ensure safety measures are properly carried out.
- Appropriate personal protective equipment (PPE) should be used; specific details are covered in Section 8.2.3.1.
- Concentrated sulphuric acid and potassium hydroxide must be handled with care.

[15] This method is adapted from the Crude Fibre – Filtration Method described in the FAO Quality Assurance for Animal Feed Analysis Laboratories Manual (FAO, 2011) including suggestions from Novotny *et al.* (2017). This method should be cited as: Crude Fibre – Filtration Method (FAO, 2011) as adapted in Velkushanova *et al.* (2021).

- Eye and hand protection must be used when preparing acids and bases. Ensure when handling concentrated acids and bases that an acid/base-proof lab coat and acid/base-proof gloves are used. For more detailed information on selecting the correct type of glove, consult the glove comparison chart (Berkeley Environment, Health & Safety[16]).
- Never add water to acid. Add acid to water (slowly and carefully).
- Be extra cautious when heating acids and bases and always use a fume hood during digestion.
- Inspect the glassware before heating acids and bases. Never use cracked glassware.
- Do not dispose of the following reagents in the same container: sulphuric acid and potassium hydroxide; sulphuric acid and petroleum; and potassium hydroxide and petroleum.

8.6.3.2.3 Required chemicals

- Dilute potassium hydroxide (0.23 M)
  - Potassium hydroxide (KOH) pellets
  - Distilled water
- Dilute sulphuric acid (0.15 M)
  - Concentrated sulphuric acid (H2SO4) (98%)
  - Distilled water
- Petroleum ether (if de-fatting is required)
- Acetone (technical quality)

8.6.3.2.4 Required apparatus and instruments

- Sintered glass Gooch crucibles (40-100 μm pore size recommended for faecal sludge samples)
- Buchner/side arm flask for the Gooch crucibles
- Beakers (500 mL)
- Volumetric flasks (1 L)
- Litmus paper (neutral)
- Desiccator with dry desiccant
- Analytical balance with four decimal places
- Vacuum pump
- Filtration manifold with variable pressure
- Fume hood
- Hot plate
- Drying oven
- Muffle furnace

8.6.3.2.5 Quality control

General information on quality assurance and quality control (QA/QC) is provided in Section 8.3. Information on standards, operating conditions and interferences that are specific to this method includes:

- A control standard should be analysed with each batch of samples. The control standard should have similar crude fibre and fat content to the samples analysed in the laboratory (*e.g.* a commercial animal feed can be used as a control standard for a laboratory analysing faecal sludge for suitability as a feed for black soldier fly larvae). A range of ± 2 standard deviations of the average value of the measurements of the control standard is acceptable (FAO, 2011).
- Duplicate measurements of each sample are recommended. Acceptable differences between duplicates are no more than ± 0.3% for samples with less than 10% crude fibre, and no more than ± 3% for samples with ≥ 10% crude fibre (FAO, 2011)
- Ensure the desiccant in the desiccator is not saturated, otherwise the samples may absorb water while cooling down in the desiccator. The desiccant should be routinely dried in the oven at 105 °C (or at a different temperature, depending on the manufacturer's instructions), prior to the colour indicating that the desiccant is nearly saturated.
- Always keep the lid of the desiccator on and use a lubricant on the rim to ensure airtight sealing. Do not overload the desiccator.
- Ensure the samples are fully cooled in the desiccator to ambient temperature prior to weighing.
- Care should be taken during quantitative transfer from the beakers following digestion to ensure that no material is lost. This is one of the greatest sources of error in the determination of crude fibre content (Novotny *et al.*, 2017).
- The pore size of the Gooch crucibles can impact the results, as coarse membranes will allow small fibre particles to escape, while fine membranes may clog during filtration. The pore sizes suggested in Section 8.6.3.2.4 are based on the method development with faecal sludge, but

---

[16] https://ehs.berkeley.edu/workplace-safety/glove-selection-guide#comparison

additional trials with specific sludge to be analysed may be necessary to determine the best pore size for a specific sample. The pore size of the Gooch crucibles should be reported with the results. Filtration of a sample should not take more than 10 min. Information about modifications to this method to deal with difficult-to-filter samples can be found in Novotny *et al.* (2017).

- Crucibles should be checked periodically to screen for cracks or clogging. It should take approximately 180 seconds for 50 mL of water to pass through a crucible without a vacuum. Times less than 120 sec indicate cracks or leaks, while times longer than 240 sec indicate clogs. Cracked crucibles should be discarded. Clogged crucibles should be cleaned by soaking in 6 N HCl for 30 min, then re-evaluated for filtration time (Novotny *et al.*, 2017).
- Follow the quality control recommendations for calibration of the analytical balance and drying oven described in Method 8.6.1.1, and for calibration of the muffle furnace described in Method 8.6.1.2.

8.6.3.2.6 Sample preservation

To preserve them, samples can be freeze-dried or oven-dried (at 80 °C if crude fat will also be analysed in the sample, otherwise dry at 103 °C) to constant weight, following Method 8.6.1.1. Freeze-dried or oven-dried samples can be stored for an extended period in a cool, dry place.

8.6.3.2.7 Sample preparation

- Determine the amount of faecal sludge required for analysis. 1 g of dried, ground sludge is required per crude fibre analysis. It is recommended to prepare at least 3 g of dried, ground sludge to allow for duplicate measurements and determination of moisture content. Ensure that a sufficient amount of faecal sludge is dried to allow for at least 3 g of dried, ground sample, taking into account that some sample may be lost during grinding and sieving. The amount of wet sludge required will be dependent on the TS of the sample.
- Before drying, uniformly mix all the faecal sludge using a stainless-steel rod (or other appropriate tool) in order to have thoroughly-mixed, representative samples. If desired, samples may also be blended (see Section 8.4.2).
- Freeze-dry or oven-dry (103 °C) samples to a constant weight, following Method 8.6.1.1.
- Grind the dried samples (*e.g.* with pestle and mortar or coffee grinder) to pass through a 1 mm sieve.
- Measure the moisture content of the dried, ground sample using either Method 8.6.1.1 or Method 8.6.15. Moisture content should not exceed 10%. If moisture content is higher than 10%, the sample should be dried until the moisture content is below 10%. Note the moisture content of the final dried sample.

8.6.3.2.8 Analysis protocol

Preparation of reagents, chemicals, and standard solutions

- Dilute potassium hydroxide (0.23 M)
  - Add 12.9042 g KOH pellets to a 1 L volumetric flask.
  - Fill to the 1 L mark with distilled water and swirl to fully dissolve the KOH pellets.
- Dilute sulphuric acid (0.15 M)
  - Add some distilled water to a 1 L volumetric flask.
  - Add 8.1582 mL concentrated sulphuric acid (98%) to the 1 L volumetric flask.
  - Top up the 1 L volumetric flask to the mark with distilled water.

Preparation of equipment

- Place a clean Gooch crucible in the furnace at 550 ±20 °C for 15 min prior to use to remove any potential residual organic material from previous measurements. Cool the Gooch crucible in the desiccator until it reaches room temperature.
- Weigh out 1 g dried and ground sample into the Gooch crucible. Record the exact weight of the sample to 4 decimal places (W_1).

Defatting pre-treatment

Perform defatting pre-treatment if the sample has more than 10% fat, and fat has not already been extracted for the crude fat analysis.

- Place the Gooch crucible on the filtration manifold.
- Add 30 mL petroleum ether.

- Filter using the vacuum pump.
- Repeat 2 more times.
- Dry the residue in the air.

Digestion and filtration

- Acid digestion and filtration:
  - Transfer the residue quantitatively to a 500 mL beaker using hot distilled water.
  - Add 150 mL 0.15 M sulphuric acid.
  - Add a glass rod to avoid bumping, see Figure 8.5.
  - Boil on a heating mantle for 30 ± 1 min under a fume hood (maintain the volume with hot distilled water).
  - Leave to cool.
  - Filter through the Gooch crucible using the vacuum pump, see Figure 8.6.
  - Wash the residue 5 times, each time with 10 mL hot distilled water (check with litmus paper for neutrality).
  - Just cover the residue with acetone, and leave for a few minutes.
  - Apply slight suction to remove the acetone.
- Base/alkaline digestion and filtration:
  - Transfer the residue quantitatively to a 500 mL beaker using hot distilled water.
  - Add 150 mL 0.23 M potassium hydroxide.
  - Add a glass rod to avoid bumping, see Figure 8.5.
  - Boil on a heating mantle for 30 ± 1 min in a fume hood (maintain the volume with hot distilled water).
  - Leave to cool.
  - Filter through the Gooch crucible using the vacuum pump, see Figure 8.6.
  - Wash the residue with hot distilled water until the rinsings are neutral (check with litmus paper for neutrality).
  - Wash with 30 mL acetone and vacuum filter, and repeat 3 times in total.

Oven drying

- Place the Gooch crucible in the drying oven at 103 ± 2 °C for 4 hr.
- Remove from the oven and place in the desiccator to cool to room temperature.
- Weigh the Gooch crucible + dry the residue immediately after removing from the desiccator. Record the weight to 4 decimal places (W_2).

Muffle furnace

- Place the Gooch crucible in the muffle furnace at 550 ± 20 °C for 2 hr.
- Remove from the furnace and place in the desiccator to cool to room temperature.
- Weigh the Gooch crucible + incinerated residue immediately after removing from the desiccator. Record the weight to 4 decimal places (W_3).

Figure 8.5 Acid-base digestion setup (photo: UKZN PRG).

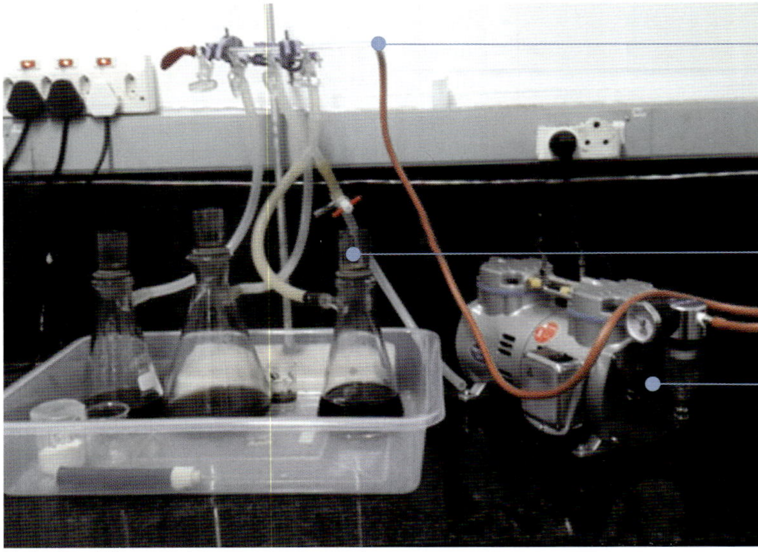

Figure 8.6 Vacuum filtration setup (photo: UKZN PRG).

8.6.3.2.9 Calculation

$$\% \text{ crude fibre} = \frac{W_2 - W_3}{W_1} \times 100\%$$

Where:

W_1 = Weight of the sample (g)

W_2 = Weight Gooch crucible and residue after drying (g)

W_3 = Weight Gooch crucible and residue after incineration (g)

8.6.3.2.10 Data set example

An example of a different acid digestion fibre method for the determination of fodder quality for plants grown on planted drying beds is presented in Gueye *et al.* (2016). Acid digestion fibre (ADF) was measured by boiling a sample of forage in a detergent under acid conditions (pH = 2) and filtering the boiled sample through filter paper. In addition to fibre, TKN was determined by the Kjeldahl digestion method, crude protein (CP) was calculated as TKN × 6.25, crude cellulose (CC) was determined by first digesting in sulfuric acid (0.26 N), and then potassium hydroxide (0.23 N). TP was extracted by dry ashing in a muffle furnace diluted with an acid mix (HCL/HNO_2) and analysed by the molybdate procedure.

8.6.4 Nitrogen

Measurements of the different forms of nitrogen indicate the stabilisation and nutrient availability in treatment and resource recovery processes. Nitrogen is present in water and wastewater in different concentrations and chemical forms, including inorganic forms (ammonia, ammonium, nitrite and nitrate) and organic nitrogen. Organic nitrogen consists of a complex mixture of compounds including amino acids, amino sugars, and proteins. Total nitrogen consists of the sum of organic and inorganic fractions, and Kjeldahl nitrogen consists of organic nitrogen and ammonia/ammonium, which can be determined together analytically. In faecal sludge, nitrogen concentrations can be 10-100 times greater than observed in influent to sewer-based, municipal wastewater treatment plants, so special care needs to be taken with sample preparation and method selection. For more information on the nitrogen cycle, refer to Strande *et al.* (2014).

Ammonia exists either as free ammonia (NH_3) or in the ionic form ammonium (NH_4^+). In aqueous solutions, a pH-dependent equilibrium exists between the two forms, and thus ammonia occurs partly in the form of ammonium and partly as free ammonia. In faecal sludge treatment processes, it is important to monitor ammonia measurements because high ammonia concentrations can inhibit biological processes, such as anaerobic digestion and stabilisation ponds. In planted drying beds, high ammonia concentrations can inhibit plant growth, which is important for treatment processes. The leachate from planted and unplanted drying beds typically also still has high concentrations of ammonia. Due to the ability to inhibit biological activity, ammonia concentrations can also be used to slow the growth of pathogens, and during alkaline stabilisation of faecal sludge, the addition of lime increases the pH, which results in greater NH_3 concentrations (than NH_4^+) (Englund and Strande, 2019). In this book, methods for determination of ammonia include test strips, and spectrophotometric and titrimetric methods. Other methods not yet included in this book are the ammonia selective electrode method and flow injection analysis.

Nitrite, which is an intermediate of the oxidation of ammonia to nitrate (nitrification), is usually present in low to non-detectable concentrations in wastewater and faecal sludge. However, in the event of an imbalance in the nitrification process, nitrite may accumulate. Nitrite is toxic and effluents with high nitrite concentration can adversely affect aquatic life when dilution in the receiving water is insufficient. In aerobic treatment processes, accumulation of nitrite may signify deficiencies in the treatment process. In this book, methods for determination of nitrite include nitrite colorimetric and spectrophotometric methods.

Nitrate and ammonium are considered to be plant-available nutrients. However, ammonia/ammonium is rapidly oxidised to nitrate in the environment, and nitrate can leach rapidly through soils. Nitrate becomes toxic and affects public health when it enters the food chain through contamination of surface or groundwater. Eutrophication and algal blooms of surface water are direct environmental impacts of pollution from excess nutrients. Nitrate management in faecal sludge treatment is thus essential to protect public and environmental health. In this book, methods for determination of nitrate concentrations include colorimetric and spectrophotometric methods.

8.6.4.1 Total nitrogen – spectrophotometric method[17]

Total nitrogen can be determined by several different methods, including in-line UV/persulfate digestion and oxidation with flow injection. The method described here is total nitrogen quantified with spectrophotometry, based on the oxidation of all the nitrogenous compounds with the persulfate digestion to nitrate, followed by a cadmium reduction process to measure the nitrate.

Commercial test kits based on standard methods for measuring total nitrogen are available, with pre-packaged individual aliquots of necessary chemical in pillows (dry chemicals) and vials (liquid chemicals). Commonly used total nitrogen test kits from manufacturers such as Hach and Merck employ a variety of methods. The example provided here is the Hach total nitrogen spectrophotometric test[D] (Hach, 2020) for samples with concentrations of 2-150 mgN/L, which is based on the manufacturer's protocol for water and wastewater using the standard method 4500-N C (Rice *et al.*, 2017). For faecal sludge, samples must be diluted to prevent the false high readings associated with turbid solutions.

8.6.4.1.1 Safety precautions

- General health and safety (H&S) procedures specific for conducting laboratory analysis of faecal sludge are presented in Section 8.2. Before conducting this method, it is important to be familiar with Section 8.2.3 to ensure safety measures are properly carried out.
- Appropriate personal protective equipment (PPE) should be used; specific details are covered in Section 8.2.3.1.

[17] This method should be cited as: Method 4500-N C (Rice et al., 2017), and if test kits are used, also as per the manufacturer's directions including any modifications.

8.6.4.1.2 Required chemicals

- Deionised or distilled water (free from nitrogen)
- Ammonia nitrogen standard solution (100 mg/L NH_3-N)
- Total nitrogen persulfate reagent powder pillow (supplied by the manufacturer)
- Total nitrogen reagent A powder pillow (supplied by the manufacturer)
- Total nitrogen reagent B powder pillow (supplied by the manufacturer).

8.6.4.1.3 Required apparatus and instruments

- Spectrophotometer (*e.g.* DR6000)
- Digester or heating block (capable of heating to 150 °C)
- 2 total nitrogen hydroxide digestion reagent vials (supplied by the manufacturer)
- TN reagent C vial (supplied by the manufacturer)
- Analytical balance with weighing boats (slurry to solid samples)
- Blender (slurry to solid samples)
- Pipettes and pipette tips
- Volumetric flask (1 L)
- Glass beakers (50 or 100 mL)
- Test tube rack
- Funnel
- Finger cot.

8.6.4.1.4 Quality control

General information on quality assurance and quality control (QA/QC) is provided in Section 8.3. Information on standards, operating conditions and interferences that are specific to this method includes:

- Perform quality control with the Hach ammonia nitrogen standard solution with every test batch (or on a daily or weekly basis, depending on the testing load).
  Note: quality control for other methods can also include nitrate standard solutions.
- Measure a blank sample for every test batch and subtract the blank values from the sample results.
- During the reaction, undissolved powder will remain at the bottom of the sample cell after the reagent dissolves. This deposit will not affect results.
- Common interferences in faecal sludge may include chloride and bromide ions. For specific concentrations refer to the manufacturer's instructions.
- Suspended solids and turbid solutions interfere with measurement, thus faecal sludge samples should be diluted with an appropriate dilution factor based on the type of sludge for accurate measurements. Always use serial dilutions.

8.6.4.1.5 Sample preservation

- Samples should be analysed as soon as possible after sampling. For analysis within 48 hr of collection, sample must be refrigerated at 4 °C (Rice *et al.*, 2017).
- Samples can be stored for a maximum of 28 days by preserving with concentrated sulphuric acid and refrigerating at 4 °C.
- For samples preserved with sulphuric acid, pH must be adjusted to 7 with sodium hydroxide solution of a known normality (5.0 N NaOH is recommended) before analysis.
- Samples must be thawed to room temperature before the analysis is performed.

8.6.4.1.6 Sample preparation

For liquid, slurry, semi-solid or solid samples:

- A turbid solution can falsely increase the spectrophotometric reading. Turbid samples should be diluted to prevent false high measurements.
- Samples containing more than 150 mgN/L must be diluted with nitrogen-free distilled water.
- The results can be reported on a mass per volume basis (gN/L).

For slurry to solid samples:

- Weigh out between 1.8 g and 2.0 g of thoroughly-mixed faecal sludge sample into a beaker.
- Dilute the sample gravimetrically and transfer to a blender, as described in Section 8.4.2.
- If necessary, adjust the dilution based on the specific sludge characteristics
- Blend for 30 sec on the highest speed.
- Transfer the contents of the blender into a 1 L volumetric flask, rinse the blender twice with 250 mL distilled water and dilute to the mark with distilled water.
- Transfer this 1 L solution into a plastic bottle and store at 4 °C until ready for testing.

- Total solids analysis should be performed on the samples so that the results of the total nitrogen measurement can be expressed on a mass per mass basis (gN/gTS).

8.6.4.1.7 Analysis protocol

Calibration

Follow the spectrophotometer manufacturer's instructions for calibration. It may be necessary to calibrate the instrument before every reading, or it may only be necessary to perform periodic calibration checks to determine when calibration is necessary. For this method, calibration can be performed as follows:

- Prepare a series of at least four different dilutions of the standard solution, making sure to include the lowest and highest concentrations of the kit testing range with serial dilutions, or dilutions with a uniform interval including the lowest and highest concentrations.
- Determine the total nitrogen concentration of the standard solutions.
- Multiply the answer by the dilution factor and report the results in mgN/L.
- Prepare a calibration curve by plotting the instrument response against the standard concentration.

Procedure

For measuring the range of 2-150 mgN/L:

- Preheat the digester or heating block to 150 °C.
- With the aid of a funnel, add the contents of one total nitrogen persulfate reagent powder pillow to each of the two total nitrogen hydroxide digestion reagent vials, ensuring that the reagent does not stick to the lips of the vials.
- Pipette 0.5mL of the sample into one of the vials.
- Pipette 0.5mL of deionised or nitrogen-free water into the second vial as a blank.
- Cover the vials with caps and shake vigorously for at least 30 sec to mix. The undissolved powder does not affect the accuracy of the test.
- Place the vials in the heating block and digest for 30 min.
- Remove the vials from the heating blocks after 30 min, place them onto a test tube rack and allow them to cool to room temperature.
- Add one total nitrogen (TN) reagent A powder pillow to each vial and shake for 30 sec.
- Allow the mixture to react for 3 min using a set timer.
- Add one TN reagent B powder pillow to each vial and shake vigorously for 15 sec. The solution will turn yellow but with some undissolved powder which does not affect the accuracy of the test.
- Allow a 2 min reaction time using a set timer and pipette 2 mL of the sample into a TN reagent C vial.
- Mix the solution by slowly inverting the vials approximately 10 times and allow a reaction time of 5 min using a set timer.
- Wipe the vials with a laboratory tissue so it is clean (*e.g.* to remove water spots, fingerprints, etc), and measure in a spectrophotometer.

8.6.4.1.8 Calculations

Liquid and slurry samples:

Result of analysis (mg/L)

$$\text{Final concentration}\left(\frac{\text{mg}}{\text{L}}\right) = \text{A}\left(\frac{\text{mg}}{\text{L}}\right) \times \text{DF}$$

$$\text{Dry basis}\left(\frac{\text{g}}{\text{g}}\right) = \frac{\text{Final concentration}\left(\frac{\text{mg}}{\text{L}}\right)}{\text{Total solids concentration}\left(\frac{\text{mg}}{\text{L}}\right)}$$

Where:

A = Spectrophotometric measurement value (mg/L)

DF = Dilution factor (F/I)

F = Final diluted volume (L)

I = Initial sample volume (L)

Slurry, semi-solid and solid samples:

$$\text{Dry basis}\left(\frac{\text{g}}{\text{g}}\right) = \frac{\frac{\text{A}\left(\frac{\text{mg}}{\text{L}}\right)}{1{,}000} \times \frac{\text{V}}{\text{M}}\left(\frac{\text{L}}{\text{g}}\right)}{\text{Total solids content}\left(\frac{\text{g}}{\text{g}}\right)}$$

Where:

A = Spectrophotometric measurement value (mg/L)

V = Volume of dilution (L)

M = Wet weight of sludge used in sample preparation (g)

8.6.4.1.9 Data set example

While developing methods to reliably estimate faecal sludge qualities and quantities, Strande *et al.* (2018) measured the total nitrogen of faecal sludge (in addition to other physicochemical parameters) from pit latrines and septic tanks located in households, non-households, and public toilets. 180 samples were analysed, with an uneven distribution and high variability. Hence, the median rather than the mean of the results were used for the statistical analysis. In spite of the high variability in the results, as illustrated in Figure 8.7 of the TN results for pit latrines and septic tanks, there were differences based on households, non-households and public toilets, and the pit latrines in general had a higher TN concentration. The complete raw data set is available at the link provided below[18].

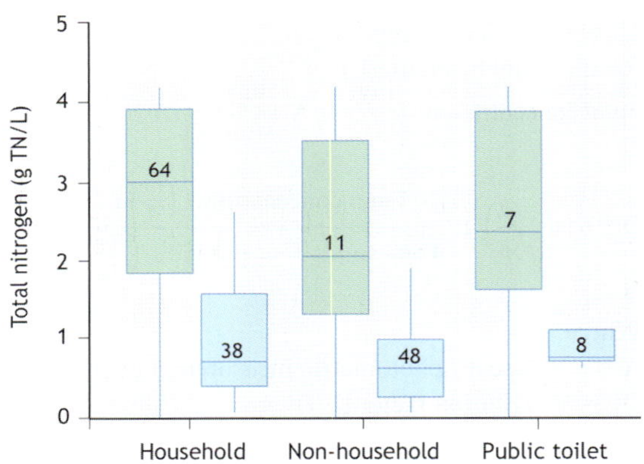

Figure 8.7 Total nitrogen for septic tanks (blue boxes) and pit latrines (green boxes) from households, non-households, and public toilets (Strande *et al.*, 2018).

8.6.4.2 Ammonium – colorimetric (test strip) method[19]

8.6.4.2.1 Introduction

Measurement of ammonium is important for assessing the impact on receiving water bodies, evaluating treatment performance, and to check for inhibitory concentrations in treatment processes. In the test strip method, ammonium ions react with Nessler's reagent to form a yellow-brown compound. The concentration of ammonium is measured semi-quantitatively by visual comparison of the reaction zone of the test strip with the fields of a colour scale. This method is based on the specific manufacturer's protocol for water and wastewater. The example provided here is one of many ammonium colorimetric test methods for samples with concentrations of 10-400 mg/L NH_4^+. Ammonium measurement results are expressed either as mg/l NH_4^+ or NH_4-N, and it is important to note how the results are expressed by the selected test method. For faecal sludge, samples must be diluted to prevent masking of the resultant colour change.

8.6.4.2.2 Safety precautions

- General health and safety (H&S) procedures specific for conducting laboratory analysis of faecal sludge are presented in Section 8.2. Before conducting this method, it is important to be familiar with Section 8.2.3 to ensure safety measures are properly carried out.
- Appropriate personal protective equipment (PPE) should be used; specific details are covered in Section 8.2.3.1.

8.6.4.2.3 Required chemicals

The required chemicals will be specific to each manufacturer's test kit; specific information on test kits can be found on the manufacturers' websites.

- Distilled water
- Reagent NH4-1 (provided with the test kit)
- Ammonium standard solution Centipur®

8.6.4.2.4 Apparatus and instruments

- Test strips
- Glass beakers (50 or 100 mL)
- Balance with weighing boats (slurry to solid samples)
- Blender (slurry to solid samples)
- Volumetric flask (250 mL)
- Absorbent paper towel
- Filter paper (adequate for removing solids from the sample, for example 0.45µm filter for liquid samples)

---

[18] https://doi.org/10.25678/0000tt

[19] This method should be cited as the specific manufacturer's method along with any modifications. The example used here is the Merck MQuant Ammonium Test (Merck, 2020j)[D].

8.6.4.2.5 Quality control

General information on quality assurance and quality control (QA/QC) is provided in Section 8.3. Information on standards, operating conditions and interferences that are specific to this method includes:

- Perform quality control with an ammonium standard solution with every test batch (or on a daily or weekly basis depending on the testing load).
- Dilute the ammonium standard solution with distilled water to 100 mgNH_4^+/L.
- The test strips are stable, and can be stored up to the date stated on the package, when stored closed at 2 to 8°C.
- The colour of the reaction zone might continue to change after the specified reaction time has elapsed. This should not be considered in the measurement, which should always be recorded at the stated time.
- Common interferences in faecal sludge include nitrate, nitrite, phosphate, potassium and magnesium. For specific concentrations of interference, refer to the manufacturer's instructions.
- For faecal sludge, samples should be diluted with an appropriate dilution factor, based on the type of sludge, to prevent interference with the colour of the test strips. Always use serial dilutions.
- The ammonium test strip method is used for qualitative to semi-quantitative measurement. For a quantitative measurement, refer to Method 8.7.5.6.

8.6.4.2.6 Sample preservation

Samples should be analysed as soon as possible after sampling. For analysis within 24 hours of collection, the sample must be refrigerated at 4 °C. Refrigeration reduces biological activity, which can increase NH_4^+ concentration.

8.6.4.2.7 Sample preparation

For liquid, slurry, semi-solid or solid samples:

- Solid samples must be dissolved in water before further dilution is performed.
- Filter samples to prevent interference in colour.
- Samples containing more than 400 mgNH_4/L must be diluted with deionised or distilled water.

For slurry to solid samples:

- Weigh out between 1.8 g and 2.0 g of thoroughly-mixed faecal sludge sample into a beaker.
- Dilute the sample gravimetrically and transfer to a blender, as described in Section 8.4.2.
- If necessary, adjust the dilution based on the specific sludge characteristics
- Blend for 30 sec on the highest speed.
- Transfer the contents of the blender into a 1 L volumetric flask, rinse the blender twice with 250 mL distilled water and dilute to the mark with distilled water.
- Filter samples through a 0.45µm filter paper and measure nitrite concentration.
- Transfer this 1 L solution into a plastic bottle and store at 4 °C until ready for testing.
- Total solids analysis should be performed on the samples so that the results of the nitrite measurement can be expressed on a mass per mass basis (gNH_4^+-N/gTS).

8.6.4.2.8 Analysis protocol

For measuring range of 10-400 mgNH_4^+/L:

- Rinse the test vessel several times with the sample. Fill the test vessel with 5 mL of the sample. Add 10 drops of the reagent NH4-1 and swirl.
- Immerse the reaction zone of the test strip in the measurement sample for 3 seconds.
- Allow excess liquid to run off via the long edge of the strip onto an absorbent paper towel and after 10 seconds determine which colour field on the label the colour of the reaction zone coincides with most closely. Read off the corresponding result in mg/L NH_4^+.

8.6.4.2.9 Calculation

Liquid and slurry samples:

Result of analysis (mg/L)

$$\text{Final concentration}\left(\frac{mg}{L}\right) = A\left(\frac{mg}{L}\right) \times DF$$

$$\text{Dry basis}\left(\frac{g}{g}\right) = \frac{\text{Final concentration}\left(\frac{mg}{L}\right)}{\text{Total solids concentration}\left(\frac{mg}{L}\right)}$$

Where:

A = Spectrophotometric measurement value (mg/L)
DF = Dilution factor (F/I)
F = Final diluted volume (L)
I = Initial sample volume (L)

Slurry, semi-solid and solid samples:

$$\text{Dry basis}\left(\frac{g}{g}\right) = \frac{\frac{A\left(\frac{mg}{L}\right)}{1{,}000} \times \frac{V}{M}\left(\frac{L}{g}\right)}{\text{Total solids content}\left(\frac{g}{g}\right)}$$

Where:

A = Spectrophotometric measurement value (mg/L)
V = Volume of dilution (L)
M = Wet weight of sludge used in sample preparation (g)

8.6.4.3 Ammonium – phenate spectrophotometric method[20]

8.6.4.3.1 Introduction

There are various methods to measure ammonia; selection of the appropriate method is based on the concentration and the presence of interferences. In this method, ammonia first reacts with hypochlorite to form monochloramine. Chloramine then reacts with phenate to form 5-aminophenate which is oxidised to indophenol, a reaction catalysed by sodium nitroprusside. Indophenol has an intense blue colour that is measured spectrophotometrically. This method measures both ammonium ions and dissolved ammonia in a concentration range of 2–150 $mgNH_4$-N/L.

It is also possible to conduct this method with commercially available test kits. Commonly used test kits from manufacturers such as Hach, Merck, and Hanna employ different methods in the kits used for ammonia measurement; the Nessler method, the salicylate method and the phenate method are common methods. For example, the Hach test kit uses the salicylate method while Merck kits are based on the phenate method. The test kits are also based on standard methods, with pre-packaged individual aliquots of necessary chemical in pillows (dry chemicals) and vials (liquid).

The example provided here is the Merck ammonium spectrophotometric test kit (Merck, 2020f)[D] for samples with concentrations of 2-150 $mgNH_4$-N/L, and uses the manufacturer's protocol for water and wastewater based on the standard method 4500 (Rice *et al.*, 2017). For faecal sludge, samples must be diluted and filtered to prevent false-high readings associated with turbid solutions.

8.6.4.3.2 Safety precautions

- General health and safety (H&S) procedures specific for conducting laboratory analysis of faecal sludge are presented in Section 8.2. Before conducting this method, it is important to be familiar with Section 8.2.3 to ensure safety measures are properly carried out.
- Appropriate personal protective equipment (PPE) should be used; specific details are covered in Section 8.2.3.1.

8.6.4.3.3 Required chemicals

- Distilled water.
- Hypochlorite, and phenate and sodium nitroprusside.
  - Reagent NH_4-1 (supplied by the manufacturer).
  - Reagent NH_4-2 (supplied by the manufacturer).
- Ammonium standard solution CRM (supplied by the manufacturer).

8.6.4.3.4 Required apparatus and instruments

- Spectrophotometer equipped with 1 cm or larger cuvette
- Analytical balance with weighing boats (slurry to solid samples)
- Blender (slurry to solid samples)
- Pipettes and pipette tips
- 10 mm cuvettes (2 packs)
- Volumetric flask (1 L)
- Glass beakers (50 or 100 mL)
- Glass storage bottle
- Blue micro spoon (supplied by the manufacturer)
- Glass test tubes (30 mL or 50 mL)
- Filter paper (adequate for removing solids from sample, for example 0.45µm filter for liquid samples)

[20] This method should be cited as: Method 4500 (Rice *et al.*, 2017), and if test kits are used, also the manufacturer's directions including any modifications.

8.6.4.3.5 Quality control

General information on quality assurance and quality control (QA/QC) is provided in Section 8.3. Information on standards, operating conditions and interferences that are specific to this method includes:

- Perform quality control with an ammonium standard solution with every test batch (or on a daily or weekly basis, depending on the testing load).
- Due to the specific temperature range dependence of the colour reaction, ensure that the temperature of the reagents is maintained between 20-30 °C.
- Ammonium-free samples turn yellow on addition of reagent NH_4-2. Thus, blank samples (distilled water) should be tested with every test batch.
- Colour of the measurement solution remains stable for up to 60 minutes after the end of the reaction time, thus the spectrophotometric measurement should be conducted within that timeframe.
- Common interferences in faecal sludge include magnesium and nitrite. For specific concentrations refer to the manufacturer's instructions.
- For faecal sludge, samples should be diluted with an appropriate dilution factor and filtered based on the type of sludge for accurate measurements. Always use serial dilutions.

8.6.4.3.6 Sample preservation

- Samples should be analysed as soon as possible after sampling to prevent losses due to volatilisation or biological activity. For analysis within 24 hours of collection, sample must be refrigerated at 4 °C.
- Samples can be frozen at -20 °C or preserved by acidifying to pH ≤ 2 and kept at 4 °C for long-term storage (Rice *et al.*, 2017).
- If acidification is used in storing samples, neutralise the samples with KOH or NaOH immediately before ammonium determination.

8.6.4.3.7 Sample preparation

For liquid, slurry, semi-solid or solid samples:

- Turbidity will falsely increase the spectrophotometric reading, thus samples might have to be filtered to prevent false-high measurements. Report any preparation steps with the results
- Samples containing more than 150 mgNH_4-N/L must be diluted with distilled water.

For slurry to solid samples:

- Weigh out between 1.8 g and 2.0 g of thoroughly-mixed faecal sludge sample into a beaker.
- Dilute the sample gravimetrically and transfer to a blender, as described in Section 8.4.2.
- If necessary, adjust the dilution based on the specific sludge characteristics
- Blend for 30 sec on the highest speed.
- Transfer the contents of the blender into a 1 L volumetric flask, rinse the blender twice with 250 mL distilled water and dilute to the mark with distilled water.
- Transfer this 1 L solution into a plastic bottle and store at 4 °C until ready for testing.
- Filter the samples through a 0.45 µm filter paper and measure ammonium concentration according to the analysis protocol.
- Total solids analysis should be performed on the samples so that the results of the ammonium measurement will be reported as gNH_4-N/gTS.

8.6.4.3.8 Analysis protocol

Calibration

Follow the spectrophotometer manufacturer's instructions for calibration. It may be necessary to calibrate the instrument before every reading, or it may only be necessary to perform periodic calibration checks to determine when calibration is necessary. For this method, calibration can be performed as follows:

- Prepare a series of at least four different concentrations of a standard solution making sure to include the lowest and highest concentration of the kit testing range. It is typical to do serial dilutions or dilutions with a uniform interval including the lowest and highest concentrations.
- Take a reading for the prepared ammonium concentrations of the dilutions of the standard.
- Multiply the answer by the appropriate dilution factor and report the results in mgNH_4-N/L.
- Prepare a calibration curve by plotting the instrument response against the standard concentration.

Procedure

For measuring range of 2.0–75 mgNH_4-N/L (2.6–96.6 mgNH_4^+/L):

- Pipette 5.0 mL of reagent NH_4-1 into a glass test tube.
- Add 0.2 mL of the diluted, filtered sample into the test tube and mix by agitating.
- Add 1 level blue micro spoon of reagent NH_4-2 and shake vigorously until completely dissolved.
- Allow the solution to stand for 15 minutes in a test tube rack and fill the 10 mm cuvette with the sample.
- Wipe the cuvette with a soft tissue to remove water spots and fingerprints and measure in the spectrophotometer.

For measuring ranges of > 75-150 mgNH_4-N/L (> 96.6-193 mgNH_4^+/L):

- Pipette 5.0 mL of reagent NH_4-1 into a test tube.
- Add 0.1 mL of the diluted, filtered sample into the test tube and mix by agitating.
- Add 1 level blue micro spoon of reagent NH_4-2 and shake vigorously until completely dissolved.
- Allow the solution to stand for 15 min in a test tube rack and fill the 10 mm cuvette with the sample.
- Wipe the cuvette with a soft tissue to remove water spots and fingerprints and measure in the spectrophotometer.

8.6.4.3.9 Calculations

Result of the analysis (mg/L NH_4-N) = measurement value, A (mg/L) × dilution factor

Liquid and slurry samples:

Result of analysis (mg/L)

$$\text{Final concentration} \left(\frac{\text{mg}}{\text{L}}\right) = \text{A} \left(\frac{\text{mg}}{\text{L}}\right) \text{ x DF}$$

$$\text{Dry basis} \left(\frac{\text{g}}{\text{g}}\right) = \frac{\text{Final concentration} \left(\frac{\text{mg}}{\text{L}}\right)}{\text{Total solids concentration} \left(\frac{\text{mg}}{\text{L}}\right)}$$

Where:

A = Spectrophotometric measurement value (mg/L)

DF = Dilution factor (F/I)

F = Final diluted volume (L)

I = Initial sample volume (L)

Slurry, semi-solid and solid samples:

$$\text{Dry basis} \left(\frac{\text{g}}{\text{g}}\right) = \frac{\frac{\text{A} \left(\frac{\text{mg}}{\text{L}}\right)}{1{,}000} \times \frac{\text{V}}{\text{M}}\left(\frac{\text{L}}{\text{g}}\right)}{\text{Total solids content} \left(\frac{\text{g}}{\text{g}}\right)}$$

Where:

A = Spectrophotometric measurement value (mg/L)

V = Volume of dilution (L)

M = Wet weight of sludge used in sample preparation (g)

8.6.4.3.10 Data set example

In evaluating the effect of drying on the physical and chemical characteristics of faecal sludge for its reuse, Septien *et al.* (2020) measured the ammonium concentration of faecal sludge from pit latrines before and after drying at different temperatures using a Merck Spectroquant®,D. Raw faecal sludge samples had ammonium concentration of 24 ± 4 g/g TS. By using the convective drying rig, the ammonia concentration reduced to 4 ± 2 g/g TS when it was dried at 40 °C.

Hach test kits using the salicylate method and a Hach-Lange DR2800 spectrophotometer, following the manufacturers' directions, were used to evaluate 60 samples in Hanoi, Vietnam, and 180 in Kampala, Uganda, for NH_4^+-N concentrations. In Hanoi concentrations ranged from 54 to 1,700 NH_4^+-N/L, and in Kampala, Uganda from 24 to 3,000 mgNH_4^+-N/L (Englund *et al.*, 2020; Strande *et al.*, 2018). The complete raw data set is available using the link below[21].

[21] https://doi.org/10.25678/0000tt

8.6.4.4 Ammonia – distillation and titration method[22]

8.6.4.4.1 Introduction

Acid-base titrations are useful in determining concentrations of acids and bases in samples. An acid or a base is neutralised with a known volume of corresponding acid or base. The exact concentration of the titrant solution must be known in order to establish the concentration of the titrated solution. The endpoint of this reaction is when the solution reaches neutral pH, where the determinant in the sample is equivalent to the analyte, and is only visible when an indicator is used. This is used to calculate the concentration of other determinants from the known ratio of the reaction.

Ammonia exists either as free ammonia (NH_3) or ammonium ions (NH_4^+). In aqueous solutions, a pH-dependent equilibrium exists between the two forms and thus ammonia occurs partly in the form of ammonium ions and partly as free ammonia. Measurement of ammonia is important for assessing the impact on receiving water bodies, evaluating treatment performance and nutrient management, and checking for inhibitory concentrations in treatment processes. The sample is adjusted to pH 9.5 using a borate buffer to allow the measurement of organic nitrogen compounds. The sample containing ammonium nitrogen is distilled in boric acid and then titrated using 0.02N H_2SO_4 as follows:

$$H_2SO_4 + 2\ NH_4OH \rightarrow 2\ H_2O + (NH_4)_2SO_4$$

This method is based on the Standard Methods for the Examination of Water and Wastewater (Rice *et al.*, 2017). Based on local experience, the method might need to be adapted for different types of faecal sludge. For example, the UKZN PRG in Durban adapted the sample volume to 70 mL, based on the characteristics of faecal sludge in Durban and the semi-automated distillation setup used in the laboratory.

8.6.4.4.2 Safety precautions

- General health and safety (H&S) procedures specific for conducting laboratory analysis of faecal sludge are presented in Section 8.2. Before conducting this method, it is important to be familiar with Section 8.2.3 to ensure safety measures are properly carried out.
- Appropriate personal protective equipment (PPE) should be used; specific details are covered in Section 8.2.3.1.
- Handle concentrated acid with care.
- Use eye and hand protection when preparing acid or handling colour reagents. Ensure when handling concentrated acid that an acid-resistant apron or lab coat and acid-resistant gloves and are used. Information on selecting the correct type of glove can be found following the link below[23].
- Prepare and keep the reagents in a fume hood during use.

8.6.4.4.3 Required chemicals

Ammonia-free distilled water

- Distilled water
- Concentrated sulphuric acid, or bromine or chlorine, depending on method used; see Section 8.6.4.4.8.

Ammonium chloride standard

- Ammonia-free distilled water
- NH_4Cl

NaOH solutions (1 N and 0.1 N)

- Ammonia-free distilled water
- NaOH

Borate buffer solution

- 0.1 N NaOH solution
- 0.025 M disodium tetraborate ($Na_2B_4O_7$) solution - (9.5 g $Na_2B_4O_7 \cdot 10H_2O$ hydrous/L or 5.0 g $Na_2B_4O_7$ anhydrous/L)

Mixed indicator solution

- Methyl red indicator
- 95% organic polar solvent (isopropyl alcohol or ethanol)
- Methylene blue

Indicator boric acid solution

[22] This method is based on Method 4500-NH3 B and C of the Standard Methods for the Examination of Water and Wastewater and should be cited as: Rice *et al.* (2017) as described in Velkushanova *et al.* (2021).

[23] https://ehs.berkeley.edu/workplace-safety/glove-selection-guide#comparison, (Berkeley Environment, Health & Safety).

- Ammonia-free distilled water
- Mixed indicator solution
- H_3BO_3

Standard sulphuric acid titrant (0.02 N)

- Ammonia-free distilled water
- Concentrated sulphuric acid
- Anhydrous sodium carbonate
- Methylene blue-methyl red mixed indicator solution.

8.6.4.4.4 Required apparatus and instruments

- Distillation unit
- pH meter
- Filter paper with 0.45 µm pore size
- Drying oven that can reach temperature 270 °C
- Analytical balance
- Blender
- Pipettes and tips
- 1 L volumetric flask
- 50 mL beakers
- Plastic bottle
- Titration setup with burette in metal clamp (or automatic titration unit).

8.6.4.4.5 Quality control

General information on quality assurance and quality control (QA/QC) is provided in Section 8.3. Information on standards, operating conditions and interferences that are specific to this method includes:

- Check the accuracy of the measurement procedure by using a standard solution of known ammonia concentration. This calibration should be done with every set of samples. Dilute and reanalyse any samples exceeding the highest standard on the calibration curve.
- Prepare and analyse blanks with every sample set.
- Urea and cyanates hydrolyse when distilled at pH 9.5, which amounts to 7% for urea and 5% for cyanates at this pH. On standing, glycine, urea, glutamic acid, cyanates and acetamide hydrolyse only very slowly in solution.

8.6.4.4.6 Sample preservation

- Samples should be analysed as soon as possible after sampling. For analysis within 24 hr of collection, samples must be refrigerated at 4 °C. For longer storage durations of up to 28 days, the pH should be reduced to below pH 2 with concentrated sulphuric acid and stored at 4 °C, or the sample should be frozen at -20 °C without acidifying. If acid preservation has been used, neutralise the samples to pH 7 with NaOH or KOH immediately before carrying out the analysis.
- Storage of reagents:
  - Ammonium-free water is very difficult to store without interference by gaseous ammonia, and storage should therefore be avoided.
  - Prepare a fresh solution of ammonium chloride monthly.
  - Prepare mixed indicator solution monthly.
  - Prepare indicator boric acid solution monthly.
  - Standard sulphuric acid titrant should not be stored longer than one week.

8.6.4.4.7 Sample preparation

This method is valid within a concentration range of 5-100 mg/L NH_3-N:

This method is valid for all sludge types (liquid, slurry, semi-solid, and solid), as long as they are pre-diluted using serial dilutions within the specified range. Ideally, dilutions should aim to be within the range of 10-25 mg/L NH_3-N. Use ammonium-free distilled water for making dilutions.

For liquid faecal sludge:

- Filter turbid samples using 0.45 µm filter.

For slurry to solid faecal sludge:

- Weigh out between 1.8 g and 2.0 g of thoroughly-mixed faecal sludge sample into a beaker.
- Dilute the sample gravimetrically and transfer to a blender, as described in Section 8.4.2.
- If necessary, adjust the dilution based on the specific sludge characteristics
- Blend for 30 sec on the highest speed.
- Transfer the contents of the blender into a 1 L volumetric flask, rinse the blender twice with 250 mL distilled water and dilute to the mark with distilled water.
- Transfer this 1 L solution into a plastic bottle and store at 4 °C until ready for testing.
- Total solids analysis should be performed on the samples so that the results of the ammonia measurement can be expressed on a mass per mass basis (gNH_3-N/gTS).

8.6.4.4.8 Analysis protocol

Preparation of reagents, chemicals and standard solutions

- Ammonia-free water

Eliminate traces of ammonia in distilled water by adding 0.1 mL concentrated sulphuric acid to 1 L distilled water and then redistill. Alternatively, treat distilled water with enough bromine or chlorine to produce a free halogen residual of 2-5 mg/L and redistill after standing for 1 hr. Discard the first 100mL of distillate, and check the ammonia concentration in the collected water before use.

- Ammonium chloride standard
  - Prepare ammonium chloride standard by weighing and dissolving 3.819 g of NH_4Cl into a 1 L volumetric flask with distilled water and dilute the flask content to the mark with ammonia-free distilled water. This makes an ammonium chloride standard solution where 1.0 mL = 1 mg N = 1.22 mg NH_3.
  - Then prepare a working standard solution by transferring 10 mL stock solution into 1 L in a volumetric flask and further dilute with ammonia-free distilled water to give a concentration of 12 mg/L NH_3.
- N NaOH
  - Weigh 4 g of NaOH and dissolve in 1 L of ammonia-free distilled water.
- 1 N NaOH
  - Weigh 40 g NaOH and dissolve in 1 L of ammonia-free distilled water.
- Borate buffer solution
  - Add 88 mL of 0.1 N NaOH solution to 500 mL of 0.025 M disodium tetraborate ($Na_2B_4O_7$) solution (9.5 g $Na_2B_4O_7 \cdot 10H_2O$ hydrous/L or 5.0 g $Na_2B_4O_7$ anhydrous/L) into a 1 L volumetric flask and dilute to the mark with ammonia-free distilled water.
- Mixed indicator solution
  - Weigh and dissolve 200 mg methyl red indicator in 100 mL of 95% organic polar solvent (isopropyl alcohol or ethanol) in a beaker. In a separate beaker, dissolve 100 mg methylene blue in 50 mL of the same solvent. Then combine the solutions.
- Indicator boric acid solution
  - Weigh 20 g H_3BO_3 and dissolve in ammonia-free distilled water. Add this to a 1 L volumetric flask, add 10 mL mixed indicator solution, and dilute to the mark.
- Standard sulphuric acid titrant (0.02 N)
  - First make a 0.1 N sulphuric acid solution by diluting 2.71 mL concentrated sulphuric acid to 1L with distilled water.
  - Dilute 200 mL 0.1 N sulphuric acid solution to 1 L with distilled water to make a 0.02 N solution.
  - Allow it to cool down to ambient temperature.
- Sulphuric acid titrant standardisation
  - Dry anhydrous sodium carbonate (Na_2CO_3) at 270 °C, cool down in a desiccator, and weigh 1.325 g. Dissolve in ammonia-free distilled water and make up to 500 mL in a volumetric flask (0.05 N solution). Do not keep longer than 1 week.
  - Titrate the sulphuric acid solution against 25 mL of sodium carbonate solution using 5 drops of methylene blue-methyl red mixed indicator.
  - Calculate the exact normality of the sulphuric acid:

Normality of H_2SO_4 solution =

$$\frac{25 \times 0.05}{\text{Volume of } H_2SO_4 \text{ used}}$$

8.6.4.4.9 Calibration

- Prepare a 1,000 mgNH_3-N/L stock solution by dissolving 3.819 g ammonium chloride in 1 L ammonia-free distilled water. The stock solution is diluted to make up four-point standard solutions covering the required range of NH_3-N concentrations.
- Prepare a calibration curve by plotting the instrument response against the standard concentration and calculating the linear regression line. Compute the sample concentration by adjusting the sample response with the offset of the standard curve. Report only those values that fall between the lowest and the highest calibration standards. Dilute and reanalyse the samples exceeding the highest standard. Report the results in mg/L.

Preparation of the setup

Add 500 mL ammonia-free water and 20 mL borate buffer to a distillation flask and adjust the pH to 9.5 with 6.0 N NaOH solution. Add a few glass beads and use this mixture to steam-clean the distillation apparatus until the distillate shows no traces of ammonia.

Procedure

- Add the sample to a distillation flask. Use the sample volume as specified here:

Expected ammonia nitrogen in the sample (mg/L)	Sample volume to be used (mL)
5 – 10	250
11 – 20	100
21 – 50	50
51 – 100	25

- Add ammonia-free distilled water to make up 500 mL.
- Add 25 mL borate buffer to the distillation flask.
- Distil at a rate of 6-10 mL/min, ensuring the tip of the delivery tube is below the surface of the acid-receiving solution, and collect the distillate into 50 mL of indicator boric acid solution.
- Titrate ammonia in the distillate with standardised 0.02 N sulphuric acid:
  - Note the starting volume of the sulphuric acid in the burette.
  - Titrate until the indicator turns pale purple.
  - Note the final volume of sulphuric acid in the burette.
- Process a blank in the same way and apply the necessary correction to the results.

8.6.4.4.10 Calculations

Liquid and slurry samples:

NH_3-N (mg/L) =

$$\frac{(A-B)\times 280}{\text{Sample volume (mL)}}\times DF$$

NH_3-N on dry basis (g/g) =

$$\frac{NH_3\text{-N (mg/L)}}{\text{Total solids concentration (mg/L)}}$$

Where:

A = Volume of H_2SO_4, titrated for sample, mL

B = Volume of H_2SO_4, titrated for blank, mL
Sulphuric acid: standard solution
(0.02 N, 1 mL = 0.28 mgNH_3-N)
1 L = 280 mgNH_3- N

DF = Dilution factor (F/I)

F = Final diluted volume (L)

I = Initial sample volume (L)

Slurry, semi-solid and solid samples:

NH_3-N on dry basis (g/g) =

$$\frac{(A-B)\times 280}{\text{Sample volume (mL)}}\times\frac{V}{M}\times\frac{1}{1{,}000}\times\frac{1}{\text{Total solids content (g/g)}}$$

Where:

V = Volume of dilution (L)

M = Wet weight of sludge used in sample preparation (g)

8.6.4.4.11 Data set example

Spit *et al.* (2014) sampled three pit latrines in Blantyre, Malawi, and determined the ammonia using the titration method. Each sample was analysed in triplicate and the results were reported volumetrically, as shown in Table 8.9.

Table 8.9 The results from the ammonia and TS analysis from three pit latrines in Blantyre, Malawi.

		NH_3-N (g/L)	TS (%)
Pit 1	1	22.0	10.5
	2	30.0	12.0
	3	26.0	26.0
	Average	26.0	16.2
	SD	4.0	8.5
Pit 2	1	36.0	19.0
	2	31.0	18.0
	3	30.7	19.3
	Average		
	SD	5.5	1.5
Pit 3	1	33	25.0
	2	36	22.0
	3	26	22.0
	Average	31.7	23.0
	SD	5.1	1.7

Two faecal sludge samples were collected from two ventilated improved pit latrines in Durban, South Africa. They were analysed in four replicates each, using Method 8.6.4.5. The average ammonia content (g/g dry sample) was 0.71 and 0.74 for samples one and two respectively. The results for ammonia are presented in Table 8.10 (source: unpublished data UKZN PRG).

Table 8.10 VIP faecal sludge analysis using the ammonia distillation and titration method.

Replicate	NH_3-N (g/g dry sample) TS = 2 g/L	NH_3-N (g/g dry sample) TS = 5 g/L
1	0.0070	0.0074
2	0.0072	0.0075
3	0.0071	0.0075
4	0.0071	0.0074
Average	0.0071	0.0074
SD	0.0002	0.0001

8.6.4.5 Nitrite – colorimetric (test strip) method[24]

8.6.4.5.1 Introduction

Nitrite is a toxic intermediate of nitrification. Wastewater effluents with high nitrite concentration can adversely affect aquatic life when dilution in the receiving water is insufficient. In aerobic faecal sludge treatment processes, accumulation of nitrite signifies deficiencies in the treatment process. This method describes the Merck 110007 MQuant Nitrite Test[D] as an example of a commercially available nitrite colorimetric test strip method for samples with concentrations of 2-80 $mgNO_2^-/L$. It is based on the manufacturer's protocol for water and wastewater. In the presence of an acidic buffer, nitrite ions react with aromatic amine to form a diazonium salt, which in turn reacts with N-(1-naphthyl)-ethylenediamine to form a red-violet azo dye. The nitrite concentration is measured semi-quantitatively by visual comparison of the reaction zone of the test strip with the fields of a colour scale. Nitrite measurement results are expressed as either $mgNO_2^-/L$ or $mgNO_2$-N/L, and it is important to note how the results are expressed by the selected test method. Depending on the expected nitrite concentration in the sample, kits with the appropriate measurement range should be selected. For faecal sludge, samples must be diluted to prevent masking of the resultant colour change.

8.6.4.5.2 Safety precautions

- General health and safety (H&S) procedures specific for conducting laboratory analysis of faecal sludge are presented in Section 8.2. Before conducting this method, it is important to be familiar with Section 8.2.3 to ensure safety measures are properly carried out.
- Appropriate personal protective equipment (PPE) should be used; specific details are covered in Section 8.2.3.1.

8.6.4.5.3 Required chemicals

The required chemicals will be specific to each manufacturer's test kit; specific information on test kits can be found on the manufacturer's websites.

- Distilled water.
- Nitrite standard solution Certipur® (1,000 mg/L NO_2^-).

8.6.4.5.4 Required apparatus and instruments

- Test strips.
- Glass beakers (50 or 100 mL).
- Balance with weighing boats (slurry to solid samples).
- Blender (slurry to solid samples).
- Volumetric flask (250 mL).
- Filter paper (adequate for removing solids from the sample, for example 0.45µm filter for liquid samples).

8.6.4.5.5 Quality control

General information on quality assurance and quality control (QA/QC) is provided in Section 8.3. Information on standards, operating conditions and interferences that are specific to this method includes:

[24] Test strip methods should be cited as the specific manufacturer's method along with any modifications. The example used here is the Merck MQuant Nitrite Test (Merck, 2020h)[D].

- Dilute the standard nitrite solution with distilled water to 20 $mgNO_2^-/L$, and analyse according to the analysis protocol.
- Perform quality control with nitrite standard solution with every test batch (or on a daily or weekly basis depending on the testing load).
- The test strips are stable up to the date stated on the pack when stored closed at +2 to +8 °C.
- The colour of the reaction zone may continue to change after the specified reaction time has elapsed. This must not be considered in the measurement, which should always be recorded at the stated time.
- Common interferences in faecal sludge include potassium, magnesium, and nitrate. For specific concentrations refer to the manufacturer's instructions.
- For faecal sludge, samples should be diluted with an appropriate dilution factor, based on the type of sludge, to prevent interference with the colour of the test strips. Always use serial dilutions.
- The nitrite colorimetric test method is used for qualitative to semi-quantitative measurement. For a quantitative measurement, refer to Method 8.6.4.6.

8.6.4.5.6 Sample preservation

- Samples should be analysed immediately after sampling to prevent bacterial conversion of NO_2^- to NO_3^- or NH_3. For short-term preservation of 1 to 2 days, samples should be stored at 4 °C or frozen at −20 °C (Rice *et al.,* 2017).
- Samples should be thawed to ambient temperature before analysis.

8.6.4.5.7 Sample preparation

For liquid, slurry, semi-solid or solid samples:

- Solid samples must be dissolved in a known volume of water before further dilution is performed.
- Filter samples through a 0.45 μm filter paper to prevent interference in colour.
- Samples containing more than 80 $mgNO_2^-/L$ must be diluted with distilled water.

For slurry to solid samples:

- Weigh out between 1.8 g and 2.0 g of thoroughly-mixed faecal sludge sample into a beaker.
- Dilute the sample gravimetrically and transfer to a blender, as described in Section 8.4.2.
- If necessary, adjust the dilution based on the specific sludge characteristics
- Blend for 30 sec on the highest speed.
- Transfer the contents of the blender into a 1 L volumetric flask, rinse the blender twice with 250 mL distilled water and dilute to the mark with distilled water.
- Filter samples through a 0.45μm filter paper and measure nitrite concentration.
- Transfer this 1 L solution into a plastic bottle and store at 4 °C until ready for testing.
- Total solids analysis should be performed on the samples so that the results of the nitrite measurement can be expressed on a mass per mass basis (gNO_2-N/gTS).

8.6.4.5.8 Analysis protocol

For measuring range of 2-80 $mgNO_2^-/L$

- Immerse the reaction zone of the test strip in the diluted, filtered sample (15 - 30 °C) for 1 second. Shake off excess liquid from the strip and after 15 sec determine which colour field on the label the colour of the reaction zone coincides with most exactly. Read off the corresponding result in $mgNO_2/L$ or $mgNO_2$-N/L.

8.6.4.5.9 Calculation

The dilution factor used must be stated.

Result of the analysis (mg/L NO_2^-) = measurement value, A (mg/L) × dilution factor (DF)

Liquid and slurry samples:

$$\text{Final concentration (mg/L)} = A\left(\frac{mg}{L}\right) \times DF$$

$$\text{Dry basis}\left(\frac{g}{g}\right) = \frac{\text{Final concentration}\left(\frac{mg}{L}\right)}{\text{Total solids concentration}\left(\frac{mg}{L}\right)}$$

Where:

A = Colorimetric measurement value (mg/L)
DF = Dilution factor (F/I)
F = Final diluted volume (L)
I = Initial sample volume (L)

Slurry, semi-solid and solid samples:

$$\text{Dry basis}\left(\frac{g}{g}\right) = \frac{\frac{A\left(\frac{mg}{L}\right)}{1{,}000} \times \frac{V}{M}\left(\frac{L}{g}\right)}{\text{Total solids content}\left(\frac{g}{g}\right)}$$

Where:

A = Colorimetric measurement value (mg/L)

V = Volume of dilution (L)

M = Wet weight of sludge used in sample preparation (g)

8.6.4.6 Nitrite – spectrophotometric method[25]

8.6.4.6.1 Introduction

Nitrite can be measured using different methods, such as chromatography or spectrophotometry. This method describes nitrite measurement by spectrophotometry. The principle is that in an acidic solution nitrite ions react with sulfanilic acid to form a diazonium salt. The diazonium salt reacts with N-(1-naphthyl) ethylenediamine dihydrochloride to form a red-violet azo dye at a pH of 2.0-2.5, which is then determined spectrophotometrically. Nitrite measurement results are expressed in different ways (*i.e.* mg/L NO_2^- or mg/L NO_2-N), and it is important to know how the results are expressed by the selected test method. Spectrophotometric measurements ranging from 0.01-1.0 mg/L is applicable for this method. However, for lower concentrations < 0.01 mg/L, spectrophotometric measurements can be made by adapting this method using a 5 cm light path and a green colour filter. The sample is filtered prior to analysis, as solids interfere with the spectrophotometric method, based on the assumption that NO_2^- will be solution, and not with the solids fraction. It is possible to conduct this method with commercially available test kits. The test kits are based on standard methods, with pre-packaged individual aliquots of the necessary chemicals in pillows (dry chemicals) and vials (liquid chemicals). Commonly used test kits from manufacturers such as Hach, Merck, and Hanna vary slightly in the methods they use for nitrite measurement. The ferrosulphate method and diazotization method are most commonly used in test kits.

[25] This method should be cited as: Method 4500-NO2–B (Rice *et al.*, 2017) and, if test kits are used, also as per the manufacturer's directions including any modifications.

The example provided here is the Merck Spectroquant Nitrite Test spectrophotometric test kit (Merck, 2020e)[D] for samples with concentrations of 0.02-1.0 mgNO_2-N/L, and it is based on the manufacturer's protocol for water and wastewater using the standard method 4500 (Rice *et al.*, 2017). For faecal sludge, samples must be diluted and filtered to prevent false high readings associated with turbid solutions.

8.6.4.6.2 Safety precautions

- General health and safety (H&S) procedures specific for conducting laboratory analysis of faecal sludge are presented in Section 8.2. Before conducting this method, it is important to be familiar with Section 8.2.3 to ensure safety measures are properly carried out.
- Appropriate personal protective equipment (PPE) should be used; specific details are covered in Section 8.2.3.1.

8.6.4.6.3 Required chemicals

- Distilled water (free from nitrogen)
- Reagent NO_2-1 (supplied by the manufacturer)
- Nitrite standard solution CRM, 0.200 mg/L NO_2-N (supplied by the manufacturer).

8.6.4.6.4 Required apparatus and instruments

- Spectrophotometer equipped with 1 cm or larger cuvettes
- Analytical balance with weighing boats (slurry to solid samples)
- Blender (slurry to solid samples)
- Pipette and pipette tips
- 10 mm cuvettes
- Volumetric flask (1 L)
- Glass beakers (50 or 100 ml)
- Glass storage bottle
- pH test strips
- Blue micro-spoon (supplied by the manufacturer)
- Glass test tubes (30 mL or 50 mL)
- Laboratory cleaning tissues
- Filter paper (adequate for removing solids from the sample, for example a 0.45μm filter for liquid samples).

8.6.4.6.5 Quality control

General information on quality assurance and quality control (QA/QC) is provided in Section 8.3. Information on standards, operating conditions and interferences that are specific to this method includes:

- Perform quality control with nitrite standard solution with every test batch (or on a daily or weekly basis depending on the testing load).
- For spectrophotometric measurements, the cuvettes must be clean. Before analysis, wipe with a laboratory cleaning tissue.
- Measurement of turbid solutions yields false high readings. For faecal sludge, samples should be diluted with an appropriate dilution factor through serial dilutions and filtered based on the type of sludge for accurate measurements.
- The pH of the measurement solution must be within the range 2.0-2.5.
- The colour of the measurement solution remains stable for up to 60 min after the end of the reaction time; thus, the spectrophotometric measurement should be conducted within that timeframe.
- Common interferences in faecal sludge include magnesium and nitrates. For specific concentrations refer to the manufacturer's instructions.

8.6.4.6.6 Sample preservation

- Samples should be analysed immediately after sampling to prevent bacterial conversion of NO_2^- to NO_3^- or NH_3. For short-term preservation of 1 to 2 days, samples should be stored at 4 °C or frozen at −20 °C (Rice *et al.,* 2017). Samples should be thawed to room temperature before analysis.

8.6.4.6.7 Sample preparation

For liquid, slurry, semi-solid or solid samples:

- Samples containing concentrations of nitrite beyond the range of the test kit must be diluted with distilled water.
- The nitrite content of samples can be estimated prior to dilution with the MQuant nitrite test kit.
- Filter the samples through a 0.45 μm filter paper.

For slurry to solid samples:

- Weigh out between 1.8 g and 2.0 g of thoroughly-mixed faecal sludge sample into a beaker.
- Dilute the sample gravimetrically and transfer to a blender, as described in Section 8.4.2.
- If necessary, adjust the dilution based on the specific sludge characteristics.
- Then blend for 30 sec on the highest speed.
- Transfer the content of the blender into a volumetric flask, rinse the blender twice with a known volume of distilled water and dilute to the mark with distilled water.
- Transfer the volumetric flask content into a plastic bottle and store in a dark cold room or refrigerator.
- Filter the samples through a 0.45 μm filter paper and measure the nitrite concentration.
- Total solids analysis should be performed on the samples so that the results of the nitrite measurement can be reported as gNO_2-N/gTS.

8.6.4.6.8 Analysis protocol

Calibration

Follow the spectrophotometer manufacturer's instructions for calibration, since calibration procedure differs between instruments. It may be necessary to calibrate the instrument before every reading, or it may only be necessary to perform periodic calibration checks to determine when calibration is necessary.

- Prepare a series of at least four different concentrations of a standard solution, making sure to include the lowest and highest concentration of the kit testing range. It is typical to do serial dilutions or dilutions with a uniform interval including the lowest and highest concentrations.
- Determine the nitrite concentration of the standard solutions.
- Multiply the answer by the dilution factor and report the results in $mgNO_2$-N/L.
- Prepare a calibration curve by plotting the instrument response against the standard concentrations.

Procedure

This method is for a measuring range of 0.02-1.0 mg/L NO_2-N

- Pipette 5ml of the sample into a test tube.
- Add 1 level blue micro-spoon of reagent NO_2-1 to the test tube.

- Shake vigorously until the reagent is completely dissolved. The pH of the solution must be between 2-2.5; check this with a pH test strip.
- Leave the solution to stand for 10 min, then fill the 10 mm cuvette with the sample.
- Wipe the cuvette with a soft tissue to remove water spots and fingerprints and then measure in a spectrophotometer.

8.6.4.6.9 Calculation

Liquid and slurry samples:

Result of analysis (mg/L)

$$\text{Final concentration}\left(\frac{mg}{L}\right) = A\left(\frac{mg}{L}\right) \times DF$$

$$\text{Dry basis}\left(\frac{g}{g}\right) = \frac{\text{Final concentration}\left(\frac{mg}{L}\right)}{\text{Total solids concentration}\left(\frac{mg}{L}\right)}$$

Where:

A = Spectrophotometric measurement value (mg/L)
DF = Dilution factor (F/I)
F = Final diluted volume (L)
I = Initial sample volume (L)

Slurry, semi-solid and solid samples:

$$\text{Dry basis}\left(\frac{g}{g}\right) = \frac{\frac{A\left(\frac{mg}{L}\right)}{1{,}000} \times \frac{V}{M}\left(\frac{L}{g}\right)}{\text{Total solids content}\left(\frac{g}{g}\right)}$$

Where:

A = Spectrophotometric measurement value (mg/L)
V = Volume of dilution (L)
M = Wet weight of sludge used in sample preparation (g)

8.6.4.6.10 Data set example

Septien *et al.* (2020) evaluated the effect of drying on the physical and chemical characteristics of faecal sludge for resource recovery. The nitrite concentration of faecal sludge from pit latrines before and after drying at different temperatures were measured using a Merck Spectroquant®,D. The sample was prepared by blending it with water, centrifugation, recovering the liquid fraction and using the commercial test kit. Nitrite in a solid sample dissolves and remains in a soluble state which can be measured. In this example, nitrite was measured in the faecal sludge sample which was further subjected to drying in infra-red and convective drying rigs. For duplicate samples, nitrite concentration of 0 g/kg dry solid was measured for untreated faecal sludge and dried faecal sludge samples after drying in a convective drying rig at 30 °C, 60 °C and 80 °C. Nitrite is usually oxidised to nitrate and thus concentrations are expected to be very low to zero in faecal sludge samples. However, nitrite concentrations in untreated faecal sludge for infra-red drying were 13 ± 0 g/kg dry solid. After drying, the concentration of nitrite decreased to 2.3 ± 0.1 g/kg dry solid. The decrease in nitrite and other nitrogen forms determined in this work did not translate into a decrease in total nitrogen, and thus drying of faecal sludge was concluded to induce changes in the chemical forms of nitrogen.

8.6.4.7 Nitrate – colorimetric (test strip) method[26]

8.6.4.7.1 Introduction

Nitrate is a major source of available nitrogen for plants and microorganisms. However, nitrate becomes toxic and affects public health when it contaminates surface or groundwater used as drinking water. Eutrophication and algal blooms of surface water are also environmental impacts of nitrate pollution. Nitrate management in faecal sludge treatment is thus essential to protect public and environmental health. This method describes the Merck MQuant® Nitrate Test[D] as one example of a commercially available nitrate colorimetric test strip method, for samples with concentrations of 10-500 mgNO_3^-/L. It is based on the manufacturer's protocol for water and wastewater. In this method, nitrate ions are reduced to nitrite ions in the presence of a reducing agent. In the presence of an acidic buffer, nitrite ions react with aromatic amine to form a diazonium salt, which in turn reacts with N-(1-naphthyl)-ethylenediamine to form a red-violet azo dye. Nitrate measurement results are expressed as

[26] Test strip methods should be cited as the specific manufacturer's method along with any modifications. The example used here is the Merck MQuant Nitrate Test (Merck, 2020g)[D].

either $mgNO_3^-/L$ or $mgNO_3$-N/L, and it is important to note how the results are expressed by the selected test method. The nitrate concentration is measured semi-quantitatively by visual comparison of the reaction zone of the test strip with the fields of a colour scale. Depending on the expected nitrate concentration in the sample, kits with the appropriate measurement range should be selected. For faecal sludge, samples must be diluted to prevent masking of the resultant colour change.

8.6.4.7.2 Safety precautions

- General health and safety (H&S) procedures specific for conducting laboratory analysis of faecal sludge are presented in Section 8.2. Before conducting this method, it is important to be familiar with Section 8.2.3 to ensure safety measures are properly carried out.
- Appropriate personal protective equipment (PPE) should be used; specific details are covered in Section 8.2.3.1.

8.6.4.7.3 Required chemicals

The required chemicals will be specific to each manufacturer's test kit; specific information on test kits can be found on the manufacturer's websites.

- Distilled water,
- Nitrate standard solution Certipur® ($1,000\ mgNO_3^-/L$),
- Amidosulfuric acid.

8.6.4.7.4 Required apparatus and instruments

- Test strips.
- Glass beakers (50 or 100 mL).
- Balance with weighing boats (slurry to solid samples).
- Blender (slurry to solid samples).
- Volumetric flask (250 mL).
- Filter paper (adequate for removing solids from sample, for example 0.45 µm filter for liquid samples).

8.6.4.7.5 Quality control

General information on quality assurance and quality control (QA/QC) is provided in Section 8.3. Information on standards, operating conditions and interferences that are specific to this method includes:

- Dilute the nitrate standard solution with distilled water to $250\ mgNO_3^-/L$, and analyse according to the analysis protocol.
- Perform quality control with nitrate standard solution with every test batch (or on a daily or weekly basis depending on the testing load).
- The test strips are stable up to the date stated on the pack when stored closed at +2 to +8 °C.
- The colour of the reaction zone may continue to change after the specified reaction time has elapsed. This must not be considered in the measurement, which should always be recorded at the stated time.
- Observe critically the second reaction zone or alert zone of the test strip, which changes colour in the presence of nitrite ions.
- Common interferences in faecal sludge include potassium, magnesium, and nitrite. For specific concentrations refer to the manufacturer's instructions.
- For faecal sludge, samples should be diluted with an appropriate dilution factor, based on the type of sludge, to prevent interference with colour of the test strips. Always use serial dilutions.
- The nitrate colorimetric test method is used for qualitative to semi-quantitative measurement. For a quantitative measurement, see Method 8.6.4.8.

8.6.4.7.6 Sample preservation

- Samples should be analysed immediately after sampling. For short-term preservation of 1 to 2 days, samples should be stored at 4 °C or frozen at -20 °C. Acid preservation for long-term storage is not encouraged because nitrate and nitrite in acid-preserved samples cannot be analysed as individual species (Rice *et al.*, 2017).
- Samples should be thawed to ambient temperature before analysis.

8.6.4.7.7 Sample preparation

For liquid, slurry, semi-solid or solid samples:

- Solid samples must be dissolved in a known volume of water before further dilution is performed.
- Filter the samples to prevent interference in colour.
- Samples containing more than $500\ mgNO_3^-/L$ must be diluted with distilled water.

For slurry to solid samples:

- Weigh out between 1.8 g and 2.0 g of thoroughly-mixed faecal sludge sample into a beaker.
- Dilute the sample gravimetrically and transfer to a blender, as described in Section 8.4.2.
- If necessary, adjust the dilution based on the specific sludge characteristics.
- Blend for 30 sec on the highest speed.
- Transfer the contents of the blender into a 1 L volumetric flask, rinse the blender twice with 250 mL distilled water and dilute to the mark with distilled water.
- Transfer this 1 L solution into a plastic bottle and store at 4 °C until ready for testing.
- Filter the samples through a 0.45 µm filter paper and measure the nitrate concentration according to the analysis protocol.
- Total solids analysis should be performed on samples so that the results of the nitrate measurement can be expressed as gNO_3-N/gTS.

8.6.4.7.8 Analysis protocol

For measuring range of 10-500 $mgNO_3/L$:

- Immerse the reaction zone of the test strip in the diluted, filtered sample (15-30 °C) for 1 second. Shake off the excess liquid from the strip and after 1 min determine which colour field on the label is closest to the colour of the reaction zone. Read off the corresponding result in $mgNO_3^-/L$ or $mgNO_3$-N/L.
- If the nitrite alert zone changes colour, eliminate nitrite interference by adding 5 drops of 10% aqueous amidosulfuric acid solution to 5 mL of each sample. Shake well and repeat the nitrate measurement.

8.6.4.7.9 Calculation

The dilution factor used must be stated.

Result of analysis (mg/L NO_3^-) = measurement value, A (mg/L) × dilution factor

Liquid samples:

Result of analysis (mg/L)

$$\text{Final concentration}\left(\frac{mg}{L}\right) = A\left(\frac{mg}{L}\right) \times DF$$

$$\text{Dry basis}\left(\frac{g}{g}\right) = \frac{\text{Final concentration}\left(\frac{mg}{L}\right)}{\text{Total solids concentration}\left(\frac{mg}{L}\right)}$$

Where:

A = Colorimetric measurement value (mg/L)
DF = Dilution factor (F/I)
F = Final diluted volume (L)
I = Initial sample volume (L)

Slurry, semi-solid and solid samples:

$$\text{Dry basis}\left(\frac{g}{g}\right) = \frac{\frac{A\left(\frac{mg}{L}\right)}{1{,}000} \times \frac{V}{M}\left(\frac{L}{g}\right)}{\text{Total solids content}\left(\frac{g}{g}\right)}$$

Where:

A = Colorimetric measurement value (mg/L)
V = Volume of dilution (L)
M = Wet weight of sludge used in sample preparation (g)

8.6.4.8 Nitrate – cadmium reduction spectrophotometric method[27]

8.6.4.8.1 Introduction

The principle of this method is that nitrate interacts with cadmium particles in aqueous solutions which converts all the nitrate in the sample into nitrite. The nitrite formed is determined by diazotising with sulfanilamide and coupling with N-(1-naphthyl)-ethylenediamine dihydrochloride to form a highly coloured azo dye that is measured spectrophotometrically. Nitrate measurements are easily influenced by interfering components such as dissolved organic matter, surfactants, nitrite and various inorganic compounds in water. The concentration ranges for the different methods used for nitrate measurement are also limited. For instance, the concentration ranges of the electrode method, the cadmium reduction spectrophotometric method and the automated cadmium reduction method are 0.14 to 1,400 mg NO_3-N/L, 0.01 to 1.0 mg NO_3-N/L and 0.001 to 10 mg NO_3-N/L, respectively. This means

[27] This method should be cited as: Method 4500-NO3-E (Rice *et al.*, 2017), and if test kits are used, also as per the manufacturer's directions including any modifications.

that higher concentrations should be diluted to the range of the selected method. Methods for nitrate determination are based on the oxidizing properties of nitrate. However, other oxidants present in water and wastewater may interfere making nitrate measurement difficult.

It is possible to conduct this method with commercially available test kits. The test kits are based on standard methods, with pre-packaged individual aliquots of necessary chemical in pillows (dry chemicals) and vials (liquid chemicals). Commonly used test kits from manufacturers such as Hach, Merck, and Hanna employ different methods in the kits used for nitrate measurement. The cadmium reduction method is used in Hach and Hanna test kits.

The example provided here is the Hach nitrate spectrophotometric test kit (Hach, 2020)[D] for samples with concentrations of 0.3-30 mgNO_3-N/L, which is based on the manufacturer's protocol for water and wastewater using the standard method 4500-NO_3-E (Rice *et al.*, 2017). For faecal sludge, samples must be diluted and filtered to prevent false high readings associated with turbid solutions.

8.6.4.8.2 Safety precautions

- General health and safety (H&S) procedures specific for conducting laboratory analysis of faecal sludge are presented in Section 8.2. Before conducting this method, it is important to be familiar with Section 8.2.3 to ensure safety measures are properly carried out.
- Appropriate personal protective equipment (PPE) should be used; specific details are covered in Section 8.2.3.1.

8.6.4.8.3 Required chemicals

- Distilled water (free from nitrogen)
- NitraVer 5 nitrate reagent powder pillow (supplied by the manufacturer)
- Nitrate nitrogen standard solution 10.0 mg/L (supplied by the manufacturer).

8.6.4.8.4 Required apparatus and instruments

- Spectrophotometer (*e.g.* DR6000)
- Analytical balance with weighing boats (slurry to solid samples)
- Blender (slurry to solid samples)
- Pipettes and pipette tips
- Sample cell
- Volumetric flask (1 L)
- Glass beakers (50 or 100 mL)
- Filter paper (adequate for removing solids from the sample, for example a 0.45µm filter for liquid samples).

8.6.4.8.5 Quality control

General information on quality assurance and quality control (QA/QC) is provided in Section 8.3. Information on standards, operating conditions and interferences that are specific to this method includes:

- Perform quality control with a nitrate standard solution with every test batch (or on a daily or weekly basis, depending on the testing load).
- Measure a blank sample for every test batch and subtract the blank values from the sample results.
- This method is technique-sensitive and thus shaking time and technique influence the colour development.
- Deposits of unoxidized metal will remain at the bottom of the sample cell after the reagent dissolves. The deposit will not affect the results.
- Common interferences in faecal sludge include chloride, ferric ion, and nitrite. For specific concentrations refer to the manufacturer's instructions.
- Suspended solids and turbid solutions interfere with measurement, thus faecal sludge samples should be diluted with an appropriate dilution factor and filtered based on the type of sludge for accurate measurements. Always use serial dilutions.

8.6.4.8.6 Sample preservation

- Samples should be analysed as soon as possible after sampling. For analysis within 48 hr of collection, sample must be filtered and refrigerated at 4 °C (Rice *et al.*, 2017).
- Samples must be thawed to room temperature before the analysis is performed.

8.6.4.8.7 Sample preparation

For liquid, slurry, semi-solid or solid samples:

- Turbid solutions falsely increase the spectrophotometric reading. Therefore, turbid samples should be filtered with a 0.45 μm filter paper to prevent false high measurements.
- Samples containing more than 30 $mgNO_3$-N/L must be diluted with distilled water.

For slurry to solid samples:

- Weigh out between 1.8 g and 2.0 g of thoroughly-mixed faecal sludge sample into a beaker.
- Dilute the sample gravimetrically and transfer to a blender, as described in Section 8.4.2.
- If necessary, adjust the dilution based on the specific sludge characteristics.
- Blend for 30 sec on the highest speed.
- Transfer the contents of the blender into a 1 L volumetric flask, rinse the blender twice with 250 mL distilled water and dilute to the mark with distilled water.
- Transfer this 1 L solution into a plastic bottle and store at 4 °C until ready for testing.
- Filter the samples through a 0.45 μm filter paper and measure the nitrate concentration according to the analysis protocol.
- Total solids analysis should be performed on samples so that the results of the nitrate measurement can be expressed as gNO_3-N/gTS.

8.6.4.8.8 Analysis protocol

Calibration

Follow the spectrophotometer manufacturer's instructions for calibration. It may be necessary to calibrate the instrument before every reading, or it may only be necessary to perform periodic calibration checks to determine when calibration is necessary. For this method, calibration can be performed as follows:

- Prepare a series of at least four different concentrations of a standard solution making sure to include the lowest and highest concentration of the kit testing range. It is typical to do serial dilutions or dilutions with a uniform interval including the lowest and highest concentrations.
- Determine the nitrate concentration of the standard solutions.
- Multiply the answer by the dilution factor and report the results in $mgNO_3$-N/L.
- Prepare a calibration curve by plotting the instrument response against the standard concentration.

Procedure

For measuring range of 0.3-30.0 $mgNO_3$-N/L:

- Pipette 10.0 mL of the sample into the sample cell.
- Add the contents of one NitraVer 5 nitrate reagent powder pillow and cover the sample cell with a stopper.
- Shake the sample cell vigorously for 1 min using a timer. It is important to note that not all the powder will dissolve. The undissolved powder will not affect the test results.
- Allow the solution to sit for 5 min making sure to use a set timer. An amber colour develops in the presence of nitrate.
- While waiting for the 5 min reaction time, prepare a blank solution.
- After the 5 min reaction time is over, fill a second cell with the sample solution.
- Wipe the sample cell with a laboratory tissue to clean it (*e.g.* to remove water spots and fingerprints) and measure in a spectrophotometer.
- Measure the nitrate concentration of the blanks and the sample.
- The sample reading should be taken within 1 min after the reaction time.

8.6.4.8.9 Calculation

Liquid and slurry samples:

Result of analysis (mg/L)

$$\text{Final concentration} \left(\frac{mg}{L}\right) = A \left(\frac{mg}{L}\right) \times DF$$

$$\text{Dry basis} \left(\frac{g}{g}\right) = \frac{\text{Final concentration} \left(\frac{mg}{L}\right)}{\text{Total solids concentration} \left(\frac{mg}{L}\right)}$$

Where:

A = Spectrophotometric measurement value (mg/L)

DF = Dilution factor (F/I)

F = Final diluted volume (L)

I = Initial sample volume (L)

Slurry, semi-solid and solid samples:

$$\text{Dry basis } \left(\frac{g}{g}\right) = \frac{\frac{A\left(\frac{mg}{L}\right)}{1{,}000} \times \frac{V}{M}\left(\frac{L}{g}\right)}{\text{Total solids content } \left(\frac{g}{g}\right)}$$

Where:

A = Spectrophotometric measurement value (mg/L)

V = Volume of dilution (L)

M = Wet weight of sludge used in sample preparation (g)

8.6.4.8.10 Data set example

Strande *et al.* (2018) determined the nitrate concentrations of faecal sludge from pit latrines and septic tanks located in households, non-households, and public toilets using the Hach nitrate test kits. As shown in Figure 8.8, 180 samples were analysed and the results showed an uneven distribution and a high variability. Hence, the median rather than mean of the results was used for the statistical analysis. Similar to other forms of nitrogen determined, nitrate concentrations in pit latrines from households, non-households and public toilets were all found to be higher compared to septic tanks. The complete raw data set is available at the link below[28].

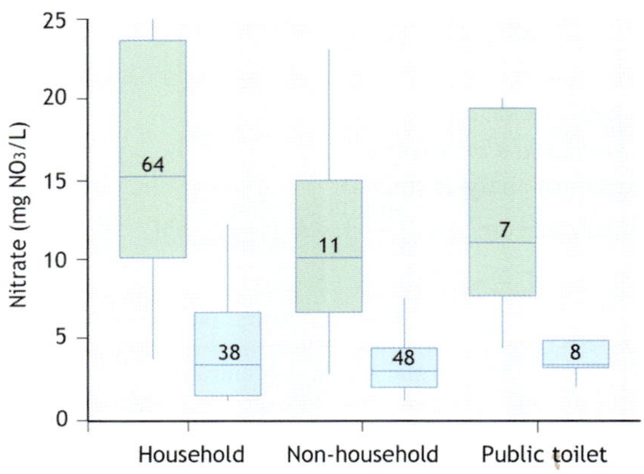

Figure 8.8 Nitrate concentrations in septic tanks (blue boxes) and pit latrines (green boxes) from households, non-households, and public toilets (Strande *et al.*, 2018).

8.6.4.9 Total Kjeldahl nitrogen – distillation and titration method[29]

8.6.4.9.1 Introduction

Kjeldahl nitrogen is defined as the total organic nitrogen together with the inorganic compounds ammonia and ammonium (NH_3/NH_4^+). In the presence of sulphuric acid, potassium sulphate, and cupric sulphate catalyst, amino nitrogen in the organic matter in the sample and free ammonia are converted into ammonium during digestion. Upon addition of a base, ammonium is converted into ammonia, and the ammonia is then distilled from an alkaline medium and absorbed in boric or sulphuric acid. The ammonia may be determined spectrophotometrically, by ammonia selective electrode or by titration with a standard mineral acid. The titration method is described here. The titrimetric and selective electrode methods of measuring ammonia in the distillate are suitable for determining a wide range of organic nitrogen concentrations.

Degradation:

$$\text{Organic N in sample} + H_2SO_4 \rightarrow (NH_4)_2SO_4(aq) + CO_2(g) + SO_2(g) + H_2O(g)$$

Liberation of ammonia:

$$(NH_4)_2SO_4(aq) + 2NaOH \rightarrow Na_2SO_4(aq) + 2H_2O(l) + 2NH_3(g)$$

Capture of ammonia:

$$B(OH)_3 + H_2O + NH_3 \rightarrow NH_4^+ + B(OH)_4^-$$

Back-titration:

$$B(OH)_3 + H_2O + Na_2CO_3 \rightarrow NaHCO_3(aq) + NaB(OH)_4(aq) + CO_2(g) + H_2O$$

All dilutions and reagents should be made with ammonia-free distilled water.

Various apparatus is available for Kjeldahl analysis, ranging from low-tech heating blocks for digestion to fully automatic Kjeldahl setups that do digestion and/or distillation and titration. This method might need to be adapted based on the type of apparatus used, and for automated systems the

---

[28] https://doi.org/10.25678/0000tt

[29] This method is adapted from Method 4500-Norg B and C of the Standard Methods for the Examination of Water and Wastewater and should be cited as: Rice *et al.* (2017), as described in Velkushanova *et al.* (2021).

manufacturer's instructions should be followed. In addition, the method might need to be adapted for different types of faecal sludge, based on local experience and the specific characteristics of faecal sludge. The total Kjeldahl nitrogen distillation and titration (TKN) method is widely accepted and has been tested for multiple substances; however, some steps in the process are affected by the composition of the sample and need to be adjusted accordingly (*e.g.* amounts of acid used, and digestion temperature and time). A useful guide with further information on considerations for Kjeldahl analysis can be found in 'A Guide to Kjeldahl Nitrogen Determination Methods and Apparatus' by Labconco Corporation (1998), and the Labconco website[30]. Methods should always be adapted according to rigorous quality control principles. For example, UKZN PRG in Durban uses a semi-automated digestion setup, and for samples with a high organic/fat content they mix samples overnight on the digestion apparatus prior to digestion.

8.6.4.9.2 Safety precautions

- General health and safety (H&S) procedures specific for conducting laboratory analysis of faecal sludge are presented in Section 8.2. Before conducting this method, it is important to be familiar with Section 8.2.3 to ensure safety measures are properly carried out.
- Appropriate personal protective equipment (PPE) should be used; specific details are covered in Section 8.2.3.1.
- Use eye and hand protection when preparing acid or handling colour reagents. Ensure when handling concentrated acid that an acid-proof lab coat and acid-proof gloves and goggles are used. For more detailed information on selecting the correct type of glove, consult the glove comparison chart provided by Berkeley Environment, Health & Safety[31].
- Prepare and keep reagents in a fume hood during use.
- Fumes are generated during the digestion step. Take care to not inhale these fumes! Ideally, this procedure is conducted in a fume hood.

8.6.4.9.3 Required chemicals

- Preparation of ammonia-free water
  - Distilled water
  - Concentrated sulphuric acid, or bromine or chlorine, depending on the method used; see Section 8.6.4.9.8.
- Digestion reagent
  - K_2SO_4
  - $CuSO_4$
  - Alternatively, commercial pre-mixed Kjeldahl tablets or powder, free of Hg and Se, are available and can be used instead of mixing digestion reagent manually.
- Boric acid – 4%
  - Boric acid
  - Ammonia-free distilled water
- Concentrated sulphuric acid – 98%
- Sulphuric acid – 0.1 N
  - Concentrated sulphuric acid
  - Ammonia-free distilled water
- Sodium hydroxide (NaOH) – 35%
  - NaOH
  - Ammonia-free distilled water
- Mixed indicator
  - Methyl red indicator
  - Bromocresol green indicator
  - Ethanol.

8.6.4.9.4 Apparatus and instruments

- Analytical balance
- (Kjeldahl) digestion apparatus (can be gas or electric, should be able to reach temperatures between 375-385 °C), or semi-automated Kjeldahl device
- Kjeldahl flasks or digestion tubes that fit the digestion apparatus used
- Boiling stones
- 100 mL volumetric flask
- 1 L volumetric flasks
- 250 mL Erlenmeyer flasks
- Plastic bottle
- Distillation setup
- Titration setup with burette in metal clamp (or automatic titration unit).

[30] https://www.labconco.com/articles/a-brief-introduction-to-kjeldahl-nitrogen-determ

[31] https://ehs.berkeley.edu/workplace-safety/glove-selection-guide#comparison

8.6.4.9.5 Quality control

General information on quality assurance and quality control (QA/QC) is provided in Section 8.3. Information on standards, operating conditions and interferences that are specific to this method includes:

- Check the accuracy of the measurement procedure by using a standard solution of known nitrogen concentration. This calibration should be done with every set of samples. Dilute and reanalyse any samples exceeding the highest standard on the calibration curve.
- Prepare and analyse blanks with every set of samples.
- Do not use $HgCl_2$ for sample preservation because it interferes with the ammonia removal liberation process.
- Avoid nitrate: during Kjeldahl digestion, nitrate in excess of 10 mg/L can oxidise a portion of the ammonia released from the digested organic nitrogen, producing N_2O and resulting in a negative interference. According to USEPA (2001), no known method exists to prevent this interference, but its effect can be predicted on the basis of preliminary nitrate determination of the sample.
- Control the system temperature, the inorganic salts and solids: the acid and salt content of the Kjeldahl digestion reagent is intended to produce a digestion temperature of approximately 380 °C.
- If the sample contains a very large quantity of salts or inorganic solids, the temperature may rise to 400 °C during digestion at which point pyrolytic loss of nitrogen occurs. To prevent this, add controlled amounts (*e.g.* 1 mL H_2SO_4/g salt) of sulphuric acid to maintain an acid-salt balance.
- Samples containing chloride should be dechlorinated prior to analysis.

8.6.4.9.6 Sample preservation

- The most reliable results are obtained from fresh samples. If immediate analysis is not possible, preserve the samples for Kjeldahl digestion by acidifying between pH 1.5 and 2.0 with concentrated sulphuric acid and storing in a refrigerator at 4 °C. Let the samples return to room temperature before starting the analysis. If acid preservation was used, neutralise the samples to pH 7 with NaOH or KOH immediately before starting the analysis.
- Storage of reagents:
  - Mixed indicator solution should be freshly prepared every month.
  - Sulphuric acid titrant should not be stored longer than one week.

8.6.4.9.7 Sample preparation

For liquid faecal sludge:

- Filter turbid samples using a 0.45 μm pore size filter.

For slurry to solid faecal sludge:

- Weigh out between 1.8 g and 2.0 g of thoroughly-mixed faecal sludge sample into a beaker.
- Dilute the sample gravimetrically and transfer to a blender, as described in Section 8.4.2.
- If necessary, adjust the dilution based on the specific sludge characteristics.
- Blend for 30 sec on the highest speed.
- Transfer the contents of the blender into a 1 L volumetric flask, rinse the blender twice with 250 mL distilled water and dilute to the mark with distilled water.
- Transfer this 1 L solution into a plastic bottle and store at 4 °C until ready for testing.
- Filter the samples through a 0.45 μm filter paper and measure the TKN concentration according to the analysis protocol.
- Total solids analysis should be performed on samples so that the results of the TKN measurement can be expressed as gTKN/gTS.

8.6.4.9.8 Analysis protocol

Preparation of reagents, chemicals and standard solutions

Ammonia-free water

- Eliminate traces of ammonia in the distilled water by adding 0.1 mL concentrated sulphuric acid to 1 L distilled water and redistill. Alternatively, treat the distilled water with enough bromine or chlorine to produce a free halogen residual of 2-5 mg/L and redistill after standing for 1 hr. Discard the first 100 mL of distillate, and check the ammonia concentration in the collected water before use.

Boric acid - 4%

- Dissolve 40g of boric acid in the distilled water in a 1L volumetric flask, and dilute up to the mark.

Sulphuric acid - 0.1 N

- Dilute 2.71 mL 98% concentrated sulphuric acid in the distilled water in a 1 L volumetric flask, and dilute up to the mark.

Mixed indicator

- Weigh 0.02 g methyl red and 0.10 g bromocresol green indicator in a 100 mL volumetric flask and fill to the mark with ethanol.

Calibration

- Prepare a 80 mg N/L stock solution by dissolving 0.4285 g glycine in 1 L ammonia-free distilled water. The stock solution is diluted to make up seven-point standard solutions covering the required range of nitrogen concentrations, including a blank.
- Test the standard solutions following the TKN procedure.
- Prepare a calibration curve by plotting the instrument response against the standard concentration and calculating the linear regression line. Compute the sample concentration by adjusting the sample response with the offset of the standard curve. Report only those values that fall between the lowest and the highest calibration standards. Dilute and reanalyse those samples exceeding the highest standard. Report the results in mg/L.

Procedure

- Include standards and a blank with every analysis batch.

Digestion

- Transfer 50 mL of the homogenised diluted sample into a 300 mL Kjeldahl flask or digestion tube (depending on the digestion apparatus used), and add 5 boiling stones.
- Use either 1 Kjeldahl tablet or mix 3.5 g K_2SO_4 and 0.5 g $CuSO_4$ together, and add to the flask. Then slowly add 10 mL of concentrated sulphuric acid and swirl to dissolve.
- Digest the samples at 380 °C for 1 hr under a fume hood. Fumes will appear above the liquid.
- Wait for the samples to cool to ambient temperature.

Distillation

- Add 50 mL distilled water and 50 mL NaOH to the flasks after digestion. The mixture should be above pH 11 before distillation.
- Prepare the absorption solution by placing 25 mL of 4% boric acid in a 250 mL Erlenmeyer flask, then insert under the condenser outlet with the tip below the surface of the boric acid.
- Distil the sample into the boric acid.

Titration

- Fill the burette with 0.1 N sulphuric acid and note the starting volume.
- Add 3 drops of mixed methyl red-bromocresol green indicator to the distillate and titrate the distillate with sulphuric acid until the colour changes from blue to pale pink.
- Read off the final volume on the burette and note down.

8.6.4.9.9 Calculation

Liquid and slurry samples:

$$\text{TKN (mg/L)} = \frac{(A - B) \times 0.1 \times 14 \times 1{,}000}{\text{Sample volume (mL)}} \times DF$$

Where:

A = Volume H_2SO_4 titrated in sample (mL)
B = Volume H_2SO_4 titrated for blank (mL)
0.1 = Normality of sulphuric acid used in titration
14 = Atomic weight of nitrogen
1,000 = Conversion from g to mg
DF = Dilution factor (F/I)
F = Final diluted volume (L)
I = Initial sample volume (L)

TKN on dry basis (g/g) =

$$\frac{\text{TKN (mg/L)}}{\text{Total solids concentration (mg/L)}}$$

Slurry, semi-solid and solid samples:

TKN on dry basis (g/g) =

$$\frac{(A - B) \times 0.1 \times 14 \times 1{,}000}{\text{Sample volume (mL)}} \times \frac{V}{M} \times \frac{1}{1{,}000} \times$$

$$\frac{1}{\text{Total solids content (g/g)}}$$

Where:

V = Volume of dilution (L)

M = Wet weight of sludge used in sample preparation (g)

8.6.4.9.10 Data set example

Faecal sludge samples were collected from a ventilated improved pit (VIP) latrine and from a urine diversion toilet (UDDT) in Durban, South Africa, and were analysed using the total Kjeldahl nitrogen distillation and titration method. Each sample was analysed in four replicates and reported gravimetrically. The average TKN values (g/g dry sample) were 0.02 and 0.03 for the VIP and UDDT, respectively. The results for TKN are presented in Table 8.11 (source: unpublished data UKZN PRG).

Table 8.11 TKN values (g/g dry sample) for a VIP and a UDDT toilet analysed by the titrimetric method.

	TKN (g/g dry sample)	
Replicate	VIP	UDDT
1	0.0212	0.0298
2	0.0204	0.0275
3	0.0188	0.0289
4	0.0200	0.0292
Average	0.0201	0.0289
SD	0.0010	0.0010

8.6.5 Phosphorus

Phosphorus is important for monitoring the nutrient availability in both faecal sludge treatment and resource recovery processes. It is an essential nutrient for the growth of plants and organisms, and also a potential environmental pollutant (Tchobanoglous *et al.*, 2003). Phosphorus is present in faecal sludge as orthophosphates, polyphosphates, and organically bound phosphates (Strande *et al.*, 2014).

There are three ways of determining phosphorus in wastewater or faecal sludge: orthophosphate, acid hydrolysable phosphate, and total phosphorus, which are used to analyse orthophosphate, polyphosphates (condensed phosphate), and organic phosphate. Orthophosphate is also called reactive or available phosphorus, as it is soluble, readily interacts with other positive elements or compounds, and is bioavailable. It is the only phosphorus form that can be determined directly without preliminary hydrolysis or oxidative digestion of the sample. To measure the concentration of polyphosphates, acid hydrolysis of samples at boiling water temperature is used to convert the dissolved and particulate phosphates into dissolved orthophosphate which is then quantified. Organic phosphates can be determined by an oxidative destruction of the organic matter in the sample, which converts organic phosphates into orthophosphates, which are then quantified. Total phosphorus concentrations will be greater than the orthophosphate concentration. In this book, a spectrophotometric method for determining total phosphorus and orthophosphate concentrations is described.

8.6.5.1 Total phosphorus and orthophosphate - spectrophotometric method[32]

8.6.5.1.1 Introduction

The determination of total phosphorus and orthophosphate is based on the principle that in sulphuric acid solution, orthophosphate ions react with molybdate ions to form phosphomolybdic acid. This molybdate compound is reduced by ascorbic acid to form an intense phosphomolybdenum blue (PMB) that is then quantified with a spectrophotometer. To measure total phosphorus, other phosphorus forms are initially converted into orthophosphate before measurement.

Commercial test kits based on standard methods for measuring total phosphorus and orthophosphates are available, with pre-packaged individual aliquots of any necessary chemicals in pillows (dry chemicals) and vials (liquid chemicals). Commonly used total phosphorus and orthophosphate test kits from manufacturers such as Hach and Merck employ a variety of methods. For example, the Merck test kit uses the ascorbic acid method. The example provided here is the Merck phosphate spectrophotometric test[D]

[32] This method should be cited as: Method 4500-P E (Rice *et al.*, 2017), and if test kits are used, also as per the manufacturer's directions including any modifications.

which is used for the determination of total phosphorus and orthophosphate for samples with concentrations of 0.05-5 mg PO_4-P/L, and it is based on the manufacturer's protocol for water and wastewater using the standard method 4500-P E (Rice *et al.*, 2017). Phosphate measurement results are expressed as either mg PO_4^{3-}/L or mg PO_4-P/L (or gTP/gTS or gPO_4-P/gTS with TS analysis) and it is important to note how the results are expressed by the selected test method. For faecal sludge, samples must be diluted (for total phosphorus) and filtered (for orthophosphate) to prevent false high readings associated with turbid solutions.

8.6.5.1.2 Safety precautions

- General health and safety (H&S) procedures specific for conducting laboratory analysis of faecal sludge are presented in Section 8.2. Before conducting this method, it is important to be familiar with Section 8.2.3 to ensure safety measures are properly carried out.
- Appropriate personal protective equipment (PPE) should be used; specific details are covered in Section 8.2.3.1.

8.6.5.1.3 Required chemicals

- Deionised or distilled water
- Total phosphorus standard solution (4 mg/L PO_4^-P)
- Reagent P-1K (supplied by the manufacturer)
- Reagent P-2K (supplied by the manufacturer)
- Reagent P-3K (supplied by the manufacturer).

8.6.5.1.4 Required apparatus and instruments

- Spectrophotometer (*e.g.* Merck, Hatch, and Hanna)
- Digester or heating block (capable of heating to 150 °C)
- Analytical balance with weighing boats (slurry to solid samples)
- Blender (slurry to solid samples)
- Pipettes and pipette tips
- Reaction cells with reagent (supplied by the manufacturer)
- Volumetric flask (1 L)
- Glass beakers (50 or 100 mL)
- Test tube rack
- Glass storage bottle
- Laboratory cleaning tissues
- Filter paper (adequate for removing solids from the sample, for example, a 0.45µm filter for liquid samples).

8.6.5.1.5 Quality control

General information on quality assurance and quality control (QA/QC) is provided in Section 8.3. Information on standards, operating conditions and interferences that are specific to this method includes:

- Perform quality control with the total phosphorus standard solution with every test batch (or on a daily or weekly basis, depending on the testing load).
- Measure a blank sample for every test batch and subtract the blank values from the sample results.
- For orthophosphate analysis, samples must be pre-treated by filtration with a 0.45 µm filter to remove most of the turbidity which interferes with spectrophotometric measurement.
- Filters may contribute a significant amount of phosphate to samples with low phosphate concentration. Thus, filters should be washed with distilled water before use.
- For faecal sludge samples, the filtration step removes struvite (magnesium ammonium phosphate) that has potentially precipitated during storage due to the reactions of urine and phosphorus in the faecal sludge.
- For total phosphorus, the sample should be diluted without filtration.
- Acid-washed glassware (cleaned with HCL) should be used for determining low concentrations of orthophosphates. Avoid using detergents containing phosphate.
- Phosphate can adsorb on glass surfaces, so glassware needs to be carefully cleaned to prevent contamination.
- Common interferences in faecal sludge for this method include ammonium, nitrites, sodium, and chemical oxygen demand. For specific concentrations of concern, refer to the manufacturer's instructions.
- Suspended solids and turbid solutions interfere with the spectrophotometric measurement; therefore, faecal sludge samples should be diluted with an appropriate dilution factor based on the

type of sludge for accurate measurements. Always use serial dilutions.

- The colour remains stable for at least 60 min following the end of the reaction time, which takes 5 min.

8.6.5.1.6 Sample preservation

- Samples must be filtered immediately after collection to prevent hydrolysis of polyphosphates. For short-term preservation of 1 to 2 days, samples should be stored at 4 °C or frozen at -20 °C. For longer storage of up to 28 days, 40 mg$HgCL_2$/L may be added. However, $HgCL_2$ is a hazardous substance and the appropriate H&S precautions must be observed.
- Samples with low phosphorus concentration must be stored in glass bottles and not plastic bottles unless they are kept frozen. Phosphates may be adsorbed onto the walls of both glassware and plastic bottles. However, washing glassware with acids prevents adsorption and thus it can be used for storage of samples. For plastic bottles, adsorption can still occur during storage under refrigerated conditions unless samples are kept frozen (Rice *et al.*, 2017).

8.6.5.1.7 Sample preparation

For liquid, slurry, semi-solid or solid samples:

- Turbid solutions falsely increase the spectrophotometric reading. Therefore, turbid samples should be diluted and/or filtered with a 0.45µm filter paper to prevent false high measurements.
- For orthophosphate analysis, the samples should be filtered and diluted where necessary using serial dilutions.
- For total phosphorus, the samples should not be filtered and appropriate dilutions should be used with serial dilutions.
- Samples containing more than 5 mgPO_4-P/L must be diluted with distilled water.

For slurry, semi-solid or solid samples:

- Weigh out between 1.8 g and 2.0 g of the thoroughly-mixed faecal sludge sample into a beaker.
- Dilute the sample gravimetrically and transfer to the blender, as described in Section 8.4.2.
- If necessary, adjust the dilution based on the specific sludge characteristics.
- Blend for 30 sec on the highest speed.
- Transfer the contents of the blender into a 1 L volumetric flask, rinse the blender twice with 250 mL distilled water and dilute to the mark with distilled water.
- Transfer this 1 L solution into a plastic bottle and store at 4 °C until ready for testing.
- Dilute with an appropriate dilution factor to achieve a concentration between 0.05 and 5.0 mg/L total phosphorus or orthophosphate and filter the samples through a 0.45µm filter paper for orthophosphate measurement.
- Total solids analysis should be performed on the samples so that the results of the total phosphorus or orthophosphate measurement can be expressed as gTP/gTS or gPO_4-P/gTS.

8.6.5.1.8 Analysis protocol

Calibration

Follow the spectrophotometer manufacturer's instructions for calibration. It may be necessary to calibrate the instrument before every reading, or it may only be necessary to perform periodic calibration checks to determine when calibration is necessary. For this method, calibration can be performed as follows:

- Prepare a series of at least four different concentrations of a standard solution, making sure to include the lowest and highest concentration of the kit testing range. This can be done with serial dilutions, or dilutions with a uniform interval throughout the range.
- Determine the total phosphorus concentration of the standard solutions.
- Multiply the spectrophotometer readings by the appropriate dilution factor and report results in mg TP/L.
- Prepare a calibration curve by plotting the instrument response against the concentration of the standards.

Procedure

- Orthophosphate measurement

For measuring range of 0.05-5.0 mg/L PO_4-P:

  - Pipette 5 mL of the sample into a test tube.
  - Add 5 drops of reagent P-2K and mix.
  - Add 1 dose of reagent P-3K and shake vigorously until the reagent is completely dissolved.
  - Leave to stand for 5 min (reaction time), then pour the sample into the cell (fill it), and measure in the spectrophotometer.

- Total P measurement

For measuring range of 0.05-5.0 mg/L PO4-P:

  - Pipette 5 mL of the sample into a reaction cell.
  - Add 1 dose of reagent P-1K, close the cell tightly, and mix.
  - Heat the cell at 120 °C in a preheated heating block for 30 min.
  - Allow the closed cell to cool to room temperature in a test tube rack.
  - Do not cool with cold water as this action will result in cracking of the glass, loss of the sample, and risk of acid spills.
  - Shake the tightly closed cell vigorously after cooling.
  - Add 5 drops of reagent P-2K, close the cell tightly, and mix.
  - Add 1 dose of reagent P-3K, close the cell tightly, and shake vigorously until the reagent is completely dissolved.
  - Leave to stand for 5 min (reaction time), and then wipe the cell with a laboratory tissue to remove water spots and fingerprints and measure in a spectrophotometer.

8.6.5.1.9 Calculation

Liquid and slurry samples:

Result of analysis (mg/L)

$$\text{Final concentration}\left(\frac{mg}{L}\right) = A\left(\frac{mg}{L}\right) \times DF$$

$$\text{Dry basis}\left(\frac{g}{g}\right) = \frac{\text{Final concentration}\left(\frac{mg}{L}\right)}{\text{Total solids concentration}\left(\frac{mg}{L}\right)}$$

Where:

A = Spectrophotometric measurement value (mg/L)

DF = Dilution factor (F/I)

F = Final diluted volume (L)

I = Initial sample volume (L)

Slurry, semi-solid and solid samples:

$$\text{Dry basis}\left(\frac{g}{g}\right) = \frac{\frac{A\left(\frac{mg}{L}\right)}{1{,}000} \times \frac{V}{M}\left(\frac{L}{g}\right)}{\text{Total solids content}\left(\frac{g}{g}\right)}$$

Where:

A = Spectrophotometric measurement value (mg/L)

V = Volume of dilution (L)

M = Wet weight of sludge used in sample preparation (g)

8.6.5.1.10 Data set example

Phosphate was measured by Septien *et al.* (2020) while evaluating the effect of drying on the physical and chemical characteristics of faecal sludge for resource recovery. Phosphate concentrations of faecal sludge from pit latrines were measured before and after drying under convective and infra-red drying rigs using a Merck Spectroquant. Different concentrations of phosphate were determined for untreated faecal sludge samples. Before drying in the convective drying rig, average phosphate in untreated faecal sludge was 2.4 ± 0.7 g PO_4^{3-}/g dry solid. The untreated faecal sludge used for infra-red drying was 13 ± 0 gPO_4^{3-}/g dry solid. Drying under the convective drying rig did not affect the phosphate concentrations because 2.3 ± 0.9, 2.5 ± 0.6 and 2.1 ± 0.7 g PO_4^{3-}/g dry solid were measured after drying at 40 °C, 60 °C and 80 °C, respectively. For infra-red drying, phosphate concentrations were slightly reduced to 11 ± 0 g PO_4^{3-}/g dry solid after drying at 80% medium infra-red for 9 min. It was concluded from the study that drying does not affect the phosphorus concentration because most phosphorus forms are bound to the solids in faecal sludge. This further suggests that dried faecal sludge pellets present an attractive nutrient composition for agricultural applications.

Phosphate was measured with the Hach test kit for 60 samples in Englund *et al.* (2020) and phosphate and total phosphorus for 180 samples in Strande *et al.* (2018) and reported as mg/L. The range of observed PO_4 in septic tanks in Hanoi, Vietnam was 3.5 to 33 mg PO_4-P/L. The range of PO_4 was 4-367 mgPO_4-P/L and TP was 6-2,040 mg TP/L for a range of household and non-household sources in pit latrines and septic tanks in Kampala, Uganda as shown in Figure 8.9. Both complete raw data sets can be downloaded using the link below[33].

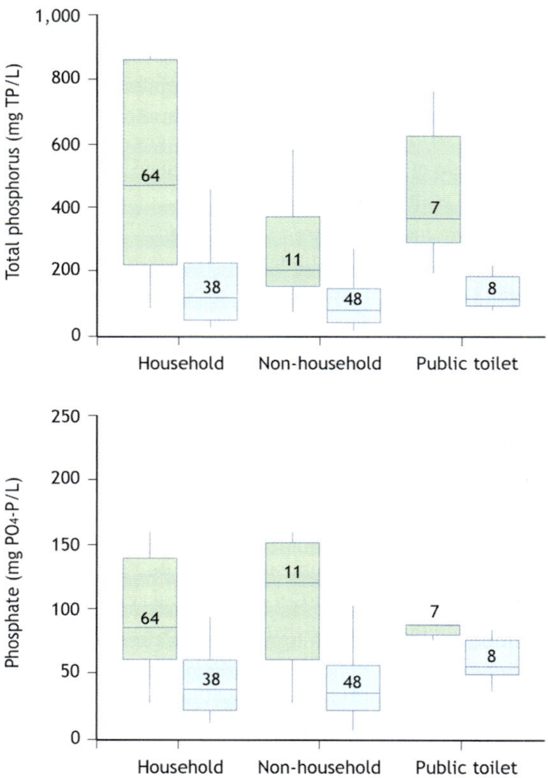

Figure 8.9 Range of total phosphorus (above) and phosphate (bellow) concentrations in faecal sludge samples from Kampala, Uganda. 180 samples consisting of pit latrines (green boxes) and septic tanks (blue boxes) (*Strande et al., 2018*).

8.6.5.2 Orthophosphate - colorimetric (test strip) method[34]

8.6.5.2.1 Introduction

Similar to nitrates, phosphorus is a major nutrient that contributes to eutrophication and it is present in aqueous solutions mainly as orthophosphate, polyphosphate or organically-bound phosphate. Concentrations of phosphorus in faecal sludge are usually 2-5 times higher than in wastewater and thus orthophosphate determination is crucial in faecal sludge treatment. The colorimetric method described here is one of many orthophosphate colorimetric test methods for samples with concentrations of 10-500 mg/L PO_4^{3-}. It is based on the Merck MQuant Phosphate Test[D] protocol for water and wastewater. In this method, orthophosphate ions (PO_4^{3-}) react with molybdate ions in the presence of a sulphuric solution to form molybdophosphoric acid, which is reduced to phosphomolybdenum blue (PMB). The phosphate concentration is measured semi-quantitatively by visual comparison of the reaction zone of the test strip with the fields of a colour scale. Phosphate measurement results are expressed as either mgPO_4^{3-}/L or mgPO_4-P/L or mgP_2O_5/L and it is important to note how the results are expressed by the selected test method. Depending on the expected phosphate concentration in the sample, kits with the appropriate measurement range should be selected. For faecal sludge, samples must be diluted and filtered to prevent masking of the resultant colour change.

8.6.5.2.2 Safety precautions

- General health and safety (H&S) procedures specific for conducting laboratory analysis of faecal sludge are presented in Section 8.2. Before conducting this method, it is important to be familiar with Section 8.2.3 to ensure safety measures are properly carried out.
- Appropriate personal protective equipment (PPE) should be used; specific details are covered in Section 8.2.3.1.

[33] https://doi.org/10.25678/0000tt

[34] Test strip methods should be cited as the specific manufacturer's method along with any modifications. The example used here is the Merck MQuant Phosphate Test (Merck, 2020i)[D].

8.6.5.2.3 Reagents, chemicals and standard solutions

The required chemicals will be specific to each manufacturer's test kit; specific information on test kits can be found on the manufacturers' websites.

- Distilled water.
- Reagent PO_4^{3-}-1.
- Standard phosphate solution Certipur® (1,000 mg/L PO_4^{3-}).

8.6.5.2.4 Apparatus and instruments

- Test strips.
- Absorbent towel.
- Glass beakers (50 or 100 mL).
- Balance with weighing boats (slurry to solid samples).
- Blender (slurry to solid samples).
- Volumetric flask (250 mL).
- Filter paper (adequate for removing solids from sample, for example 0.45 µm filter for liquid samples).

8.6.5.2.1 Quality control

General information on quality assurance and quality control (QA/QC) is provided in Section 8.3. Information on standards, operating conditions and interferences that are specific to this method includes:

- Dilute the phosphate standard solution with distilled water to 100 mg/L PO_4^{3-}, and analyse according to the analytical protocol.
- Perform quality control with phosphate standard solution with every test batch (or on a daily or weekly basis depending on the testing load).
- The test strips are stable up to the date stated on the pack when stored closed at +15 to +25 °C.
- The colour of the reaction zone may continue to change after the specified reaction time has elapsed. This must not be considered in the measurement, which should always be recorded at the stated time.
- Common interferences in faecal sludge include nitrite and nitrate. For specific concentrations refer to the manufacturer's instructions.
- For faecal sludge, samples should be diluted with an appropriate dilution factor, based on the type of sludge, to prevent interference with the colour of the test strips. Always use serial dilutions.
- The orthophosphate colorimetric test method is used for qualitative to semi-quantitative measurement. For a quantitative measurement, refer to Method 8.7.3.10 (the total phosphorus and orthophosphate spectrophotometric method).

8.6.5.2.6 Sample preservation

- Samples must be filtered immediately after collection to prevent hydrolysis of polyphosphates. For short-term preservation of 1 to 2 days, samples should be stored at 4 °C or frozen at -20 °C. For longer storage up to 7 days, 40 mg$HgCL_2$ /L may be added. However, $HgCL_2$ is a hazardous substance and the appropriate H&S precautions must be observed.
- Samples with low phosphorus concentration must be stored in glass bottles and not plastic bottles unless they are kept frozen. Phosphates may be adsorbed onto the walls of the plastic bottles if used for storage under refrigerated conditions (Rice *et al*, 2017).
- Samples should be thawed to ambient temperature before analysis.

8.6.5.2.7 Sample preparation

For liquid, slurry, semi-solid or solid samples:

- Solid samples must be dissolved in a known volume of water before further dilution is performed.
- Filter the samples through a 0.45 µm filter paper to prevent interference in the colour.
- Samples containing more than 500 mgPO_4^{3-}/L must be diluted with distilled water.

For slurry, semi-solid or solid samples:

- Weigh out between 1.8 g and 2.0 g of the thoroughly-mixed faecal sludge sample into a beaker.
- Dilute the sample gravimetrically and transfer to the blender, as described in Section 8.4.2.
- If necessary, adjust the dilution based on the specific sludge characteristics.
- Blend for 30 sec on the highest speed.
- Transfer the contents of the blender into a 1 L volumetric flask, rinse the blender twice with 250 mL distilled water and dilute to the mark with distilled water.

- Transfer this 1 L solution into a plastic bottle and store at 4 °C until ready for testing.
- Filter the samples through a 0.45 μm filter paper and measure the orthophosphate concentration according to the analytical protocol.
- Filter the samples through a 0.45 μm filter paper and measure the PO_4^{3-} concentration.
- Total solids analysis should be performed on the samples so that the results of the total phosphorus or orthophosphate measurement can be expressed as g PO_4^{3-}/gTS or gPO_4-P/gTS.

8.6.5.2.8 Analysis protocol

This method is valid for a measuring range of 10-500 mgPO_4^{3-}/L.

- Immerse the reaction zone of the test strip in the diluted, filtered sample (15-30 °C) for 1 sec. Allow the excess liquid to run off via the long edge of the strip onto an absorbent paper towel.
- Add 1 drop of reagent PO_4^{3-}-1 and place on the reaction zone and allow to react for 15 sec.
- Allow the excess liquid to run off via the long edge of the strip onto an absorbent paper towel and after 1 min determine with which colour field on the label matches the colour of the reaction zone most closely. Read off the corresponding result in mgPO_4^{3-}/L or mgPO_4-P/L.

8.6.5.2.9 Calculation

Liquid samples:

Result of analysis (mg/L)

$$\text{Final concentration}\left(\frac{mg}{L}\right) = A\left(\frac{mg}{L}\right) \times DF$$

$$\text{Dry basis}\left(\frac{g}{g}\right) = \frac{\text{Final concentration}\left(\frac{mg}{L}\right)}{\text{Total solids concentration}\left(\frac{mg}{L}\right)}$$

Where:

A = Colorimetric measurement value (mg/L)

DF = Dilution factor (F/I)

F = Final diluted volume (L)

I = Initial sample volume (L)

Slurry, semi-solid and solid samples:

$$\text{Dry basis}\left(\frac{g}{g}\right) = \frac{\frac{A\left(\frac{mg}{L}\right)}{1{,}000} \times \frac{V}{M}\left(\frac{L}{g}\right)}{\text{Total solids content}\left(\frac{g}{g}\right)}$$

Where:

A = Colorimetric measurement value (mg/L)

V = Volume of dilution (L)

M = Wet weight of sludge used in sample preparation (g)

8.6.6 pH and electrical conductivity

pH, electrical conductivity (EC), and alkalinity are measurements of solution properties in faecal sludge samples. pH is a measure of how acidic or basic an aqueous solution is. The pH scale is logarithmic, and is inversely related to the concentration of hydrogen ions in the solution. A pH of 7 is neutral, an acidic solution has a lower pH (< 7) and a high concentration of hydrogen ions, while a basic solution has a high pH (>7) and a low concentration of hydrogen ions. Biological treatment processes and conditioning for improved dewatering commonly require a pH between 6 and 9, and therefore pH is a standard design and operational parameter for faecal sludge treatment. pH values outside of this range may indicate contamination by other solid or liquid wastes, and can contribute to issues with biological processes or problems with corrosion. EC is a measure of dissolved salts, and can be used as a proxy for salinity. The concentration of dissolved salts influences the flocculation and dewatering properties of faecal sludge, and high ECs may inhibit biological treatment processes. EC is also an important parameter in resource recovery from faecal sludge as irrigation or soil conditioner, as high concentrations of dissolved salt may be harmful to crops and contribute to the accumulation of salinity in soils. Measurements of pH and EC are temperature-dependent, and thus temperature should always be recorded together with pH and EC. Alkalinity is also an important solution property of faecal sludge, and is important for biological reactions as it is a measure of the capacity of the solution to resist acidification. A method for measuring alkalinity in faecal sludge is not included in this section, but a thorough explanation of how to measure alkalinity can be found in Method 2320 of the Standard Methods for the Examination of Water and Wastewater (Rice *et al.*, 2017).

8.6.6.1 pH – electrode method[35]

8.6.6.1.1 Introduction

This method is an electrometric measurement procedure to determine the pH of faecal sludge samples with different TS contents. This method is preferred over the colorimetric test strip method for faecal sludge due to its increased accuracy and precision, and the fact that pH readings obtained with this method are not influenced by the colour of the sample and can be made at higher TS contents than the pH test strip method. In this method, different sample preparation and measurement steps may be necessary depending on the TS of the sample. If the electrode does not produce a stable reading when inserted into a sample, the sample needs to be diluted before measuring the pH. For most types of faecal sludge (liquid, slurry, semi-solid, and some solid samples), dilution will not be necessary. However, for some solid samples or samples that have been dewatered or dried, dilution will be required.

8.6.6.1.2 Safety precautions

- General health and safety (H&S) procedures specific for conducting laboratory analysis of faecal sludge are presented in Section 8.2. Before conducting this method, it is important to be familiar with Section 8.2.3 to ensure safety measures are properly carried out.
- Appropriate personal protective equipment (PPE) should be used; specific details are covered in Section 8.2.3.1.

8.6.6.1.3 Required chemicals

- Distilled water
- pH standard buffer solutions.

8.6.6.1.4 Required apparatus and instruments

- pH meter including a potentiometer
- Glass electrode and reference electrode with a temperature sensor and compensation (should be accurate to 0.1 pH unit with pH range of 0 -14)
- Beakers
- Magnetic stirrer with stir bar
- Analytical balance capable of weighing 0.1 g (if pre-dilution of the sample is required).

8.6.6.1.5 Quality control

General information on quality assurance and quality control (QA/QC) is provided in Section 8.3. Information on standards, operating conditions and interferences that are specific to this method includes:

- The pH meter and electrode must be calibrated with minimum two points that encompass the expected pH of the samples and are at least 3 pH units apart (USEPA 2004a). For faecal sludge samples, it is recommended to calibrate with three buffer solutions at pH 4, 7, and 10.
- Only fresh buffer solutions should be used. Changing all the solutions daily is good practice.
- If the measured pH of a sample is outside of the calibration range (*e.g.* below 4 or above 10), a calibration should be performed with an additional buffer to extend the calibration range, and the sample should be measured again.
- USEPA recommends repeat measurements on successive aliquots of sample (duplicate or triplicate measurements). Replicates should differ by no more than 0.1 pH units (USEPA 2004a).
- Typically, pH values should be reported to the nearest 0.1 pH unit (Rice *et al.*, 2017).
- Samples should always be measured at room temperature, and the actual temperature should always be measured and recorded together with pH.
- If the sample temperature is more than 2 °C different from the buffer solutions, the measured pH values must be corrected. The mode of correction depends on the type of instrument, and is either done automatically or adjusted manually (see the manufacturer's instructions) (USEPA 2004a).
- The electrode should not touch the stir bar, bottom, or sides of the beaker during the pH measurement.

[35] This method is adapted from USEPA Method 9040C (USEPA 2004a), USEPA Method 9045D (USEPA 2004b), Method 4500 from the Standard Methods for the Examination of Water and Wastewater (Rice *et al.*, 2017), and the UK Method for Measuring the pH of Food Products (Vijayakumar and Adedeji, 2017). This method should be cited as described in Velkushanova *et al.* (2021).

- Allow the electrodes sufficient time to stabilise while calibrating or measuring. A stability indicator on most meters prompts the user when readings should be taken.
- Electrodes must be rinsed thoroughly (and be fully cleaned) between measurements of different samples. Errors will occur if the electrode is coated with oily material or particulate matter. To clean electrodes, first rinse with distilled water. If coated with an oily material, the electrodes may not rinse free and should instead:
  - be cleaned with an ultrasonic bath, or,
  - be washed with detergent, rinsed several times with water, placed in 1:10 HCl so that the lower third of the electrode is submerged, and then thoroughly rinsed with water, or,
  - be cleaned as per the manufacturer's instructions.
  - It should be noted that adding distilled water to the sample may change the pH. If the pH of a sample can be measured without adding water, it should not be diluted.

8.6.6.1.6 Sample preservation

Samples should be analysed as soon as possible.

8.6.6.1.7 Sample preparation

For liquid and slurry samples:

- Homogenise the sample with a blender.
- Pour the sample into a beaker, ensuring the liquid level is high enough for complete immersion of the sensing elements of the electrode while allowing for enough space at the bottom of the beaker for the magnetic stir bar to avoid colliding with the electrode, as shown in Figure 8.10.
- Allow the sample to reach room temperature.

For semi-solid and solid samples:

- Homogenise the sample with a blender.
- If blending the sample produces a paste-like consistency, add the sample to a beaker, ensuring the level is high enough for complete immersion of the sensing elements of the electrode.
- Allow the sample to reach room temperature.
- If the sample is too dry to form a paste after blending, water addition may be necessary - follow the sample preparation steps for very dry samples.

For dry samples (e.g. very 'thick' or 'dry' solid samples, dried end products such as pellets):

- Method development to establish the appropriate dilution will be required, and dilution should be reported with the results. Presented here is an example of dilutions used by the UKZN PRG laboratory.
- If the sample is too dry to produce a paste when blended, make a 1:1 dilution with distilled water by weighing equal masses of the sample and distilled water into a beaker (common masses are 20 g sample + 20 g distilled water).
- Cover the beaker and continuously stir the suspension for 5 min.
- If the sample absorbs all of the added distilled water, begin the sample preparation again using a 2:1 dilution (*e.g.* 20 g sample + 40 g distilled water). Report the dilution with the results.
- After stirring, let the diluted suspension stand for about 15 min to allow most of the suspended solids to settle out, or filter or centrifuge to isolate the aqueous phase for pH measurement.
- If there is an oily layer floating at the top of the supernatant, decant the oily phase before measuring the pH of the liquid phase. The electrode will need to be cleaned if it becomes coated with oily material.

8.6.6.1.8 Analysis protocol

Calibration

- Pour fresh buffer solutions into separate beakers with magnetic stir bars: for typical faecal sludge samples, pH buffer solutions at pH 4, 7, and 10 are used.
- Buffer solutions should be stirred gently with the magnetic stirrer during calibration.
- Calibrate the pH meter according to the manufacturer's instructions using standard buffer solutions.

Measurement

For liquid and slurry samples:

- Rinse the electrode with distilled water. Dry the electrode with a laboratory tissue.
- Turn on the magnetic stir bar and place the electrode into the beaker containing the sample. Ensure that the sensing elements of the electrode are completely immersed, and do not allow the

magnetic stir bar to collide with the electrode (see Figure 8.10).

- If not using a magnetic stir bar, create a stirring motion with the electrode to ensure movement of the sample across the sensing element and to homogenise the sample.
- Once the reading has stabilised, record the pH and temperature of the sample.
- Thoroughly rinse the electrode with distilled water prior to measuring the pH of the next sample.

For semi-solid and solid samples:

- Rinse the electrode with distilled water. Dry the electrode with a laboratory tissue.
- Dip the electrode into the paste-like sample, ensuring that the sensing elements of the electrode are completely immersed.
- Allow the reading to stabilise, then record the pH and temperature.
- Take two additional readings at different locations in the sample, and record. Report the average of the three readings.
- Thoroughly rinse the electrode with distilled water prior to measuring the pH of the next sample.

For dry samples requiring water addition:

- Rinse the electrode with distilled water. Dry the electrode with a laboratory tissue.
- Place the electrode into a beaker that contains the settled diluted sample. Ensure that the sensing elements of the electrode are completely immersed in the supernatant.
- Once the reading has stabilised, record the pH and temperature of the sample.
- Thoroughly rinse the electrode with distilled water prior to measuring the pH of the next sample.

8.6.6.1.9 Calculation

No calculation required - direct reading.

8.6.6.1.10 Data set example

Ward *et al.* (2021) measured the pH of 465 faecal sludge samples collected from pit latrines and septic tanks in Lusaka, Zambia. pH was measured either on the day of collection or on the following day, with samples stored at 4 °C. Samples were homogenised with a blender before measuring the pH. The median pH value for all the samples was 7.7. TS values of the samples in this study ranged from < 0.5% ds to approximately 20% DS. None of the samples required dilution prior to pH measurement. The entire raw data set is included with publication[36].

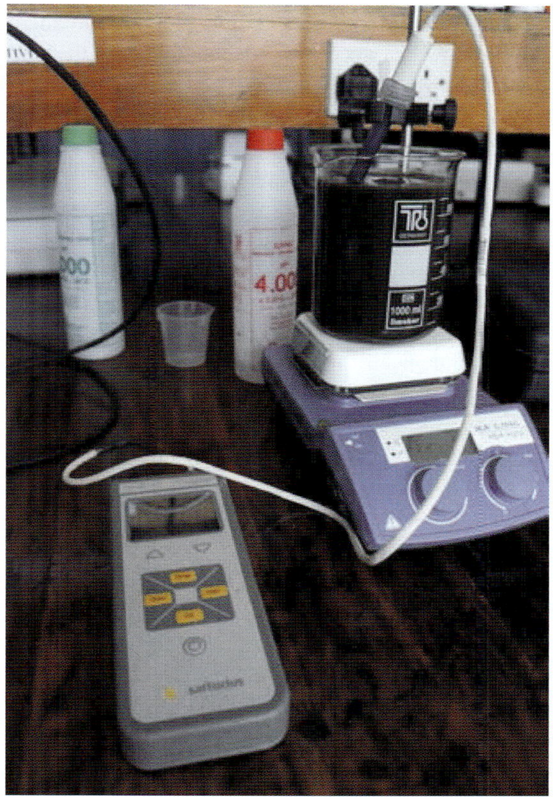

Figure 8.10 Analysis of pH for liquid faecal sludge sample. Note position of pH electrode: submerged in the well-mixed sample, but with enough clearance above the magnetic stir bar (photo: Eawag).

[36] https://doi.org/10.25678/00037X

8.6.6.2 Electrical conductivity – electrode method[37]

8.6.6.2.1 Introduction

Electrical conductivity (EC) is a measure of the ability of an aqueous sample to conduct electric current. Samples with higher concentrations of ions have higher EC, and EC can be used as a proxy measurement for dissolved salt concentrations. EC in faecal sludge is reported in units of millisiemens per centimetre or microsiemens per centimetre. The symbol for siemens is written either as 'S' or 'mho' so EC measurements for faecal sludge may be reported as any of the following: mS/cm, mmhoS/cm, µS/cm, µmho/cm.

Different sample preparation and measurement steps may be necessary depending on the TS concentration of the sample. If the sample does not have high enough water content for a reading to be obtained, dilution will be required. For most types of faecal sludge (liquid, slurry, semi-solid, and some solid samples), dilution will not be necessary. However, for some solid samples or samples that have been dewatered or dried, dilution will be required.

8.6.6.2.2 Safety precautions

- General health and safety (H&S) procedures specific for conducting the laboratory analysis of faecal sludge are presented in Section 8.2. Before conducting this method, it is important to be familiar with Section 8.2.3 to ensure safety measures are properly carried out.
- Appropriate personal protective equipment (PPE) should be used; specific details are covered in Section 8.2.3.1.

8.6.6.2.3 Required chemicals

- Reagent-grade water with low conductivity compared to the samples being measured (see Section 1080 in Rice *et al.* (2017) for instructions on how to prepare reagent-grade water)
- Standard potassium chloride (KCl) solution, 0.01 M (or molarity specified by the manufacturer).

8.6.6.2.4 Required apparatus and instruments

- Conductivity meter (capable of measuring conductivity with an error no more than 1% or 1 µS/cm, whichever is greater
- Conductivity electrode with a temperature sensor
- Beakers
- Analytical balance capable of weighing 0.1 g (if pre-dilution of the sample is required)

8.6.6.2.5 Quality control

General information on quality assurance and quality control (QA/QC) is provided in Section 8.3. Information on standards, operating conditions and interferences that are specific to this method includes:

- Multi-point calibrations are generally not required for conductivity meters. The conductivity cell is calibrated by determining the cell constant using a one-point calibration, commonly with 0.01 M KCl, which gives an EC reading of 1412 µS/cm at 25 °C.
- Conductivity is highly temperature-dependent. It is important to understand whether the conductivity meter compensates for temperature differences, or whether readings must be manually adjusted.
- Samples should be allowed to reach room temperature before measuring EC, and the actual temperature should always be recorded along with the EC measurements. See Method 2510B in Rice *et al.* (2017) for instructions on manual adjustment of EC readings based on temperature.

8.6.6.2.6 Sample preservation

Samples should be analysed as soon as possible.

8.6.6.2.7 Sample preparation

For liquid and slurry samples:

- Homogenise the sample with a blender.
- Pour the sample into a beaker, ensuring the liquid level is high enough for complete immersion of the sensing elements of the electrode.
- Allow the sample to reach room temperature.

---

[37] This method is based on Method 2510B of the Standard Methods for the Examination of Water and Wastewater (Rice *et al.*, 2017), with specific adaptations to deal with the extended range of TS present in faecal sludge samples. This method should be cited as: Method 2510B (Rice *et al.*, 2017), as adapted in Velkushanova *et al.* (2021).

For semi-solid and solid samples:

- Homogenise the sample with a blender.
- If blending the sample produces a paste-like consistency, add the sample to a beaker, ensuring the level is high enough for complete immersion of the sensing elements of the electrode.
- Allow the sample to reach room temperature.
- If the sample is too dry to form a paste after blending, water addition may be necessary - follow the sample preparation steps for very dry samples.

For dry samples (e.g. very 'thick' or 'dry' solid samples, dried end products such as pellets):

- Method development to establish the appropriate dilution will be required, and dilution should be reported with the results. Presented here is the example of dilutions used by the UKZN PRG laboratory.
- If the sample is too dry to produce a paste when blended, make a 1:1 dilution with reagent water by weighing equal masses of the sample and distilled water into a beaker (common masses are 20 g sample + 20 g distilled water).
- Cover the beaker and continuously stir the suspension for 5 min to allow the salts to solubilise.
- If the sample is hygroscopic and absorbs all of the added reagent water, begin the sample preparation again using a 2:1 dilution (*e.g.* 20 g sample + 40 g distilled water). Report the dilution with the results.
- After stirring, let the diluted suspension stand for approximately 15 min to allow most of the suspended solids to settle out, or filter or centrifuge to isolate the aqueous phase for EC measurement.
- If there is an oily layer floating at the top of the supernatant, decant the oily phase before measuring the EC of the liquid phase. The electrode will need to be cleaned if it becomes coated with an oily material.
- Note: the suggested dilutions are based on methods developed for faecal sludge at UKZN PRG, and may not be appropriate for every sample. Method development to establish the appropriate dilution will be required, and dilution should be reported with the results.

8.6.6.2.8 Analysis protocol

Calibration

- Rinse the electrode with reagent water, then rise three times with 0.01 M KCl standard solution.
- Pour fresh room-temperature 0.01 M KCl standard solution into a beaker, ensuring the liquid level is high enough for complete immersion of the sensing elements of the electrode.
- Calibrate the conductivity meter according to the manufacturer's instructions.

Measurement

For liquid and slurry samples:

- Rinse the electrode with distilled water. Dry the electrode with a laboratory tissue.
- Place the electrode into the beaker containing the sample. Ensure that the sensing elements of the electrode are completely immersed.
- Once the reading has stabilised, record the EC and temperature of the sample.
- Thoroughly rinse the electrode with distilled water prior to measuring the EC of the next sample.

For semi-solid and solid samples:

- Rinse the electrode with distilled water. Dry the electrode with a laboratory tissue.
- Dip the electrode into the paste-like sample, ensuring that the sensing elements of the electrode are completely immersed.
- Allow the reading to stabilise, then record the EC and temperature.
- Take two additional readings at different locations in the sample, and record. Report the average of the three readings.
- Thoroughly rinse the electrode with distilled water prior to measuring the EC of the next sample.

For dry samples requiring water addition:

- Rinse the electrode with distilled water. Dry the electrode with a laboratory tissue.
- Place the electrode into the beaker that contains the settled diluted sample. Ensure that the sensing elements of the electrode are completely immersed in the supernatant.
- Once the reading has stabilised, record the EC and temperature of the sample.
- Thoroughly rinse the electrode with distilled water prior to measuring the EC of the next sample.

8.6.6.2.9 Calculations
No calculation required - direct reading.

8.6.6.2.10 Data set example
Ward *et al.* (2021) measured the EC of 465 faecal sludge samples collected from pit latrines and septic tanks in Lusaka, Zambia. EC was measured either on the day of collection or on the following day, with samples stored at 4 °C. Samples were homogenised with a blender before measuring the EC. The median EC value for pit latrine samples was 14.5 mS/cm, and the median value for septic tank samples was 1.8 mS/cm. The TS values of samples in this study ranged from < 0.5% DS to approximately 20% DS. None of the samples required dilution prior to the EC measurement. The entire raw data set is included with publication.

8.6.7 Elemental analysis

Elemental analysis is a process where a sample is analysed for chemical elements. Examples of metal analysis include macro- and micro-nutrients that are necessary for treatment performance and plant and animal growth (*e.g.* boron, chlorine, copper, iron, manganese, molybdenum, and zinc), and heavy metals for compliance in land application or incineration (*e.g.* arsenic, beryllium, cadmium, chromium, copper, lead, mercury, nickel, selenium, and zinc). Examples of ultimate analysis include total carbon to estimate energy content or carbon sequestration in biochar, and total carbon and nitrogen for stabilisation of compost. Methods for chlorine and chloride are also covered in this section.

8.6.7.1 Metals – overview
Although some laboratories have methods that are routinely used for analysis of metals in faecal sludge, in general, there is a need for methods to be further developed based on types of faecal sludge, different objectives for the analysis, and available laboratory capacity and analytical machines. Provided here is a general overview of the analysis of metal concentrations, some examples of methods, and references to refer to when developing methods for these specific needs. General considerations for laboratory method development are covered in Section 8.1.

Appropriate forms of sample preparation and analysis will depend on the objectives of the analysis. For example, metals in the effluent of a treatment plant prior to discharge in a receiving water body, in contrast to metal concentrations in compost prior to land application. Preliminary treatment prior to analysis of metals will typically be required. As discussed in *Section 3000 Metals* of the Standard Methods for the Examination of Water and Wastewater, metals can be operationally defined as dissolved, suspended, total, or acid-extractable (Rice *et al.*, 2017). Sample preparation will depend on which of these is being analysed (*e.g.* a 0.45 µm membrane filter for suspended metals; a filtered solution for dissolved metals). Acid digestions are required for all samples other than very clear samples such as drinking water with a turbidity < 1 NTU. The USEPA 3050B method for acid digestion is most commonly used for slurry to solid faecal sludge samples, and in general sediments, sludge and soil samples including readily oxidisable organic matter. A nitric acid digestion can be adequate for relatively more 'clean' samples (*e.g.* metals loosely adsorbed on particulate matter), whereas further digestion with additional acids may be required with more difficult to oxidise samples, or if quantifying total metals (*i.e.* dissolved and particulate, organic and inorganically bound).

To evaluate the accuracy and precision of the digestion, it is important to use standards as positive controls that have similar organic characteristics and metal concentrations to evaluate the percentage recovery of total metals (*e.g.* National Institute of Standards and Technology (NIST) standards for soils and sludge, and standard biochars (UK Bioresearch Center (UKBRC)). It is also important to use consistent conditions for each sample (weights and volumes), and report the type of acid digestion along with the results. Great care needs to be taken during the sampling and laboratory preparation and analysis in order not to introduce metals into the sample.

Metals can then be quantified; Table 3010:I in Rice *et al.* (2017) summarises which methods can be used for the quantification of different metals. The table includes colorimetric and instrument methods, including atomic absorption (AA), flame photometry, and inductively coupled plasma mass spectrometry (ICP-MS)). The choice of analytical instrument will depend on what instruments are available, the required detection limits and range of concentrations, and the sample matrix and potential interferences.

Metals - instrument methods

Instrument methods are not covered in more detail, as they will be specific to the available laboratory equipment and the manufacturer's operating directions. The instrument used to quantify results following sample preparation and digestion will depend on availability, type of metals, required level of detection, and sample matrix. As mentioned above, the corresponding method for extraction must also be matched to the type of instrument. Table 3010:I in Rice *et al.* (2017) summarises which instrument methods can be used for the quantification of different metals. In general, the AES and MS detectors can detect lower concentration ranges than AA detectors. For example, the optimal concentration range for lead for FLAA with direct aspiration atomic absorption is 1-20 mg/L (Method 3111), whereas for the ICP-MS it is 42-4,700 µg/L (Method 3125 (Rice *et al.*, 2017)). It is important to note that the given concentrations are for what is analysed with the instrument (*e.g.* following digestion), and their translation to corresponding concentrations in faecal sludge needs to be back-calculated based on the method and moisture content.

General considerations for the operation of instruments includes running calibration curves covering the entire range of the analysed concentrations, quantifying MDLs using a selected set of standards appropriate to the metals being analysed, spiked samples (*e.g.* known concentrations of standards in HNO_3), and instrument blanks. The mean concentration and standard deviation for each sample need to be calculated (a calculation of the sample variation, with replicates), and the percentage recovery of standards, spiked digestion, and spiked instrument samples need to be reported. All the blanks must be below the method detection limit.

Metals - colorimetric and spectrophotometric methods

Some elements can be quantified with more simple instrumentation, using colorimetric methods and quantification with a spectrophotometer (see the methods in Section 3500 and Table 3010:I) (Rice *et al.*, 2017). Commercially available test kits are also available for many of these methods. However, the applicability for faecal sludge is limited, as the methods are not applicable with high organic content samples, which reduces the digestion capacity. Possible applications are if metal concentrations are high enough that following dilution to reduce organics prior to digestion they are still above the MDL, relatively clear liquid samples such as effluents of treatment processes, or samples that are filtered for analysis of dissolved metal concentrations. Also of concern are interference from colour and turbidity, and chemicals used in sample preparation that could complex with metals (*e.g.* EDTA, citric acid). Preliminary tests and sample digestions are always required in advance to validate the method for the specific type of sample, along with using the appropriate controls and blanks, and suspended solids have to be uniformly distributed in the sample prior to digestion. Concentration ranges for the Hach Crack Set digestion method include: iron (0.24-7.2 mg/L), lead (0.12-2.40 mg/L), nickel (0.12-7.2 mg/L), cadmium (0.02-0.3 mg/L) and zinc (0.24-7.2 mg/L). If using commercially available test kits, it is always important to follow the manufacturer's directions and to document how the method has been validated and adapted for each use.

8.6.7.2 Metals – acid digestion for environmentally available metals[38]

8.6.7.2.1 Introduction

The USEPA Method 3050B for metal analysis of sediments, sludge, and soil samples is commonly used for analysis of metals in faecal sludge. It is not considered to be a total digestion; the results are considered to be representative of metal concentrations that could become environmentally available (USEPA, 1996). The method can be coupled with instrumental methods for analysis, either flame atomic absorption spectrometry (FLAA) or

[38] This method should be cited as USEPA Method 3050B (1996).

inductively-coupled plasma atomic emission spectrometry (ICP-AES), *OR* Graphite Furnace AA (GFAA) or inductively-coupled plasma mass spectrometry (ICP-MS), depending on the modifications in the protocol. Hence, it is important to follow each step in the method specific for the different types of instrumental analysis. For GFAA or ICP-MS the final digest is diluted to 100 mL, whereas for ICP-AES or FLAA hydrochloric acid is added for a final reflux. Typical faecal sludge sample sizes for analysis are 1-2 g wet weight or 1 g dry weight. The complete method can be downloaded using the link below[39]. This method is intended for acid-extractable, environmentally-available metals in solid, semi-solid and slurry faecal sludge samples. For more liquid samples (*e.g.* liquid and slurry, depending on characteristics), it is necessary to determine if suspended and/or dissolved metals are most relevant to quantify, and then carry out extractions on the filter and/or filtered sample (see Section 3030 in Rice *et al.*, 2017).

8.6.7.2.2 Safety precautions

- General health and safety (H&S) procedures specific for conducting laboratory analysis of faecal sludge are presented in Section 8.2. Before conducting this method, it is important to be familiar with Section 8.2.3 to ensure safety measures are properly carried out.
- Appropriate personal protective equipment (PPE) should be used; specific details are covered in Section 8.2.3.1.
- When working with concentrated acid special protective measures need to be taken, for example working in a fume hood, and wearing an acid-safe laboratory coat, goggles, and gloves. Spills need to be neutralised prior to cleaning, per established laboratory protocols.

8.6.7.2.3 Required chemicals

The reagent blank must be less than the MDL.

- Reagent water.
- Nitric acid (concentrated HNO_3).
- Hydrochloric acid (concentrated HCl).
- Hydrogen peroxide (30% H_2O_2).

[39] https://www.epa.gov/sites/production/files/2015-06/documents/epa-3050b.pdf.

8.6.7.2.4 Required apparatus and instruments

- Digestion vessels - 250 mL.
- Vapour recovery device (*e.g.* watch glasses, solvent handling system, refluxing device).
- Drying oven.
- Thermometer - accurate measurement to at least 125 °C.
- Filter paper - Whatman No. 41 or equivalent.
- Centrifuge and centrifuge tubes.
- Analytical balance - accurate to 0.01 g.
- Heating source - able to maintain 90-95 °C (*e.g.* hot plate, block digester, microwave).
- Funnel.
- Graduated cylinder.
- Volumetric flasks - 100 mL.
- Standard laboratory glassware and utensils.

8.6.7.2.5 Quality control

General information on quality assurance and quality control (QA/QC) is provided in Section 8.3. Information on standards, operating conditions and interferences that are specific to this method includes:

- Important measures include acid washing of all the glassware to prevent background contamination of metals, and special consideration to not include metal contamination during the sample preparation and analysis.
- To ensure precision and accuracy, ongoing laboratory and analytical control measures are necessary, including ensuring that reagent blanks do not have background metal concentrations greater than the MDL.
- A method blank must be taken through the entire digestion and analytical procedure to ensure no background contamination above the MDL is introduced during the process.
- Duplicate spiked samples in the sample matrix should be analysed periodically, and always when analysing a new sample matrix/type of faecal sludge.
- In addition to standard quality control measures such as calibration, standards, duplicates and blanks, it is also important to use standards as positive controls in the digestion that have similar organic characteristics and metal concentrations

(*e.g.* NIST standards for soils and sludge). Examples of recoveries from NIST standards for hot-plate and microwave heat sources, and total digestion values are provided in tables 3, 4, and 5 in the USEPA method (1996).

8.6.7.2.6 Sample preservation

Samples should be refrigerated upon arrival at the laboratory (kept on ice during collection and transport), and analysed as soon as possible. Prior to analysis, the samples need to be air-dried. Dried samples can be stored for the longer term, if they can be stored in conditions that ensure they remain dry; if samples are damp or moist, biological degradation will continue.

8.6.7.2.7 Sample preparation

It can also be difficult to obtain representative results with wet and damp faecal sludge from the extraction process, so sub-samples are oven-dried, crushed, and ground prior to analysis to reduce sub-sample variability.

Solid, semi-solid and slurry samples (TS > 5%):

- Mix the dried sample well and take a representative sub-sample.
- Oven-dry the sub-sample at 105 °C to dryness (*i.e.* weight does not change), and then grind the sample (*e.g.* rolling pin, pestle and mortar, or coffee grinder), and then sieve to 2 mm (*e.g.* USS #10).
- If debris, rubbish, rocks, etc. are removed by sieving, this should be noted in the results.
- Weigh out the thoroughly-mixed samples of 1-2 g in the digestion vessel.

8.6.7.2.8 Analysis protocol

For a step-by-step protocol and a flow chart of modifications depending on the instrument method, refer to USEPA (1996)[40]. In general, the following steps are carried out.

- A series of reflux reactions with HNO_3 are repeated until brown fumes are no longer produced (*i.e.* samples heated without boiling for specified times, with the vapour condensed and returned to the flask).
- As an alternative to using hot plates or block digesters, the method can be modified for microwave digestion of samples for analysis with GFAA or ICP-MS.
- Samples are then warmed with aliquots of H_2O_2 until there is minimal effervescence.
- The H_2O_2 is then reduced by heating without boiling for 2 hours (or microwave modification).
- After cooling, dilute to 100 mL with water and remove particulates with filtration or centrifugation for analysis with GFAA or ICP-MS.
- For analysis with FLAA or ICP-AES, an additional reflux with HCl is required (or microwave modification) and filter with Whatman No. 41 paper (or equivalent).

Modifications to the method need to be carefully developed and documented, and will depend on the available resources. For example, the UKZN PRG laboratory has a microwave digestion system (Milestone Ethos One, Italy)[D], which allows for a specific mass of a prepared sample (between 0.1 and 1 g, depending on TS content and digestion time) to be digested with 10 mL of Aqua regia (9 mL concentrated nitric acid + 3 mL concentrated hydrochloric acid) in a closed polytetrafluoroethylene (PTFE) vessel for 105 min with the following pre-settings:

05:00 min at 1,000 w and at 90 °C
60:00 min at 1,000 w and at 130 °C
10:00 min at 1,000 w and at 40 °C
30:00 min at 1,000 w and at 30 °C.

8.6.7.2.9 Calculations

The dilution factor must be taken into consideration for the total recovery of the analyte. An example calculation for the determination of the mass fraction of cadmium (mg/kg TS, also called 'dry mass') in a slurry, semi-solid, or solid faecal sludge sample is provided here:

---

[40] https://www.epa.gov/sites/production/files/2015-06/documents/epa-3050b.pdf

Slurry, semi-solid and solid (mg/kg TS):

$$\frac{5\text{ mg Cd}}{\text{L}} \times \frac{100\text{ mL extracted}}{1\text{ g dried sludge}} \times \frac{1000\text{ g}}{\text{kg}} \times \frac{\text{L}}{1000\text{ mL}} = 5\,\frac{\text{mg Cd}}{\text{kg DS}}$$

8.6.7.2.10 Data set example

Examples in the literature of reported concentrations of metals in faecal sludge include 60 samples from Hanoi, Vietnam with a range of 1.7-64 g/L TS. Samples were analysed with inductively-coupled plasma (ICP) based on Standard method 3120 Metals by plasma emission spectroscopy (Rice *et al.*, 2017), and concentrations of 0.1-41 mg/L Ni, 0.1-3-3 mg/L Pb, 2-2,000 mg/L Fe, and 1-118 mg/L Zn were reported (Englund *et al.*, 2020, the complete raw data set is available using the link below[41]).

Examples of reported mass fractions of metals in biochar from faecal sludge in composting toilets include 188.8 (± 3.1) mg Zn/kg DS, < 5 mg/kg DS Cd, and < 12.5 mg/kg DS for Cu, Cr, Pb and Ni. Samples were analysed with ICP-OES following a microwave digestion (ultraCLAVE 4, MLS GmbH, Germany) at 250 °C and 120 bar for ten minutes, with 0.2 g sample, 5 mL HNO_3, 1 mL H_2O_2 and 0.3 mL hydrofluoric acid (HF) (Bleuler *et al.*, 2020).

Examples of total metals in faeces and faecal sludge from pit latrines in Colorado, USA and Kampala, Uganda have been reported gravimetrically as a percentage of metal oxides in the ash (% ash). Samples were dried at 105 °C, pulverised and homogenised, then incinerated at 550 °C and the resulting ash was digested in aqua regia and hydrofluoric acid using a microwave digester prior to analysis with ICP-OES, following ASTM D6357-11 (ASTM 2011). P, Mn, Fe, Mg, Si, Al, Ca, Ti, Na, and K were analysed (Hafford *et al.*, 2018).

8.6.7.3 Ultimate analysis – total carbon, hydrogen, nitrogen, oxygen, and sulphur[42]

8.6.7.3.1 Introduction

Ultimate analysis is the quantification of the major organic elemental composition of a sample or material. It includes elemental carbon, hydrogen, nitrogen and oxygen, and sometimes sulphur, halogens or ash. Samples are combusted in an ultimate analyser at a range of high temperatures in a pure oxygen environment, and then quantified by the subsequent release of gasses (*e.g.* CO_2, SO_2, NO_2). Ultimate analysis is used for compliance with standards for fuel, to routinely analyse plants, soil samples, food and feed, or to evaluate the stability of organic matter based on the carbon-to-nitrogen ratio (C:N) (*e.g.* compost). The step-by-step procedure will vary depending on the laboratory and available equipment. One example is the total carbon, nitrogen and sulphur analysis used by the UKZN PRG laboratory in Durban, South Africa, that is described here. The LECO-TruMac-CNS Series 928 analyser[D] is a carbon, nitrogen and sulphur analyser utilising a pure oxygen environment in a ceramic horizontal furnace regulated at high temperatures (1,100 to 1,450 °C). The combustion gas collection and handling system uses a helium carrier gas and a thermal conductivity cell for the detection of nitrogen.

8.6.7.3.2 Safety precautions

- General health and safety (H&S) procedures specific for conducting laboratory analysis of faecal sludge are presented in Section 8.2. Before conducting this method, it is important to be familiar with Section 8.2.3 to ensure safety measures are properly carried out.
- Appropriate personal protective equipment (PPE) should be used; specific details are covered in Section 8.2.3.1.
- Check gas pressures before opening the gas cylinders. Exercise standard safety protocols for working with pressurised gas cylinders (*e.g.* keep them safely tethered at all times, and never transport without the gas cap in place).

---

[41] https://doi.org/10.25678/0000tt

[42] This method should be cited as the specific method that is carried out in each laboratory, including the manufacturer's make and model of analyser, and the exact method of the sample preparation. Existing standard methods for coal and coke include ISO 17247:2020 and ASTM D3176–15

- Ensure the furnace door seals during combustion to avoid risk of fire or burns.

8.6.7.3.3 Required chemicals

Gas cylinders

- Pure oxygen (35 psi, 241 kPa).
- Helium (35 psi, 241 kPa).
- Air (40 psi, 276 kPa).

8.6.7.3.4 Required apparatus and instruments

- CNS analyser equipped with furnace, computer and auto-sampler.
- Analytical balance accurate to four decimal places.
- Flow meter and regulators.
- Nickel liners (size specific to the instrument).
- Ceramic boats or crucibles (size specific to the instrument).

8.6.7.3.5 Quality control

General information on quality assurance and quality control (QA/QC) is provided in Section 8.3. Information on standards, operating conditions and interferences that are specific to this method includes:

- Calibrate the instruments daily using a certified acetanilide/EDTA standard containing nitrogen (6.48 ± 0.09), carbon (72.48 ± 0.25) and sulphur (7.47 ± 0.05), measured in g for solid and mL for liquid samples.
- Test replicates of 3 different masses ranging from 0.1 g to 0.5 g.
- No sample dilution is required, but it should be ensured that all solid waste is removed, and the sample must be thoroughly mixed.
- Nickel liners should be used for liquid and slurry samples with high moisture content (TS < 5%) as the ceramic boats are porous.
- Use NIST standards 'Montana soil' and 'sludge standard' and/or UKBRC standard biochars depending on the analysis, along with acetanilide and benzoic acid as internal standards, and report percentage recoveries.

8.6.7.3.6 Sample preservation

Samples should be refrigerated upon arrival at the laboratory (and kept on ice during collection and transport), and analysed as soon as possible. Some types of ultimate analysers require samples to be air-dried prior to analysis, while other analysers are able to characterise wet samples. If an analyser requires air-dried samples, it is recommended to dry samples before storage. Dried samples can be stored for the longer term, if they can be stored in conditions that ensure they remain dry; if samples are damp or moist, biological degradation will continue.

8.6.7.3.7 Sample preparation

As discussed in Chapter 3, it is difficult to obtain representative samples of faecal sludge; hence, a carefully designed plan needs to be followed to ensure representativeness. It can also be difficult to obtain representative results with wet and damp faecal sludge, so sub-samples are frequently oven-dried, crushed, and ground prior to analysis to reduce sub-sample variability. The analyser used by the UKZN PRG is able to take wet samples, but requires a nickel liner for the crucible as it is permeable.

8.6.7.3.8 Analysis protocol

Instrument setup

- Switch on the instrument and auto-sampler.
- Switch on the computer and software.
- Set the furnace temperature to 1,350 °C.
- Turn on the gasses.
- Perform an instrument check.
- Check the furnace temperature (1,350 °C)
- A system check is done through the software according to the instructions of the supplier.
- Check for leaks (oxygen and helium (1,263 Hg) independently).

Analysis parameters

- Furnace temperature = 1,350 °C.
- Cooler temperature = 5 °C.
- Dehydration time = 0 sec.
- Purge cycles = 2 sec.
- Equilibration time = 30 sec.

Blank analysis

- Condition system by analysing 3-5 blanks of empty crucibles.

Solid, semi-solid and slurry samples (TS > 5%)

- Weigh 0.1 g of sample into the crucible; enter the mass and sample name into the software.
- Place the crucible into the auto-sampler.
- Start the analysis, according to the manual supplied by the manufacturer.

Liquid (TS < 5%)

- Place a nickel boat liner on the crucible (crucible is porous).
- Weigh 0.1–0.15 g of the liquid sample into the cover of the crucible; enter the mass and sample name into the software.
- Place the crucible into the auto-sampler.
- Start the analysis following the manual supplied by the manufacturer.

8.6.7.3.9 Calculations

There are no required calculations; the results are automatically converted by the software programme, including conversions of units (ppm, %, mg/L).

Report values as % based on dry weight basis, and the C:N as ratio as weight:weight.

8.6.7.3.1 Data set example

Example 1

Carbon, nitrogen and sulphur were analysed in duplicate on dried, pulverised samples with a HEKAtech Eurovector and a Leco TruSpec CHNS Marco Analyser[D], to evaluate the use of dried faecal sludge as a dry combustion fuel in kilns. Elevated concentrations of nitrogen, sulphur and chlorine indicate potential for dioxin, furan, NOx, N_2O, SO_2, HCl, HF and CxHy formation during combustion (Gold *et al.*, 2017). Results from this study are summarised in Table 8.12.

Table 8.12 Ultimate analysis of dried faecal sludge (Gold *et al.*, 2017).

	Kampala, Uganda		Dakar, Senegal	
	Average	SD	Average	SD
Carbon %	27.8	3.1	28.8	3.4
Hydrogen %	4.2	0.5	4.2	0.4
Nitrogen %	3.2	0.4	3.0	0.6
Sulphur %	0.7	0.1	1.7	0.0

Example 2

Three samples were collected – two of them were faecal sludge samples from ventilated improved pit latrines and urine diversion toilets in Durban, South Africa, the third sample was fresh faeces also from donors in Durban. These samples were analysed in six replicates each, using Method 8.6.7.3. The values were similar to those obtained by Gold *et al.*, (2017).

The average CNS content for the VIP was: 21.05% for carbon, 2.06% for nitrogen and 0.65% for sulphur (Table 8.13). For the UDDT it was: 5.25% for carbon, 0.46% for nitrogen and 0.15% for sulphur (Table 8.14), and for the fresh faeces: 18.47% for carbon, 5.25% for nitrogen and 3.97% for sulphur (Table 8.15) (source: unpublished data UKZN PRG).

Table 8.13 Example of CNS values for VIP faecal sludge (UKZN PRG).

Samples	Mass	Carbon (%)	Nitrogen (%)	Sulphur (%)
1	0.2095	19.180	1.9100	0.6120
2	0.2009	19.513	1.9629	0.5990
3	0.2025	20.878	1.9743	0.6600
4	0.2056	24.753	2.3859	0.7770
5	0.2018	21.093	2.0852	0.6510
6	0.2020	20.907	2.0305	0.6270
Average	0.2037	21.0540	2.0581	0.6543
SD	0.0032	1.9811	0.1714	0.0643

Table 8.14 Example of CNS values for UDDT faecal sludge (UKZN PRG).

Samples	Mass	Carbon (%)	Nitrogen (%)	Sulphur (%)
1	0.2051	4.0186	0.3549	0.1320
2	0.2071	4.8903	0.4033	0.1360
3	0.2077	5.0185	0.5597	0.1370
4	0.2012	5.0963	0.4172	0.1440
5	0.2076	5.3617	0.3953	0.1540
6	0.2093	7.095	0.6611	0.2150
Average	0.2063	5.2467	0.4653	0.1530
SD	0.0028	1.0138	0.1187	0.0313

Table 8.15 Example of CNS values for fresh faeces (UKZN PRG).

Samples	Mass	Carbon (%)	Nitrogen (%)	Sulphur (%)
1	0.1075	16.3180	2.3198	7.6700
2	0.1135	18.0580	2.5307	5.5500
3	0.1085	18.0510	2.4342	8.9100
4	0.1036	19.3000	2.2742	0.1320
5	0.1137	19.1050	2.8293	1.5200
6	0.1085	20.0080	2.5805	0.0656
Average	0.1092	18.4733	2.4947	3.9746
SD	0.00351	1,1856	0.1841	3.5722

8.6.7.4 Chlorine – colorimetric (test strip) method[43]

8.6.7.4.1 Introduction

Chlorine is used as a disinfectant to reduce microbial load in drinking water and wastewater. Excessive chlorine in wastewater effluents cam form carcinogenic chloro-organics that affect aquatic organisms. In faecal sludge treatment, chlorine measurement is essential as it influences electrical conductivity in the effluent. In this method, chlorine oxidizes an organic compound to a violet dye. The chlorine concentration is measured semi-quantitatively by visual comparison of the reaction zone of the test strip with the fields of a colour scale. The example provided here is the is the Merck MQuant Chlorine - Test[D], one of many chlorine colorimetric test methods for samples with concentrations of 25-500 $mgCl_2/L$. For faecal sludge, samples must be filtered and/or diluted to prevent masking of the resultant colour change by the faecal sludge sample.

8.6.7.4.2 Safety precautions

- General health and safety (H&S) procedures specific for conducting laboratory analysis of faecal sludge are presented in Section 8.2. Before conducting this method, it is important to be familiar with Section 8.2.3 to ensure safety measures are properly carried out.
- Appropriate personal protective equipment (PPE) should be used; specific details are covered in Section 8.2.3.1.

8.6.7.4.3 Required chemicals

- Distilled water
- Dichloroisocyanuric acid sodium salt dehydrate (for the quality control procedure)

8.6.7.4.4 Apparatus and instruments

- Test strips
- Glass beakers (50 or 100 mL)
- Balance with weighing boats (slurry to solid samples)
- Blender (slurry to solid samples)
- Volumetric flask (250 mL)
- Filter paper (adequate for removing solids from the sample, for example 0.45μm filter for the liquid samples)

8.6.7.4.5 Quality control

General information on quality assurance and quality control (QA/QC) is provided in Section 8.3. Information on standards, operating conditions and interferences that are specific to this method includes:

- To make a standard solution, dissolve 1.85 g of dichloroisocyanuric acid sodium salt dihydrate in distilled water, make up to 1,000 mL with distilled water, and mix. This corresponds to approximately 1,000 mg/L free chlorine.
- Perform quality control on the standard on a daily or weekly basis (depending on the testing load).
- The test strips are stable up to the date stated on the pack when stored closed at +2 to +8 °C.
- The colour of the reaction zone can continue to change after the specified reaction time has elapsed. This gives an incorrect measurement.
- Common interferences in faecal sludge include sodium and nitrites (for specific concentrations refer to the manufacturer's instructions).
- For faecal sludge, filter and/or dilute the samples based on the type of sludge to prevent interference in colour with the test strips. Always use serial dilutions.
- The chlorine test strip method is used for qualitative to semi-quantitative measurement. For a quantitative measurement, see Method 8.6.7.5.
- The colorimetric test kit must be selected based on the expected range of chlorine concentration in the sample.

8.6.7.4.6 Sample preservation

- Samples for chlorine analysis must not be stored. Samples must be analysed immediately after sampling without exposure to excessive light and agitation. In aqueous solutions, chlorine is unstable and its concentration decreases with storage, agitation, and exposure to light (Rice *et al.*, 2017).

[43] This method should be cited as the specific manufacturer's method along with any modifications. The example used here is the Merck MQuant Chlorine Test Kit (Merck, 2020l)[D]

8.6.7.4.7 Sample preparation

For liquid, slurry, semi-solid or solid samples:

- Solid samples must be dissolved in a known volume of water before further dilution is performed.
- Filter the samples to prevent interference in the colour.
- Samples containing more than 500 $mgCl_2/L$ must be diluted with distilled water.

For slurry to solid samples:

- Weigh out between 1.8 g and 2.0 g of thoroughly-mixed faecal sludge sample into a beaker.
- Dilute the sample gravimetrically and transfer to a blender, as described in Section 8.4.2.
- If necessary, adjust the dilution based on the specific sludge characteristics
- Blend for 30 sec on the highest speed.
- Transfer the contents of the blender into a 1 L volumetric flask, rinse the blender twice with 250 mL distilled water and dilute to the mark with distilled water.
- Filter samples through a 0.45µm filter paper and measure nitrite concentration.
- Transfer this 1 L solution into a plastic bottle and store at 4 °C until ready for testing.
- Total solids analysis should be performed on the samples so that the results of the nitrite measurement can be expressed on a mass per mass basis (gCl_2/gTS).

8.6.7.4.8 Analysis protocol

The measuring range of this method is 25-500 $mgCl_2/L$:

- Immerse the reaction zone of the test strip in the diluted sample (15-25 °C) for 2 seconds. Shake off excess liquid from the strip after exactly 10 seconds to determine with which colour field on the label the colour of the reaction zone coincides most closely. Read off the corresponding result in $mgCl_2/L$.
- Always analyse samples together with the standard solution to ascertain the reliability of the results.

8.6.7.4.9 Calculation

Liquid samples:

Result of analysis (mg/L Cl_2)

$$\text{Final concentration}\left(\frac{\text{mg Cl}_2}{\text{L}}\right) = A\left(\frac{\text{mg}}{\text{L}}\right) \times DF$$

$$\text{Dry basis}\left(\frac{\text{g}}{\text{g}}\right) = \frac{\text{Final concentration}\left(\frac{\text{mg}}{\text{L}}\right)}{\text{Total solids concentration}\left(\frac{\text{mg}}{\text{L}}\right)}$$

Where:

A = Colorimetric measurement value (mg/L)
DF = Dilution factor (F/I)
F = Final diluted volume (L)
I = Initial sample volume (L)

Slurry, semi-solid and solid samples:

$$\text{Dry basis}\left(\frac{\text{g}}{\text{g}}\right) = \frac{\frac{A\left(\frac{\text{mg}}{\text{L}}\right)}{1{,}000} \times \frac{V}{M}\left(\frac{\text{L}}{\text{g}}\right)}{\text{Total solids content}\left(\frac{\text{g}}{\text{g}}\right)}$$

Where:

A = Colorimetric measurement value (mg/L)
V = Volume of dilution (L)
M = Wet weight of sludge used in sample preparation (g)

8.6.7.5 Chlorine – spectrophotometric method[44]

8.6.7.5.1 Introduction

This method for quantifying chlorine in faecal sludge is based on the principle that in a weakly acidic buffer solution, free chlorine reacts with dipropyl-p-phenylenediamine (DPD) in the presence of potassium iodide to form a red-violet dye. The degree of the colour that is formed in solution is then measured spectrophotometrically. Commercially available test kits for measuring chlorine based on standard methods are available with pre-packaged individual aliquots of the necessary chemicals in pillows (dry chemicals) and vials (liquid chemicals). Commonly used test kits from manufacturers such as Hach, Merck, and Hanna vary slightly in the methods they use for chlorine measurement. The example provided here is the Merck Spectroquant® Chlorine Cell Test (Merck, 2020b)[D] for samples with

[44] This method should be cited as: Method 4500 CL2-G (Rice *et al*., 2017), and if test kits are used, also as per the manufacturer's directions including any modifications.

concentrations of 0.03-6.0 $mgCl_2/L$, and it is based on the manufacturer's protocol for water and wastewater using the standard method 4500 (Rice *et al.*, 2017). For faecal sludge, samples must be diluted and filtered to prevent false high readings associated with turbid solutions.

8.6.7.5.2 Safety precautions

- General health and safety (H&S) procedures specific for conducting laboratory analysis of faecal sludge are presented in Section 8.2. Before conducting this method, it is important to be familiar with Section 8.2.3 to ensure safety measures are properly carried out.
- Appropriate personal protective equipment (PPE) should be used; specific details are covered in Section 8.2.3.1.

8.6.7.5.3 Required chemicals

- Distilled water (free from chlorine)
- Bottle of reagent Cl_2-1 (supplied by the manufacturer)
- Chlorine standard solution (3.00 mg/L Cl_2) can be purchased from the manufacturer or prepared as described in Rice *et al.* (2017) using the chlorine DPD colorimetric method.
- 0.1 N sodium hydroxide
- 0.1 N sulphuric acid.

8.6.7.5.4 Required apparatus and instruments

- Spectrophotometer
- Analytical balance with weighing boats (slurry to solid samples)
- Blender (slurry to solid samples)
- Pipette and pipette tips
- Sample cell (supplied by the manufacturer)
- Volumetric flask (1 L)
- Glass beakers (50 or 100 mL)
- Glass storage bottle
- pH test strips
- Blue micro spoon (supplied by the manufacturer)
- Laboratory cleaning tissues
- Filter paper (adequate for removing solids from the sample, for example a 0.45μm filter for liquid samples).

8.6.7.5.5 Quality control

General information on quality assurance and quality control (QA/QC) is provided in Section 8.3. Information on standards, operating conditions and interferences that are specific to this method includes:

- Perform quality control with chlorine standard solution with every test batch (or on a daily or weekly basis depending on the testing load).
- For spectrophotometric measurements, the sample cells must be clean. Before analysis, wipe with a laboratory cleaning tissue.
- Measurement of turbid solutions yields false high readings. For faecal sludge, samples should be diluted with an appropriate dilution factor through serial dilutions and filtered for accurate measurements. Recommended dilution factors and filtration protocols will need to be developed based on the characteristics of the specific sludge.
- The pH of the measurement solution must be within the range 4.5-5.5.
- The colour of the measurement solution remains stable for up to 30 min after the end of the reaction time; thus the spectrophotometric measurement should be conducted within that timeframe.
- Common interferences in chlorine measurement include bromine and iodine. For specific concentrations, refer to the manufacturer's instructions.

8.6.7.5.6 Sample preservation

- Samples for chlorine analysis must not be stored. Samples must be analysed immediately after sampling without exposure to excessive light and agitation. In aqueous solutions, chlorine is unstable and its concentration decreases with storage, agitation and exposure to light (Rice *et al.*, 2017).

8.6.7.5.7 Sample preparation

For liquid samples:

- Samples containing more than 6 $mgCl_2/L$ must be diluted with distilled water to within the range of 0.03-6.0 $mgCl_2/L$
- Turbid solutions falsely increase spectrophotometric readings. Therefore, turbid samples should be filtered with a 0.45 μm filter paper to prevent false high measurements.

For slurry to solid samples:

- Weigh out between 1.8 g and 2.0 g of the thoroughly-mixed faecal sludge sample into a beaker.
- Dilute the sample gravimetrically and transfer to the blender, as described in Section 8.4.2.
- If necessary, adjust the dilution based on the specific sludge characteristics.
- Blend for 30 sec on the highest speed.
- Transfer the contents of the blender into a 1 L volumetric flask, rinse the blender twice with 250 mL distilled water and dilute to the mark with distilled water.
- Transfer this 1 L solution into a plastic bottle and store at 4 °C until ready for testing.
- Filter the samples through a 0.45 µm filter paper.
- The sample pH must be within the range of 4-8; if necessary, adjust with sulphuric acid or sodium hydroxide solution.
- Total solids analysis should be performed on the samples so that the results of the chlorine measurement can be expressed as gCl_2/gTS.

8.6.7.5.1 Analysis protocol

Calibration

Follow the spectrophotometer manufacturer's instructions for calibration. It may be necessary to calibrate the instrument before every reading, or it may only be necessary to perform periodic calibration checks to determine when calibration is necessary. For this method, calibration can be performed as follows:

- Prepare a series of at least four different concentrations of a standard solution making sure to include the lowest and highest concentration of the kit testing range. It is typical to perform serial dilutions or dilutions with a uniform interval including the lowest and highest concentrations.
- Determine the chlorine concentration of the standard solutions.
- Multiply the spectrophotometric reading by the dilution factor and report the results in $mgCl_2/L$.
- Prepare a calibration curve by plotting the instrument response against the standard concentration.

Procedure

For measuring range of 0.03-6 mg/L free chlorine

- Pipette 5 mL of the sample into a sample cell.
- Add 1 level blue micro spoon of reagent Cl_2-1.
- Shake vigorously until the reagent is completely dissolved.
- Leave to stand for 1 min (reaction time).
- Wipe the sample cell with a laboratory tissue to clean it and then measure in a spectrophotometer.

8.6.7.5.9 Calculations

Liquid and slurry samples:

Result of analysis (mg/L)

$$\text{Final concentration}\left(\frac{mg}{L}\right) = A\left(\frac{mg}{L}\right) \times DF$$

$$\text{Dry basis}\left(\frac{g}{g}\right) = \frac{\text{Final concentration}\left(\frac{mg}{L}\right)}{\text{Total solids concentration}\left(\frac{mg}{L}\right)}$$

Where:

A = Spectrophotometric measurement value (mg/L)
DF = Dilution factor (F/I)
F = Final diluted volume (L)
I = Initial sample volume (L)

Slurry, semi-solid and solid samples:

$$\text{Dry basis}\left(\frac{g}{g}\right) = \frac{\frac{A\left(\frac{mg}{L}\right)}{1{,}000} \times \frac{V}{M}\left(\frac{L}{g}\right)}{\text{Total solids content}\left(\frac{g}{g}\right)}$$

Where:

A = Spectrophotometric measurement value (mg/L)
V = Volume of dilution (L)
M = Wet weight of sludge used in sample preparation (g)

8.6.7.6 Chloride – colorimetric (test strip) method[45]

8.6.7.6.1 Introduction

Chloride is classified as a pollutant because of the impact on organisms and plants in aquatic ecosystems. In faecal sludge, high levels of chloride can influence microbial cells activity and the characteristics of the

[45] This method should be cited as the specific manufacturer's method along with any modifications. The example used here is the Merck MQuant Chloride Test Kit (Merck, 2020k)[D].

sludge such as dewatering and settling. In this method, chloride ions react with silver ions, decolorising red-brown silver chromate. The chloride concentration is measured semi-quantitatively by visual comparison of the reaction zones of the test strip with the colour rows of a colour scale provided by manufacturers. The example provided here is the MQuant Chloride -Test[D], one of many chloride colorimetric test methods for samples with concentrations of 500-3,000 mgCl^-/L. It is recommended for determination of the chloride content and interference level in conjunction with COD analysis.

8.6.7.6.2 Safety precautions

- General health and safety (H&S) procedures specific for conducting laboratory analysis of faecal sludge are presented in Section 8.2. Before conducting this method, it is important to be familiar with Section 8.2.3 to ensure safety measures are properly carried out.
- Appropriate personal protective equipment (PPE) should be used; specific details are covered in Section 8.2.3.1.

8.6.7.6.3 Required chemicals

- Distilled water
- 0.1 M sodium hydroxide or 0.1 M nitric acid
- Chloride standard solution (typically supplied with a test strip kit)

8.6.7.6.4 Apparatus and instruments

- Test strips
- Glass beakers (50 or 100 mL)
- pH strips (0-14)
- Balance with weighing boats (slurry to solid samples)
- Blender (slurry to solid samples)
- Volumetric flask (250 mL)
- Filter paper (adequate for removing solids from a sample, for example 0.45μm filter for liquid samples)

8.6.7.6.5 Quality control

General information on quality assurance and quality control (QA/QC) is provided in Section 8.3. Information on standards, operating conditions and interferences that are specific to this method includes:

- To make a standard solution, dilute the 1,000 mg/L Cl^- chloride standard solution with distilled water to 10-100 mg/L Cl^-.
- Perform quality control on a daily or weekly basis (depending on the testing load).
- The test strips are stable up to the date stated on the pack when stored closed at +2 to +8 °C.
- The colour of the reaction zones can continue to change after the specified reaction time has elapsed. This must not be considered in the measurement. The colour after the specified reaction time is the correct reading.
- Common interferences in faecal sludge include calcium, potassium, magnesium, sodium, ammonium, nitrates and nitrites.
- For faecal sludge, samples should be diluted with an appropriate dilution factor, based on the type of sludge, to prevent interference with the colour of the test strips. Always use serial dilutions.
- The chloride test strip method is used for qualitative to semi-quantitative measurement. For a quantitative measurement, refer to Method 8.6.7.7.
- The colorimetric test kit must be selected based on the expected range of chlorine concentration in the sample.

8.6.7.6.6 Sample preservation

- Samples should be analysed as soon as possible after sampling. For analysis within 24 hours of collection, sample must be refrigerated at 4 °C. Samples should be analysed immediately to prevent the reduction of residual chlorine to chloride, depending on the redox potential.

8.6.7.6.7 Sample preparation

For liquid, slurry, semi-solid or solid samples:

- Solid samples must be dissolved in water before further dilution is performed.
- Samples containing more than 3,000 mg/L Cl must be diluted with distilled water. The pH must be within the range 5-8. Adjust, if necessary, with 0.1 N sodium hydroxide solution or nitric acid.

For slurry to solid samples:

- Weigh out between 1.8 g and 2.0 g of the thoroughly-mixed faecal sludge sample into a beaker.

- Dilute the sample gravimetrically and transfer to the blender, as described in Section 8.4.2.
- If necessary, adjust the dilution based on the specific sludge characteristics.
- Blend for 30 sec on the highest speed.
- Transfer the contents of the blender into a 1 L volumetric flask, rinse the blender twice with 250 mL distilled water and dilute to the mark with distilled water.
- Transfer this 1 L solution into a plastic bottle and store at 4 °C until ready for testing.
- Filter the samples through a 0.45 µm filter paper.
- The sample pH must be within the range of 4-8; if necessary, adjust with sulphuric acid or sodium hydroxide solution.
- Total solids analysis should be performed on the samples so that the results of the chlorine measurement can be expressed as gCl^-/gTS.

8.6.7.6.8 Analysis protocol

For measuring range of 500-3,000 $mgCl^-/L$

- Immerse all the reaction zones of the test strip in the sample (15-25 °C) for 1 second. Shake off the excess liquid from the strip and after 1 min determine with which colour row on the label the colours of the reaction zones coincide most closely. Read off the corresponding result in $mgCl^-/L$.

8.6.7.6.9 Calculation

Result of analysis ($mgCl^-/L$) = measurement value × dilution factor

Liquid samples:

$$\text{Final concentration}\left(\frac{mg}{L}\right) = A\left(\frac{mg}{L}\right) \times DF$$

$$\text{Dry basis}\left(\frac{g}{g}\right) = \frac{\text{Final concentration}\left(\frac{mg}{L}\right)}{\text{Total solids concentration}\left(\frac{mg}{L}\right)}$$

Where:

A = Colorimetric measurement value (mg/L)
DF = Dilution factor (F/I)
F = Final diluted volume (L)
I = Initial sample volume (L)

[46] This method should be cited as: Method 4500 CL-E (Rice *et al.*, 2017) and, if test kits are used, also as per the manufacturer's directions including any modifications.

Slurry, semi-solid and solid samples:

$$\text{Dry basis}\left(\frac{g}{g}\right) = \frac{\frac{A\left(\frac{mg}{L}\right)}{1{,}000} \times \frac{V}{M}\left(\frac{L}{g}\right)}{\text{Total solids content}\left(\frac{g}{g}\right)}$$

Where:

A = Colorimetric measurement value (mg/L)
V = Volume of dilution (L)
M = Wet weight of sludge used in sample preparation (g)

8.6.7.7 Chloride – spectrophotometric method[46]

8.6.7.7.1 Introduction

This method for quantifying chloride in faecal sludge is based on the principle that chloride ions react with mercury (II) thiocyanate to form slightly dissociated mercury (II) chloride. The thiocyanate released reacts with iron (III) ions to form red iron (III) thiocyanate which is quantified spectrophotometrically. Commercially available test kits for measuring chloride based on the standard methods are available with pre-packaged individual aliquots of the necessary chemicals in pillows (dry chemicals) and vials (liquid chemicals). Commonly used test kits from manufacturers such as Hach, Merck, and Hanna vary slightly in the methods they use for chloride measurement. The example provided here is the Merck Spectroquant® Chloride Cell Test (Merck, 2020c)[D] for samples with concentrations of 0.5-15.0 $mgCl^-/L$, and it is based on the manufacturer's protocol for water and wastewater using the standard method 4500 (Rice *et al.*, 2017). For faecal sludge, samples must be diluted and filtered to prevent false high readings associated with turbid solutions.

8.6.7.7.2 Safety precautions

- General health and safety (H&S) procedures specific for conducting laboratory analysis of faecal sludge are presented in Section 8.2. Before conducting this method, it is important to be familiar with Section 8.2.3 to ensure safety measures are properly carried out.

- Appropriate personal protective equipment (PPE) should be used; specific details are covered in Section 8.2.3.1.

8.6.7.7.3 Required chemicals

- Distilled water (free from chloride)
- Bottle of reagent Cl-1K (supplied by the manufacturer)
- Chloride standard solution (2.50 mg/L Cl) can be purchased from the manufacturer or prepared as described in Rice *et al.* (2017) using the chloride automated ferricyanide method.

8.6.7.7.4 Required apparatus and instruments

- Spectrophotometer
- Analytical balance with weighing boats (slurry to solid samples)
- Blender (slurry to solid samples)
- Pipette and pipette tips
- Sample cell (supplied by the manufacturer)
- Volumetric flask (1 L)
- Glass beakers (50 or 100 mL)
- Glass storage bottle
- Laboratory cleaning tissues
- Filter paper (adequate for removing solids from sample, for example a 0.45 µm filter for liquid samples).

8.6.7.7.5 Quality control

General information on quality assurance and quality control (QA/QC) is provided in Section 8.3. Information on standards, operating conditions and interferences that are specific to this method includes:

- Perform quality control with chloride standard solution with every test batch (or on a daily or weekly basis depending on the testing load).
- For spectrophotometric measurements, the sample cells must be clean. Before analysis, wipe with a laboratory cleaning tissue.
- Measurement of turbid solutions yields false high readings. For faecal sludge, samples should be diluted with an appropriate dilution factor through serial dilutions and filtered based on the type of sludge for accurate measurements.
- The colour of the measurement solution remains stable for up to 60 min after the end of the reaction time; thus the spectrophotometric measurement should be conducted within that timeframe.
- Common interferences in chloride measurement include calcium, potassium, magnesium, sodium, ammonium, nitrates and nitrites. For specific concentrations refer to the manufacturer's instructions.

8.6.7.7.6 Sample preservation

- Samples for chloride analysis must not be stored. Chloride samples must be analysed immediately to prevent reduction of residual chloride to chloride depending on the redox potential (Rice *et al.*, 2017).

8.6.7.7.7 Sample preparation

Liquid samples:

- Samples containing more than 15 mgCl/L must be diluted with distilled water to within the range of 0.5-15.0 $mgCl^-/L$.
- Turbid solutions falsely increase the spectrophotometric reading. Therefore, turbid samples should be filtered with a 0.45 µm filter paper to prevent false high measurements.

Slurry to solid samples:

- Weigh out between 1.8 g and 2.0 g of the thoroughly-mixed faecal sludge sample into a beaker.
- Dilute the sample gravimetrically and transfer to the blender, as described in Section 8.4.2.
- If necessary, adjust the dilution based on the specific sludge characteristics.
- Blend for 30 sec on the highest speed.
- Quantitatively transfer the contents of the blender into a 1 L volumetric flask, rinse the blender twice with 250 mL distilled water and dilute to the mark with distilled water.
- Transfer this 1 L solution into a plastic bottle and store at 4 °C until ready for testing.
- Filter the samples through a 0.45 µm filter paper.
- Total solids analysis should be performed on the samples so that the results of the chloride measurement can be expressed as mgCl/gTS.

8.6.7.7.8 Analysis protocol

Calibration

Follow the spectrophotometer manufacturer's instructions for calibration. It may be necessary to calibrate the instrument before every reading, or it may only be necessary to perform periodic calibration checks to determine when calibration is necessary. For this method, calibration can be performed as follows:

- Prepare a series of at least four different concentrations of a standard solution making sure to include the lowest and highest concentration of the kit testing range. It is typical to do serial dilutions or dilutions with a uniform interval including the lowest and highest concentrations.
- Determine the chloride concentration of the standard solutions.
- Multiply the answer by the dilution factor and report the results in mgCl/L.
- Prepare a calibration curve by plotting the instrument response against the standard concentration.

Procedure

For measuring range of 0.5-15.0 mg/L chloride

- Pipette 10 mL of the sample into a reaction cell.
- Pipette 0.25 mL of reagent Cl-1K and add to the sample in the reaction cell and mix.
- Leave to stand for 10 min (reaction time).
- Wipe the sample cell with a laboratory tissue to clean it and then measure in a spectrophotometer.

8.6.7.7.9 Calculations

Result of analysis (mg/L)

Liquid and slurry samples:

$$\text{Final concentration}\left(\frac{\text{mg}}{\text{L}}\right) = \text{A}\left(\frac{\text{mg}}{\text{L}}\right) \times \text{DF}$$

$$\text{Dry basis}\left(\frac{\text{g}}{\text{g}}\right) = \frac{\text{Final concentration}\left(\frac{\text{mg}}{\text{L}}\right)}{\text{Total solids concentration}\left(\frac{\text{mg}}{\text{L}}\right)}$$

Where:

A = Spectrophotometric measurement value (mg/L)

DF = Dilution factor (F/I)

F = Final diluted volume (L)

I = Initial sample volume (L)

Slurry, semi-solid and solid samples:

$$\text{Dry basis}\left(\frac{\text{g}}{\text{g}}\right) = \frac{\frac{\text{A}\left(\frac{\text{mg}}{\text{L}}\right)}{1{,}000} \times \frac{\text{V}}{\text{M}}\left(\frac{\text{L}}{\text{g}}\right)}{\text{Total solids content}\left(\frac{\text{g}}{\text{g}}\right)}$$

Where:

A = Spectrophotometric measurement value (mg/L)

V = Volume of dilution (L)

M = Wet weight of sludge used in sample preparation (g)

8.6.8 Colour and turbidity

Colour and turbidity in faecal sludge are related to concentrations of suspended particles, dissolved organic matter, and inorganic compounds. Colour imparted by dissolved compounds is referred to as 'true colour', whereas the 'apparent colour' of a sample refers to the colour contributed from soluble and suspended material. Turbidity is the cloudiness or haziness of a fluid resulting from suspended and colloidal material. Colour and turbidity can be used as indicators of concentration of particulate matter in faecal sludge, and colour can further indicate concentrations of soluble material. Colour and turbidity can be used to characterise both untreated faecal sludge and effluent from a treatment process. Colour can also be used as an indicator for the level of stabilisation.

Colour can be measured by visual comparison with standards, which works for both apparent and true colour. Other standard methods for colour analysis are spectrophotometric, and are only applicable for filtered samples (2120C, D, E, F in Rice *et al.*, 2017). A method using digital image analysis of colour-corrected photographs for monitoring colour and texture of faecal sludge is also in development by Eawag (Ward *et al.*, 2021). The preferred method for measuring turbidity in faecal sludge, water, and wastewater samples is the nephelometric method, due to its precision and sensitivity over a wide range of values of turbidity.

8.6.8.1 Colour – visual comparison method[47]

8.6.8.1.1 Introduction

In a visual comparison method, the colour of a sample is determined by comparison to a set of colour standards. The standards are typically either a set of calibrated concentrations of coloured solutions or coloured glass discs. A comparison between the colour of a sample and the colour standards are made under identical conditions of illumination. This method can be used to measure true colour and apparent colour.

Standard methods exist for quantifying the colour of water and wastewater samples, which are typically much more dilute and lower in particulate matter than faecal sludge. Effluent from faecal sludge treatment processes will likely be relatively uncomplicated to characterise using existing methods for water and wastewater; however, untreated faecal sludge may be outside the range of the platinum-cobalt standard colour scale (*e.g.* untreated faecal sludge can be green or light grey as well as shades of brown, and higher TS samples may be heterogeneous in colour). Existing standard methods may not be applicable for semi-solid or solid sludges, as they were designed for liquid samples. Another method under development by Eawag is the digital image analysis of colour-corrected photographs of 10 mL aliquots of faecal sludge (Ward *et al.*, 2021). The method presented in Ward *et al.* (2021) does not rely on the platinum-cobalt colour scale, but instead uses a colour checker chart to standardise colours for image analysis, as shown in Figure 8.11. This method enables the characterisation of faecal sludge with any TS, and is able to accommodate heterogeneous colours along with texture analysis.

Step-by-step procedures for colour characterisation will vary depending on the laboratory, the available equipment, and the characteristics of the incoming faecal sludge samples. One example is the visual colour comparison method used by the UKZN PRG laboratory in Durban, South Africa that is described here, which is based on the Hach colour test field method (Hach, 2016)[D]. In this method, the colour is determined by visual comparison of the sample with calibrated glass discs, which represent the colours of specific concentrations of platinum-cobalt standard solutions. The results are reported in PtCo units.

8.6.8.1.2 Safety precautions

- General health and safety (H&S) procedures specific for conducting the laboratory analysis of faecal sludge are presented in Section 8.2. Before conducting this method, it is important to be familiar with Section 8.2.3 to ensure safety measures are properly carried out.
- Appropriate personal protective equipment (PPE) should be used; specific details are covered in Section 8.2.3.1.

8.6.8.1.3 Required chemicals

- Deionised water.

8.6.8.1.4 Required apparatus and instruments

- Colour disc, 0-100 colour units, platinum-cobalt scale
- Colour comparator box
- Long path adapter
- Glass viewing tubes, 18 mm.

8.6.8.1.5 Quality control

General information on quality assurance and quality control (QA/QC) is provided in Section 8.3. Information on standards, operating conditions and interferences that are specific to this method includes:

- The colour value of water can be pH-dependent. Some laboratories measure the colour value at the natural pH of the sample, while some laboratories adjust samples to neutral pH before measuring colour. In cases where pH is adjusted before colour measurement, pH adjustment and the original sample pH should be reported along with the colour value. In every case, the pH of the

[47] This method should be cited as the specific method that is carried out in each laboratory, including the type of standards, the make and model of equipment, and the exact method of sample preparation. Existing standard methods for the measurement of colour in liquid samples can be found in the methods 2120B-2120E from the Standard Methods for the Examination of Water and Wastewater (Rice *et al.*, 2017).

sample during colour measurement should be reported with the colour value.

- When measuring true colour, pre-treatment has to be carried out to remove turbidity. When reporting the true colour value, specify the details of the pre-treatment method to ensure comparability with other results.
- In practice, an apparent colour test will be applied prior to filtration/centrifugation and a true colour test will be applied after filtration/centrifugation. In this way 'apparent colour' tests measure all the colour in a sample, irrespective of how it is caused. Even slight turbidity causes the measured colour to be noticeably higher (or different) to the same sample without turbidity.
- Samples are typically contained in some type of vessel and the perception of colour may be influenced by the background colour of the containing surface. The perception of colour against a white or light background can be quite different from a black or dark background. These issues should be minimised by always characterising colour with a consistent background (*e.g.* using a colour comparator box) and consistent lighting conditions. As the light source can affect the perception of colour, information about the light source should be reported with the results. Important information to include could be time of day and season if using sunlight, and colour temperature and light intensity if using a lamp.

8.6.8.1.6 Sample preservation

It is recommended that colour is measured immediately after sample collection. If immediate measurement is not possible, the samples should be stored at 4 °C for no longer than 24 hr before analysis, and storage time and temperature noted and reported. The pH of the samples should not be altered for preservation, as this may affect the colour.

8.6.8.1.7 Sample preparation

If measuring apparent colour:

- Homogenise the sample thoroughly (see Section 8.4.2).

If measuring true colour:

- Homogenise the sample thoroughly (see Section 8.4.2).
- Filter (using a 0.45 μm filter) or centrifuge the sample to remove suspended matter.
- Note: centrifugation and filtration affect the colour, so it is important to report the sample preparation procedures along with the results.

8.6.8.1.8 Analysis protocol

- Measure and record the pH of the sample. Report the pH at which the colour was determined with the results.
- Load the colour disc and long path adapter into the colour comparator box.
- Fill a glass viewing tube to the top line with deionised water as a blank, and load the blank into the left opening of the colour comparator box.
- Fill a second glass viewing tube to the top line with the homogenised sample, and load the sample tube into the second opening in the colour comparator box.
- Hold the colour comparator box up to a light source (*e.g.* sunlight or lamp for a consistent light source). Turn the colour disc to find the colour match.
- Read and record the result displayed in platinum cobalt colour units in the scale window.

8.6.8.1.9 Calculation

No calculation required – direct reading.

8.6.8.1.10 Data set example

Meng *et al.* (2020) measured the colour intensity of effluent from anaerobic digesters by comparison with platinum-cobalt standard solutions within the range of 5-300 PtCo units, following Method 2120B from Rice *et al.* (2017). The effluent colour was influenced by the treatment technology employed. In this case, the effluent was slightly darker if a free nitrous acid pre-treatment was used before anaerobic digestion (1,667 ± 27 mg PtCo/L for pre-treated sludge, compared with 1,433 ± 27 mg PtCo/L for a control). A darker colour corresponded to higher concentrations of soluble COD in the effluent.

Ward *et al.* (2021) presented a new method for characterising the colour and texture of faecal sludge using a standard colour checker chart with subsequent colour correction assuming standard natural lighting (CIE Standard Illuminant D65) before digital image analysis. The average RGB (red, green, blue) colour of a faecal sludge sample in a colour-corrected image was calculated by averaging the R, G, and B values in a selected sample. RGB colours were then converted into HSV (hue, saturation, value). Examples of colour-corrected images are shown in Figure 8.11. In general, samples from pit latrines were more saturated in colour than samples from septic tanks. More saturated colour corresponded to higher COD concentrations, and colour and turbidity measurements were strongly correlated. The entire raw data set will be included with publication.

Figure 8.11 Example of photographs with standard colour checker chart used for colour characterisation (Ward *et al.*, 2021).

8.6.8.2 Turbidity – nephelometric method[48]

8.6.8.2.1 Introduction

Turbidity is used in faecal sludge management as an indicator of suspended solids concentration, for example in liquid fractions after solid-liquid separation processes (*e.g.* filtrate from drying beds, supernatant from a settling tank). Turbidity of supernatant or filtrate is sometimes used as a metric to indicate how well a treatment process is removing solids from the liquid stream. Examples of evaluation of supernatant turbidity are given in Section 2.3.4.1. Turbidity measured using the nephelometric method should be reported in nephelometric turbidity units (NTU). A consideration for measuring the turbidity of faecal sludge samples is that there are upper limits to accurate turbidity measurements. If the sample is too concentrated, turbidity readings may not be accurate. More concentrated samples with turbidity exceeding the measurement range of the instrument can be measured, but they must first be diluted. Dilution is generally not recommended, as this can affect the behaviour of the suspended solids in unanticipated ways, and introduce error to the measurement. For this reason, turbidity is mostly relevant for samples that have already undergone an initial solid-liquid separation step.

8.6.8.2.2 Safety precautions

- General health and safety (H&S) procedures specific for conducting the laboratory analysis of faecal sludge are presented in Section 8.2. Before conducting this method, it is important to be familiar with Section 8.2.3 to ensure safety measures are properly carried out.
- Appropriate personal protective equipment (PPE) should be used; specific details are covered in Section 8.2.3.1.
- If using formazine calibration standards: exercise caution when working with hydrazine sulfate. It is carcinogenic. Do not inhale, ingest, or let it contact skin. Take care when handling formazine suspensions, as they may contain residual hydrazine sulfate.

[48] This method is based on Method 2130B from the Standard Methods for Examination of Water and Wastewater (Rice *et al.*, 2017). This method should be cited as: Method 2130 (Rice *et al.*, 2017).

8.6.8.2.3 Required chemicals

- Chemicals for calibration standards (0)
- Low-turbidity dilution water.

8.6.8.2.4 Required apparatus and instruments

- Nephelometer
- Sample cells (clear glass or plastic)
- Lint-free tissue
- Ultrasonic bath (not necessary, but helpful for dissipation of bubbles).

8.6.8.2.5 Quality control

General information on quality assurance and quality control (QA/QC) is provided in Section 8.3. Information on operating conditions and interferences that are specific to this method includes:

- The type, concentration and number of standards required to calibrate the nephelometer is dependent on the instrument, as is the frequency of calibration. General recommendations from USEPA Method 180.1 state that determination of the linear calibration range must use at least three standards and one blank. At a minimum, a calibration check with one standard and one blank should be analysed with every batch of samples. If a calibration check is not within ± 10% of the expected value, the instrument must be recalibrated. If a portion of the calibration range is nonlinear, a sufficient number of standards must be used to define the nonlinear relationship (USEPA, 1993).
- If a sample reading is outside of the calibration curve, it should be diluted 1:1 and re-characterised.
- Air bubbles in the sample will result in incorrectly high readings. Bubbles should be dissipated before measurement.
- Allowing particles to settle to the bottom will result in incorrectly low readings. Samples should be homogenised and particles resuspended directly before measurement.
- The presence of particles that absorb light (*e.g.* activated carbon) can result in low readings.
- Smudged, dirty, or scratched sample cells will result in incorrect measurements. To avoid damage, sample cells should not be cleaned with solvents. Cells must be discarded if scratched or not clear.
- Sample cells should not be touched where the light beam will pass through (to avoid getting dirty fingerprints on the sample cell).

8.6.8.2.6 Sample preservation

It is recommended that turbidity is measured immediately after sample collection. If immediate measurement is not possible, the samples should be stored at 4 °C for no longer than 24 hr before analysis, and the storage time and temperature noted and reported. The pH of the samples should not be altered for preservation, as this may affect turbidity.

8.6.8.2.7 Sample preparation

Before a measurement is taken, allow the sample to reach room temperature. Ensure the samples are thoroughly homogenised by stirring or shaking.

8.6.8.2.8 Analysis protocol

Preparation of standards

Standards used for calibration of the nephelometer can either be made in the laboratory, or procured from the manufacturer of the nephelometer or a manufacturer of chemical standards. The manufacturer's instructions should always be consulted as to which calibration standards to use for a specific instrument. See Rice *et al.* (2017) Section 2130B for a detailed explanation of the different types of turbidity standards available. The following instructions on preparation of low-turbidity dilution water and formazine standards are from Rice *et al.* (2017), Section 2130B:

Low-turbidity dilution water (≤ 0.02 NTU)

- Filter laboratory-grade distilled water through a filter capable of removing particles larger than 0.1 µm.
- Rinse a collection flask at least twice with filtered water and discard the next 200 mL.
- Alternatively, if low-turbidity dilution water cannot be produced in the laboratory, some commercial bottled demineralised waters can be used, provided the turbidity is ≤ 0.02 NTU.

Formazine standard stock suspension (4,000 NTU)

- Add 1.00 g hydrazine sulfate, (NH2)2 H2SO4 to low-turbidity dilution water and dilute to 100 mL.

- Add 10.00 g hexamethylenetetramine (CH2)6N4 to low-turbidity dilution water and dilute to 100 mL.
- Mix 5.0 mL of the hydrazine sulfate solution + 5.0 mL of the hexamethylenetetramine solution in a flask. Let stand for 24 hr at 25 ± 3 °C. The resulting suspension has a turbidity of 4,000 NTU.
- Store 4,000 NTU stock solution in an amber or opaque bottle. Stock suspension will remain stable for up to 1 year if properly stored.

Dilute formazine standard suspensions

- Dilute 4,000 NTU formazine standard stock suspension with low-turbidity dilution water to produce standard suspensions of various turbidities. Prepare immediately before use and discard directly after use. Typical standard suspension values could be *e.g.* 1, 10, 100, 500, 1,000, 2,000 NTU, depending on the nephelometer.

Calibration

Follow the nephelometer manufacturer's instructions for calibration. Calibration procedures vary between instruments. Certain instruments will require calibration with specific standards provided by the manufacturer, with a self-prepared formazine standard used only when no other options are available. It may be necessary to calibrate the instrument before every reading, or it may only be necessary to perform periodic calibration checks to determine when calibration is necessary (Section 8.3.4.2). Depending on the instrument, it may automatically integrate the calibration measurements, or a calibration curve may need to be manually prepared.

Procedure

- Follow the nephelometer manufacturer's instructions for measurement.
- Fill a sample cell with a homogenised, representative sample. Take care not to touch parts of the sample cell through which light will pass (*e.g.* only handle the top of the cell). Ensure that no bubbles are suspended in the sample. If possible, sonicate the cell in an ultrasonic bath for 1-2 sec to dissipate bubbles.
- Wipe the cell with a soft, lint-free cloth or laboratory tissue to remove water spots and fingerprints.
- Follow the instrument instructions to measure turbidity. Allow time for the displayed turbidity value to stabilise. Record the turbidity value (NTU) from the instrument display.

8.6.8.2.9 Calculation

No calculation – direct reading.

If the sample was diluted, multiply the reading by a dilution factor to calculate the turbidity of the undiluted sample.

Report turbidity measurements with the following precision (from Rice *et al.*, 2017 and USEPA, 1993):

Turbidity reading (NTU)	Record to the nearest (NTU)
0.0-1.0	0.05
1-10	0.1
10-40	1
40-100	5
100-400	10
400-1,000	50
> 1,000	100

8.6.8.2.10 Data set example

Ward *et al.* (2021) evaluated the turbidity of supernatant following the centrifugation of faecal sludge samples from septic tanks and pit latrines in Lusaka, Zambia. A Hach 2100N nephelometer with an upper limit of 2,000 NTU was used, following AWWA standard method 2130B (Rice *et al.*, 2017). Supernatant turbidity for 179 samples of faecal sludge from septic tanks had a median turbidity of 100 NTU, an average of 180 NTU, and a standard deviation of 230 NTU. Supernatant turbidity for 46 samples of faecal sludge from pit latrines had a median of 650 NTU, an average of 850 NTU, and a standard deviation of 800 NTU. Only three supernatant samples initially measured were above 2,000 NTU. These were diluted 1:1 and remeasured, but were ultimately excluded from the dataset due to concerns about error introduced by dilution. Examples of measurement error (average values ± standard deviation) of triplicate measurements of three individual samples are: 21 ± 1 NTU, 90 ± 0 NTU, and 340 ± 10 NTU. The entire raw data set is included with publication.

Junglen *et al.* (2020) measured the turbidity of faecal sludge samples from pit latrines in Nairobi, Kenya following AWWA standard method 2130 (Rice *et al.*, 2017). This study evaluated the use of turbidity as an indicator of TSS in influent sludge. Turbidity readings were highly variable and were very high, requiring dilution before measurement. The turbidity of six pit latrine sludge samples from different containments had a median value of 13,300 NTU, an average of 14,600 NTU, and a standard deviation of 11,000 NTU.

8.6.9 Settleability and dewaterability

Settleability and dewaterability describe different aspects of the process of separating liquids and solids during faecal sludge treatment. Settleability is a description of how well sludge settles, and can be described by multiple characteristics, including how fast it settles, how compact the settled sludge is, and how many unsettled solids remain. Dewaterability is a description of how well liquid can be removed from sludge, and includes characteristics such as how quickly it can be filtered, the characteristics of the filtrate, the moisture content of the sludge cake after it has been dewatered, and how strongly water is bound to the sludge solids. More information about why and when to characterise metrics of settleability and dewaterability is included in Chapter 4.

This section includes a method for evaluating the selection and dose of conditioners for improved settling and dewatering (the jar test method), two methods for characterising dewaterability (the capillary suction time method and the water activity method), and one method for characterising settleability (the sludge volume index method).

8.6.9.1 Jar test[49]

8.6.9.1.1 Introduction

The jar test method is used to test the efficacy of a treatment process for removal of suspended solids from faecal sludge. This method can be used to identify required mixing time and intensity, and optimal conditioner and dose. Conditioners can be inorganic chemicals such as lime, ferric chloride, or aluminium sulphate, or they can be polymers. The aim of conditioners is to improve settling and dewatering performance by destabilising suspended particles in faecal sludge to form larger aggregates. Objectives of conditioning can include: supernatant turbidity reduction, compact settled sludge cake formation, reduced filter clogging and faster filtration, and lower moisture content in dewatered sludge. Jar tests enable controlled testing of different types and doses of conditioners to evaluate which yields the optimal settling or dewatering performance.

8.6.9.1.2 Safety precautions

- General health and safety (H&S) procedures specific for conducting laboratory analysis of faecal sludge are presented in Section 8.2. Before conducting this method, it is important to be familiar with Section 8.2.3 to ensure safety measures are properly carried out.
- Appropriate personal protective equipment (PPE) should be used; specific details are covered in Section 8.2.3.1.

8.6.9.1.3 Required chemicals

- Conditioner standard solution (also called 'makedown' solution). Follow the manufacturer's instructions for the dilution factor of the standard solution. If no manufacturer's recommendations are available, typically 0.1-1.0% dilutions work best.

8.6.9.1.4 Required apparatus and instruments

- Jar test apparatus with gang stirrers with variable rpm settings.
- 1,000 mL beakers (one for each stirrer in jar test apparatus).
- Syringes or pipettes (10 mL).
- Analytical balance (for weighing out conditioners for standard solutions).

[49] This method is expanded from the jar test procedure outlined in Chapter 5.2 Conditioning in Faecal Sludge Management: Highlights and Exercises (Ward and Strande, 2019), and includes input from several industrial jar testing protocols for water and wastewater (SNF (2015), Microdyn Nadir (2020), and Christophersen (2000). This method should be cited as: Ward and Strande, 2019, as described in Velkushanova *et al.* (2021).

- Blender/homogeniser, or bottle with secure cap (for mixing polymer standard solutions).
- Timer.

8.6.9.1.5 Quality control

General information on quality assurance and quality control (QA/QC) is provided in Section 8.3. Information on standards, operating conditions and interferences that are specific to this method includes:

- Always include a blank faecal sludge sample with no added conditioner.
- Store the conditioners in accordance with the manufacturer's instructions. Polymer conditioners are often sold as concentrated polymer emulsions that are subsequently diluted to make conditioner standard solutions. Concentrated polymer emulsions may separate over time and will need to be re-mixed before making standard solutions.
- Cationic polymer emulsions are unstable in water with high levels of hardness or alkalinity. Standard solutions may need to be re-made if they are more than a few hours old.
- Before addition to the sample, ensure thorough mixing of conditioner standard solutions, especially polymer conditioners. This can be done using a blender or homogeniser, or by shaking thoroughly inside a capped bottle. Keep homogenising until all the visible droplets of polymer are gone.
- In cases when both a coagulant and a flocculant are being tested for use together, determine the correct selection and optimal dose of coagulant first. Dose the faecal sludge with the optimal dose of coagulant before evaluating flocculants, then repeat the test procedure with different flocculant doses.
- Avoid homogenising a faecal sludge sample in a way that would destroy particles (*e.g.* with a blender or homogeniser), instead mix with a spoon or stirrer, or pour back and forth between beakers to combine. Good practice is to not mix any faster than the highest rpm setting of the jar test.
- Ensure that the jar tests are performed on faecal sludge samples that are at the same temperature. Allow the samples to reach room temperature before proceeding with the jar test.
- Jar tests are only appropriate for use with liquid and slurry samples.

8.6.9.1.6 Sample preservation

It is recommended that jar tests are performed immediately after sample collection. If this is not possible then samples should be stored at 4 °C for no longer than 7 days before analysis, and storage time noted and reported.

8.6.9.1.7 Sample preparation

- Allow all the samples to reach room temperature.
- Homogenise the faecal sludge sample well by carefully stirring or pouring back and forth between beakers.
- Characterise TS (Method 8.6.1.1) or TSS (Method 8.6.1.3) of the faecal sludge sample, as these values will be needed to calculate conditioner doses.

8.6.9.1.8 Analysis protocol

- Fill each beaker with 1 L (or consistent volume) of the faecal sludge to be evaluated. Record the volume.
- Set the gang stirrers to 100 rpm.
- Add conditioner doses to each beaker, leaving one beaker unconditioned as a blank (example: from left to right, 0, 20, 40, 60 mL). If working with polymer solutions, ensure complete mixing by using a syringe or pipette submerged in the liquid sample to inject the polymer solution about halfway between the beaker wall and the stirrer.
- High-speed mixing: mix at 100 rpm for 2 min. This is a suggestion based on typical operation: the mixing speed and duration can be adjusted to mimic the conditions in a specific treatment process.
- Observe the jars during the high-speed mixing. Note which dose formed the first flocs. Note also which dose produces the clearest supernatant and/or largest floc size.
- After the high-speed mixing time has elapsed, reduce the stirring speed to 30-40 rpm (slow-speed mixing). Continue mixing for 5 to 20 min. It is recommended to adjust the mixing speed and duration in this step to better match the treatment process configuration.
- After the slow speed mixing time has elapsed, turn off the mixer. Let the conditioned samples settle for 15-20 min, or select an amount of time to

reflect the settling time in a specific treatment process.

- After the settling time has elapsed, note the supernatant and floc appearance. Taking pictures may be helpful. If further characterisation is desired, now is the time to sample the supernatant for turbidity (Method 8.6.8.2), TSS (Method 8.6.1.3), or COD measurements (Method 8.6.2.1 or 8.6.2.2), and/or to conduct further quantification of the SVI (Method 8.6.9.4), CST (Method 8.6.9.2), or specific dewaterability of the conditioned sludge.

8.6.9.1.9 Calculation

$$\text{Conditioner dose}\left(\frac{g}{kg\ TS}\right) =$$

$$\frac{V_c\,(mL) \times C_c\left(\frac{g}{mL}\right)}{TS_{FS}\left(\frac{g}{L}\right) \times V_{FS}(L)} \times \frac{1{,}000\ g}{kg}$$

Where:

V_c = Volume of conditioner solution added to beaker (mL)

Cc = Concentration of conditioner solution (g/mL)

TS_{FS} = Total solids in the faecal sludge sample (g/L)

V_{FS} = Volume of the faecal sludge sample in the beaker (L)

8.6.9.1.10 Data set example

Gold *et al.* (2016) used jar tests to evaluate the performance of different conditioners at a range of doses for faecal sludge samples from Dakar, Senegal. A Velp Scientifica FC6S^{D} jar test apparatus was used. A high-speed mixing step at 200 rpm for 2 min was selected, with no follow-up slow-speed mixing step.

Moto *et al.* (2018) used jar tests to identify the optimal doses of chitosan and Moringa seed powder for faecal sludge from Dar es Salaam, Tanzania. A high-speed mixing step at 100 rpm for 2 hr was selected to mimic mixing conditions at a pilot scale treatment facility. More details about this study can be found in Case Study 4.1 in Chapter 4.

8.6.9.2 Capillary suction time[50]

8.6.9.2.1 Introduction

Capillary suction time (CST) is a measure of the rate of water release from sludge. This measurement is used as an indicator for the performance of many faecal sludge dewatering processes, for example dewatering time on drying beds and performance with geotextile filtration and mechanical presses. CST measurements can be used to evaluate different doses and types of conditioners, and are often used in combination with jar testing (Method 8.6.9.1). CST is measured by pouring sludge into a small reservoir placed on top of a sheet of chromatography paper. The water in the sludge is drawn into the chromatography paper via capillary action. The time it takes the water to travel a certain distance along the chromatography paper is recorded by a set of electrodes in contact with the paper. CST is reported in seconds or normalised by the sample solids content and reported as s L/gTS or s L/g TSS.

8.6.9.2.2 Safety precautions

- General health and safety (H&S) procedures specific for conducting laboratory analysis of faecal sludge are presented in Section 8.2. Before conducting this method, it is important to be familiar with Section 8.2.3 to ensure safety measures are properly carried out.
- Appropriate personal protective equipment (PPE) should be used; specific details are covered in Section 8.2.3.1.

8.6.9.2.3 Required chemicals

- Distilled water.

8.6.9.2.4 Required apparatus and instruments

- CST Apparatus with 18 mm reservoir (s) (Figure 8.12). Commercially available from Triton Electronics Ltd., Essex, England.
- CST paper (supplied by the manufacturer CST, or Whatman No. 17 chromatography paper cut into 7 × 9 cm sections, with grain parallel to 9 cm side).
- Thermometer (accuracy of ±0.5 °C).
- Beaker with pourable spout.

[50] This method is based on Method 2710 G from the Standard Methods for the Examination of Water and Wastewater (Rice *et al.*, 2017) with adaptations based on specific experience with faecal sludge. This method should be cited as: Method 2710 G (Rice *et al.*, 2017), as adapted in Velkushanova *et al.* (2021).

8.6.9.2.5 Quality control

General information on quality assurance and quality control (QA/QC) is provided in Section 8.3. Information on operating conditions and interferences that are specific to this method includes:

- Measure the CST of distilled water, and record the measurement. Repeat this with every new box of filter paper. CST results are normalised by subtracting the CST of distilled water, which should account for the differences in performance due to variability in the batches of filter paper.
- Avoid homogenising the sample in a way that would destroy flocs (*e.g.* with a blender or homogenizer), instead mix with a spoon or stirrer, or pour back and forth between beakers to combine.
- Variations in sludge temperature can affect CST results. Ensure that CST tests are performed on faecal sludge samples that are at the same temperature. Allow the samples to reach room temperature before proceeding with the CST measurement.
- Sludge suspended solids concentration has a significant effect on the test results. This effect can be mitigated by homogenising well when performing replicates of the same sample. When comparing different samples, a rough correction can be made for sludges with different solid concentrations by dividing the CST value by the TS (Method 8.6.1.1) or TSS (Method 8.6.1.3) concentration of the sludge.
- CST values typically have a high variability for replicates of the same faecal sludge samples. It is suggested to perform a minimum of 5 replicates for each faecal sludge sample characterised. Up to a 10% relative standard error between replicates is considered acceptable.
- CST apparatus typically comes with reservoirs of two different diameters (10 mm and 18 mm). It is conventional to use the 18 mm diameter reservoir with faecal sludges. Reservoir diameter has a significant impact on the CST result, which is important to note when comparing results from other studies. Reservoir diameter should always be reported with experimental results.
- CST analysis is only appropriate for use with liquid and slurry samples.

8.6.9.2.6 Sample preservation

It is recommended that CST measurements are performed immediately after sample collection. If this is not possible then samples should be stored at 4 °C for no longer than 7 days before analysis, and the storage time noted and reported.

8.6.9.2.7 Sample preparation

- Allow the sample to reach room temperature.
- Homogenise the faecal sludge sample well by stirring or pouring back and forth between the beakers.
- Characterise the TS (Method 8.6.1.1) or the TSS (Method 8.6.1.3) of the faecal sludge sample, as these values will be used to normalise the effect of the solid concentration on the CST.

8.6.9.2.8 Analysis protocol

- Turn on and reset the CST apparatus. Plug the test block(s) into the CST apparatus (See Figure 8.12).
- For each test block: place new filter paper on the lower test block, with the rough side up. Add the upper test block, then insert an 18-mm funnel into the test block and seat it using light pressure and a quarter turn to prevent leaks where it meets the CST paper.
- Measure and record the temperature of the faecal sludge.
- Ensure the sludge is homogenised by stirring or pouring back and forth. Then pour a representative sludge sample into the reservoir until the liquid level reaches the top (approximately 7 mL). The CST apparatus will begin time measurement when the liquid being drawn into the CST paper reaches the inner pair of electrical contact points. Timing ends when the liquid reaches the outer contact points.
- When timer on digital display stops, record the CST value (in seconds).
- Empty the remaining sludge from the reservoir and remove and discard the used CST paper. Rinse and dry the test block and reservoir.

8.6.9.2.9 Calculations

$$CST_{measured}(s) - CST_{Distilled\ water}(s) = CST_{adjusted}(s)$$

$$\frac{CST_{adjusted}(s)}{TS\ of\ faecal\ sludge\ (\frac{g}{L})} = CST_{normalized}(\frac{s \cdot L}{gTS})$$

$$\frac{CST_{adjusted}(s)}{TSS\ of\ faecal\ sludge\ (\frac{g}{L})} = CST_{normalized}(\frac{s \cdot L}{gTSS})$$

8.6.9.2.10 Data set example

Ward *et al.* (2021) measured the CST of 217 faecal sludge samples from septic tanks and pit latrines in Lusaka, Zambia using a Triton 319 Multi-CST apparatus with 18-mm funnels. CST values were adjusted by subtracting the CST of distilled water. As the objective of this study was to compare the time it took different samples to filter, CST was reported in seconds and not normalised by TS or TSS. Four replicates were performed for each sample, and the average relative standard error of the replicate CST measurements was 5%. The entire raw data set is included with publication[51].

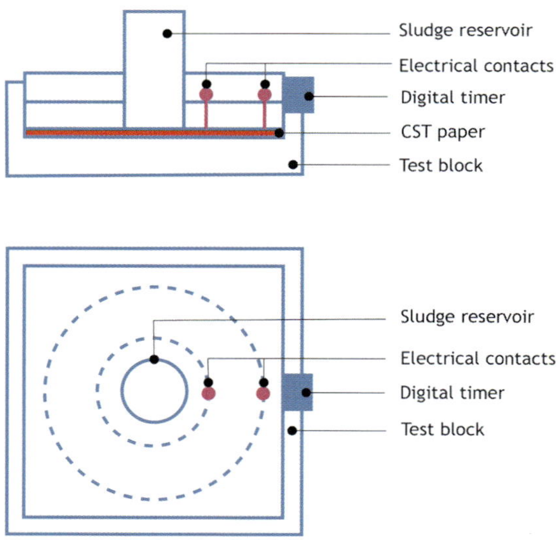

Figure 8.12 CST apparatus in side view (above) and top-down view (below).

8.6.9.3 Water activity[52]

8.6.9.3.1 Introduction

Water activity is a thermodynamic parameter that is defined as the vapour pressure of water in a sample divided by the vapour pressure of pure water in the same conditions. Water activity is an indicator of the binding strength of moisture within a faecal sludge sample. Water activity (often referred to as a_w) is unitless, and values can range from 0 to 1. A water activity of 1 would indicate that none of the water in the sample is bound to the solids, and is easily removable, whereas a water activity of 0 would mean that all of the water in the sample is strongly bound to the solids (Stringel, 2020). Water activity is also an indicator of microbial activity within the sample, as microorganisms cannot survive in environments with low levels of free water (typically no growth occurs at water activities less than 0.62, and most pathogenic organisms are inactivated at water activities less than 0.86) (Barbosa-Cánovas *et al.*, 2003). Measuring water activity can provide information for the design of faecal sludge and faeces treatment technologies that incorporates drying and dewatering processes, and can provide information on the required level of dryness in order to limit microbial activity during storage of dried solids.

There are a number of methods for measuring water activity; however, only two have been reported in the literature for the characterisation of faecal sludge: 1. the static gravimetric saturated salts method; and 2. the chilled mirror dew point hygrometer method using a water activity meter. For the static gravimetric saturated salts method, sludge samples are sealed in containers with selected saturated salt solutions to produce a range of relative humidity conditions (Bourgault *et al.*, 2019). Samples are stored for several weeks at a constant temperature to reach equilibrium conditions. Once the samples have equilibrated, the equilibrium moisture content is determined by measuring TS. Equilibrium moisture content values are used to calculate water activity, and can be plotted with relative humidity to produce a sorption isotherm. This method can be challenging

[51] https://doi.org/10.25678/00037X

[52] This method should be cited as adapted in Velkushanova *et al.* (2021), together with the specific analytical equipment (if different), and any manufacturer's modifications

due to the time it takes to reach equilibrium and special precautions that need to be taken to ensure that bacterial growth does not occur. An alternative method for measuring water activity is the use of an automated water activity meter (Stringel, 2020; Getahun *et al.*, 2020). In contrast to the saturated salts method, water activity can be quantified in several minutes. One of the most popular water activity meters is the AquaLab Series produced by Meter, because of its relative accuracy, precision, rapidity and ease of use. AquaLab water activity meters use the chilled mirror dew point hygrometer method to characterise water activity (Barbosa-Cánovas *et al.*, 2003). This method works by sealing a sample inside an equilibration chamber and equilibrating the liquid phase water in the sample with the vapour phase water in the headspace and calculating the relative humidity in the headspace (using the dew point temperature of the air and sample temperature). When the sample is at equilibrium moisture content, the relative humidity of the headspace is equal to the water activity of the sample.

For a description of the saturated salts method for faecal sludge and faeces, see Bourgault *et al.* (2019) and Remington *et al.* (2020) (available open access). In this section, the example of a water activity meter with a dew point hygrometer method (AquaLab TDL)[D] is provided. The step-by-step procedure will vary depending on the laboratory and available equipment.

8.6.9.3.2 Safety precautions

- General health and safety (H&S) procedures specific for conducting laboratory analysis of faecal sludge are presented in Section 8.2. Before conducting this method, it is important to be familiar with Section 8.2.3 to ensure safety measures are properly carried out.
- Appropriate personal protective equipment (PPE) should be used; specific details are covered in Section 8.2.3.1.

8.6.9.3.3 Required chemicals

- Standard salt solutions (LiCl, NaCl, KCl) with a specific molality and water activity constant (provided by manufacturer).
- Distilled water or USP purified water.

8.6.9.3.4 Required apparatus and instruments

- Water activity meter (e.g, Meter AquaLab, Rotronic AwTherm, or Novasina Lab).
- Sample cup with lid (comes with a water activity meter).

8.6.9.3.5 Quality control

General information on quality assurance and quality control (QA/QC) is provided in Section 8.3. Information on operating conditions and interferences that are specific to this method includes:

- Calibration with standard salt solutions of specific molality and water activity constant is necessary. Frequency of calibration is dependent on the water activity meter and the manufacturer's instructions should be followed. A general guideline is to check the readings daily with a standard (or before each use if the meter is not used daily) to ensure measurement falls within an acceptable range.
- It is recommended to use calibration standards provided by the instrument manufacturer to reduce preparation error and ensure the highest accuracy. Calibration standards should span the range of water activities to be measured.
- Temperature heavily affects measurements of water activity.
    - Ensure that the water activity meter is located in a laboratory where the temperature is relatively stable (*e.g.* away from air conditioning, heating vents or open windows).
    - Allow sufficient time for the water activity meter to warm up and reach a stabile temperature before commencing with the measurements.
    - Samples that are more than 4 °C colder or warmer than the instrument chamber temperature need to equilibrate to the instrument temperature for accurate reading. Previously refrigerated samples should be allowed to reach room temperature in a sealed container before analysis.
- Proper cleaning and maintenance of the instrument is crucial for obtaining accurate and repeatable measurements. Follow the manufacturer's instructions.
- It is possible to measure water activity for liquid, slurry, semi-solid, and solid faecal sludge samples.

8.6.9.3.6 Sample preservation

It is recommended that water activity measurements are performed immediately after the sample collection. If this is not possible then samples should be stored at 4 °C for no longer than 7 days before analysis, and the storage time noted and reported.

8.6.9.3.7 Sample preparation

- Uniformly mix the sample using a stainless steel rod (or other appropriate tool) in order to have a thoroughly-mixed representative sample.
- Fill the sample cup about halfway full with the thoroughly-mixed sample. Ensure that the bottom of the cup is completely covered by the sample (this increases the efficiency of the instrument). Do not fill the sample cup more than halfway to prevent contamination of the sensors in the sampling chamber.
- Ensure the rim and outside walls of the sample cup are clean.
- Cover the sample cup immediately as evaporation can affect the characteristics of the sample. Analyse the sample directly after adding to the sample cup, or cover the sample cup if it must stand on the counter prior to analysis.

8.6.9.3.8 Analysis protocol

Calibration

Calibrate the water activity meter according to the manufacturer's instructions using standard salt solutions. Molalities and water activities of calibration standards used for the AquaLab TDL are displayed in Table 8.14.

Table 8.14 Water activities of calibration standards for the AquaLab TDL water activity meter.

Verification Standard at 25 °C	Water Activity
17.18 mol/kg LiCl	0.150 ±0:005
13.41 mol/kg LiCl	0.250 ±0:005
8.57 mol/kg LiCl	0.500 ±0:005
6.00 mol/kg NaCl	0.760 ±0:005
2.33 mol/kg NaCl	0.920 ±0:005
0.50 mol/kg KCl	0.984 ±0:005
USP purified water	1.000 ±0:005

Measurement

- Place the prepared sample cup in the instrument chamber.
- Carefully close the chamber to avoid spills and contamination of the chamber.
- Seal the chamber and start the reading. The instrument will take some time (several minutes) to take the measurement.
- At the end of the measurement, the instrument will display the water activity reading.
- Write down the reading.
- It is advisable to conduct at least two replicate measurements of each sample. Some extremely dry, dehydrated, highly viscous, high fat, or glassy samples may require multiple measurements before readings stabilise due to their slow water-emitting properties.

8.6.9.3.9 Calculation

No calculation is required, as the water activity meter reports the water activity directly.

8.6.9.3.10 Data set example

Practical example of the moisture content *versus* water activity for faecal sludge is presented in Figure 8.13.

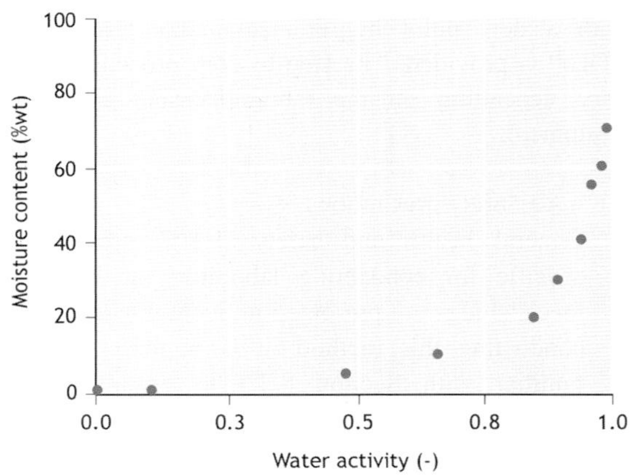

Figure 8.13 Moisture content *versus* water activity for faecal sludge from urine-diverting dry toilets dried to different final moisture contents (source: UKZN PRG).

8.6.9.4 Sludge volume index[53]

8.6.9.4.1 Introduction

The sludge volume index (SVI) is a measurement of sludge settling performance, based on the amount of suspended solids that settle within a specified amount of time. SVI measurements can be used to monitor settling performance at faecal sludge treatment plants, to observe changes in settling that could lead to process upsets, such as sludge bulking, or to assess settling performance to aid in the design of new settling-thickening tanks. By definition, the SVI is the volume (mL) of settled sludge occupied by 1 g of sludge after 30 min of settling in a 1,000 mL graduated cylinder or an Imhoff cone. The reading is expressed in terms of mL/g.

Modifications to these methods for adaption to faecal sludge include using either a graduated cylinder or an Imhoff cone, depending on what equipment is available and the anticipated settled sludge volume, as described below. The settling time has also been modified from 30 minutes (specified in Rice *et al.*, 2017) to 30-60 minutes, depending on the study objective (for example, mimicking the residence time of a specific treatment facility). This method is for liquid and slurry samples, with adaptations depending on the settleability.

8.6.9.4.2 Safety precautions

- General health and safety (H&S) procedures specific for conducting laboratory analysis of faecal sludge are presented in Section 8.2. Before conducting this method, it is important to be familiar with Section 8.2.3 to ensure safety measures are properly carried out.
- Appropriate personal protective equipment (PPE) should be used; specific details are covered in Section 8.2.3.1.

8.6.9.4.3 Apparatus and instruments

- 1 L graduated Imhoff cone or 1 L Class A graduated cylinder
- Timer
- TSS and VSS analytical equipment, see sections 8.6.1.3.3 and 8.6.1.4.3.
- Standard laboratory glassware and utensils

8.6.9.4.4 Quality control

General information on quality assurance and quality control (QA/QC) is provided in Section 8.3. Information on operating conditions and interferences that are specific to this method includes:

- Settling behaviour is influenced by temperature. For comparable results, SVI should be measured on sludge that is at room temperature. To ensure a consistent temperature, sludge should be kept out of direct sunlight during the course of settling tests.
- Jostling the settling vessel during the settling period can affect the results. Sludge should be allowed to settle in a place where it will not get bumped or disturbed.
- Imhoff cones should be used for more dilute sludge, while graduated cylinders should be used for sludge with higher solids content. Following the recommendations from Rice *et al.* (2017), Imhoff cones should be used for samples with less than approximately 100 mL settled sludge per L. For samples more than approximately 100 mL settled sludge per L, it is recommended to use a Class A graduated cylinder. This is because Imhoff cones are typically only graduated at the bottom, so if settled sludge volumes are higher than 100 mL, graduated cylinders will offer better resolution.
- It is advised to always state whether a graduated cylinder or Imhoff cone was used in determining SVI, as results can be affected by the choice of settling vessel.

[53] This method is based on methods 2540 F, 2710C, and 2710 D from Rice *et al.* (2017), with minor changes to adapt to use with faecal sludge. The first steps of adapting this method to faecal sludge are described in Chapter 6 of Faecal Sludge Management: Systems Approach for Implementation and Operation (Dodane and Bassan, 2014), and further adaptations are presented here. This method should be cited as Method 2710 D (Rice *et al.*, 2017) as adapted in Velkushanova *et al.* (2021).

- The use of graduated cylinders with a high aspect ratio (height to diameter ratio) should be avoided. The friction created by the walls can reduce settling velocities, which can cause discrepancies in SVI results.
- If the settled sludge contains large pockets of liquid within the settled layer, the volume of trapped liquid should be estimated and subtracted from the volume of settled sludge, or the test should be repeated. If this is a recurring problem, it should be reported with the results.
- Floating material should not be included as part of the settled sludge.
- A settling time should be selected based on the objectives of the study (for example, to mimic the conditions of a specific treatment process). Typical settling times in SVI tests of faecal sludge range from 30-60 minutes. Longer settling times have also been selected in the past for studies investigating settling performance after long residence times, designed to mimic residence times at a specific treatment facility (*e.g.* 100 minutes, Dodane and Bassan, 2014). Because the duration of settling affects the volume of settled sludge, the settling time should always be reported with the SVI results. SVI results obtained using different settling times might not be comparable.

8.6.9.4.5 Sample preservation

- It is recommended that the sludge volume index is measured immediately after the sample collection; if not possible then samples should be stored at 4 °C for no longer than 7 days before analysis, and the storage time noted and reported.

8.6.9.4.6 Sample preparation

- Allow the sample to reach room temperature.
- Determine the total suspended solids concentration of a thoroughly-mixed sample (see Method 8.6.1.3).

8.6.9.4.7 Analysis protocol

- Uniformly mix the sample.
- Fill the Imhoff cone or graduated cylinder to the 1,000 mL mark.
- Allow the sludge to settle for a predetermined settling time.
- After settling time is complete, record the volume of the settled sludge and settling time.

8.6.9.4.8 Calculation

Calculate the sludge volume index (SVI) as follows:

$$\text{SVI}\left(\frac{\text{mL}}{\text{g}}\right)=\frac{\text{Settled sludge volume}\left(\frac{\text{mL}}{\text{L}}\right)}{\text{Suspended solids}\left(\frac{\text{g}}{\text{L}}\right)}$$

8.6.9.4.9 Data set example

Ward *et al*. 2019 evaluated SVI in liquid faecal sludge samples collected from 20 septic tanks in Dakar, Senegal using Imhoff cones and a 30-minute settling time. The median SVI was 32 mL/g, and mean and standard deviation were 45 and 51 mL/g, respectively. Several samples underwent no visible settling during the settling period, thus the settled sludge volume was recorded as the total volume of sample. In wastewater sludge literature, an SVI less than or equal to 100 mL/g is designated as good settling performance (Dodane and Bassan, 2014). The publication and entire raw data set are available at the link below[54].

Gold *et al.* (2016) evaluated the settling performance of liquid faecal sludge samples from Dakar, Senegal using Imhoff cones before and after the addition of conditioners. The settling time was 60 minutes. The SVI results were highly variable, so the researchers chose to report the volume of settled sludge, along with the TSS of the supernatant after settling, as metrics of the settling performance.

8.7 PHYSICAL PROPERTIES

8.7.1 Physical and mechanical properties

Physical properties are characteristics that do not change the chemical composition of a material. Examples of physical properties are density, particle size, and mechanical properties. Mechanical properties are the physical properties of a material that are measured by the application of force. These can include, for example, shear strength, viscosity,

[54] https://doi.org/10.1016/j.watres.2019.115101

plasticity, etc. The physical and mechanical properties of faecal sludge are important for developing emptying and treatment technologies, especially when these technologies involve the application of force on the sludge; for example, in the case of pumping, compressing, or extruding faecal sludge or solid end products from its treatment (see Chapter 4). Methods included in this section include density, particle size, rheology, liquid limit, plastic limit, and texture analysis.

8.7.1.1 Density – mass and volume measurement method[55]

8.7.1.1.1 Introduction

Density is the relationship between the mass and volume of a substance. As explained in Chapter 2, density is important to convert concentrations between weight/volume and weight/weight. It is therefore recommended to measure density when possible, but especially when the faecal sludge to be analysed spans a range of sludge types. In the following method, the mass of a known volume of faecal sludge is measured and density is determined by direct calculation. This method measures wet bulk density, which is a commonly used parameter for faecal sludge, for example to convert between units. If dry bulk density and/or particle density or pore space are required, Method 8.7.1.2 should be used. This method provides sufficient accuracy for most of the applications where density for faecal sludge is needed. However, because the risk of potential (human) error is relatively high, it should not be used for measurements that require an accuracy of several decimal places. If a higher level of accuracy is required, a digital density meter can be used. A useful guide with more information on how to make accurate density measurements can be found on the Anton Paar website[56].

8.7.1.1.2 Safety precautions

- General health and safety (H&S) procedures specific for conducting the laboratory analysis of faecal sludge are presented in Section 8.2. Before conducting this method, it is important to be familiar with Section 8.2.3. to ensure safety measures are properly carried out.
- Appropriate personal protective equipment (PPE) should be used; specific details are covered in Section 8.2.3.1.

8.7.1.1.3 Required apparatus and instruments

- 5 mL measuring spoon or scoop (or other appropriate volume)
- Knife, to trim excess sludge from the measuring spoon
- Analytical balance
- Glass weighing dish
- Distilled water (for the quality control procedure).

8.7.1.1.4 Quality control

General information on quality assurance and quality control (QA/QC) is provided in Section 8.3. Information on standards, operating conditions, and interferences that are specific to this method include:

- The analytical balance must be calibrated regularly, following the method outlined in Method 8.6.1.1.
- Inaccuracies may arise for the following reasons:
    - If the measuring spoon is not completely filled with the sample.
    - If the sample is compressed in the process of filling the spoon.
    - If the sample is not levelled completely.

  These manual errors should be reduced as much as possible, to increase the accuracy of the measurement.
- Before every series of density measurements, do a check with distilled water. Follow the measuring procedure with distilled water, and compare the density with the density of water: $\rho_{water} = 0.998203$ g/cm^3 for T = 20 °C. A common tolerance limit is 0.0001 g/cm^3. If the results are not within the tolerance limit, clean the spoon thoroughly and try again.

8.7.1.1.5 Sample preparation

- Homogenise the faecal sludge sample thoroughly by stirring with a spoon or stirring rod.
- It is important to prepare the sample for density in the same way as other analysis that is being conducted, especially if the results will be used to

[55] This method should be cited as described in Velkushanova *et al.* (2021).

[56] https://wiki.anton-paar.com/en/density-and-density-measurement/

convert between weight/weight or weight/volume concentrations (*e.g.* if a blended sample is used for TS measurement and that is the parameter of interest for the density measurement, then density should be measured on the blended sample).

- Exclude larger, inconsistent or floating particles from the sample if it is determined that their inclusion may affect the final result (*e.g.* hair, stones, glass, and maggots).

8.7.1.1.6 Analysis protocol

Liquid, slurry, semi-solid and solid samples:

- Place the measuring spoon and the glass dish on the balance, and tare the balance.
- Use the spoon to scoop a sample of faecal sludge, such that the sample completely fills the spoon. Avoid compressing the sample as much as possible.
- Wipe the bottom of the spoon with a laboratory tissue, removing any excess sample.
- Level the sample by removing any excess above the surface of the edge of the spoon with a knife, to leave a flat surface that is flush with the top of the spoon.
- Place the measuring spoon on the glass dish on the scale, and record the mass of the sample contained in the spoon.

8.7.1.1.7 Calculation

$$\text{Density}\ (\frac{\text{g}}{\text{cm}^3}) = \frac{\text{Mass (g)}}{\text{Volume (mL)}}$$

8.7.1.2 Density – volume displacement method[57]

Bulk density is a measure of mass per unit volume. It is used as a measure of wetness, volumetric water content, and porosity. Factors that influence the measurement include the organic matter content, porosity, and material structure. Particle density, or solid density, represents only the weight of dry material per unit volume of the material solids; the pore space is not included in the volume measurement. The porosity of a material is the pore space portion of the material volume occupied by air and water. Both density parameters, bulk and particle (solid), are commonly used, depending on the purpose of the measurement. For example, particle density might be more suitable for calculations on drying beds, while the bulk density will have more relevance for emptying and transportation.

8.7.1.2.1 Introduction

Wet bulk density is determined using the same techniques as presented in Method 8.7.1.1. Dry bulk density is determined by oven-drying a known volume of sample and measuring the mass of the dry sample. Particle density is determined using the volume displacement technique. Pore space is then calculated from these values.

8.7.1.2.2 Safety precautions

- General health and safety (H&S) procedures specific for conducting the laboratory analysis of faecal sludge are presented in Section 8.2. Before conducting this method, it is important to be familiar with Section 8.2.3 to ensure safety measures are properly carried out.
- Appropriate personal protective equipment (PPE) should be used; specific details are covered in Section 8.2.3.1.
- Wear gloves suitable for withstanding high temperatures when placing and removing crucibles from the oven.
- Use appropriate mechanical tools, such as metal tongs, to remove crucibles and trays after drying in the oven to avoid direct contact with hot surfaces.

8.7.1.2.3 Required apparatus and instruments

- Porcelain crucibles
- Desiccator with dry desiccant
- Drying oven
- Analytical balance with four decimal places
- 100 mL measuring cylinder
- 7.5 mL measuring scoop or 10 mL measuring cylinder (depending on sludge type)
- Tube to hold the sample that fits inside the 100 mL measuring cylinder
- Glass weighing dish
- Laboratory tissue

[57] This method follows British Standard 812-2:1995, determination of density for testing aggregates (1995), and should be cited as BS 812-2 (1995) as adapted in Velkushanova *et al.* (2021).

- Knife, to trim excess sludge from the measuring scoop
- Heat-resistant gloves
- Thermometer (for the quality control procedure)
- Set of standard calibration weights (for the quality control procedure)
- Distilled water (for the quality control procedure).

8.7.1.2.4 Quality control

General information on quality assurance and quality control (QA/QC) is provided in Section 8.3. Information on standards, operating conditions, and interferences that are specific to this method include:

- The analytical balance and oven must be checked and calibrated weekly.
  - Check the temperature throughout the oven area by placing a calibrated thermometer on each shelf. After 30 min, check the temperature at each level against the oven setting. Using the same method, also check for temperature differences between the front and back of the oven. Adjust the oven setting if necessary. If temperatures are uneven on the shelves, check the insulation.
  - To calibrate the analytical balance, place a standard calibration weight on the balance and weigh. Adjust the balance manually if necessary. Do this with the whole range of weights from the calibration set. Make sure to include a standard weight of a mass similar to the mass of the expected sample + crucible.
- Make sure the desiccant in the desiccator is not saturated, otherwise samples may absorb water while cooling down in the desiccator. Routinely dry the desiccant in the oven at 105 °C (or at a different temperature, depending on the manufacturer's instructions), prior to the colour indicating that the desiccant is nearly saturated.
- Always keep the lid of the desiccator on and use a lubricant on the rim to ensure airtight sealing. Do not overload the desiccator.
- Before every series of density measurements, do a check with distilled water. Follow the measuring procedure with distilled water, and compare the density with the density of water: $\rho_{water} = 0.998203$ g/cm^3 for T = 20 °C. A common tolerance limit is 0.0001 g/cm^3.

8.7.1.2.5 Sample preparation

- Homogenise the faecal sludge sample thoroughly by stirring with a spoon or stirring rod.
- It is important to prepare the sample for density in the same way as other analysis that is being conducted, especially if the results will be used to convert between weight/weight or weight/volume concentrations (*e.g.* if a blended sample is used for TS measurement and that is the parameter of interest for the density measurement, then density should be measured on the blended sample).
- Exclude larger, inconsistent or floating particles from the sample if it is determined that their inclusion may affect the final result (*e.g.* hair, stones, glass, and maggots).

8.7.1.2.6 Analysis protocol

- Pre-heat the oven to 103-105 °C.
- Place a clean crucible in the oven at a temperature of 103-105 °C for 1 hr prior to use (to remove any moisture). After drying, place the crucible in the desiccator and allow it to cool down to room temperature. Keep the crucible in the desiccator until the next step.
- Weigh the crucible and record the mass (W_1).
- Place the measuring scoop and the glass dish on the balance, and tare the balance. If required, for liquid and slurry sludge types a measuring cylinder might also be used.
- Use the scoop to measure 7.5 mL of the sample, such that the sample completely fills the scoop. Avoid compressing the sample as much as possible.
- Wipe the bottom of the scoop with a laboratory tissue, removing any excess sample.
- Trim any sample from the top of the scoop with the knife, to leave a flat surface flush with the top of the scoop.
- Place the measuring scoop on the glass dish on the scale, and record the mass of the sample contained in the scoop (W_2).
- Transfer all the sample from the scoop into a dried crucible. Rinse the scoop with small volumes of distilled water to dislodge heavy particles. Make sure that all the particles are transferred to the crucible. Add the washings to the crucible.
- Oven-dry the sample at 103-105 °C for at least 24 hr.

- Take the sample out of the oven, and place it in the desiccator to reach room temperature.
- Weigh the dry mass of the sample + crucible using an analytical balance and record the weight (W_3).
- Fill the 100 mL measuring cylinder with 50 mL water.
- Suspend an empty sample-holding tube inside the 100 mL measuring cylinder filled with 50 mL water and record the volume level of water (V_1).
- Carefully transfer all the dry sample from the crucible into the holding tube, ensuring that all the particles are transferred.
- Suspend the tube with the sample in the measuring cylinder with water and record the new level of the water (V_2).

8.7.1.2.7 Calculation

Bulk density

$$\text{Bulk density (wet)} \left(\frac{g}{mL}\right) = \frac{(W_2\text{-}W_1)\ (g)}{Vt\ (mL)}$$

Where:

W_1 = Mass of the crucible (g)
W_2 = Wet mass of sample
V_t = Total volume of sample (7.5 mL).

$$\text{Bulk density (dry)} \left(\frac{g}{mL}\right) = \frac{(W_3\text{-}W_1)\ (g)}{Vt\ (mL)}$$

Where:

W_1 = Mass of the dried crucible (g)
W_3 = Dry residue + crucible after drying at 103-105 °C (g)
V_t = Total volume of the sample, pore volume + solid volume (7.5 mL).

Particle density

Particle density values represents only the weight of dry sample per unit volume of the sample solids; the pore space is not included in the volume measurement.

$$\text{Particle density} \left(\frac{g}{mL}\right) = \frac{(W_3\text{-}W_1)\ (g)}{Vs\ (mL)}$$

Where:

W_1 = Mass of the dried crucible (g)
W_3 = Dry residue + crucible after drying at 103-105 °C (g)
V_1 = Volume in measuring cylinder with holding tube (mL)
V_s = Volume of the solids (ONLY) = V_t-V_1 (mL).

Pore space

Pore space (g/mL) =

Bulk density (g/mL) − Particle density (g/mL)

8.7.1.3 Particle size - laser light scattering method[58]

8.7.1.3.1 Introduction

Characterising particle size distribution can help in designing treatment processes and monitoring process effectiveness. Particle size influences how much organic material is available organic material is for degradation by microorganisms, and how the particle size distribution changes over the course of stabilisation and treatment. Particle size distribution affects settling and dewatering performance, and is also an important characteristic of end products from faecal sludge treatment (*e.g.* dried sludge solid fuels or feedstock for larvae rearing).

Several standard methods for characterising the particle size of water and wastewater exist, and these are discussed in Method 2560 Particle Counting and Size Distribution in the Standard Methods for the Examination of Water and Wastewater (Rice *et al.*, 2017). These include manual sequential sieving and filtration, the use of electronic measurement devices (including electronic sensing zone instruments, light-blockage instruments, and light-scattering instruments), and direct sizing and counting using microscopy. Manual sieving and filtration are slow, labour-intensive, and has a lower level of accuracy, but does not require expensive instrumentation. Electronic measurement of particle size is typically the method of choice if instruments are accessible. However, when large aggregates of particles (> 500 µm) are to be analysed, direct microscopic methods are advised (Rice *et al.*, 2017).

[58] This method should be cited as adapted in Velkushanova *et al.* (2021), together with the specific analytical equipment (if different), and any manufacturer's modifications.

Step-by-step procedures for measuring particle size will vary depending on the selected method, the available equipment, and the characteristics of the incoming faecal sludge samples. One example of electronic measurement of particle size is the laser light scattering method used by the UKZN PRG laboratory in Durban, South Africa that is described here. This method is specifically written to be used with a Malvern Mastersizer 3000 particle size analyser[D], and follows the Malvern Mastersizer 3000 User Manual (Malvern Instruments, 2017) and Method 2560D in Rice *et al.* (2017). The Malvern Mastersizer 3000 measures particle size by shooting a laser beam through a dispersed sample, and measuring the angle and intensity of light scattered off the particles. Mie and Frauhofer theories are used to calculate the particle sizes based on the scattering pattern. A wet dispersion unit is used with faecal sludge samples to circulate samples through the measurement cell. The size range of the Mastersizer 3000 is 0.01-3,500 µm.

8.7.1.3.2 Safety precautions

- General health and safety (H&S) procedures specific for conducting the laboratory analysis of faecal sludge are presented in Section 8.2. Before conducting this method, it is important to be familiar with Section 8.2.3 to ensure safety measures are properly carried out.
- Appropriate personal protective equipment (PPE) should be used; specific details are covered in Section 8.2.3.1.

8.7.1.3.3 Required chemicals

- Particle size standards (for the Mastersizer 3000, Malvern recommends the Malvern QAS3002 Quality Audit Standard).

8.7.1.3.4 Required apparatus and instruments

- Mastersizer 3000
- Mastersizer wet dispersion unit
- Beaker.

8.7.1.3.5 Quality control

General information on quality assurance and quality control (QA/QC) is provided in Section 8.3 .Information on standards, operating conditions, and interferences that are specific to this method include:

- Calibration is performed using standard suspensions or dry powders of spherical particles of known size (*e.g.* standards provided by the manufacturer or NIST standard particles). Rice *et al.* (2017) recommend using at least three different-sized particle standards to calibrate a particle sensor. Follow the manufacturer's instructions to set up a calibration strategy.
- Sample blanks, handled identically to the faecal sludge samples, should be analysed daily. Generally, blanks should not show more than 5% of the counts in any size channel compared to the samples. See Section 7 Quality Control in 2560A (Rice *et al.,* 2017) for a detailed discussion of quality control for particle size analysis, or refer to the manufacturer's instructions.
- Large particles, solid waste, stones, and hair should be removed before testing, as they can harm the instrument. This can be achieved by passing the sample through a sieve before analysis. Sieve size and other pre-treatment steps should be selected based on the upper measurement limit of the specific instrument and the manufacturer's instructions.
- Minimise particle contamination (*e.g.* from airborne particles, contaminated dilution water, or contaminated glassware). Keep the samples in a closed container, ensure the dilution water is particle-free and run blanks to ensure this, and ensure the glassware is thoroughly cleaned and particle-free before use. For information about producing particle-free dilution water, see 2560A in Rice *et al.* (2017).
- Faecal sludge samples may require dilution prior to analysis. It is important to avoid breaking up aggregates or flocs during the sample preparation, so dilutions should be made carefully using pipettes with wide openings. Wide openings can be made by cutting off the tips of the pipettes. The sample should be added to the dilution water (not water added to the sample) in order to reduce shear on the sample. Be careful to use slow, low-intensity mixing. Avoid mechanical stirring and ultrasonication. For more sample preparation tips, see Section 3 Sample collection and handling in 2560A, Rice *et al.* (2017).
- Minimise the time between sample collection and measurement, as particles may agglomerate over

- time, changing the particle size distribution. Dilution can also influence agglomeration – make dilutions immediately before analysis.
- If samples must be stored before analysis, refrigerate them (4 °C), but make sure that they are brought back to room temperature before analysis.

8.7.1.3.6 Sample preservation

Samples should be analysed as soon as possible after collection, to prevent changes in particle size distribution due to agglomeration. If the samples must be stored before analysis, store them in a refrigerator (4 °C) and do not dilute them before storage.

8.7.1.3.7 Sample preparation

- Remove all the particles larger than the upper limit of the instrument by sieving.
- If the sample is semi-solid or solid, dilute the sample in particle-free water and gently mix to produce a slurry. Add to a beaker.
- If the sample is liquid or slurry, dilution may not be necessary. Gently mix the sample to ensure homogeneity, then add a portion of the sample to a beaker.

8.7.1.3.8 Analysis protocol

Instrument setup

- Switch on the instrument.
- Switch on the computer and start the Mastersizer software.
- Wait 30 min for the instrument to stabilise before using the instrument.

Measurement

- Select the instrument protocol for measuring the specific sample type (*e.g.* faecal sludge from VIP latrines) and allow the instrument to initialise. A background light measurement will then be taken.
- When prompted, add the sample a small amount at a time until the obscuration is within the correct range (displayed on the computer screen). Note: if the sample is too concentrated, it will immediately exceed the obscuration range - if this happens, the sample will need to be diluted and measured again.
- Run the sample measurement protocol.
- After measurement is completed, clean the system by following the prompts on the user interface.

8.7.1.3.9 Calculation

No calculation required – direct reading is based on the overall percentage of particle volume and does not require adjustment based on dilution.

8.7.1.3.10 Data set example

Faecal sludge at UKZN PRG was analysed (unpublished data, Figure 8.14), with the results interpreted as follows:

- Weighted residual - an indication of how well the calculated data was fitted to the measurement data. A good fit is indicated by a residual of less than 1%, while a residual over 1% may indicate the use of an incorrect refractive index and adsorption values for the sample.
- Dv 50, Dv 10 and Dv 90 are standard percentile readings from the analysis.
    - Dv 50 - the particle diameter in µm at which 50% of the sample volume is smaller and 50% is larger. This value is also known as the Mass Median Diameter (MMD) or the median of the volume distribution. The v in the expression Dv 50 shows that this refers to the volume distribution. Following the same naming convention, Ds refers to the surface area distribution, Dl is the length distribution, and Dn is the number distribution.
    - Dv 10 - the particle diameter below which 10% of the sample volume lies.
    - Dv 90 - the particle diameter below which 90% of the sample volume lies.
- D [4, 3] - the volume-weighted mean or Mass Moment Mean Diameter.
- D [3, 2] - the surface-weighted mean, also known as the Surface Area Moment Mean Diameter.
- Span - is the measurement of the width of the distribution. The narrower the distribution, the smaller the span becomes.
- Concentration - the volume concentration. This is calculated using the Beer-Lambert law.
- Obscuration - an ideal range of obscuration is usually between 3 and 20%, depending on the sample and dispersion unit used.
- Distribution - shows the type of distribution the analysis has used. Options include volume, surface area, length, or number. The Mastersizer 3000 measurement is fundamentally a

measurement of volume distribution; the transformation of the results into a surface area, length, or number distribution may amplify any error in the original result, especially at the fine end of the size distribution.

Analysis

Created by: Malvern Instruments Ltd
Last edited: 2012/04/13 05:12:51 PM

Malvern

Measurement Details

Sample Name Average of 'synthetic_1 '
Operator Name MS3000
SOP File Name SOP_vip dry_example.msop
Measurement Date Time 2013/10/15 10:34:50 AM
Analysis Date Time 2013/10/15 10:34:50 AM
Result Source Averaged

Analysis

Particle Name Default
Dispersant Name Water
Particle Absorption Index 0.100
Weighted Residual 0.75 %
Analysis Model General Purpose
Particle Refractive Index 1.520
Dispersant Refractive Index 1.330
Laser Obscuration 19.01 %
Scattering Model Mie
Analysis Sensitivity Normal

Result

Concentration 0.0431 %
Uniformity 0.841
Specific Surface Area 195.8 m²/kg
D [3,2] 17.0 µm
D [4,3] 958 µm
Span 2.487
Result Units Volume
Dv 10 4.16 µm
Dv 50 898 µm
Dv 90 2240 µm

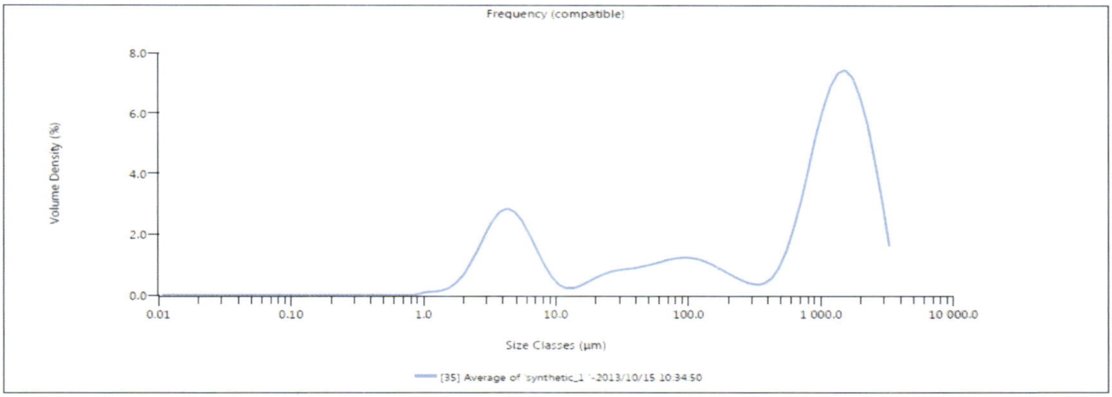

Size (µm)	% Volume In	Size (µm)	% Volume In	Size (µm)	% Volume In	Size (µm)	% Volume In	Size (µm)	% Volume In	Size (µm)	% Volume In	Size (µm)	% Volume In
0.0100	0.00	0.0679	0.00	0.460	0.00	3.12	2.03	21.2	0.56	144	0.81	976	5.14
0.0114	0.00	0.0771	0.00	0.523	0.00	3.55	2.28	24.1	0.62	163	0.70	1110	5.77
0.0129	0.00	0.0876	0.00	0.594	0.00	4.03	2.38	27.4	0.67	186	0.59	1260	6.15
0.0147	0.00	0.0995	0.00	0.675	0.00	4.58	2.29	31.1	0.69	211	0.48	1430	6.24
0.0167	0.00	0.113	0.00	0.767	0.00	5.21	2.03	35.3	0.72	240	0.38	1630	6.02
0.0189	0.00	0.128	0.00	0.872	0.03	5.92	1.65	40.1	0.75	272	0.31	1850	5.50
0.0215	0.00	0.146	0.00	0.991	0.08	6.72	1.22	45.6	0.79	310	0.27	2100	4.72
0.0244	0.00	0.166	0.00	1.13	0.10	7.64	0.81	51.8	0.84	352	0.30	2390	3.73
0.0278	0.00	0.188	0.00	1.28	0.12	8.68	0.49	58.9	0.90	400	0.44	2710	2.57
0.0315	0.00	0.214	0.00	1.45	0.18	9.86	0.27	66.9	0.96	454	0.73	3080	1.32
0.0358	0.00	0.243	0.00	1.65	0.31	11.2	0.18	76.0	1.00	516	1.18	3500	
0.0407	0.00	0.276	0.00	1.88	0.54	12.7	0.18	86.4	1.02	586	1.81		
0.0463	0.00	0.314	0.00	2.13	0.86	14.5	0.26	98.1	1.01	666	2.59		
0.0526	0.00	0.357	0.00	2.42	1.25	16.4	0.36	111	0.97	756	3.46		
0.0597	0.00	0.405	0.00	2.75	1.66	18.7	0.47	127	0.90	859	4.34		

Figure 8.14 Example of data generated during the particle size analysis (source: unpublished data, UKZN PRG).

8.7.1.4 Rheological properties – rheometer method[59]

8.7.1.4.1 Introduction

Rheology is related to the measurement of the response of soft solid materials or liquids to an applied force, such as sheering, where the deformation is a plastic flow in contrast to elastic deformation. Rheological properties for faecal sludge for example, give an estimation of the 'pumpability' of a sample by the change in viscosity under applied shear stress. A rheometer is used to carry out a number of rheological tests on faecal sludge and faeces samples. These include flow curves, amplitude and frequency sweeps, variable temperature tests and stress recovery tests. A number of different measuring systems exist (cone-cup, plate-plate, building material cell), and each is suited to different types of samples.

Step-by-step procedures for measuring rheological properties will vary depending on the available equipment and the characteristics of the incoming faecal sludge samples. One example of rheology measurement is the method used by the UKZN PRG laboratory in Durban, South Africa that is described here.

8.7.1.4.2 Safety precautions

- General health and safety (H&S) procedures specific for conducting the laboratory analysis of faecal sludge are presented in Section 8.2. Before conducting this method, it is important to be familiar with Section 8.2.3 to ensure safety measures are properly carried out.
- Appropriate personal protective equipment (PPE) should be used; specific details are covered in Section 8.2.3.1.

8.7.1.4.3 Required apparatus and instruments

- Rheometer
- 27 mm cone-cup attachment
- 32 mm cone-cup attachment
- Plate-to-plate attachment
- Building material cell
- Spatula.

8.7.1.4.4 Quality control

General information on quality assurance and quality control (QA/QC) is provided in Section 8.3. Information on standards, operating conditions, and interferences that are specific to this method include:

- A motor calibration and inertia calibration service must be performed every 90 days.
- Using a cone-cup attachment with a larger diameter (hence a larger surface area) produces more accurate results.
- Fluid samples with low viscosity tested at low shear rates may produce inaccurate results due to surface tension effects.
- For a long duration of testing (*e.g.* more than 1 hr), it is recommended to cover the cup containing the sample to avoid loss of moisture due to evaporation.
- Measurements should be conducted at a standard temperature of 25 °C. The temperature at which measurements were taken should be reported with the results.
- Blending or intensive mixing of the sample prior to measurement will change the structure of the faecal sludge material and affect the results. These homogenization techniques must never be applied for this method, and gentle mixing is preferable if homogenization is required.

8.7.1.4.5 Sample preservation

It is recommended that samples are analysed as soon as possible after collection. If immediate measurement is not possible, the samples should be stored at 4 °C for no longer than 14 days before analysis.

8.7.1.4.6 Sample preparation

- The sample should preferably not be homogenised as this will change its structural properties. The sample should be used as received and undisturbed. Only in case of differential moisture content within the sample, homogenise gently prior to the measurement. Never use a blender to homogenise.

[59] This method should be cited as adapted in Velkushanova *et al.* (2021), together with the specific analytical equipment (if different), and any manufacturer's modifications.

8.7.1.4.7 Analysis protocol

Calibration

- The instrument must be switched on for at least an hour before analysis to perform motor adjustment (refer to specific manufacturer's information).
- For the motor adjustment procedure, connect the measuring system and leave a 1 mm gap from the top, then select the start service device, and then 'Start motor adjustment' from the software programme menu. This takes approximately 10 min to complete.
- Once the motor adjustment is completed, move the measuring system to the loading position.

Measurement

- Switch on the instrument.
- Remove the cap protecting the instrument's coupler.
- Open the rheometer software programme and control panel tab to initialise the instrument.
- Attach the cone cup to the rheometer plate.
- Manually enter 25 °C for the rheometer plate.
- Manually enter 23 °C for the VT-2 tower.
- Load the sample into the cone cup and then attach the cone cup to the base plate (Figure 8.15 A).
- Attach the measuring system to the adapter (Figure 8.15 B).
- Open the control panel and lower the measuring system to a gap of 0.00 mm.
- Close the control panel and select the measuring method (*e.g.* flowability: flow curve logarithmic).
- Select Start, enter the sample details, then select Continue.
- The test begins after reaching the set temperature (approximately 1 min).
- A generated report appears when testing is complete that can be saved as a PDF document.
- Once the test is complete, uncouple the measuring system, open the control panel and raise the measuring system.
- Remove the cone cup and discard the sample.
- Close the program and switch off the computer.

Note: these values are provided as an example, the actual measurement procedure will differ due to model type and software.

8.7.1.4.8 Calculation

No calculation required - direct reading.

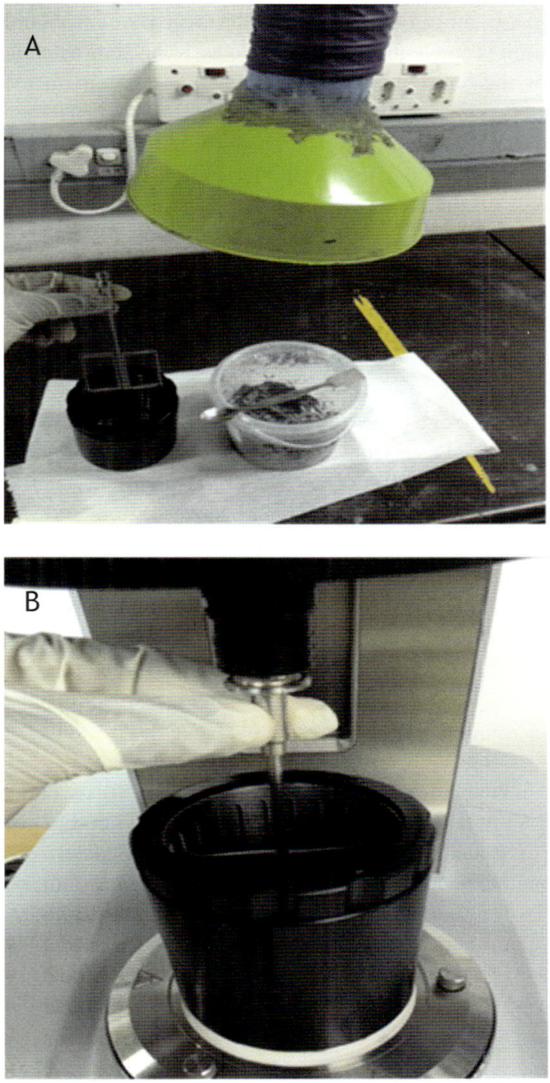

Figure 8.15 Loading the sample into the cone cup and then to the plate (A), and attaching the measuring system to the adapter (B) (source: UKZN PRG).

8.7.1.5 Liquid limit – cone penetrometer method[60]

8.7.1.5.1 Introduction

The liquid limit is the moisture content at which a material passes from the liquid phase into the plastic phase, and is determined experimentally. The liquid limit is conventionally used to classify the consistency of the soil. Measurements of liquid limit in faecal sludge are useful when determining optimal methods of sludge emptying from onsite systems, or when determining the TS at which to pelletise dried sludge end products for resource recovery (Septien *et al.*, 2018). The liquid limit and plastic limit (methods 0 and 0.) can be used to calculate the consistency index, liquidity index, and plasticity index, which are helpful in characterising the consistency of a material over a range of TS.

The cone penetrometer provides a static test using a material's resistance to penetration to determine its liquid limit. The cone is dropped from a specific height into the sample and the penetration into the sample is recorded and correlated to the moisture content of the sample. The moisture content is increased continuously and the test repeated, until an approximate linear graph can be produced. The water content corresponding to a penetration of 20 mm is the liquid limit of the sample.

8.7.1.5.2 Safety precautions

- General health and safety (H&S) procedures specific for conducting the laboratory analysis of faecal sludge are presented in Section 8.2. Before conducting this method, it is important to be familiar with Section 8.2.3 to ensure safety measures are properly carried out.
- Appropriate personal protective equipment (PPE) should be used; specific details are covered in Section 8.2.3.1.

8.7.1.5.3 Required chemicals

- Distilled water.

8.7.1.5.4 Required apparatus and instruments

- Cone penetrometer with a standard stainless steel cone (50 g), fitted with a 150 mm-diameter dial indicator for direct reading of the penetration
- Sample cup of 40 mm diameter and 55 mm deep (standard size)
- A stopwatch readable to 1 sec
- Two palette knives or spatulas
- Laboratory spoons for loading the penetrometer cup with the sample
- Rubber spatula to scrape any sample out of the cup
- A knife or flat sharp object to scrape excess material off the top of the cup.

8.7.1.5.5 Quality control

General information on quality assurance and quality control (QA/QC) is provided in Section 8.3. Information on standards, operating conditions, and interferences that are specific to this method include:

- Ensure that the penetrometer base level is horizontal using a bubble leveller.
- Measurements should be conducted at a standard temperature of 25 °C. The temperature at which measurements were taken should be reported with the results.

8.7.1.5.6 Sample preservation

It is recommended that samples are analysed as soon as possible after collection. If immediate measurement is not possible, the samples should be stored at 4 °C for no longer than 14 days before analysis.

8.7.1.5.7 Sample preparation

- Select a representative sample.
- Remove any foreign objects that might be mixed with the sludge (*e.g.* solid waste, stones, hair, and maggots).
- Do not allow the sample to dry before testing.

[60] This method is based on the British Standards Institution methods of test for soils for civil engineering purposes: Classification tests. This method should be cited as BS 1377:2, as adapted in Velkushanova *et al.* (2021).

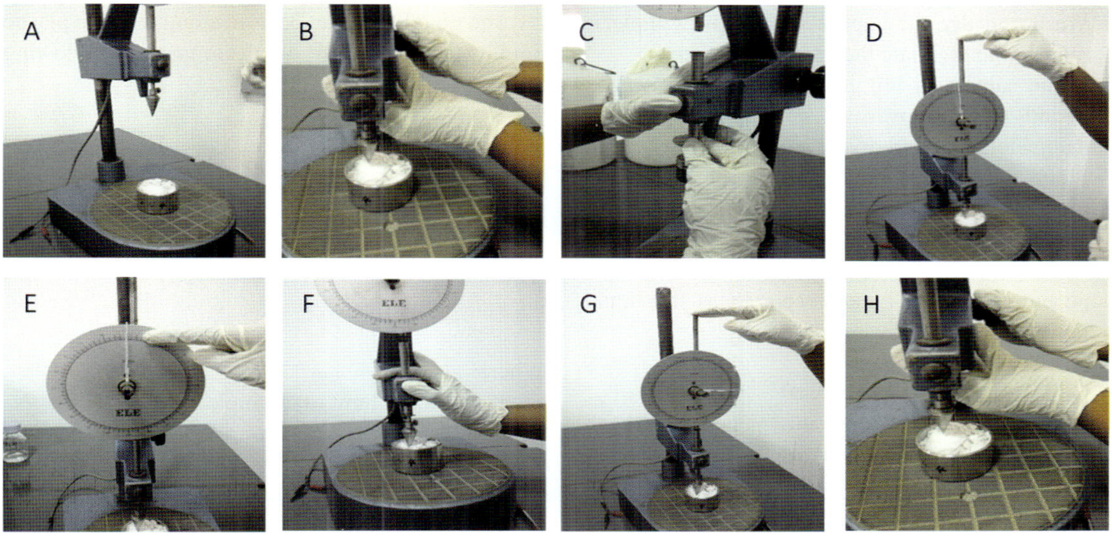

Figure 8.16 Illustration of testing stages (Source: UKZN PRG).

8.7.1.5.8 Analysis Protocol

1. Place 300 g of the prepared sample on the glass plate.
2. Mix the sample with the two palette knives for approximately 10 min and adjust with distilled water to achieve a first cone penetration reading of 15 mm.
3. Transfer the mixed sample to a metal cup using the palette knife and use a knife or long, straight edge to scrape the excess sludge from the top, creating a smooth, level surface flush with the top of the metal cup.
4. Place the metal cup in the designated position on the base of the instrument, ensuring the penetration cone is locked in a raised position (Figure 8.16 A).
5. Lower the penetration cone carefully until it just touches the surface of the sample; the correct position is indicated if the cone just scratches the surface when the cup is moved (Figure 8.16 B). Move the clamp to lower the cone.
6. When the cone has been placed in the correct position, lower the stem of the dial gauge until it just touches the cone shaft.
7. Set the dial gauge to zero (to the nearest 0.1 mm) (Figure 8.16 C).
8. Release the cone for 5 sec ± 1 sec. Lock the cone into position after the 5 sec have lapsed and lower the stem of the dial gauge again to touch the cone shaft. Read the dial gauge to the nearest 0.1 mm; this value is recorded as the cone penetration (Figure 8.15 G).
9. Lift the cone from the cup and clean it carefully.
10. Replace the wet sample in the cup, ensuring no air is trapped and repeat steps 3 through 9.
11. If the difference between the first and second penetrations is less than 0.5 mm, report the average value. If the difference is greater than 0.5 mm but less than 1 mm, repeat the test a third time and if the overall range is no greater than 1 mm, report the average of the three values. If the overall range is greater than 1 mm, remove the sample from the cup and repeat the procedure from step 2.
12. Take approximately 10 g of the sample where the cone penetrated the cup and measure the moisture content (see Method 8.6.1.1).
13. Repeat the entire procedure at least 3 times using the same sample to which increments of distilled water have been added.
14. Go from drier to wetter samples, until a cone penetration range of approximately 15 mm to 25 mm has been reached over the course of at least 4 test runs and the values are evenly distributed.

15. Wash and dry the cup each time the sample is removed to facilitate the addition of water.
16. If the sample is left in the open for extended periods of time, cover with an evaporating dish or damp cloth to avoid drying.
17. Plot the moisture content against cone penetration to obtain a linear regression that best fits the plotted points.
18. The moisture content corresponding to a cone penetration of 20 mm is reported as the liquid limit, to the nearest whole number.

8.7.1.5.9 Calculation (if necessary)

Plot the water content (%) against penetration depth (mm). The water content corresponding to 20 mm penetration is the liquid limit (w_L). If the liquid limit cannot be measured, report the sample as non-plastic (see Figure 8.17).

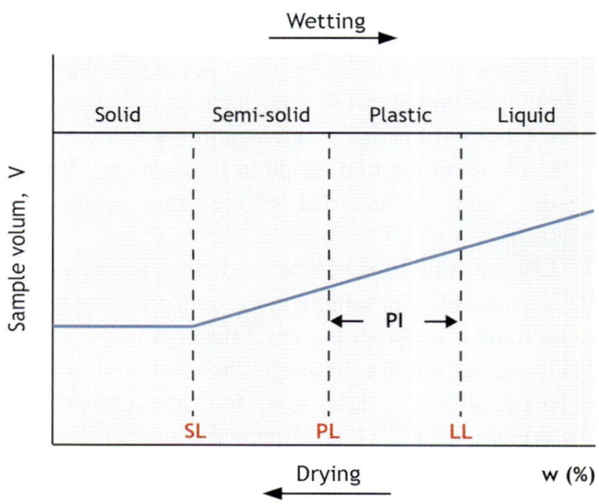

Figure 8.17 Relationship between liquid limit (w_L), plastic limit (w_P) and volume of the sample (adapted from Das and Sobhan, 2013).

The liquid limit can be used along with the plastic limit (Method 8.7.1.6) to determine the following indices:

$$\text{Consistency index [-]} = \frac{w_L - w}{w_L - w_P}$$

$$\text{Liquidity index [-]} = \frac{w - w_P}{w_L - w_P}$$

$$\text{Plasticity index [-]} = w_L - w_P$$

Where:

w_L = Liquid limit of a sample

w_P = Plastic limit of a sample (see Method 8.7.1.6)

w = Water content of a sample (see Method 8.6.1.1).

8.7.1.6 Plastic limit – thread-rolling method[61]

8.7.1.6.1 Introduction

The plastic limit of faecal sludge is the experimentally determined moisture content at which the sample is too dry to behave as a plastic. It indicates the plasticity of a material at a given moisture content. It is used in conjunction with the liquid limit to calculate the Plasticity Index in order to classify the consistency of faecal sludge. The sample is moulded from a ball shape into a thin thread of approximately 3 mm until cracks appear in the thread, both longitudinally and transversely. The moisture content at which the cracks appear and the thread cannot be rolled anymore without breaking is the plastic limit.

8.7.1.6.2 Safety precautions

- General health and safety (H&S) procedures specific for conducting the laboratory analysis of faecal sludge are presented in Section 8.2. Before conducting this method, it is important to be familiar with Section 8.2.3 to ensure safety measures are properly carried out.
- Appropriate personal protective equipment (PPE) should be used; specific details are covered in Section 8.2.3.1.

[61] This method is based on the British Standards Institution Methods of test for soils for civil engineering purposes: Classification tests. This method should be cited as BS 1377:2, as adapted in Velkushanova *et al.* (2021).

8.7.1.6.3 Required apparatus and instruments

- Flat glass plate, to mix and roll samples (10 mm thick, 300 mm square)
- Two pallet knives or spatulas
- Rod (3 mm in diameter and 100 mm long).

8.7.1.6.4 Quality control

General information on quality assurance and quality control (QA/QC) is provided in Section 8.3. Information on standards, operating conditions, and interferences that are specific to this method include:

- This method assumes that heat from the hands contributes to drying out the faecal sludge during handling, contributing to transverse and longitudinal shearing. The length of time taken to dry out the sludge may be extended due to the necessity of wearing latex gloves.
- If the sample is too wet, a hair dryer can be used to evaporate extra moisture and save time on hand rolling.
- The results are subject to the interpretation of the person performing the test, leading to variations in the results. These differences should be quantified as part of a comprehensive quality control procedure (see Section 8.3).

8.7.1.6.5 Sample preservation

It is recommended that the samples are analysed as soon as possible after collection. If immediate measurement is not possible, the samples should be stored at 4 °C for no longer than 14 days before analysis.

8.7.1.6.6 Sample preparation

- Select a representative sample.
- Remove any foreign objects that might be mixed with the sludge (*e.g.* solid waste, stones, hair, and maggots).
- Ensure the sludge sample is thoroughly mixed.

8.7.1.6.7 Analysis protocol

1. Weigh approximately 20 g of the sample on to the glass plate for mixing.
2. Allow the sample to dry until it can be shaped into a ball.
3. Mould the sample into a ball between the fingers and then roll it between the palms until the heat of the hands has made it dry enough that small cracks appear on the surface.
4. Divide the sample into 2 subsamples of approximately 10 g, carrying out a separate determination for each subsample.
5. Divide each subsample into 4 more samples of approximately equal size and apply the following steps to each sample.
6. Mould the sample between the fingers to equally distribute the moisture and then roll the sample into a thread of approximately 6 mm between the thumb and first finger.
7. Roll the thread on the glass plate with the fingers, from their tip to the second knuckle using enough pressure to reduce the diameter to approximately 3 mm in 5 to 10 forward and backward rolls. It is important to maintain a constant rolling pressure.
8. Pick up the sample and mould between the fingers, reproduce a thread shape and repeat step 7.
9. Continue step 8 until the thread shears both longitudinally and transversely when it is rolled to a 3-mm diameter, which is determined using the rod. After the thread has crumbled, do not reproduce the thread, as the first crumbling point is the plastic limit.
10. Place the pieces of the thread in a container and seal with a lid.
11. Place the pieces of all four threads in the one container and determine the moisture content (Method 8.6.1.1).
12. Repeat steps 5 through 11 for the second set of 4 subsamples.

If the moisture content of the 2 sample replicates differs by more than 0.5% the entire test must be repeated.

8.7.1.6.8 Calculation (if necessary)

Calculate the average of the two moisture content values and round to the nearest whole number. This is the plastic limit (w_P).

Plastic limit can be used along with liquid limit (Method 8.7.1.5) to determine the following indices:

$$\text{Consistency index [-]} = \frac{w_L - w}{w_L - w_P}$$

$$\text{Liquidity index [-]} = \frac{w - w_P}{w_L - w_P}$$

$$\text{Plasticity index [-]} = w_L - w_P$$

Where:

w_L = Liquid limit of a sample (see Method 8.7.1.5)

w_P = Plastic limit of a sample

w = Water content of a sample (see Method 8.6.1.1).

8.7.1.7 Compressibility and stickiness – texture analyser method[62]

8.7.1.7.1 Introduction

Texture analysis refers to a technique for evaluating the mechanical and physical properties of materials. A texture analyser can be used to carry out a variety of tests for characterising the properties of faecal sludge and treated end products. Different mechanical properties of the sample can be measured with a texture analyser, including: compression, tension, flexure, penetration, extrusion, and adhesion, which are used to measure hardness, crispiness, crunchiness, softness, springiness, stickiness, tackiness and other material properties. An assortment of probes is available, and changing the probe enables different properties to be characterised in a variety of samples. Compression tests, the focus of this method, are specifically used to measure compressibility, compactibility, springiness, stress relaxation, crust strength, firmness, and elastic recovery. Characterisation of these properties of solid and semi-solid faecal sludge and end products is informative for the design of new treatment and resource recovery technologies.

8.7.1.7.2 Safety precautions

- General health and safety (H&S) procedures specific for conducting the laboratory analysis of faecal sludge are presented in Section 8.2. Before conducting this method, it is important to be familiar with Section 8.2.3 to ensure safety measures are properly carried out.
- Appropriate personal protective equipment (PPE) should be used; specific details are covered in Section 8.2.3.1.

8.7.1.7.3 Required apparatus and instruments

- Texture Analyser (Stable Micro Systems TA.XT Express)
- Petri dish (> 75 mm diameter, glass)
- Probe - p/75 compression platen attachment.

8.7.1.7.4 Quality control

General information on quality assurance and quality control (QA/QC) is provided in Section 8.3. Information on standards, operating conditions, and interferences that are specific to this method include:

- Major calibration of the equipment should be carried out by the service company at regular intervals, according to the manufacturer's instructions.
- Routine calibration of the equipment should be performed when starting up the equipment, following the manufacturer's instructions.
- Do not use solvents or cleaners with the instrument or probes. Clean with a nonabrasive cloth using water.
- Do not apply excessive upward, downward or sideways force to the probe while connected to the texture analyser, as damage may occur to the load cell.
- Use a consistent mass for all the replicates.
- Remove any foreign objects that might be mixed with the sludge (*e.g.* solid waste, stones, hair, and maggots).

8.7.1.7.5 Sample preservation

It is recommended that the samples are analysed as soon as possible after collection. If immediate measurement is not possible, the samples should be stored at 4 °C for no longer than 14 days before analysis.

[62] This method should be cited as adapted in Velkushanova *et al.* (2021), together with the specific analytical equipment (if different), and any manufacturer's modifications.

8.7.1.7.6 Sample preparation

- Ensure the sample is thoroughly mixed.
- After mixing, if the samples are not to be analysed immediately, cover them with foil or a lid to minimise water loss due to evaporation.

8.7.1.7.7 Analysis protocol

Start up

- Turn on the computer and texture analyser. Start the texture analysis software.
- Release the safety button on the texture analyser by pressing and rotating it clockwise.
- Attach the probe to the texture analyser.

Calibration

- Place the petri dish containing the sample in the designated place and secure its position with clamps.
- For height calibration before each set of experiments, select 'Calibrate' (set up for the specific petri dish in use) in the software programme and calibrate the height.
- In the software window, ensure the contact force is 1 g and the height is 25 mm return distance, and the return speed is 10.
- The probe should go down to touch the base of the petri dish and return to 25 mm above the surface.
- Click 'OK' after successfully calibrating the height.

Run a test

- Select 'Run a test' in the software programme.
- Set the test mode ('Compression'), pre-test speed (1 mm/sec), test speed (2 mm/sec), post-test speed (10 mm/sec), target mode ('Distance'), distance (5 mm), trigger type ('Auto-force') and 'Trigger force' (5 g) in Newtons, and click 'OK'. For tension experiments, set the test mode (tension), pre-test speed (3 mm/sec), post-test speed (10 mm/sec), target mode (distance), distance (10 mm) and trigger type (button).
- Do not change the parameters; these include sample shape which is undefined as default, and the data acquisition. Use the default settings. These have the sample width, sample length, sample height, temperature and stress area without any values as the sample is not rigid and the temperature is not automatically regulated. The data acquisition tab has the data acquisition rate (pps) as 200 and the typical test time (sec) at 150 as the default settings.
- Weigh 100 g of the thoroughly-mixed sample into the petri dish, and ensure the sample surface is level for optimal contact with the probe.
- Secure the petri dish with clamps.
- Lower the probe using the stylus or the software.
- Start the test by clicking 'Start'.
- The compression platen on the texture analyser moves until it reaches the top surface of the sludge sample then it exerts a force of 0.098N (10 g) on the sample and compresses the sludge at a speed of 2 mm/s until the probe is in tight contact with the bottom of the petri dish. Afterwards, the probe is raised back up to its initial position at a speed of 10 mm/s, thus exerting pulling force on the faecal sludge.

Check results and save

- Once the analysis is done, click on the 'Results' tab to view the tabulated results such as peak positive and negative forces.
- To view the maximum peak negative and positive force, click on the 'Go to' tab then view and select the required force.
- To save data, export and save the raw data as an Excel spreadsheet to the required directory.
- After saving a graph and raw data (combined), close the software program.
- Move the probe to 25 cm above the platform.

Shut down

- Press the emergency stop button.
- Switch off the texture analyser.
- Remove the sample and clean the probe.

8.7.1.7.8 Calculation

No calculation required - direct reading. An example of a typical graphical test result is presented in Figure 8.18

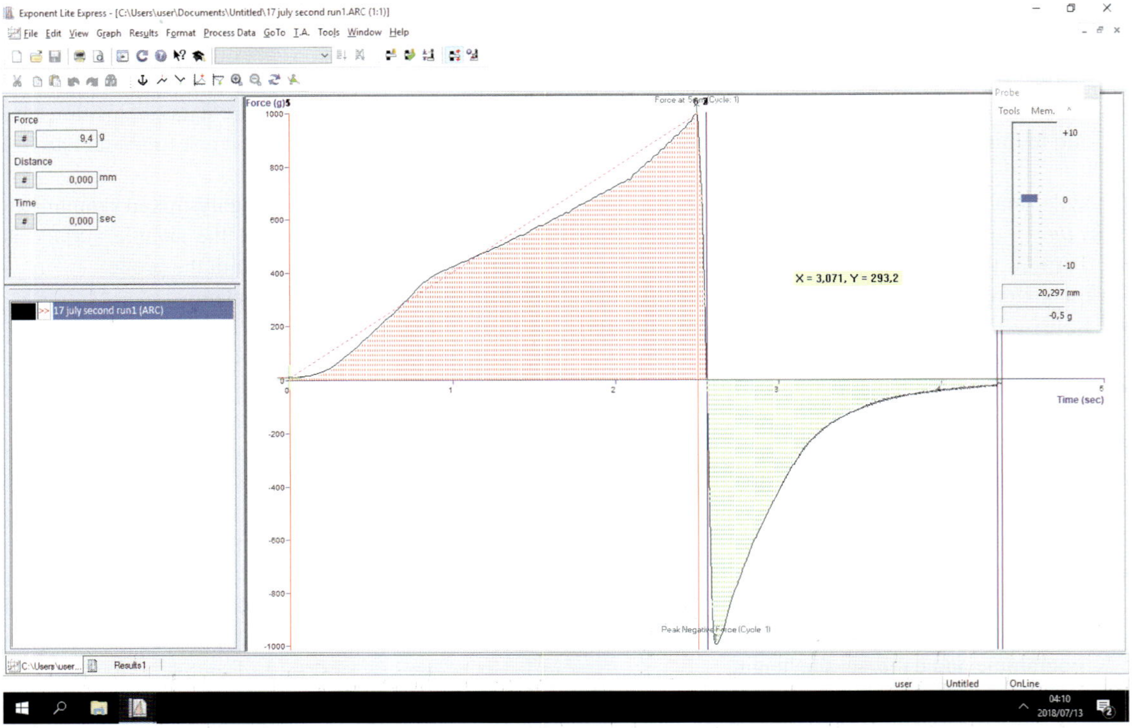

Figure 8.18 Typical graph of a test with the red part showing the compression forces and the green part showing the tension forces.

8.7.2 Physical and thermal properties

The evaluation of thermal properties of faecal sludge is important for treatment and resource recovery applications, such as the production of heat-treated pellets and the combustion of solid fuels or biofuels from faecal sludge. Thermal properties include thermal conductivity and diffusivity, specific heat, and calorific value. The calorific value of a material is the quantity of heat produced from its combustion, which is important in evaluating the suitability of faecal sludge end products as solid fuels. The thermal conductivity is the ability of a material to conduct heat, and heat capacity is the amount of heat energy required to change the temperature of an object by a certain amount. Thermal conductivity and heat capacity are important in the heat treatment and drying of faecal sludge for pathogen reduction and resource recovery.

8.7.2.1 Thermal conductivity – thermal conductivity analyser method[63]

8.7.2.1.1 Introduction

Thermal conductivity is the thermal property of a material that describes its ability to conduct heat. Thermal conductivity, along with several other thermal characteristics, can be measured using a thermal conductivity analyser. The method presented here is written for a C-Therm TCi Thermal conductivity analyser[D]. This instrument uses the modified transient plane source method, which involves monitoring the temperature increase in the

[63] This method should be cited as adapted in Velkushanova *et al.* (2021), together with the specific analytical equipment (if different), and any manufacturer's modifications.

sample during and after exposure to a short heat stimulus. The thermal conductivity of the sample is inversely proportional to the rate of temperature increase in the sample. This method usually requires small amounts of sample and can accommodate solid, semi-solid, slurry, and liquid faecal sludge samples, as well as dried end products. In addition to thermal conductivity, this method can be used to determine thermal effusivity, thermal diffusivity, heat capacity, the R value, and depth of heat penetration. Characterisation of these thermal properties of faecal sludge and end products is helpful in the development and design of drying technologies.

8.7.2.1.2 Safety precautions

- General health and safety (H&S) procedures specific for conducting laboratory analysis of faecal sludge are presented in Section 8.2. Before conducting this method, it is important to be familiar with Section 8.2.3 to ensure safety measures are properly carried out.
- Appropriate personal protective equipment (PPE) should be used; specific details are covered in Section 8.2.3.1.

8.7.2.1.3 Required apparatus and instruments

- Thermal conductivity analyser (C-Therm TCi)
- Small volume test kit (SVTK) to measure liquid and powder samples
- Measuring spoons (1/8 or 0.63 mL and ¼ or 1.25 mL).

8.7.2.1.4 Quality control

General information on quality assurance and quality control (QA/QC) is provided in Section 8.3. Information on standards, operating conditions, and interferences that are specific to this method include:

- The sensor is factory-calibrated. See the manufacturer's instructions for routine calibration and maintenance.
- Check the R2 value for each measurement. If the R2 value is less than 0.995, the measurement is not valid and should be repeated.

8.7.2.1.5 Sample preservation

It is recommended that samples are analysed as soon as possible after collection. If immediate measurement is not possible, the samples should be stored at 4 °C for no longer than 14 days before analysis.

8.7.2.1.6 Sample preparation

- Remove any foreign objects that might be mixed with the sludge (*e.g.* solid waste, stones, hair, and maggots).
- Ensure the sample is thoroughly mixed.
- After mixing, if samples are not to be analysed immediately, cover the samples with foil or a lid to minimise water loss due to evaporation.
- To perform this method, the density of the sample must be known. Before starting the analysis, measure the density of the sample using Method 8.7.1.1.

8.7.2.1.7 Analysis protocol

Measurement - solid and dry faecal sludge samples:

- In the TCi software, select 'New test'.
- Choose 'Liquids and powders' as the group/material.
- Wait for the instrument and sensor to be detected.
- Enter the density of the sample (m/v) alongside the material.
- Scoop 3 × 1/8 teaspoons, or 1.8 mL of the sample onto the test cell.
- Place the quick clamp cap on the test cell.
- Monitor the sensor temperature with the TCi software until it is stable and the sensor, sample and environment have all reached a state of thermal equilibrium.
- Initiate the test sequence with the TCi software.
- Alter 'Test method' to enter the number of measurements required (usually five measurements are required).

Measurement - liquid or slurry faecal sludge samples:

- In the TCi software, select 'New test'.
- Choose 'Liquids and powders' as the group/material.
- Wait for the instrument and sensor to be detected.
- Enter the density of the sample (m/v) alongside the material.

- Measure 1.25 mL (1/4 teaspoon) of the total liquid volume of the specimen.
- Transfer this volume directly to the test cell.
- Place the quick clamp cap on the test cell.
- Monitor the sensor temperature with the TCi software until it is stable and the sensor, sample and environment have all reached a state of thermal equilibrium.
- Initiate the test sequence with the TCi software.
- Alter 'Test method' to enter the number of measurements required (usually five measurements are required).

Cleaning after testing

- Pour out the contents of the sample from the test cell or remove it with a paper towel.
- Place the sensor upside down and remove the test cell by gradually unfastening the three screws.
- Remove the sensor test and clean with either soap and water, water, or propyl alcohol.
- To test again, place the test cell on the sensor and place upside down in order to have easy access to the screws.
- Tighten gradually until the test cell seats perfectly flat against the sensor-housing surface.

8.7.2.1.8 Calculation (if necessary)

No calculation is required, as the analyser calculates and reports the thermal properties.

8.7.2.1.9 Data set example

Table 8.15 contains experimental data from five replicate measurements of thermal properties of a faecal sludge sample from VIP latrines in Durban, South Africa, using a C-Therm TCi thermal conductivity analyser. Effusivity, thermal conductivity, diffusivity, heat capacity, and the R value are thermal properties of the material. R2 is an indicator of smoothness of the curve generated for each measurement. Measurements are deemed valid when R2 > 0.995 (unpublished data, UKZN PRG).

Table 8.15 Experimental data for faecal sludge from VIP latrines in Durban, South Africa.

Replicate number	Sensor	Effusivity ($Ws^{½}/m^2K$)	Thermal conductivity (W/mK)	Diffusivity (m^2/s)	Heat capacity (J/kg.K)	R value (m^2K/W)	R2
1	T298	488.58	0.1816	0.0000	1,205,649.27	0.0037	0.9990
2	T298	489.81	0.1821	0.0000	1,208,939.75	0.0037	0.9988
3	T298	507.26	0.1880	0.0000	1,255,668.41	0.0035	0.9985
4	T298	517.42	0.1915	0.0000	1,282,799.11	0.0035	0.9986
5	T298	523.54	0.1936	0.0000	1,299,144.06	0.0034	0.9986

8.7.2.2 Calorific value – bomb calorimeter method[64]

8.7.2.2.1 Introduction

Calorific value is defined as the amount of heat energy released by the mass of a sample when combusted in an enclosure of constant volume. It is a measure of the energy content of a sample. Calorific value is an important metric for evaluating the suitability of faecal sludge end products as biofuels, for example, dried sludge, pellets, and char. The calorific value of faecal sludge is affected by multiple factors (*e.g.* type of onsite sanitation technology, level of stabilization, and sand content), and a range of variation in calorific values has been reported for faecal sludge end products worldwide, as summarised in Andriessen *et al.* (2019). Faecal sludge end products can have calorific values comparable to wood and waste biomass (Andriessen *et al.*, 2019; Murray Muspratt *et al.*, 2014; Diener *et al.*, 2014). Calorific value is

[64] This method should be cited as adapted in Velkushanova *et al.* (2021), together with the specific analytical equipment (if different), and any manufacturer's modifications.

expressed as energy/mass. Common units for calorific value of fuels are MJ/kg or BTU/lb.

The bomb calorimeter method is commonly used for measurement of calorific value. The calorific value obtained with a bomb calorimeter represents the higher heating value (HHV), or the gross heat of combustion per unit mass of sample. This is the heat produced when the sample burns, plus the heat given up when the newly formed water vapour condenses and cools to the temperature of the calorimeter. This method is intended for use with dry end products or oven-dried samples. Calorific value can be determined for oven-dried moisture-free samples, or for as-received samples containing some moisture. A thorough introduction to the theory and use of bomb calorimetry is available in Parr Instrument Company (2013).

Step-by-step procedures for measuring calorific value will vary depending on the available equipment and the characteristics of the incoming faecal sludge samples. One example of calorific value measurement using a bomb calorimeter is the method used by the UKZN PRG laboratory in Durban, South Africa using a Parr 6200 Oxygen Bomb Calorimeter[D], described here.

8.7.2.2.2 Safety precautions

- General health and safety (H&S) procedures specific for conducting the laboratory analysis of faecal sludge are presented in Section 8.2. Before conducting this method, it is important to be familiar with Section 8.2.3 to ensure safety measures are properly carried out.
- Appropriate personal protective equipment (PPE) should be used; specific details are covered in Section 8.2.3.1.
- Exercise caution and follow the instrument's instructions carefully when charging and handling the oxygen bomb. Never over-pressurise the bomb. Maximum filling pressure may vary with the equipment being used, for example, for the Parr 6200 Oxygen Bomb Calorimeter the bomb must never be filled to more than 600 psi (40 atm).
- Work carefully when moving the pressurised bomb after filling with oxygen.
- During firing and for at least 15 sec after firing, stand back from the calorimeter and keep clear of the top of the calorimeter. If the bomb does explode, it is likely that the force of the explosion will be directly upward.

8.7.2.2.3 Required chemicals

- Oxygen cylinder (≥ 99.5% pure oxygen)
- Standard benzoic acid pellets
- Ethylene glycol as a combustion aid (for samples that are difficult to combust).

8.7.2.2.4 Required apparatus and instruments

- Oxygen bomb calorimeter, including an oxygen combustion bomb
- Bomb head support stand
- Analytical balance with four decimal places.

8.7.2.2.5 Quality control

General information on quality assurance and quality control (QA/QC) is provided in Section 8.3. Information on standards, operating conditions, and interferences that are specific to this method include:

- Benzoic acid is used as a calibration standard of known calorific value to determine the heat capacity of the calorimeter.
- The heat capacity of the calorimeter should be checked at least once a month, and also after changing any part on either the calorimeter or the oxygen canister. See ASTM D5865 (ASTM 2004) for a thorough description of the heat capacity checks, and refer to the manufacturer's instructions.
- The temperature measurement in the calorimeter needs to be accurate to 0.0001 °C, and should be calibrated by a recognised certifying agency (ASTM 2004).
- The analytical balance must be calibrated regularly, following the method outlined in Method 8.6.1.1.
- After filling the bomb with oxygen, check that the bomb is not leaking by submerging it in water. Do not fire the bomb if gas bubbles are leaking from the lid.
- If leaking, depressurise and then open it, then clean the seals and O rings, and then re-seal, re-fill, and re-check.

- The fuse used to ignite the sample is made of cotton thread. Ensure the fuse thread stays dry, as a wet fuse will prevent the sample from igniting. The fuse should not be immersed in the sample; instead, it must be placed above the sample.
- The test should be operated at room temperature (20-25 °C).
- Regular maintenance must be carried out after every 30 tests, including replacement of the O ring and wire.
- Allow at least 20 min for the calorimeter to warm up and the jacket temperature to reach standard operational temperature.
- If an as-received sample containing some moisture is to be characterised, it may be difficult to achieve complete combustion of the sample. In this case, ethylene glycol or another combustion aid may be used. According to ASTM D5865, a minimum of 0.4 g of combustion aid is used, and its weight recorded to the nearest 0.0001 g. Calorific value results must be corrected for the use of the combustion aid by subtracting the heat of combustion of the aid multiplied by the mass of the combustion aid from the overall calorific value obtained.
- Other corrections to the measured calorific value may also be required, although this process is automated in more automated instruments. See the manufacturer's instructions for specific calibration calculations and protocols.
- Do not use too much sample. The standard bomb cannot withstand the effects of combustible charges which liberate more than 8,000 calories. This generally limits the total weight of combustible material (the sample plus the combustion aid) to no more than 1.1 g.

8.7.2.2.6 Sample preservation

When this analysis is to be performed on moisture-free samples, faecal sludge can be dried immediately after collection (in a drying oven at 105 °C), then ground to powder. Dried powder samples can be stored long-term prior to analysis, in a cool dry place. When samples are to be analysed as-received, still containing moisture, they should be refrigerated (4 °C) until analysis and stored for no longer than 30 days. The preparation steps and duration of storage should be reported with the results.

8.7.2.2.7 Sample preparation

- If the sample is to be analysed as a moisture-free sample, dry at 105 °C in a crucible for 24 hr or until completely dry, following the method outlined in Method 8.6.1.1 for assuring complete sample dryness. Even if the sample has been dried as a sample preservation step, it should be dried again immediately before characterisation to ensure it is entirely free of any moisture that could have been absorbed during storage.
- If the sample is to be analysed as-received, do not dry.
- Grind or pulverise moisture-free and as-received samples, and sieve to ensure the particle size of the sample is less than 250 μm (*e.g.* with a 250 μm sieve or No. 60 sieve).

8.7.2.2.8 Analysis protocol

Calibration

- Use a 1-g benzoic acid pellet for calibration. Always record the exact weight of benzoic acid (to 0.0001 g) used for the calibration even if using the pellets, as weight may vary.
- The heat of combustion of benzoic acid is ~26 MJ/kg, and the exact heat capacity will be listed on the certificate that comes with the benzoic acid standard.
- Run the benzoic acid standard using the same method used for a sample measurement in the following section.

Measurement

- Open the oxygen gas cylinder and set the flow rate to constant pressure (*e.g.* 400 psi, 3,000 kPa).
- Fill the water chamber with distilled water to the mark, and then turn on the calorimeter, pump and heater. When the calorimeter is ready the 'Start' key will appear and begin testing.
- Fill the calorimeter bucket with 2 L of distilled water and place the bucket in the calorimeter. It is important that the water level is exactly at the 2 L mark to maintain accuracy. It is recommended to replace the water every day in case there is ionization/adulteration of the distilled water during the experiments.
- Record the moisture content of the sample. If analysing a moisture-free sample, note 0% moisture. If analysing as-received samples,

measure the moisture content of a representative sample (see Method 8.6.1.1).

- Weigh between 0.5 and 0.7 g of sample into a capsule, and record the sample weight to the nearest 0.0001g.
- If the moisture content of the sample is higher than 80%, add ethylene glycol to the sample before combustion to facilitate ignition. Adjust the sample mass accordingly so that the combined weight of ethylene glycol and sample is no more than 1 g. Record the mass of ethylene glycol added to the nearest 0.0001 g.
- Follow the manufacture's instruction to set up the bomb.
- The Parr 6200 calorimeter will conduct the test automatically.
- Read and record the calorific value from the screen.
- Remove the bomb from the chamber after 3 min and depressurise the bomb by opening the valve knob slowly. After all the pressure has been released, unscrew the cap and lift the head straight out.
- Remove the chamber containing the ash.
- Wipe the inside of the bomb with a clean laboratory tissue and proceed with the next sample.
- Clean the bomb and soak in a citric acid solution, either overnight or whenever the bomb is dirty.
- Aqua regia can also be used to clean the bomb; remove the wire if aqua regia is used.

8.7.2.2.9 Calculation

No calculation required - direct reading. If using a basic calorimeter instead of an automated calorimeter such as the Parr 6200, follow the instructions in the manual, or use the equations in ASTM D5865 to calculate the heat capacity of the calorimeter and the calorific value of the sample.

8.8 BIOLOGICAL PROPERTIES

8.8.1 Pathogens

The current methods presented in this section are for pathogens, specifically the detection of total and viable Helminth eggs, and the qualitative and quantitative enumeration of coliform bacteria and coliphage as indicators of pathogens. Biological activities related to the production and consumption of organic matter, or respiration, are included under Section 8.6. Further types of analytical methods for biological examinations include identifying specific pathogens (*e.g.* viruses, bacteria, protozoans, and helminths), metrics of toxicity (*e.g.* use of bioassays), enumeration (*e.g.* plate counts, flow cytometry, and MPN), and types and functions of organisms (*e.g.* DNA/RNA analysis).

The methods of detection for pathogens are important to ensure adequate protection of public health and are commonly used for compliance, evaluation of treatment performance, and research purposes. Isolating specific microorganisms can be tedious, so indicator organisms that send a signal that samples could be contaminated with pathogens are commonly used (Madigan *et al.,* 2018). Indicator organisms are selected based on their ease of detection, similar behaviour in the environment, and greater resistance to die off than other pathogens of concern. For a detailed overview of disease-causing organisms and enteric pathogens of concern in wastewater treatment, and the pathways of contamination, see Gerba (2020) and Cairncross and Feacham (2019).

Coliform bacteria are commonly used as indicators, as they live in the intestines of humans and other warm-blooded animals and so are considered to indicate faecal contamination. Coliforms are operationally defined as 'aerobic and facultatively aerobic, gram-negative, non-spore-forming, rod-shaped bacteria that ferment lactose with gas formation within 48 hr at 35 °C' (Madigan *et al.,* 2018). In general, most coliforms are not harmful, and only some cause signs of infection. One caveat on using coliforms as indicators is that they are not specific to humans. *E coli* can be distinguished from other types of coliforms (*e.g.* the absence of urease and the presence of B-glucuronidase) and is more likely to indicate faecal pollution (Gerba, 2020).

The coliform test is commonly carried out with the most-probable-number (MPN) method or the membrane filter procedure. In the MPN method, samples are serially diluted in liquid culture medium in test tubes until no growth is observed. In the membrane filter procedure, samples are passed through a filter that captures all of the bacteria, and the filter is then incubated on a plate of eosin-methylene blue (EMB) culture media. However, microbial methods can be challenging when resources are limited, requiring a wide range of laboratory equipment and skilled personal. Presented in Bain *et al.* (2012) is a summary of commercially available test kits for coliform bacteria, including a cost breakdown, and which type of settings they are appropriate for (no laboratory, a basic laboratory, or a highly resourced laboratory). The summary covers the types of presence/absence tests, and quantitative tests using colony counts and MPN. Colony counts use plating, filtration, or immobilization of the bacteria within a media, whereas MPN tests rely on sample division or dilution and a statistical method for estimating the number of organisms (Bain *et al.,* 2012). This book presents examples of the 3M Petrifilm colony count method[D], and the Colilert presence/absence[D] and MPN method. For a more detailed review of available kits, refer to Bain *et al.* (2012). Reporting of MPN and CFU are equivalent, both are units to measure the estimated number of bacteria in a water sample. CFU/100 mL is based on counting colonies on a plate, whereas MPN/100 mL is the statistical probability of the number of organisms. The membrane filtration and the multi-enzyme substrate methods are included in the Standard Methods for the Examination of Water and Wastewater, Section 9060A (Rice *et al.,* 2017).

Bacteriophage are a type of virus that are able to infect and grow in bacteria. Coliphage are a type of bacteriophage that specifically infect *E. coli*, so their presence suggests the presence of *E. coli*. They are used as indicators due to their continual presence in wastewater. Coliphage are similar to many human viruses, they cannot replicate in the environment without a host, are relatively resistant to disinfection processes, and they can be detected with simple and inexpensive methods that yield results in 8-18 hr, which make them good candidates for indicators (Rice *et al.,* 2017; Gerba, 2020). Presented in this book is the example of a coliphage plaque assay. Step-by-step procedures for variants of the double-agar-layer method, single-agar-layer format, and filter-adsorption technique, are presented in Rice *et al.* (2017).

In faecal sludge, helminths are commonly used as an indicator of the effectiveness of pathogen reduction due to their prevalence in low- and middle-income countries, and their persistence following treatment. Helminths are important pathogens to monitor, especially *Ascaris lumbricoides*, whose eggs are one of the pathogens most resistant to inactivation in treatment processes. Presented in this chapter is a method developed by the UKZN PRG specifically for the detection of total and viable helminth eggs in faecal sludge.

8.8.1.1 *E. coli* and total coliforms – colony forming unit method[65]

8.8.1.1.1 Introduction

Coliform bacteria are present in the faeces of warm-blooded animals, and are used as common indicators of faecal pollution. Escherichia coli (*E. coli)* is one type of coliform bacteria; most *E. coli* and coliforms are harmless, but some cause signs of infection in humans. Coliform bacteria are relatively easy to culture, and are used as indicators of the possible contamination by or presence of faecal matter. Coliforms and *E. coli* can be measured by pour plate or spread plate counts, the membrane filtration method or the multi-well enzyme substrate method results. The plate count test for *E. coli* and total coliforms is based on the principle that microbial cells quickly grow into visible colonies when they are provided with a suitable growth media and growth

[65] This method should be cited as the specific method that is carried out in each laboratory, including the manufacturer's make and model (where necessary) for the total coliforms and *E. coli* test, and the exact method of sample preparation.

conditions. This method has been used frequently to culture different microorganisms on a nutrient media by adding substances that enhance the growth of organisms of interest or inhibit unwanted species. Colony-forming methods for analysing *E. coli* and total coliforms give an estimate of the population density of *E. coli* and total coliforms in the sample. To ensure that plates with countable colonies are produced, diluted samples are used for this method. Decreasing concentrations of the original sample are made using serial dilutions to plate a range of dilutions. This ensures that the plates contain reduced numbers of *E. coli* and total coliform that are distinct and can be counted as individual colonies. The number of colonies counted on a plate gives the colony-forming units, which is divided by the volume of sample used to get the CFU/volume (CFU/mL).

The example provided here is the 3M Petrifilm *E. coli*/coliform test[D], which is based on the manufacturer's protocol and can be found on the 3M website[66]. For a more detailed review of available kits, refer to Bain *et al.* (2012). The 3M Petrifilm *E. coli*/coliform count plate is a sample-ready culture-medium system, which contains Violet Red Bile (VRB) nutrients, a cold-water-soluble gelling agent, an indicator of glucuronidase activity (BCIG), and a tetrazolium indicator that facilitates colony enumeration in microbiological samples. These plates provide both *E. coli* and total coliform count information with confirmed results in 24-48 hr. UKZN PRG in Durban has adapted this method for analysis of total coliforms and *E. coli* in faecal sludge.

8.8.1.1.2 Safety precautions

- General health and safety (H&S) procedures specific for conducting the laboratory analysis of faecal sludge are presented in Section 8.2. Before conducting this method, it is important to be familiar with Section 8.2.3 to ensure safety measures are properly carried out.
- Appropriate personal protective equipment (PPE) should be used; specific details are covered in Section 8.2.3.1.

8.8.1.1.3 Required chemicals

- Ringer's solution
- Saline solution (0.85-0.90%)
- Phosphate buffer water or distilled water.
- 0.1 N NaOH
- 0.1 N HCl.

8.8.1.1.4 Required apparatus and instruments

- 3M Petrifilm *E. coli* /coliform count plates
- Laminar flow hood
- Autoclave
- 1 mL micropipette and tips
- 50 or 100 mL beaker
- Blender
- Analytical balance
- Incubator for temperatures up to 45 °C.

8.8.1.1.5 Quality Control

General information on quality assurance and quality control (QA/QC) is provided in Section 8.3. Information on standards, operating conditions, and interferences that are specific to this method include:

- Use appropriate sterile buffered dilution water, Ringer's solution, saline solution (0.85-0.90%), phosphate buffer or distilled water.
- Do not use diluents containing citrate, bisulphite or thiosulphate with 3M Petrifilm plates as these may inhibit the growth of organisms.
- The samples should be adjusted to a pH of 6.6-7.2 to ensure optimum growth and recovery of the organisms.
- Opened 3M Petrifilm should be sealed by folding the end of the pouch over and applying adhesive tape.
- Storage of 3M Petrifilm plates: store unopened pouches at temperatures lower than 8 °C or frozen. For opened pouches, store sealed pouches in a cool dry place for no longer than four weeks and prevent exposure to moisture.
- If the sample is too concentrated, serial dilutions can be done using distilled water or appropriate buffer solutions using sterile equipment to prevent errors when counting the colonies.
- The samples should be mixed thoroughly as analytical results depend on adequate sample

[66] https://multimedia.3m.com/mws/media/701951O/product-instructions-3m-petrifilm-e-coli-coliform-count-plate.pdf

mixing. If the sample is not adequately shaken before the aliquots are removed, the actual bacterial density could be underestimated.

8.8.1.1.6 Sample preservation

- The samples should be analysed immediately after sampling if possible. Samples can be stored for up to 24 hr between 1-4 °C (Rice *et al.,* 2017). The samples should be allowed to reach room temperature before analysis.

8.8.1.1.7 Sample preparation

For liquid samples:

- In general, no preparation for liquid samples is required. Appropriate dilutions should be made using a serial dilution procedure. For faecal sludge samples, a minimum of four sample dilutions are required. However, five or more dilution are preferred (USEPA, 2006).

For slurry to solid samples:

- Weigh out between 1.8 g and 2.0 g of the thoroughly-mixed sample into a beaker.
- Dilute the sample gravimetrically and transfer to the blender, as described in Section 8.4.2.
- If necessary, adjust the dilution based on the specific sludge characteristics.
- Blend for 30 sec on the highest speed.
- Transfer this solution into a plastic bottle for testing.
- The sample pH must be within the range of 6.6-7.2. Adjust the pH with sulphuric acid or sodium hydroxide solution if necessary.
- Total solids analysis should be performed on the samples so that the results of the measurement can be expressed as CFU/gTS.

8.8.1.1.8 Analysis protocol

- Place the 3M Petrifilm *E. coli*/coliform count plate on a flat surface (Figure 8.19 A)
- Lift the top film of the plate and dispense 1 mL of the sample onto the centre of the bottom film (Figure 8.19 B).
- Slowly roll the top film down onto the sample, making sure there is no entrapment of air bubbles (Figure 8.19 C).
- Distribute the sample evenly within the circular well using a gentle downward pressure on the centre of the plastic spreader (flat side down) (Figure 8.19 D). Do not slide the spreader across the film.
- Remove the spreader and leave the plate undisturbed for one min to permit solidification of the gel.
- Incubate the plates in a horizontal position with the clear side up.
- Incubate the plates at 35 °C for 24 + 2 hr and examine for coliforms and *E. coli* growth. The incubation times and temperature can be selected based on the current local reference methods.
- Some *E. coli* colonies require additional time to form the blue precipitate. Re-incubate the plates an additional 24 + 2 hr to detect any additional *E. coli* growth.
- The count plates can be counted on a standard colony counter (Figure 8.19 E)
- Enumerate the total coliforms as the sum of the red colonies plus the blue colonies associated with entrapped gas. Enumerate *E. coli* as the sum of the blue colonies with entrapped gas.
- Report the results as a colony count cfu/mL.
- Figure 8.19 F and Figure 8.19 G display an example where colonies are too numerous to count.
- Colonies may be isolated for further culturing. Lift the top film and pick the colony from the gel (Figure 8.19 H).

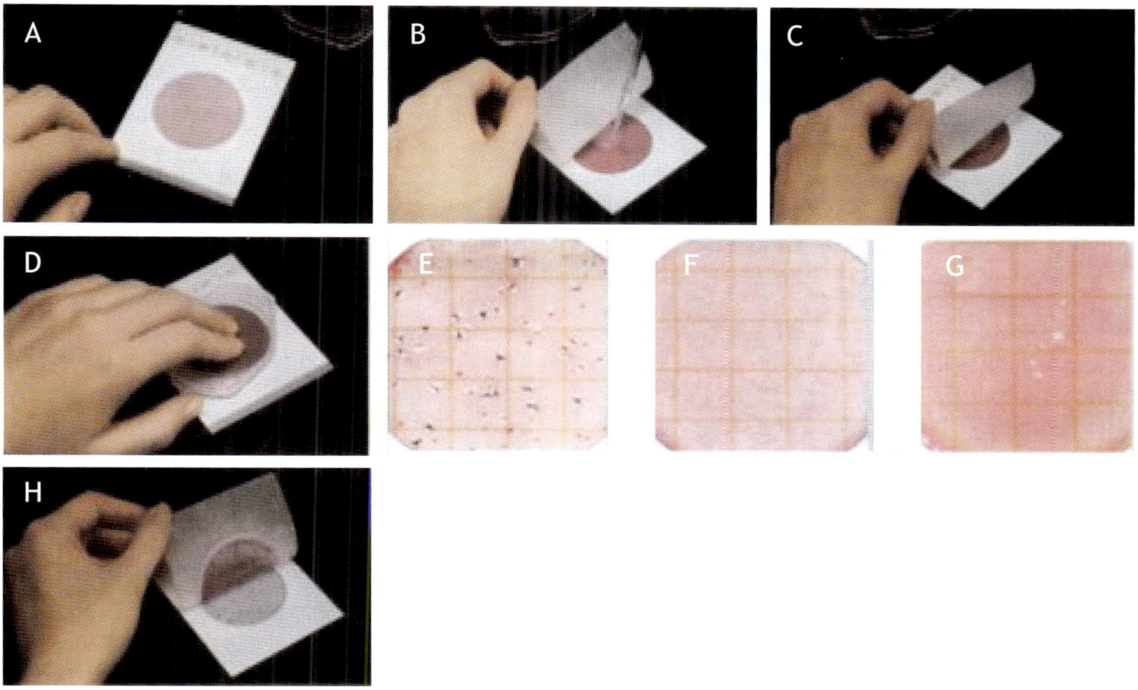

Figure 8.19 *E. coli* count – stages of the analysis protocol (www.multimedia.3M.com).

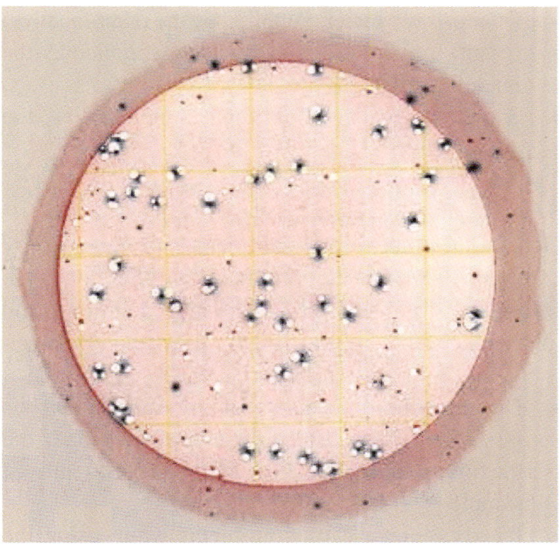

Figure 8.20 *E. coli* count = 49 (blue colonies with gas); Total Coliforms count = 87 (red colonies and blue colonies with gas). Source: UKZN PRG.

8.8.1.1.9 Calculation

$$\text{Coliform forming unit} = \frac{A}{V} \times DF$$

Where:

A = Number of counted colonies
V = Volume plated (mL)
DF = Dilution factor

8.8.1.2 *E. coli*, faecal coliforms, and total coliforms – the most probable number method[67]

8.8.1.2.1 Introduction

The most probable number (MPN) method is a very common quantitative technique, used to estimate numbers of viable cells in water, soils and sediment. The test is based on diluting the microbial content in the sample to a point where samples might not contain any microbial unit that can be cultured. In the MPN technique, replicate sample dilutions are made in an appropriate growth medium and incubated. In the

[67] This method is designed for water analysis and has been adapted by UKZN PRG in Durban for analysis of total coliform and *E. coli* in faecal sludge. This method should be cited as the specific method that is carried out in each laboratory, including the manufacturer's version and model for the total coliform and *E. coli* test, and the exact method of sample preparation.

dilutions, some will contain a single viable cell (which will grow), whereas others will not. Usually, the growth of cells is indicated by changes in the medium. By counting the number of positive and negative tubes for each dilution, and referring to statistical tables, the MPN can be determined. The use of MPN in the quantification of cells is particularly useful for samples with low cell densities (Duncan and Horan, 2003; Oblinger and Koburger, 1975).

Presented here is the example of the Colilert 18®D commercially available test kit. For a more detailed review of available kits, refer to Bain *et al.* (2012). The Colilert 18® is a proprietary test designed for the qualitative presence/absence detection and/or quantification by MPN of total coliforms and *E. coli* in water, and faecal coliforms in wastewater. It uses what the manufacturer terms as 'Defined Substrate Technology (DST)'. The method is based on β-d-galactosidase activity, an essential enzyme that is possessed by both coliforms and *E. coli* and used for lactose fermentation. Two nutrient indicators, o-nitrophenyl-β-d-galactopyranoside (ONPG) and 4-methyl-umbellfieryl-β-d-glucuronide (MUG) are used to detect coliforms and *E. coli*, respectively. Coliforms use β-d-galactosidase enzyme to metabolise ONPG to o-nitrophenol, a yellow-coloured product. In addition, β-glucuronidase, expressed by the majority of *E. coli*, can hydrolyse MUG, forming the fluorescent product 4-methylumbelliferone. These enzymes do not usually occur in non-coliforms. Thus, the growth of non-coliforms is eliminated by this method. Nevertheless, the growth of non-coliforms that possess these enzymes is inhibited by the specially formulated Colilert 18 test matrix. The Colilert 18 test is therefore able to determine *E. coli* and single viable coliforms without interference from non-targeted organisms as compared to conventional media methods. The test can be qualitative following the presence/absence procedure or quantitative following the Quanti-tray® or Quanti-tray 2000® analysis procedure.

8.8.1.2.2 Safety precautions

- General health and safety (H&S) procedures specific for conducting the laboratory analysis of faecal sludge are presented in Section 8.2. Before conducting this method, it is important to be familiar with Section 8.2.3 to ensure safety measures are properly carried out.
- Appropriate personal protective equipment (PPE) should be used; specific details are covered in Section 8.2.3.1.

8.8.1.2.3 Required chemicals

- Colilert 18 reagent 'snap packs' for 18-22 hr incubation (supplied by the manufacturer)
- Colilert sterile trays (supplied by the manufacturer)
- Sterile deionised or distilled water (supplied by the manufacturer)
- Antifoam
- Thiosulphate.

8.8.1.2.4 Required apparatus and instruments

- Autoclave
- Incubator (35-37 °C)
- Pipette and pipette tips
- 50 or 100 mL beaker
- Test tube (sterile)
- Fridge or cold room (8-15 °C)
- Water bath (35-44.5 °C)
- Fluorescent UV lamp 6 watts, 365 nm
- Quanti-tray® sealer (supplied by the manufacturer)
- Quanti-tray/2,000® rubber inserts (supplied by the manufacturer)
- Sampling bottle, 100 mL (supplied by the manufacturer).

8.8.1.2.5 Quality control

General information on quality assurance and quality control (QA/QC) is provided in Section 8.3. Information on standards, operating conditions, and interferences that are specific to this method include:

- Detection limit: the detection limit for this analysis is 1 MPN per 100 mL of sample, and the maximum limit is 2,000 MPN per 100 mL of sample.
- If excess foam causes problems while using the Quanti-Tray®, IDEXX antifoam solution can be used.
- In samples with excessive chlorine, a blue flash may be seen when adding Colilert 18. If observed the sample should be considered invalid and discarded.

8.8.1.2.6 Sample preservation

- The samples should be analysed immediately after sampling, if possible. Samples can be stored for up to 24 hr between 1-4 °C (Rice *et al.*, 2017). The samples should be thawed to room temperature before analysis.
- Store Colilert 18 reagent kits at room temperature (between 20-25 °C) and preferably in a cool dry place.

8.8.1.2.7 Sample preparation

For liquid samples:

- Collect the samples in 100 mL autoclaved sampling bottles.
- Ensure the sampling bottles have 1 mg thiosulphate to neutralise chlorine interferences.
- Dilute highly turbid, concentrated samples with sterile distilled water to concentrations within the range of the test.
- Test the samples at room temperature.
- For 10-fold serial dilutions, fill 9 mL of distilled water into ten tubes labelled as 10-1 through 10-10. Pipette 1 mL of the prepared sample to the first tube titled 10-1 and mix very well. Transfer 1 mL from the first tube labelled 10-1 to the next tube, labelled 10-2. Mix this tube as well. Continue this pattern to create a serial dilution series. You will end up with 9 tubes of 9 mL and 1 tube of 10 mL. Each tube represents a ten-fold dilution of the sample.

For slurry to solid samples:

- Weigh out between 1.8 g and 2.0 g of thoroughly-mixed faecal sludge sample into a beaker.
- Dilute the sample gravimetrically and transfer to the blender, as described in Section 8.4.2.
- If necessary, adjust the dilution based on the specific sludge characteristics.
- Blend for 30 sec in a blender on the highest speed.
- Transfer this solution into bottles with 1 mg thiosulphate to neutralise chlorine interferences.
- Dilute highly turbid, concentrated samples with sterile distilled water.
- Test the samples at room temperature.
- Do not dilute with buffers to avoid interferences with the Colilert reagent.
- Mix the sample by inverting 3 times.

8.8.1.2.8 Analysis protocol

Procedure - Quanti-Tray 2000® total coliforms and E. coli and faecal coliforms (quantitative):

- Switch on the Quanti-Tray® sealer and allow 15 min to warm up.
- Add a sachet of Colilert reagent to the 100 mL sample and mix. Shake the sample until the reagent is completely dissolved.
- Allow the sample to sit and the foam to settle. Add antifoam if necessary.
- Hold the tray upright with the well side facing the palm and bend the ends inwards to open up the tray.
- Pour the sample into the Quanti-tray® once the foam settles.
- Avoid contamination. Do not touch the foil.
- Hold the black rubber frame upright with one hand and place the tray against the frame to fit it into the slots.
- Feed the unit with tray side up into the sealer and collect the sealed tray at the bottom end.
- Press the reverse button on the sealer if jamming occurs.
- Incubate trays at 35 °C for 18 hr ± 4 hr for total coliforms and *E. coli.* Trays can be stacked during incubation.
- Incubate trays at 44 °C for 18 hr ± 4 hr for faecal coliforms (thermo-tolerant coliforms).

Figure 8.21 Quanti-Tray 200 total coliforms and *E. coli* procedure (source: UKZN PRG).

Reading

- Yellow wells = total coliforms.
- Yellow/fluorescent wells = *E. coli.*
- Multiply by the dilution factor for MPN/100 mL if necessary (IDEXX/MPN table is available at the link below[68])
- Enumerate the total coliform by counting the number of large yellow wells (including the single large well at the top) and the small wells on each tray.
- Record the numbers on the relevant data sheet and refer to the table for the total coliform MPN value.
- Enumerate *E. coli* by counting the number of large and small wells that fluoresce under UV illumination in a dark room (Figure 8.22).
- Count the positive wells and refer to the MPN table provided by the manufacturer using the link below.
- Multiply by the dilution factor for MPN/100 mL if necessary.
- Calculate and then record the MPN/100 mL and the upper and lower boundaries of the 95% confidence interval. Record the QA sample results (including zero) the same as the routine results (using IDEXX/ MPN table).

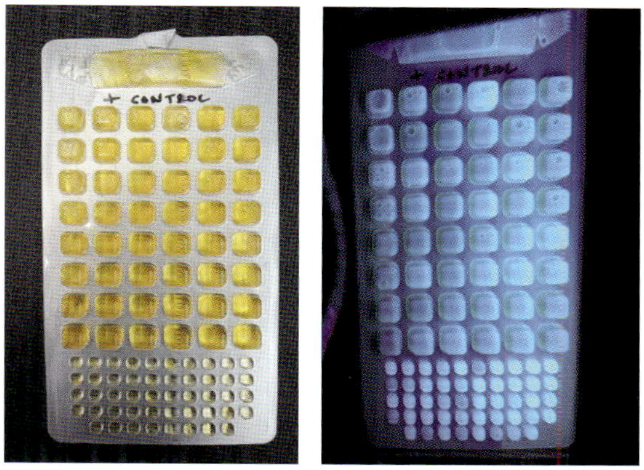

Figure 8.22 Control tray for Total Coliforms (left) and fluorescent control tray of *E. coli* (right) (source: UKZN PRG).

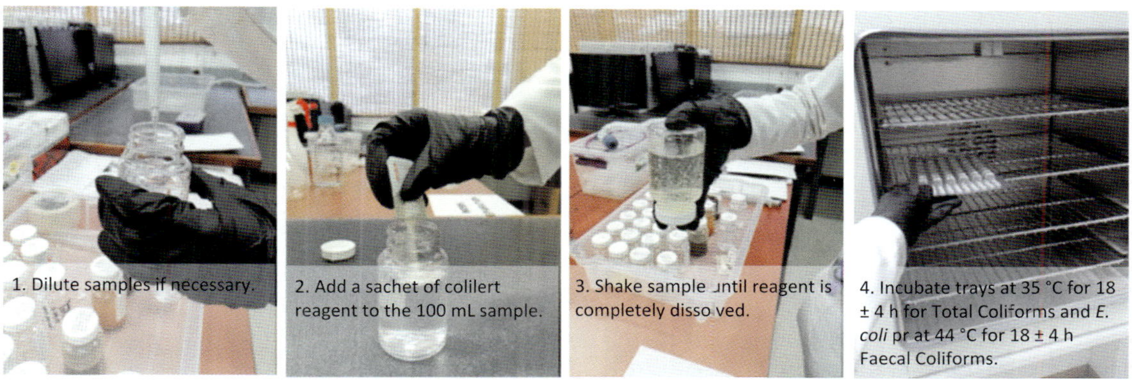

Figure 8.23 Procedure for Presence/Absence (P/A) test for total coliforms and *E. coli* and Faecal Coliforms (source: UKZN PRG).

[68] https://www.idexx.co.za/en-za/water/resources/mpn-generator/

Procedure - presence/absence (P/A) test for total coliforms and E. coli and faecal coliforms (qualitative):

- Dilute the samples with distilled water if necessary.
- Add a sachet of Colilert 18 reagent to the 100 mL sample and mix. Shake the sample until the reagent is completely dissolved.
- Allow the sample to sit and the foam to settle. Add antifoam if necessary.
- Incubate the bottles at 35 °C for 18 hr ± 4 hr for total coliforms and *E. coli*. The bottles can be stacked during incubation.
- Incubate the trays at 44 °C for 18 hr ± 4 hr for faecal coliforms (thermo-tolerant coliforms).

Reading

- Total coliforms - samples turn yellow after incubation at 35 °C.
- *E. coli* - fluorescent.
- Faecal coliforms - samples turn yellow after incubation at 44 °C.

Disposal of used Quanti-trays

Dispose of the trays in a biohazard waste box and arrange collection by a relevant hazardous waste management company.

Table 8.16 Comparison of total coliforms, E. coli and faecal coliform procedure.

E. coli (positive control)			Yellow wells and fluorescence	
Blank sample (sterile water) (negative control)			Clear wells and no florescence	
	Total coliforms	*E. coli*	Faecal coliforms	P/A
Volume	100 mL	100 mL	100 mL	100 mL
Incubation temperature	35 °C	35 °C	44 °C	35 °C
Incubation time	18 hr	18 hr	18 hr	18 hr
Reading	Counting positive wells	UV light	UV light	UV light

Figure 8.24 Control sample for Total Coliforms (left) and fluorescent control sample of *E. coli* (right) (source: UKZN PRG).

8.8.1.3 Bacteriophage - plaque assay method[69]

8.8.1.3.1 Introduction

Bacteriophages are viruses that are able to infect and replicate inside of bacteria and archaea, and they are pervasive in the environment. They could potentially be an alternative to using faecal indicator bacteria, due to their close morphological and biological properties (McMinn *et al.*, 2017). The problems with the use of faecal coliforms as indicators are that they are present in many different types of animals and are not specific to humans, and they could also have different environmental fates than other pathogens (*e.g.* viruses and protozoa). Coliphage are specifically a type of bacteriophage that infect and replicate in *E. coli.* These organisms have been found to be persistent in sewage systems and resistant to treatment. Additionally, they are relatively easy to enumerate as compared to enteric viruses (Jofre *et al.*, 2016). These characteristics make coliphages potentially significant in monitoring the effluent quality of wastewater and faecal sludge treatments. Coliphage can be used to track the origin of faecal contamination in the environment, and also in laboratory experiments, such as spiking into reactors to evaluate treatment performance.

Plaque assays are used to quantify bacteriophage. The theory behind the method is that a cloudy layer of bacterial cells called a 'lawn' grows across a petri dish. If a bacteriophage grows inside a bacterial cell and lyses it, then this results in a clear spot representing no growth, which is called a plaque. Since the cells are lysed, they no longer scatter light and therefore no longer look cloudy. Examples of step-by-step procedures for variants of the double-agar-layer method, single-agar-layer format, and filter-adsorption technique are presented in Rice *et al.* (2017).

8.8.1.3.2 Safety precautions

- General health and safety (H&S) procedures specific for conducting the laboratory analysis of faecal sludge are presented in Section 8.2. Before conducting this method, it is important to be familiar with Section 8.2.3 to ensure safety measures are properly carried out.
- Appropriate personal protective equipment (PPE) should be used; specific details are covered in Section 8.2.3.1.

8.8.1.3.3 Required chemicals

- Tryptone agar
- Tryptone broth
- Calcium chloride
- Phosphate buffer saline (PBS)
- Reagent grade water
- Yeast extract
- Glucose
- NaCl
- Beef extract
- Glycerol.

8.8.1.3.4 Required apparatus and instruments

- Laminar flow hood
- Autoclave
- Incubator (37 °C)
- Refrigerator
- Petri dishes (sterile)
- Water bath with adjustable temperature (45-100 °C)
- Pipettes and pipette tips
- 50 or 100 mL beaker
- 1 or 2 mL Eppendorf tubes
- Bunsen burner
- Shaker (optional)
- Magnetic stirrer and stirrer bars
- Mass balance
- Test tube racks and Eppendorf holders
- Thermometer

[69] This method is based on Method 9224 E of the Standard Methods for the Examination of Water and Wastewater and should be cited as Rice *et al.* (2017) as described in Velkushanova *et al.* (2021).

8.8.1.3.5 Quality control

General information on quality assurance and quality control (QA/QC) is provided in Section 8.3. Information on standards, operating conditions, and interferences that are specific to this method include:

- Coliphage positive controls (Coliphage ΦX174) must be prepared according to the procedure described in the standard method 9224B (Rice *et al.,* 2017).
- Non-sterile/contaminated apparatus such as test tubes, pipette tips or pipette can result in false results.
- Molten agar should not be too hot (> 45 °C). Insufficient cooling can result in the death of host cells put on the top of the soft agar and growth of *E. coli* host cells and plaques will not occur. However, if the agar is cooled too much (< 45 °C) it will solidify before it is dispensed.
- If the sample is too diluted no plaques may appear. Ensure the appropriate dilution factor is used.

8.8.1.3.7 Sample preservation

- The samples should be analysed immediately after sampling if possible. The samples can be stored for up to 24 hr between 1-4 °C (Rice *et al.,* 2017). The samples should be thawed to room temperature before analysis.

8.8.1.3.8 Sample preparation

For liquid samples:

- No preparation for liquid samples is required. If necessary, appropriate dilutions should be made using the serial dilution procedure. For faecal sludge samples, a minimum of four sample dilutions is required. However, five or more dilution are preferred.

For slurry to solid samples:

- Weigh out between 1.8 g and 2.0 g of thoroughly-mixed faecal sludge sample into a beaker.
- Dilute the sample gravimetrically using PBS buffer and transfer to the blender, as described in Section 8.4.2.
- If necessary, adjust the dilution based on the specific sludge characteristics.
- Blend for 30 sec in a blender on the highest speed.
- Dilution of the sample: prepare the dilution of the sample in 9 mL PBS by taking 1 mL from the 50 mL diluted sample and add to 9 mL PBS. Vortex for 30 sec and repeat the process until 6 tenfold dilutions are made (10^{-1} to 10^{-6}). Since viruses can grow to incredibly high concentrations, they need to be diluted in order to count them effectively.
- The same sample preparation method applies to all replicate measurements (if applicable).

Serial dilution for host culture

- Put 9 mL of tryptone broth in each of ten culture tubes labelled as 10^{-1} through 10^{-10}. These tubes will be used for viral serial dilutions.
- Take 1 mL of the coliphage culture stock from the freezer and let it thaw in the laminar flow. Transfer 1 mL of it to the tube labelled 10^{-1} with a pipette. Mix the tube well. This is the first ten-fold dilution (*i.e.* a 1 in 10 dilution)
- Take 1 mL of the mixed culture from your tube labelled 10^{-1} and transfer it with a new pipette to the next tube, labelled 10^{-2}. Mix this tube as well.
- Continue this pattern to create a serial dilution series. You will end up with 9 tubes of 9 mL and 1 tube of 10 mL. The viral loads in your tubes will be diluted anywhere from 10 times (your first tube) or 100 times (your second tube) to ten billion times (your final tube).
- Label all the dilution tubes and media as follows. Each tube represents a ten-fold dilution of the virus.
  - Four tryptone soft agar tubes: 10^{-3}, 10^{-4}, 10^{-5}, 10^{-6}.
  - Four tryptone hard agar plates: 10^{-3}, 10^{-4}, 10^{-5}, 10^{-6}.
  - Six PBS tubes: 10^{-1} through 10^{-6}.

8.8.1.3.8 Analysis protocol

Preparation of media

- Tryptone top (soft) agar: add 10 g tryptone, 0.2 g calcium chloride, 5 g sodium chloride, 1 g yeast extract, 1 g glucose and 7 g agar in 1 L of water in a conical flask. Using a stirrer, mix until most of the solute is dissolved. Sterilise and dissolve by autoclave at 121 °C for 20 min. Let the agar cool while it is still molten and dispense 7 mL of the molten agar in the test tubes or 15 mL centrifuge tubes.

- Tryptone bottom (hard) agar: add 10 g tryptone, 0.2 g calcium chloride, 5 g sodium chloride, 1 g yeast extract, 1 g glucose and 15 g agar in 1 L of water in a conical flask. Using a stirrer, mix until most of the solute is dissolved. Sterilise and dissolve by autoclave at 121 °C for 20 min. While the agar is cool pour it into bottom Petri dishes.
- Tryptone broth: add 10 g tryptone, 0.2 g calcium chloride, 5 g sodium chloride, 1 g yeast extract and 1 g glucose in 1 L of water in a conical flask. Using a stirrer, mix until the solute is dissolved. Sterilise and dissolve by autoclaving at 121 °C for 20 min. Store at room temperature.
- Phosphate buffer saline (PBS): add 5 mL of $MgCl_2 \cdot 6H20$ + 1.25 mL KH_2PO_4 into 1,000 mL dH_2O. Sterilise at 121 °C for 20 min. Store the sample in a cold room or refrigerator at 4 °C. The phosphate buffer will be used as diluent for the sludge sample.
- PBS dilution tubes: aseptically dispense 9 mL of PBS into sterile test tubes or 15 mL centrifuge tubes pre-sterilised by autoclaving at 121 °C for 15 min.

Storage of E. coli C host culture

- Grow *E. coli* overnight in 10 mL of tryptone broth in a centrifuge tube and incubate at 36.5 ± 2 °C. Once grown, add 10 mL of 50% glycerol and then vortex until thoroughly mixed. Once mixed, aliquot 200 µL into vials or 1 mL Eppendorf tubes. Store at -80 °C in the freezer. Autoclave at 121 °C for 10 min. Store at 4 °C.

Preparation of E. coli C host culture

- The day prior to the assay, inoculate 5ml of tryptone broth with the stored *E. coli* host culture using an inoculating loop and incubate at 36.5 ± 2 °C overnight. Transfer 1.5 mL of the incubated culture into 30 mL tryptone broth and incubate for 4 hr at 36.5 ± 2 °C with gentle shaking. Ensure that the lid is loosely closed to ensure enough oxygen for growth. Label the tube as *E. coli*/date/temperature.
- Prepare tryptone agar plates, tryptone soft agar tubes (7 mL/tube) and tryptone broth tubes (10-15 mL/tube). Label appropriately and store in the cold room for use the next day.

Assay procedure

- Warm 100 mL of the sample in a water bath at 44.5 ± 1 °C for 3 min.
- Add 5 mL of $CaCl_2$ solution and 5 mL of appropriate host bacterium (*E. coli* C) preparation to the warmed sample.
- Mix the inoculated sample with 100 mL melted tryptone agar at 44.5 ± 1 °C and distribute to eight 150 × 15 mm petri dishes.
- For the positive control, mix 1 mL of the appropriate positive control preparation (30-80 PFU/mL) and 1 mL host bacterium, *E. coli* C with 12.5 mL warmed agar that has been diluted with an equal volume of warm sterile water.
- Pour into a single 150 × 15 mm Petri dish.
- Repeat for a negative control without the 1 mL of phage preparation.
- Incubate the dishes at 36.5 ± 2 °C overnight and examine for plaques the following day.
- The petri dishes will be covered with a cloudy area of bacteria cells and clear spots indicating regions where plaques have formed.
- Count the number of plaques on the eight dishes that received the sample.

8.8.1.3.9 Counting and calculation

Plaque visualisation, counting and calculating of viral titer (pfu/mL):

- Count the plaques on each plate, taking the average for any technical replicates of the same dilution.
- Determine the viral titer of the stock sample by taking the average number of plaques for a dilution and the inverse of the total dilution factor.

  NOTE: As an example, 30 and 32 plaques counted for replicates of the 1×10^{-7} dilution [31 (average)/10^{-7} (dilution) × 0.4 mL (inoculum)] would yield a titer of 7.75 × 108 pfu/mL.

$$\text{PFU/mL} = \frac{\text{Average number of plaques}}{\text{DF} \times \text{V}}$$

DF = Dilution factor
V = Volume of diluted virus added to the plate.

8.8.1.4 Helminth – microscopy method[70]

8.8.1.4.1 Introduction

The prevalence of helminth infections in people living with rudimentary water and sanitation in low-income countries is generally high. Due to the extreme hardiness of the eggs of the roundworm, *Ascaris lumbricoides*, they are used in the waste and sanitation field as a 'marker' or 'indicator' for the safe end-use of resource recovery products from faecal sludge. Since *Ascaris* eggs are so difficult to inactivate, if treatment of faecal sludge is successful in killing *Ascaris* eggs, then it is likely that other pathogens are also inactivated (viruses, bacteria, protozoa and helminths). Other commonly found helminths are *Trichuris trichiura*, *Taenia* spp. and in areas with very sandy soils, hookworm spp. (*Necator americanus* and *Ancylostoma duodenale*) and *Strongyloides stercoralis*. Various animal parasites are also commonly encountered. In countries where piped water is not chlorinated, the presence of free-living soil and water organisms are encountered and need to be differentiated from pathogens.

Helminth eggs are thought to adhere to soil particles, possibly as a result of charge interactions with, or adsorption of, eggs to the particles. Faecal sludge samples are often contaminated with silica particles, hence the use of ammonium bicarbonate as a wash solution. Liquid samples that have a high fat content need to be treated differently. Here, it is suggested that a surfactant such as Tween 20[D], TritonX-100[D], or 7X[D] is used to break up the fats, rather than ammonium bicarbonate.

Laboratory testing for helminths is based on four main principals: washing, filtration, centrifugation and flotation of the eggs to remove them from the various waste media:

1. Ammonium bicarbonate is used as both a wash solution and also to dissociate the eggs from the soil particles (surfactants should be used for fatty samples).
2. Filtration, using 100 µm and/or 20 µm sieves is used to separate larger and smaller particles from the eggs both after washing and after flotation.
3. Centrifugation is used to sediment the deposit so water can be discarded after washing, to aid the separation process during flotation, and to sediment the washed eggs after flotation.
4. Flotation, using a solution of zinc sulphate at a specific gravity (SG) of 1.3 is used to separate eggs (with a relative density of < 1.3) out of the matter retained (retentate) with them on the 20 µm sieve.

8.8.1.4.2 Safety precautions

- General health and safety (H&S) procedures specific for conducting the laboratory analysis of faecal sludge are presented in Section 8.2. Before conducting this method, it is important to be familiar with Section 8.2.3 to ensure safety measures are properly carried out.
- Appropriate personal protective equipment (PPE) should be used; specific details are covered in Section 8.2.3.1.

8.8.1.4.3 Required chemicals

- Physiological saline (8.5 g/L NaCl)
  Dissolve 8.5 g sodium chloride in 1,000 mL deionised water. If this amount will not be utilised in less than a week, it is preferable to decant it into smaller containers and autoclave for 15 min at 121 °C. Cool to room temperature and store.
- Ammonium bicarbonate (AmBic)
  Dissolve 119 g of ammonium bicarbonate in 1 L de-ionised water (use a magnetic stirrer and bar magnet) - store in a glass jar.
- Tween 80, TritonX100, or 7X
  Note: use neat - see 'Test procedure' below.
- Zinc sulphate ($ZnSO_4 \cdot 7H_2O$)
  Dissolve 500 g zinc sulphate in approximately 800 mL deionised water (use the magnetic stirrer and bar magnet) and adjust SG using more of the chemical or water to raise or lower the SG to 1.3.
- 0.1 N sulphuric acid (H_2SO_4)
  Add 500 mL de-ionised water to a 1 L plastic bottle, pour 3 mL concentrated sulphuric acid into

[70] This method is based on Pebsworth *et al.* (2012), Belcher *et al.* (2015), Naidoo et al., (2016), Grego *et al.* (2018), Naidoo *et al.* (2019), and Naidoo *et al.* (2020), and should be cited as the UKZN PRG Helminth Method

a 10 mL graduated cylinder, then pour the H_2SO_4 into the plastic bottle containing the water, then re-cap and shake. Uncap, add 497 mL of deionised water to the plastic bottle, recap and shake.

8.8.1.4.4 Required apparatus and instruments

- Compound microscope with 10× and 40× objectives (and preferably, a camera)
- Bench-top centrifuge with a swing-out rotor that can spin a minimum of 8 × 15 mL plastic conical test tubes (*e.g.* Falcon tubes) and, if possible, buckets that can also spin a minimum of 4 × 50 mL plastic conical test tubes
- 15 mL plastic conical test tubes (Falcon tubes)
- 50 mL plastic conical test tubes (Falcon tubes)
- Sink with hose attached to tap for washing using strong water pressure
- Top-pan balance (for weighing up to 1.20 g and accurate to 2 decimal places)
- Magnetic stirrer and bar magnets
- Vortex mixer
- Hydrometer that can measure SG between 1.2 and 1.3
- 100 μm mesh stainless steel pan sieve, diameter 200 mm, height 50 mm
- 20 μm mesh stainless steel pan sieve, diameter 200 mm, height 50 mm
- 20 μm mesh stainless steel pan sieve, diameter 100 mm, height 45 mm
- Plastic conical test tubes (Falcon tubes), 15 mL or 50 mL
- Plastic test tube racks to hold the 15 mL Falcon tubes (and if using 50 mL tubes, one for these)
- Plastic 200 mL beakers
- Plastic 'hockey-stick' shaped spreaders
- Plastic 3 mL Pasteur pipettes (non-sterile)
- Non-sterile gloves
- Applicator sticks and wooden tongue depressors
- Microscope slides (76 × 26 × 1.2 mm)
- Cover glasses (22 × 40 mm).

8.8.1.4.5 Quality control

General information on quality assurance and quality control (QA/QC) is provided in Section 8.3. Information on standards, operating conditions and interferences that are specific to this method includes:

- Source an uninfected slurry or semi-solid sludge sample (preferably of a consistency very similar to the samples being tested) and uninfected liquid sludge sample, to make up QA/QC samples. Add 10% formalin (in a 1:10 ratio of formalin to sludge) and store at 4 °C for up to 6 months.
- Make one NEGATIVE control per sample type being tested, *i.e.* one uninfected sample weighed (in grams) and one measured (in litres).
- Make up one POSITIVE control per sample type being tested, *i.e.* spike a known number of *A. suum* eggs into one weighed solids sample and another known number of *A. suum* eggs into a measured liquid sample (egg stocks for preparing the positive controls should be purchased from a reputable supplier).
- Run a negative and a positive sample in parallel with a batch of similar consistency samples per day.
- Once the technician has processed and examined the slides made from the controls, they should be re-examined by a senior, experienced analyst as a control for the microscopy part of the analysis.
- Most sludge and wastewater methods consider recovering > 80% of spiked eggs to be extremely satisfactory.

8.8.1.4.6 Sample preservation

After collection, the samples should be stored at approximately 4-10 °C. Processing is always best carried out as soon after sampling as possible, but providing that there is sufficient moisture and the samples are fairly large (≥ 100 mL/100 g), the eggs should be unharmed and development will be arrested at these low temperatures.

8.8.1.4.7 Analysis protocol

Procedure for slurry, semi-solid and solid faecal sludge samples (TS > 5%):

1. Place a 200 mL plastic beaker (labelled with the sample number) on the top-pan balance, zero the balance, and weigh 10 g or 20 g of the sample into the beaker.
   Note: if waste material is very dry (*e.g.* pelletised or completely desiccated), then soak the weighed sample for 12 – 24 hr in ± 80 mL physiological saline to soften. Next, break up and mix the sample well in the saline. Allow to stand to sediment the

solids for 4 hr. Remove as much supernatant as possible without disturbing the deposit, and continue with the next step below.

2. Add 50-80 mL AmBic and a magnetic stirring bar, and mix on the magnetic stirrer for 10 min.
3. Pour this mixture over the 100 µm mesh sieve placed on top of the 20 µm sieve (wet the sieves with tap water first).
4. Rinse the beaker with tap water and pour over the sieves.
5. Wash the magnet over the sieves and remove, wash the 100 µm sieve well (using a 'hockey stick'-shaped spreader, or preferably, a gloved hand) over the 20 µm filter, regularly checking the bottom sieve for fluid build-up. Use the same hockey stick spreader to stir the sample on the 20 µm sieve while holding the 100 µm sieve directly above so as not to lose any sample. When the 20 µm sieve has drained sufficiently, place the 100 µm sieve back on top and continue washing. Repeat this until the sample on the 100 µm sieve is sufficiently well washed.
6. Separate the sieves and then rinse the 20 µm sieve well. Use water pressure to wash the material to one side of the sieve to make collection easier.
7. Rinse all the material off the 20 µm sieve into the original rinsed-out, labelled beaker.
8. Pour the beaker contents into 4 × 15 mL conical test tubes labelled with a sample number or if the retentate is large, use 50 mL labelled tubes. Rinse the beaker with a small volume of water and add this to the test tubes until all the sample is collected. (The aim after the next step is to have ≤1 mL deposit in a 15 mL tube and ≤5 mL in a 50 mL tube.)
9. Centrifuge at 3,000 rpm (1,512 g) in the centrifuge with a swing-out rotor for 10 min.
10. Pour off the supernatant, leaving deposits in the test tubes.
11. Place the test tubes in the rack with the applicator stick in each (as a stirring rod) and pipette in $ZnSO_4$, 3 mL at a time, vortexing in between the addition of the chemical, until the tubes are filled to the 14 mL mark for the 15 mL tubes/45 mL mark for the 50 mL tubes.
12. Centrifuge at 2,000 rpm (672 g) for 10 min.
13. Pour the supernatant flotation fluid over the 100 mm diameter 20 µm sieve. Wash the remaining deposits out of the test tubes and keep one aside for re-use.
14. Wash the material on the sieve well with tap water and rinse it down to one side of the sieve for collection. Using a 3 mL plastic pipette, transfer the material back into the test tube kept aside.
15. Centrifuge at 3,000 rpm (1,512 g) for 10 min to obtain the final deposit.
16. Pour off the supernatant water and pipette up the deposit, place it on one or more microscope slides (but make one slide at a time so they do not stand for long periods and dry out), place a 22 × 40 mm coverslip on top, examine and count every *Ascaris* egg, classifying them as viable, potentially viable or dead. *Trichuris*, *Taenia,* and hookworm spp. eggs must also be counted and assessed simply as potentially viable or dead (Figure 8.25).

Procedure for liquid faecal sludge samples (TS <5%):

1. Select an appropriate volume of the sample.
   Note: If the water is effluent from a wastewater treatment plant and is fairly clean with low suspended solids, then it is preferable to use a large sample of 5-10 L, measured out using a 1 L measuring cylinder. If the sample is dirty water, but with low to moderate suspended solids, use a 1-5 L sample.
   Note: If the sample is black water with a high concentration of solids (3-5%), then use amounts of 200-500 mL.
   Note: If the sample is fatty, then measure out a selected sample size (from 200-500 mL). For all types of liquid samples, pour into a plastic beaker large enough to contain the sample with at least 5-10 cm above the surface, so that it does not spill when mixing on the magnetic stirrer.
2. Add 1 mL per litre of neat Tween 80 or 7X directly into the sample (so as to make a ± 0.1% solution). Mix well using the magnetic stirrer and magnet in the beaker for 20 min.
3. The measured sample is poured slowly through a 100 µm sieve which fits on top of a 20 µm sieve and is well washed, regularly checking the bottom sieve for fluid build-up.
4. Separate the sieves and then rinse the 20 µm sieve well and wash the material to one side of the sieve for collection.

5. Rinse all the material off the 20 µm filter into 2-4 × 15 mL or 50 mL conical test tubes.
6. Centrifuge at 3,000 rpm (1,512 g) in the centrifuge with a swing-out rotor for 10 min.
7. Step 7. Pour off the supernatant and retain the deposits left in 2-4 ×15 mL or 50 mL test tubes.
8. Place the test tubes in the rack with an applicator stick in each (as a stirring rod) and pipette in $ZnSO_4$, 3 mL at a time, vortexing in between the addition of the chemical, until the tubes are filled to the 14 mL mark.
9. Centrifuge at 2,000 rpm (672 g) for 10 min.
10. Pour the supernatant flotation fluid over the 100 mm diameter 20 µm sieve. Wash out the test tubes and keep one aside for re-use.
11. Wash the material on the sieve well with tap water and rinse down to one side of the sieve for collection. Using a 3 mL plastic pipette, transfer the material back into the test tube kept aside.
12. Centrifuge at 3,000 rpm (1,512 g) for 10 min to obtain the final deposit.
13. Pour off the supernatant water. There should be about 0.2-0.3 mL water left on top of the pellet. If the pellet is small, you may be able to pipette up everything and make one slide to examine. If the deposit is thick, dilute it with water (0.1 mL at a time) until it is of a thinner consistency, then pipette up enough to make one slide at a time, place it on a microscope slide, place a 22 × 40 mm coverslip on top, and examine and count every *Ascaris* egg, classifying them as viable, potentially viable or dead. *Trichuris*, *Taenia* and hookworm spp. eggs must also be counted and assessed simply as potentially viable or dead. Continue making more slides until the entire sample has been examined and all the helminth eggs have been counted and assessed.

8.8.1.4.8 Calculation

Once all the eggs have been counted, the results should be calculated to report the number of eggs per litre or per gram for each species of helminth and within each species, and those that are viable and non-viable.

Example 1

If 2.5 L of liquid sample was analysed and there were 500 *Ascaris* eggs found, then use simple proportions: 500/2.5 : X/1 = 500 × 1/2.5 = 200 eggs/L.

Example 2

If 15 g of solid sample was analysed and 3,450 *Ascaris* eggs were counted, then using proportions again:

3,450/15 : X/1 = 3,450 × 1/15 = 230 eggs/g of sample (wet mass) (abbreviated EPG).

Note: Adjust the egg counts to per gram dry mass (or per gram TS) if a sample of the sludge has been tested for moisture content.

8.8.1.4.9 Data set example

Table 8.171 Data set of faecal sludge samples.

ID No.	Sample type	Sample quantity	*Ascaris* - DEAD			*Ascaris* -p otentially VIABLE				Trich	Trich	Taen	Taen	Other
			Inf	Dead	Nec	Imm	Mot	Devel	Undev	Pot vi	Dead	Pot vi	Dead	
0.1	Effluent	10 L	1	261	12	3	9	15	8	2	26	1	7	0
	Results per litre		<1	26.1	1.2	<1	<1	1.5	<1	<1	2.6	<1	<1	0
0.2	Sludge	10 g	6	543	28	23	19	267	399	88	54	49	9	1 E v
	Results per gram		<1	54.3	2.8	2.3	1.9	26.7	39.9	8.8	5.4	4.9	<1	<1 E v

ID No. = Sample identification number
Inf = Infertile, *i.e.* eggs that were not fertilised
Nec = Necrotic, *i.e.* egg contains a dead, shrivelled larva
Imm = Immotile larva, healthy looking, but not moving
Mot = A motile larva
Devel = Embryo in egg in ≥ 2-cell stage of development
Undev = Embryo in single cell stage
Trich = *Trichuris* sp.
Taen = *Taenia* sp.
Pot vi = Potentially viable, i.e for the same developmental stages as described under "*Ascaris* - potentially viable": Imm, Mot, Devel, Undev
Dead = The developmental stages described under "*Ascaris* - DEAD": Inf, Dead, Nec
Other = Any other helminth eggs found (count and record number of eggs only)
E v = *Enterobius vermicularis*.

Interpretation of results

The term helminths encompasses round worms (nematodes), tapeworms (cestodes) and flatworms (trematodes). The nematodes that are a concern in sanitation are those that lay eggs or produce larvae in an undeveloped stage and require time in the soil to develop into infective larvae (geohelminths) *e.g. Ascaris* spp., *Trichuris* spp., hookworm spp. and *Stongyloides stercoralis*. Other nematode eggs that rapidly develop larvae and are infective for humans in a few hours, *e.g. Enterobius vermicularis*, may also be a concern.

Geohelminth eggs are considered as potentially viable (and thus infective) if they are in the undeveloped stage, developing stages [a 2-, 4-, 8-, 16, 32-cell stage or more (blastula stages) to an immature larva (gastrula)], then to a developed larva (L1 and finally an infective L2 larva). If an L2 larva is moving in the egg, then it is viable and infective. When eggs that are undeveloped or in early cleavage (one of the blastula stages) die, they may become globular, have broken shells or collapsed walls, or appear empty inside - these are termed dead. If a formed larva dies inside the egg, it appears shrivelled and occupies much less space than a plump, healthy viable larva, and is termed necrotic. Eggs that have never been fertilised (infertile) cannot develop and are therefore classified under the NON-viable eggs.

Cestodes such as *Taenia* spp. contain an oncosphere within the egg that does not develop further, thus it is described as potentially viable if it looks in good condition and the hooklets are visible, and dead if the contents are globular or have no structure. Cestode eggs (except for *Hymenolepis nana*) require an intermediate host (such as the pig) to ingest them before that host passes on the infection to humans; however, ingesting *Taenia solium* eggs poses a serious risk for humans to become the intermediate host and develop cysts in the brain resulting in neurocysticercosis (*T. saginata* poses no risk). The eggs of these two species are indistinguishable and therefore all *Taenia* eggs are counted, assessed for potential viability and reported as *Taenia* sp.

Some trematode eggs are excreted in the undeveloped stage and a miracidium develops in approximately two weeks, while others contain a miracidium when laid - these eggs are only a concern if fresh sanitation waste is dumped into water bodies as they require aquatic plants to encyst on or an intermediate host to develop within, for transmission to occur. All trematode and cestode eggs (except for *Taenia* spp.) are thus counted and recorded, however recording their viability status is optional and not a requirement unless specifically requested by the client.

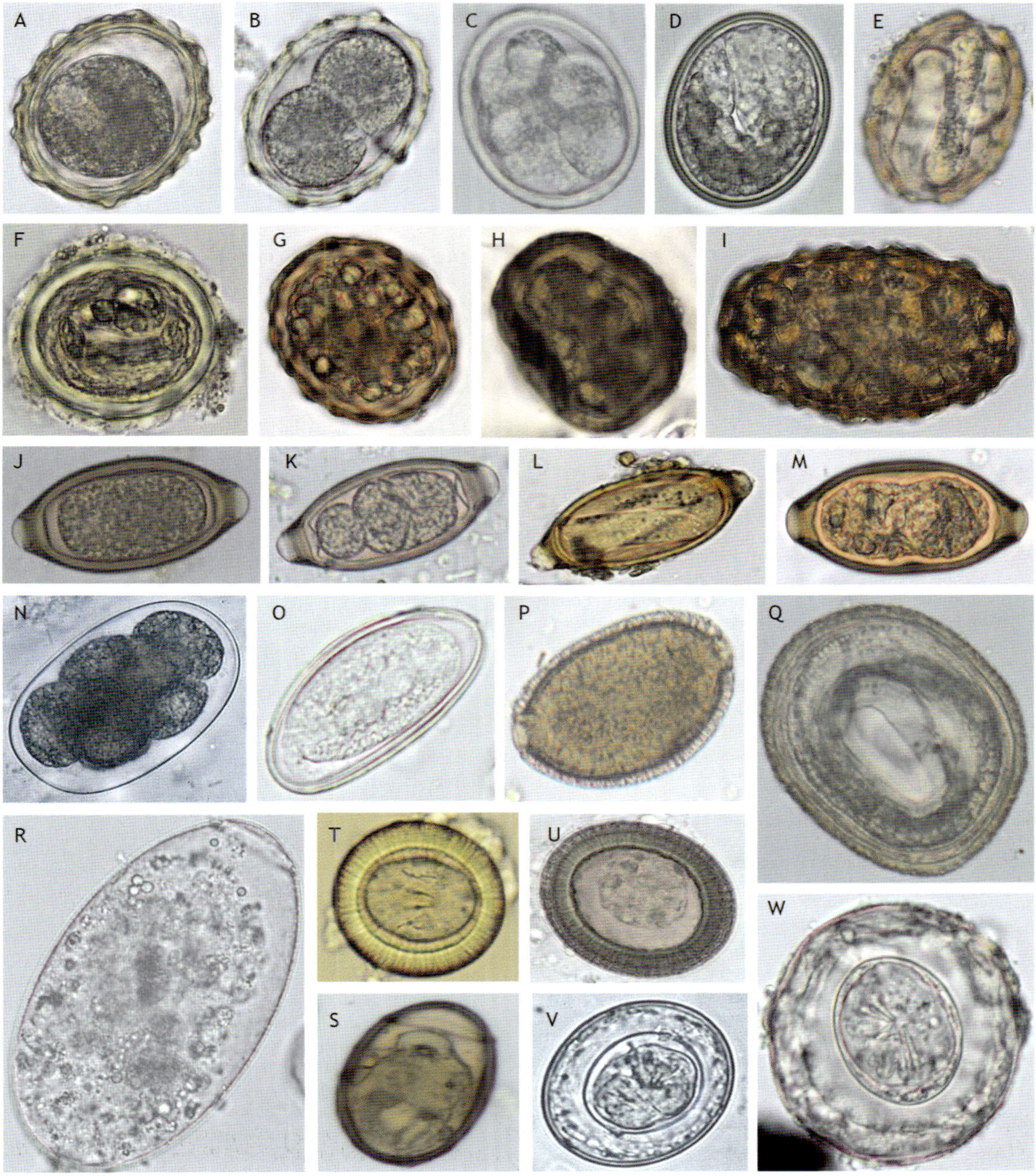

Figure 8.25 Photographs of some helminth eggs A) Undeveloped *Ascaris* egg, B) *Ascaris* 2-cell developing, C) *Ascaris* multiple cells developing, D: *Ascaris* gastrula, E) *Ascaris* viable larva, F) *Ascaris* egg containing necrotic (dead) larva, G) dead *Ascaris* egg containing globules, H) dead *Ascaris* egg, empty with collapsing wall, I) infertile *Ascaris* egg, J) undeveloped *Trichuris* egg, K) developing *Trichuris* egg, L) *Trichuris* egg containing a viable, motile larva, M) dead *Trichuris* egg, N) Hookworm sp. developing, O) *Enterobius vermicularis* dead, P) *Capillaria* sp. dead, Q) *Toxocara* sp. developed larva, R) *Fasciola* sp. dead, S) *Dicrocoelium dendriticum* possibly viable, T) *Taenia* sp. egg in good condition (probably viable), U) dead *Taenia* sp. egg, V) *Hymenolepis nana* possibly viable; and W) *Hymenolepis diminuta* possibly viable (source: UKZN PRG).

REFERENCES

3M petrifilm, product instruction, *E. coli*/coliform plate count https://multimedia.3m.com/mws/media/701951O/product-instructions-3m-petrifilm-e-coli-coliform-count-plate.pdf (accessed 10 December 2020).

American Chemical Society (ACS), https://www.acs.org/content/acs/en/chemical-safety/basics/glassware-and-equipment.html (accessed December 2020).

Andriessen N., Ward B.J. and Strande L. (2019). To char or not to char? Review of technologies to produce solid fuels for resource recovery from faecal sludge. *Journal of Water, Sanitation and Hygiene for Development*, *9*(2), 210-224.

Anton Parr GmbH. (2020). Instruction Manual and Safety Information MCR x2 Series. https://www.anton-paar.com/za-en/products/details/rheometer-mcr-72-and-mcr-92

Anton Paar GmbH. Density and density measurement. https://wiki.anton-paar.com/en/density-and-density-measurement/ (accessed December 2020).

AOAC G, MD U. (2000). Fat (crude) or ether extract in animal feed. Official methods of analysis. Gaithersburg, MD, USA.

ASTM (2004) Standard Test Method for Gross Calorific Value of Coal and Coke (D5865-04), ASTM International.

ASTM (2011) Test Methods for Determination of Trace Elements in Coal, Coke, & Combustion Residues from Coal Utilization Processes by Inductively Coupled Plasma Atomic Emission, Inductively Coupled Plasma Mass, & Graphite Furnace Atomic Absorption Spectrometry (D6357 – 11). ASTM International.

ASTM (2015a). Standard Test Method for Determination of Total Solids in Biomass (E1756–08), Method A. ASTM International

ASTM (2015b). Standard Test Method for Determination of Total Solids in Biomass (E1756–08), Method B. ASTM International.

ASTM (2015c) Standard Practice for Ultimate Analysis of Coal and Coke (D3176–15). ASTM International.

Bain R., Bartram J., Elliott M., Matthews R., McMahan L., Tung R., Chuang P. and Gundry S. (2012). *Int J Environ Res Public Health*, May; 9(5):1609-25.

Barker K. (2005). *At the Bench: A Laboratory Navigator*. CSHL Press.

Barbosa-Cánovas G.V., Fernández-Molina J.J., Alzamora S.M., Tapia M.S., López-Malo A. and Chanes J.W. (2003). *Handling and preservation of fruits and vegetables by combined methods for rural areas: Technical Manual* (No. 149). Food & Agriculture Organization of the United Nations. *ISBN 92-5-104861-4.*

Belcher D., Foutch G.L., Smay J., Archer C. and Buckley C.A. (2015). Viscous heating effect on deactivation of helminth eggs in ventilated improved pit sludge. *Water Science & Technology,* 72 (7): 1119-1126.

Berkeley Environment, Health & Safety. Glove comparison chart. https://ehs.berkeley.edu/workplace-safety/glove-selection-guide#comparison (Accessed November 2020).

Bleuler M., Gold M., Strande L. and Schönborn A. (2020). Pyrolysis of Dry Toilet Substrate as a Means of Nutrient Recycling in Agricultural Systems: Potential Risks and Benefits. *Waste and Biomass Valorization.*

Bourgault C., Lessard P., Remington C. and Dorea C.C. (2019). Experimental determination of moisture sorption isotherm of fecal sludge. *Water*, *11*(2), 303.

British Standards Institution (1990). British Standard 1377:2. Methods of test for soils for civil engineering purposes: Classification tests, British Standards Institution.

British Standards Institution (1995). British Standard 812:20. Testing aggregates; Methods of determination of density, British Standards Institution.

British Standards Institution (2000). British Standard 1016:102. Methods for analysis and testing of coal and coke. Determination of total moisture of coke, British Standards Institution.

Byron Inouye https://manoa.hawaii.edu/exploringourfluidearth/physical/world-ocean/map-distortion/practices-science-precision-vs-accuracy (accessed 2020)

C-Therm Technologies. C-Therm TCi Operator Manual, TH130041 Rev. H. https://www.manualslib.com/products/C-Therm-Tci-9217324.html (Accessed December 2020).

Cairncross S. and Feacham R. (2019). *Environmental Health Engineering in the Tropics*, 3rd edition, Routledge Earthscan, London and New York.

Christophersen D. (2000) Jar Testing Procedures. Handbook of Industrial Water Treatment, GE Power and Water, pp. 369-377.

Coleman, D., and Vanatta, L. (2010). Statistics in Analytical Chemistry: Part 36: Significant Digits (Concluded). *American laboratory*, *42*(1), 31-33.

Das B.M. and Sobhan K. (2013). *Principles of geotechnical engineering*, Nelson Education.

Diener S., Semiyaga S., Niwagaba C.B., Murray Muspratt A., Gning J.B., Mbéguéré M., Ennin J.E., Zurbrugg C. and Strande L. (2014). A value proposition: Resource recovery from faecal sludge—Can it be the driver for improved sanitation? *Resources, Conservation and Recycling*, *88*, 32-38.

Dodane P.H. and Bassan M. (2014). Chapter 6: Settling-thickening tanks. *Faecal sludge management: systems approach for implementation and operation. London: IWA*, 123-139.

Duncan M. and Horan N. (2003) *Handbook on water and wastewater microbiology*. First Edition, Academic Press.

Englund M. and Strande L. (2019). Faecal Sludge Management: Highlights and Exercises. Eawag: Swiss Federal Institute of Aquatic Science and Technology. Duebendorf, Switzerland

Englund M., Carbajal J.P., Ferré A., Bassan M., Vu T.H.A., Nguyen V.A. and Strande L. (2020). Modelling quantities and qualities (Q&Q) of faecal sludge in Hanoi, Vietnam and Kampala, Uganda for improved management solutions. *Journal of Environmental Management,* 261, 110202.)

European Commission (EC) (2009) Determination of crude oils and fats. Commission Regulation (EC) No 152/2009. https://eur-lex.europa.eu/legal-content/EN/TXT/?uri=CELEX:32009R0152

FAO (2011) *Quality assurance for animal feed analysis laboratories.* FAO Animal Production and Health Manual No. 14. Rome.

Gerba C.P. (2020). Pathogen Removal. Book chapter in Biological Wastewater Treatment: 2nd edition. G.-H. Chen, M.C.M. van Loosdrecht, G.A. Ekama and D. Brdjanovic (editors), IWA Publishing.

German Association for Water, Wastewater and Waste (DWA) (2008). Merkblatt DWA-M 383 Kennwerte der Klärschlammentwässerung (leaflet DWA - M 383 Characteristic values for wastewater sludge dewatering), Hennef, Germany.

Gold M., Cassar C.M., Zurbrügg C., Kreuzer M., Boulos, S., Diener S. & Mathys A. (2020). Biowaste treatment with black soldier fly larvae: Increasing performance through the formulation of biowastes based on protein and carbohydrates. *Waste Management*, *102*, 319-329.

Gold M., Dayer P., Faye M.C.A.S., Clair G., Seck A., Niang S., Morgenroth E. and Strande L. (2016). Locally produced natural conditioners for dewatering of faecal sludge. *Environmental Technology*, 37(21), 2802-2814.

Gold M., Harada H., Therrien J.D., Nishida T., Cunningham M., Semiyaga S., Fujii S., Dorea C., Nguyen V.A. and Strande L. (2018). Cross-country analysis of faecal sludge dewatering. *Environmental Technology*, 39:23, 3077-3087, DOI: 10.1080/09593330.2017.1374472.

Gold M., Sekigongo P., Ddiba D., Seck A., Niwagaba C., Diene A., Diaw S., Niang S. and Strande L. (2017). Faecal sludge as a solid industrial fuel: A pilot-scale study. *Journal of Water, Sanitation and Hygiene for Development* 7:2, 243-251 2017.

Grego S., Barani V., Hegarty-Craver M., Raj A., Perumal P., Berg A.B. and Archer C. (2018). Soil-transmitted helminth egg assessment in wastewater in an urban area in India. *Journal of Water and Health,* 16(1): 34-43.

Gueye A., Mbeguere M., Niang S., Diop C. and Strande L. (2016). Novel plant species for faecal sludge drying beds: Survival, biomass response and forage quality. *Ecological Engineering*, 94, 617-621, 2016.

Mertens D.R. (2003). Challenges in measuring insoluble dietary fibre. *Journal of Animal Science*, *81*(12), 3233-3249.

Getahun S., Septien S., Mata J., Somorin T., Mabbett I. and Buckley C. (2020). Drying characteristics of faecal sludge from different on-site sanitation facilities. *Journal of Environmental Management*, *261*, 110267.

Hach (2016). Colour Test Kit CO-1 (223400) https://www.hach.com/asset-get.download.jsa?id=7639982666. (November 2020).

Hach Nitrate cadmium reduction method_8039 www.hach.com https://www.hach.com/total-nitrogen-reagent-set-hr-tnt/product-details?id=7640209860.

Hach Lange GMBH LCW 902 Crack-Set, DOC312.53.94370.

Hafford L.M., Ward B.J., Weimer A.W. and Linden K. (2018). Fecal sludge as a fuel: characterization, cofire limits, and evaluation of quality improvement measures. *Water, Science & Technology,* 78:12 pp. 2437-2448.

IARC Working Group on the Evaluation of Carcinogenic Risks to Humans. (2012). Chemical agents and related occupations. IARC monographs on the evaluation of carcinogenic risks to humans, 100(PT F), 9.

International Electrotechnical Commission (2017). ISO/IEC 17025: 2017. General requirements for the competence of testing and calibration laboratories. International Organization for Standardization and International Electrotechnical Commission.

ISO (2020). 17247: Coal and coke – Ultimate analysis, International Organization for Standardization. Geneva, Switzerland.

ISO, I. (1999) 6492: Animal Feeding Stuffs —Determination of fat content. International Organization for Standardization. Geneva, Switzerland.

Jofre J., Lucena F., Blanch A.R. and Muniesa M. (2016). Coliphages as Model Organisms in the Characterization and Management of Water Resources. *Water*, 8, 199

Junglen K., Rhodes-Dicker L., Ward B.J., Gitau E., Mwalugongo W., Stradley L. and Thomas E. (2020). Characterization and prediction of fecal sludge parameters and settling behavior in informal settlements in Nairobi, Kenya.

Labconco Corporation (1998). A Guide to Kjeldahl Nitrogen Determination Methods and Apparatus.

http://www.expotechusa.com/Catalogs/Labconco/PDF/KJELDAHLguide.PDF

Leco operations manual https://eu.leco.com/product/928-series (accessed 2020)

Lewis S.J. and Heaton K.W. (1997). Stool form scale as a useful guide to intestinal transit time. *Scandinavian Journal of Gastroenterology*, 32(9), 920-924.

Macrina F.L. (2005). Scientific integrity text and cases in responsible conduct of research, 3rd, ASM Press, Washington DC.

Madigan M.T., Bender K.S., Buckley D.H., Sattley W.M. and Stahl D.A. (2018). Brock Biology of Microorganisms, 15th Edition. Boston, US: Benjamin Cummins.

Maffo P.D, Kengne E.S., Djumyom Wafo G.V., Letah Nzouebet D.A., Zuitchou M.N., Liegui G.S., Wanda C., Fotso F. and Noumsi I.M.K. (2019). Quantification and characterization of faecal sludge from a tropical urban area: the case study of Douala, Cameroon. *Int. J. Biol. Chem. Sci.*, 13(5): 105-113.

Malvern Panalytical (2017). Mastersizer 3000 User Manual MAN0474-08-EN-00, https://www.malvernpanalytical.com/en/learn/knowledge-center/user-manuals/MAN0474EN. (Accessed November 2020)

McMinn B.R., Ashbolt N.J. and Korajkic A. (2017). Bacteriophages as indicators of faecal pollution and enteric virus removal. *Letters in Applied Microbiology*, 65(1), 11–26.

Meng J., Duan H., Li H., Watts S., Liu P., Shrestha S., Zheng M., Yu W., Chen Z., Song Y., Dwyer J., Hu S. and Yuan Z. (2020). Free nitrous acid pre-treatment enhances anaerobic digestion of waste activated sludge and rheological properties of digested sludge: A pilot-scale study. *Water Research*, *172*, 115515.

Merck (2020a) - Spectroquant® phosphate cell test – 114543 https://www.merckmillipore.com/ZA/en/product/Phosphate-Cell-Test,MDA_CHEM-114543 (Accessed December 2020).

Merck (2020b) - Spectroquant® chlorine cell test – 100595 https://www.merckmillipore.com/ZA/en/product/Chlorine-Test, MDA_CHEM-100595. (Accessed December 2020).

Merck (2020c) - Spectroquant® chloride cell test - 101804. https://www.merckmillipore.com/ZA/en/product/Chlorine-Test, MDA_CHEM-101804. (Accessed December 2020).

Merck (2020d) - Spectroquant® COD cell test -114895. https://www.merckmillipore.com/ZA/en/product/COD-Cell-Test, MDA_CHEM-114895(Accessed December 2020).

Merck (2020e) - Spectroquant® nitrite test -114776.

https://www.merckmillipore.com/ZA/en/product/Nitrite-Test, MDA_CHEM-114776(Accessed December 2020).

Merck (2020f) - Spectroquant® Ammonium Test -100683 https://www.merckmillipore.com/ZA/en/product/Ammonium-Test, MDA_CHEM-100683

Merck (2020g) - MQuant Nitrate Test – 110020 https://www.merckmillipore.com/CH/de/product/Nitrate Test,MDA_CHEM-110020 (Accessed December 2020).

Merck (2020h) MQuant Nitrite Test -110007 https://www.merckmillipore.com/CH/de/product/Nitrite-Test,MDA_CHEM-110007 (Accessed December 2020).

Merck (2020i) MQuant® Phosphate Test -110428 https://www.merckmillipore.com/ZA/en/product/Phosphate-Test,MDA_CHEM-110428 (Accessed December 2020).

Merck (2020j) - MQuant® Ammonium Test – 110024 https://www.merckmillipore.com/ZA/en/product/Ammonium-Test,MDA_CHEM-110024 (Accessed December 2020).

Merck (2020k) - MQuant® Chloride Test Kit -110079 https://www.merckmillipore.com/ZA/en/product/Chloride-Test,MDA_CHEM-110079 (Accessed December 2020).

Merck (2020l) - MQuant® (2020) Chlorine Test Kit – 117924 https://www.merckmillipore.com/ZA/en/product/Chlorine-Test,MDA_CHEM-117924 (Accessed December 2020).

Microdyn Nadir (2017). Trouble Shooting Jar Test Procedure. Technical Service Guide. Form No: TSG-T-008, 02/05/2020. Retrieved from: https://www.microdyn-nadir.com/wp-content/uploads/TSG-T-008-Troubleshooting-Jar-Testing-Procedure.pdf

Moodley C.L., Archer C., Hawksworth D. and Leibach L. (2008). Standard methods for the recovery and enumeration of helminth ova in wastewater, sludge, compost and urine diversion waste in South Africa. WRC TT Report No. 322/08, Water Research Commission, Pretoria.

Moto N., Esanju M., Andriessen N., Kimwaga R. and Strande L. (2018). Use of chitosan and Moringa oleifera as conditioners for improved dewatering of faecal sludge. In *Transformation towards sustainable and resilient WASH services: Proceedings of the 41st WEDC International Conference, Nakuru, Kenya*.

Murray Muspratt A., Nakato T., Niwagaba C., Dione H., Kang J., Stupin L., Regulinski J., Mbéguéré M. & Strande L. (2014). Fuel potential of faecal sludge: calorific value results from Uganda, Ghana and Senegal. *Journal of Water, Sanitation and Hygiene for Development*, *4*(2), 223-230.

Naidoo D., Archer C., Louton B. and Rodda N. (2016). Testing household disinfectants for the inactivation of helminth eggs on surfaces and in spills during pit latrine emptying. *Water SA,* 42(4): 560-570.

Naidoo D., Appleton C.C., Archer C.E. and Foutch G.L. (2019). The inactivation of Ascaris suum eggs by short exposure to high temperatures. *Journal of Water, Sanitation and Hygiene for Development,* 9(1): 19-27.

Naidoo D., Archer C.E., Appleton C.C., Septien S., and Buckley C.A. (2020). Inactivation of *Ascaris* for thermal treatment and drying applications in faecal sludge. *Journal of Water, Sanitation and Hygiene for Development,* 18(2): 209-217.

National Research Council (2011). Prudent Practices in the Laboratory: Handling and Management of Chemical Hazards National Research Council (US) Committee on Prudent Practices in the Laboratory. Washington DC: National Academies Press (US).

National Research Council (2020). Promoting Chemical Laboratory Safety and Security in Developing Countries and Security. https://www.nap.edu/catalog/12857/promoting-chemical-laboratory-safety-and-security-in-developing-countries

Novotny L.F.G., Layton B. and Walter M. (2017). Critical Factors in Determining Fibre in Feeds and Forages A. s. L. M. a. S. Committee, Fibre Best Practices Working Group (Revision 1).

Nyakeri E.M., Ayieko M.A., Amimo F.A., Salum H. & Ogola H.J.O. (2019). An optimal feeding strategy for black soldier fly larvae biomass production and faecal sludge reduction. *Journal of Insects as Food and Feed*, *5*(3), 201-213.

Oblinger J.L. and J.A. Koburger (1975). Understanding and Teaching the Most Probable Number Technique. *J. Milk Food Technol.,* 38(9): 540-545.

Osibote B.A., Osibote I.A., Bolaji O.M. & Ana G.R.E.E. (2016). Effect of Thermal Treatment on Microbial Load of Faecal Sludge from Some Faecal Sludge Collection Sites in Oyo State, South Western, Nigeria. *Jordan Journal of Biological Sciences*, *9*(4).

Park Y.W. (2008). Moisture and water activity. *Handbook of processed meats and poultry analysis*, 35-67.

Parr Instrument Company (2010). Operating Instruction Manual 6200 Oxygen Bomb Calorimeter. Moline IL, USA. https://www.parrinst.com/products/oxygen-bomb-calorimeters/6200-isoperibol-calorimeter/documents/

Parr Instrument Company (2013). Introduction to Bomb Calorimetry. Moline IL, USA. https://www.parrinst.com/wp-content/uploads/downloads/2013/07/483M_Parr_Intro-to-Bomb-Calorimetry.pdf

Pebsworth P.A., Archer C.E., Appleton C.C. and Huffman M.A. (2012). Parasite transmission risk from geophagic and foraging behaviour in Chacma baboons. *American Journal of Primatology,* 74: 940–947.

Prichard L. & Barwick V. (2003). *Preparation of calibration curves: A guide to best practice*. LGC Group Ltd., TR LGC/VAM/2003/032, Teddington, Middlesex, UK.

Remington C., Bourgault C. & Dorea C.C. (2020). Measurement and Modelling of Moisture Sorption Isotherm and Heat of Sorption of Fresh Feces. *Water*, *12*(2), 323.

Rice E.W., Baird R.B., Eaton A.D. and Clesceri L.S. (2017). Standard Methods for the Examination of Water and Wastewater, 23rd edition. Washington: APHA, AWWA, WPCR.

Semiyaga S., Okure M.A.E., Niwagaba C.B., Nyenje P.M. & Kansiime F. (2016). Dewaterability of faecal sludge and its implications to faecal sludge management in urban slums. *International Journal of Environmental Science and Technology*.

Septien S., Singh A., Mirara S.W., Teba L., Velkushanova K. and Buckley C.A. (2018). 'LaDePa'process for the drying and pasteurization of faecal sludge from VIP latrines using infrared radiation. *South African Journal of Chemical Engineering*, 25, 147-158.

Septien S., Mirara S.W., Makununika B.S.N., Singh A., Pocock J, Velkushanova K. and Buckley C.A. (2020). Effect of drying on the physical and chemical properties of faecal sludge for its reuse. *J. Environ. Chem. Eng*., 8 (1):103652. doi:10.1016/j.jece.2019.103652.

SNF Holding Company (2015). Official Procedure for Laboratory Jar Test. Retrieved from: https://www.mi-wea.org/docs/Official%20Proced.%20for%20Laboratory%20Jar%20Test%204-15.pdf

Spit J., Malambo D., Gonzalez M.E., Nobela H., de Pooter L. and Anderson K. (2014). Emergency Sanitation: Faecal Sludge Treatment Fieldwork Summary. WASTE: Gouda, the Netherlands.

Stable Microsystems Ltd (2020). Surrey, Englandhttps://www.stablemicrosystems.com/ManualRequest.html https://www.stablemicrosystems.com/TextureAnalysisAttachments.html, https://www.stablemicrosystems.com/TAXTplus.html (Accessed November 2020).

Strande L., Ronteltap M. and Brdjanovic D. (2014). Faecal sludge management: Systems approach for implementation and operation., London: IWA Publishing.

Strande L., Schoebitz L., Bischoff F., Ddiba D., Okello F., Englund M., Ward B.J. and Niwagaba C.B. (2018). Methods to Reliably Estimate Faecal Sludge Quantities and Qualities for the Design of Treatment Technologies and Management Solutions. *Journal of Environmental Management*, 223: p. 898-907.

Stringel S.S. (2020). Final technical report Project OPP1164143 - Characterization of faecal material behaviour during drying. *Gates Open Res*, *4*. http://library.metergroup.com/Manuals/14638_TDL_Web.pdf

Tchobanoglous, G., Burton, F. L., & Stensel, H. D. (2003). Metcalf & Eddy Wastewater Engineering: Treatment and Reuse. International Edition. McGrawHill, 4, 361-411.

Trönnberg L., Hawksworth D., Hansen A., Archer C. and Stenstrom T.A. (2010). Household-based prevalence of helminthes and parasitic protozoa in rural KwaZulu-Natal, South Africa, assessed from faecal vault sampling. *Transactions of the Royal Society of Tropical Medicine & Hygiene*, 104: 646-654.

USEPA (1993). Method 180.1 Determination of Turbidity by Nephelometry. Rev. 2.0. *Methods for Chemical Analysis of Water and Wastes*.

USEPA (2001). Method 1687 Total Kjeldahl Nitrogen in Water and Biosolids by Automated Colorimetry with Preliminary Distillation/Digestion. https://www.epa.gov/sites/production/files/2015-10/documents/method_1687_draft_2001.pdf

USEPA (2004a). SW-846 Test Method 9040C: pH Electrometric Measurement. United States Environmental Protection Agency: https://www.epa.gov/sites/production/files/2015-12/documents/9040c.pdf

USEPA (2004b). SW-846 Test Method 9045D: Soil and Waste pH. United States Environmental Protection Agency: https://www.epa.gov/sites/production/files/2015-12/documents/9045d.pdf

USEPA (2006). Method 1681: Fecal Coliforms in sewage sludge (Biosolids) by multiple Tube fermentation using A-1 medium. https://www.epa.gov/sites/production/files/2015-08/documents/method_1681_2006.pdf

USEPA (2017). 40 CFR Part 136 – Guidelines establishing test procedures for the analysis of pollutants, Appendix B Definition and Procedure for the Determination of the Method Detection Limit.

USEPA (1996). Method 3050B: Acid Digestion of Sediments, Sludges, and Soils, Revision 2. Washington DC.

Vijayakumar P.P. and Adedeji A.A. (2017). Measuring the pH of Food Products. University of Kentucky College of Agriculture, Food and Environment Cooperative Extension Service. http://www2.ca.uky.edu/agcomm/pubs/ID/ID246/ID246.pdf

Ward, B.J., Andriessen, N., Tembo, J.M., Kabika, J., Grau, M., Scheidegger, A., Morgenroth, E., Strande, L. (2021) Predictive models using "cheap and easy" field measurements: Can they fill a gap in planning, monitoring, and implementing faecal sludge management solutions? *Water Research*. https://doi.org/10.25678/00037X

Ward B.J. and Strande L. (2019). Chapter 5.2 Conditioning in Faecal Sludge Management: Highlights and Exercises. Eawag: Swiss Federal Institute of Aquatic Science and Technology, Duebendorf, Switzerland. ISBN: 978-3-906484-70-9.

Ward B.J., Traber J., Gueye A., Diop B., Morgenroth E. and Strande L. (2019). Evaluation of conceptual model and predictors of faecal sludge dewatering performance in Senegal and Tanzania. *Water Research*, *167*, 115101.

Wolf J., Prüss-Ustün A., Ivanov I., Mugdal S., Corvalán C., Bos R., Neira M. and World Health Organization (2018). Preventing disease through a healthier and safer workplace.

World Bank (1984). Occupational health and safety guidelines. Office of Environmental Affairs. Washington DC.

World Bank (2019). Health, Safety and Dignity of Sanitation Workers an Initial Assessment.

World Health Organization (2011). *Laboratory quality management system: handbook*. World Health Organization.

Annex 1

Global Partnership of Laboratories for Faecal Sludge Analysis

■ Analysis
■ Teaching
■ Experimentation

	University of KwaZulu-Natal	CSE	CDD / CASS	AIT	eawag aquatic research	IHE Delft
	Analysis, Teaching, Experimentation	Analysis, Experimentation	Analysis, Experimentation	Analysis, Teaching, Experimentation	Analysis, Teaching, Experimentation	Analysis, Teaching, Experimentation
PARAMETERS						
Helminths – number and viability	■	■	■	■	■	■
E. coli, total coliforms, coliforms	■	■	■	■	■	■
Organic matter (COD, BOD, TOC)	■	■	■		■	■
Solids (TS, VS, TSS, VSS, moisture)	■	■	■	■	■	■
Nutrients (nitrogen: total / nitrate / nitrite / ammonium) and phosphate: total / orthophosphate)	■	■	■	■	■	■
Ions (selective cations / anions)	■		■	■	■	■
Heavy metals	■		■	■	■	■
Osmotic pressure	■				■	
Respirometric tests	■				■	■
Calorific value	■	■	■		■	■
Thermal conductivity	■					
Heat capacity	■					
VFA	■			■	■	■
Pyrolysis / combustion	■				■	■
Drying energy potential	■					■
Particle size distribution	■		■		■	
Rheology properties (shear strength / viscosity)	■				■	■
EQUIPMENT						
Specialist microbiology laboratory	■	■	■	■	■	■
Rheometer	■				■	■
Differential scanning calorimeter / thermogravimetric analysis	■					■
Calorimeter	■	■	■		■	■
Spectrophotometer	■	■	■	■	■	■
Moisture balance and analyser	■		■		■	
Penetrometer	■					
Particle size analyser	■		■		■	
Thermal conductivity analyser	■					
Chloride analyser	■				■	
Osmometer	■				■	
Gas chromatograph	■			■	■	■
Respirometer	■				■	■
Microwave plasma / atomic emission spectrometer	■				■	

■ Analysis
■ Teaching
■ Experimentation

	Columbia University	2iE	BITS Pilani	ENPHO	Institut Teknologi Bandung
	Analysis, Teaching, Experimentation	Analysis, Teaching, Experimentation	Analysis, Teaching, Experimentation	Analysis, Experimentation	Analysis, Teaching, Experimentation
PARAMETERS					
Helminths – number and viability		■	■	■	■
E. coli, total col forms, coliforms	■	■	■	■	■
Organic matter (COD, BOD, TOC)	■	■	■		■
Solids (TS, VS, TSS, VSS, moisture)	■	■	■	■	■
Nutrients (nitrogen: total / nitrate / nitrite / ammonium) and phosphate: total / orthophosphate)	■	■	■	■	■
Ions (selective cations / anions)	■				■
Heavy metals	■		■	■	■
Osmotic pressure	■				■
Respirometric tests	■				
Calorific value	■		■		
Thermal conductivity					
Heat capacity					
VFA	■	■			■
Pyrolysis / combustion	■				
Drying energy potential	■				
Particle size distribution	■		■		
Rheology properties (shear strength / viscosity)	■		■		■
EQUIPMENT					
Specialist microbiology laboratory	■	■	■	■	■
Rheometer	■				
Differential scanning calorimeter / thermogravimetric analysis	■				
Calorimeter	■		■		
Spectrophotometer	■	■	■	■	■
Moisture balance and analyser	■	■	■		
Penetrometer					
Particle size analyser	■				
Thermal conductivity analyser					
Chloride analyser	■	■			
Osmometer					
Gas chromatograph	■		■		■
Respirometer	■				
Microwave plasma / atomic emission spectrometer	■				

Unlike wastewater, there are very few laboratories that specialize in faecal sludge analysis. In addition, due to the lack of standard methods for sampling and analysing faecal sludge, standard methods from other fields, such as water, wastewater and soil science, are usually applied. This is why the experts on faecal sludge analysis recently established the Global Partnership of Laboratories for Faecal Sludge Analysis to address these challenges and to work towards standardized methods for the characterization and quantification of faecal sludge from onsite sanitation technologies, including sampling techniques and health and safety procedures for faecal sludge handling.

The Partnership also delivers on-campus courses and training and aims to improve communication between sanitation practitioners, provide a comparative faecal sludge database, and improve confidence in the methods and obtained results. The Partnership currently consists of eleven laboratories in Durban, New Delhi, Bangalore, Bangkok, Zurich, Delft, New York, Ouagadougou, Goa, Kathmandu and Bandung.

IHE Delft

IHE Delft Institute for Water Education

The new Laboratory for Faecal Sludge Analysis was opened at IHE Delft on 19 November 2018. It was equipped using funds from the 'Global Sanitation Graduate School' grant, provided by the Bill & Melinda Gates Foundation (BMGF). In this facility sanitation professionals and academics from all over the world can analyse, research and learn about the characteristics, use and re-use of human excreta, for the benefit of improving people's health and quality of life. The lab, initiated in the framework of the new Master of Science in Sanitation program at IHE Delft, was designed for analytical work, teaching and training, as well as to support experimentation as part of research by master's and doctoral students. The laboratory is equipped with facilities for video recording and offers tailor-made training courses to third parties.

Contact
Dr. Claire Furlong
c.furlong@un-ihe.org
+31 15 2151 724
IHE Delft Institute for Water Education
Westvest 7
2611 AX Delft
The Netherlands
www.un-ihe.org

WASH Research & Development Centre

University of KwaZulu-Natal

UNIVERSITY OF KWAZULU-NATAL
INYUVESI YAKWAZULU-NATALI

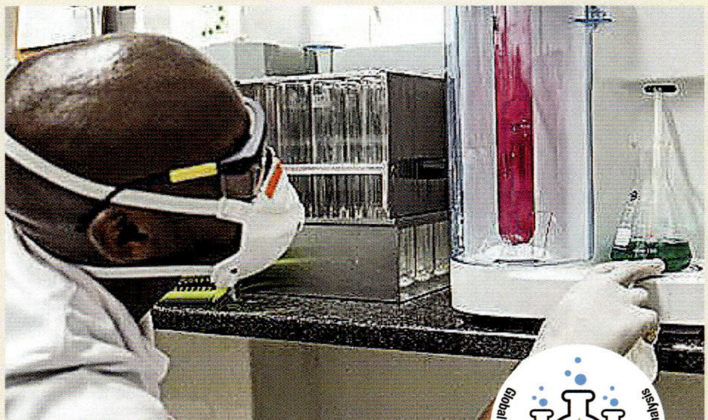

WASH Research & Development Centre operates a recently modernized and fully equipped reference sanitation laboratory, primarily used to investigate human excreta, faecal sludge from different on-site sanitation facilities, and wastewater. WASH Research & Development Centre also provides support to other sanitation laboratories in Africa and Asia for their set-up and improvement. It offers a range of facilities and activities to support research and education activities: access to different sanitation systems, mechanical workshops, field and laboratory testing and sampling, technology and prototype testing, specialized training, and sharing of data.

For more details visit
https://washcentre.ukzn.ac.za/

Contact

Merlien Reddy
Reddym5@ukzn.ac.za
+27 31 260 1584
Wash Research & Development Centre
University of KwaZulu-Natal
South Africa

Eawag

Swiss Federal Institute of Aquatic Science and Technology

eawag
aquatic research

Contact
Dr. Linda Strande
linda.strande@eawag.ch
Sandec / Eawag
Überlandstrasse 133 8600
Dübendorf Switzerland
www.sandec.ch

Eawag (the Swiss Federal Institute of Aquatic Science and Technology) started in 1936 as a water and wastewater treatment research institute, and is part of the ETH Domain (Swiss Federal Institutes of Technology). The mandate of the Sandec (Sanitation, Water, and Solid Waste for Development) Department at Eawag is to develop and test methods and technologies that help the world's poorest to access sustainable water and sanitation services. Sandec has been conducting research into faecal sludge management for 25 years. Faecal sludge analysis is conducted at the Eawag and ETH laboratories in Switzerland and at partner-institution laboratories in many countries throughout Asia and Africa. Numerous resources are available free of charge on the Sandec website www.sandec.ch/fsm_tools, including publications, books, online courses, workshops, newsletters and reference materials.

CSE

Centre for Science and Environment

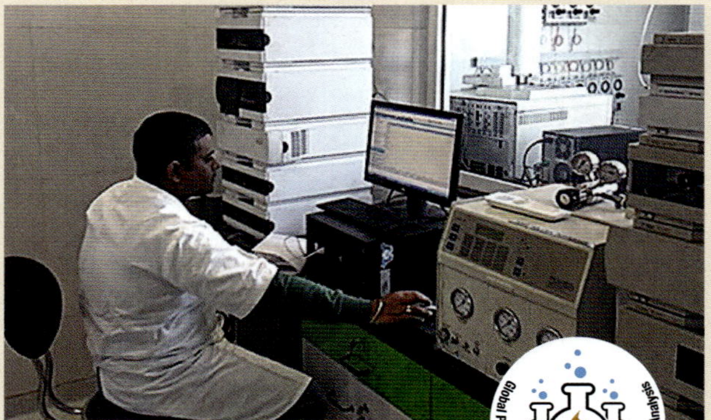

The Environment Monitoring Laboratory (EML) was established to support CSE's specific research activities and it now undertakes independent research in a variety of fields. The EML has partnered with BMGF to undertake research in the field of faecal sludge and septage management, which includes the collection and analysis of data related to wastewater and septage, and the assessment of novel technologies for treating human excreta. As a part of CSE, EML is committed to teaching and training stakeholders at their residential training facility, and disseminating knowledge and information through its outreach platforms.

Find out more about CSE's work on www.cseindia.org.

Contact
Chandra Bhushan
chandra@cseindia.org
+91 11 4061 6000
Anil Agarwal Environment Training Institute (AAETI) Vill.
Nimli, Tijara
Alwar, Rajasthan
India

CDD/CASS

Consortium for DEWATS Dissemination / Centre for Advanced Sanitation Solutions

The CDD-CASS Analytical Laboratory was set up in 2010 as a water and wastewater-testing laboratory. As CDD's work extended into faecal sludge management, the laboratory was expanded in 2017 with the support of the BMGF to include faecal sludge testing capabilities. These include testing for heavy metals, calorific value, e-coli, helminth eggs etc. Primarily the laboratory provides BORDA-CDD researchers with testing services to support the monitoring of decentralised wastewater treatment systems (DEWATS) and faecal sludge treatment plants. It also supports research and development activities in this field.

Find out more about CDD-CASS's work at www.cddindia.org/CASS.

Contact

Rohini Pradeep
rohini.p@cddindia.org
+91 8064532829
+91 8028482144

CDD Society|Bangalore Survey No. 205
(Opp. Beedi Workers Colony)
Kommaghatta Road, Bandemath
Kengeri Satellite Town Bangalore 560
060, Karnataka, India

AIT
Asian Institute of Technology

The Asian Institute of Technology (AIT) laboratory was established in 1964 in support of its postgraduate program in Sanitary Engineering. Supported by a number of donors, the laboratory has been upgraded and improved to comply with ISO17025 standards. As well as being an advanced analytical laboratory, the AIT laboratory offers facilities for lab- and pilot-scale experimentation and an environmental research station for field testing. The lab provides a platform for capacity building of postgraduate researchers, practitioners, and professionals in environmental-related fields including faecal sludge management and it is particularly equipped to support laboratory-, pilot- and field-scale research with an emphasis on faecal sludge and sanitation systems.

Contact
Dr. Atitaya Panuvatvanich
atitaya_p@ait.asia
Asian Institute of Technology
P.O. Box 4
58 Moo 9, Km. 42, Paholyothin Highway Klong Luang, Pathum Thani 12120 Thailand

Columbia University
Kartik Chandran Laboratories

Columbia University
IN THE CITY OF NEW YORK

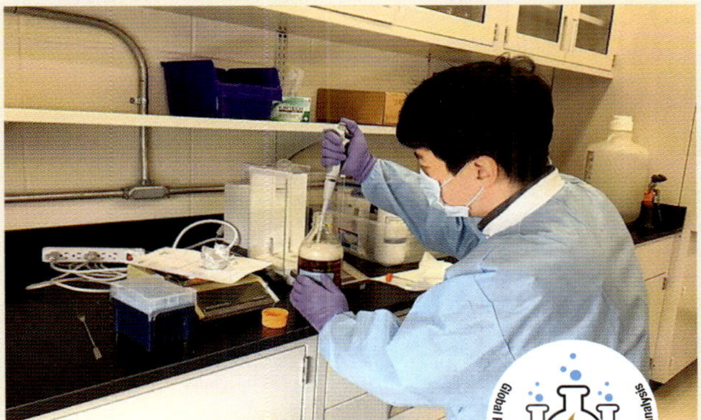

The Kartik Chandran Laboratories at Columbia University provide state-of-the-art facilities for the chemical and biological interrogation and characterization of fecal sludge and other human waste streams. These techniques and protocols are standardized and often applied within the framework of innovative FSM technologies including non-sewered sanitation systems. We are also a leading laboratory engaged in the global surveillance of fecal sludge prokaryotic, eukaryotic and viral microbiomes. This is accomplished using advanced multi-oimcs and bioinformatics techniques. Our laboratory welcomes FSM researchers and professionals from around the world towards mutually beneficially capacity-building efforts.

Contact
Prof. Kartik Chandran
kc2288@columbia.edu
+1 212 854 9027
Columbia University
500 West 120th Street, Mudd 918
New York, NY, 10027, USA
www.columbia.edu/~kc2288

2iE

International Institute for Water and Environmental Engineering

The laboratory for faecal sludge analysis at 2iE was opened on 2005 to promote the development of scientific knowledge on faecal sludge management. The laboratory has been gradually equipped using funds from the Switzerland Agency for Development and Cooperation, the Japan International Cooperation Agency through the project entitled Improving Sustainable Water and Sanitation Systems in Sahel Region in Africa: Case of Burkina Faso, and the Bill & Melinda Gates Foundation through project Stimulating Local Innovation on sanitation for the Urban Poor in Sub-Saharan Africa and South-East Asia. The laboratory is used for both academic and research activities for masters and PhD students, capacity building of sanitation professionals and building the expertise on faecal sludge characterization.

Contact
Pr. Yacouba Konate
Yacouba.konate@2ie-edu.org
+226 78 88 08 61
Institut 2iE, 1 rue de la Science 01 BP 594
Ouagadougou 01, Burkina Faso
www.2ie-edu.org

BITS

Birla Institute of Technology & Science

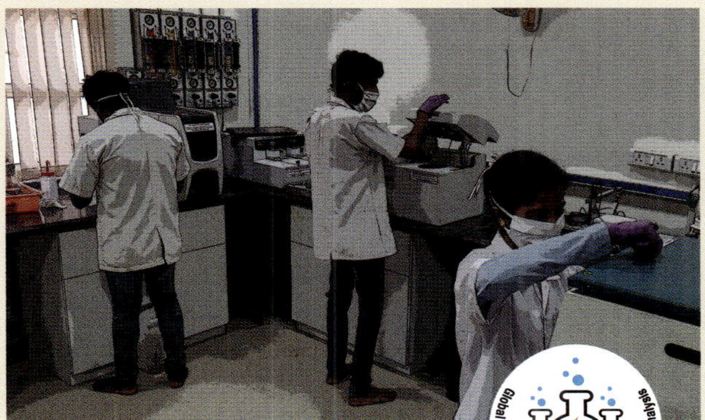

Faecal sludge management laboratory was set up with the generous support of BITS Pilani alumni - class of 1966-1971. This laboratory has variety of equipment and can analyse physical, chemical and biological characteristics of faecal sludge. The laboratory adds value to the Applied Environmental Biotechnology Laboratory of Department of Biological Sciences at K K Birla Goa campus. We had recently started a Master's program in Sanitation Science Technology and Management with support from Bill & Melinda Gates Foundation where our faecal sludge laboratory will play an important role by providing hands-on training to the students.

Contact
Prof. Dr. Srikanth Mutnuri
Birla Institute of Technology & Science, Pilani
K K Birla Goa Campus , Near NH 17B, Bypass Road
Zuarinagar 403726, Goa, India
+91 0832 2580125
srikanth.mutnuri@gmail.com
srikanth@goa.bits-pilani.ac.in

ENPHO
Environment and Public Health Organization

Global Partnership of Laboratories for Faecal Sludge Analysis
MEMBER

Contact
Pramina Nakarmi
pramina.nakarmi@enpho.org
+977-1-5244641
ENPHO, 110/25 Adarsha
Marg-1,Thapagaon, New Baneswor,
Kathmandu, Nepal

Environment and Public Health Organization (ENPHO) is an NGO established in 1990 with a vision of creating eco-societies by providing quality services on water, sanitation and hygiene (WASH), environment and public health. Since its establishment, ENPHO laboratory, as a division under ENPHO, has been providing analytical services in water, wastewater, air, food, soil quality, and recently in faecal sludge analysis. ENPHO laboratory is equipped with trained professionals and advanced equipment. ENPHO laboratory is accredited by Nepal Bureau of Standards and Metrology (NBSM), Government of Nepal. It has been providing a platform for research and monitoring activities on faecal sludge through its characterization in terms of nutrients, solids, organic matter, heavy metals and microbiological parameters including helminths eggs. For more details about ENPHO please visit http://enpho.org/research-and-development/.

ITB

Institut Teknologi Bandung

As the oldest water laboratory in Indonesia (founded in 1935), the Water Quality Laboratory (WQL) of the Faculty of Civil and Environmental Engineering at Institut Teknologi Bandung (ITB) provides services in water and water quality analysis and wastewater treatment. WQL has been a water and wastewater quality assessment laboratory since 2003 which was accredited by the National Accreditation Committee (KAN) and ISO/IEC 17025:2017. Since 2007, WQL has two divisions which are Water Laboratory and Environmental Microbiology Laboratory. This laboratory is conducteding various analysis regarding faecal sludge and accepting faecal sludge samples for both quality testing and research purposes.

Contact
Prof. Dr-Ing. Ir. Prayatni Soewondo, MS
+62-22-2534176
prayatni@ftsl.itb.ac.id
Water Quality Laboratory
Faculty of Civil and Environmental Engineering – ITB
Jl. Ganesa No. 10, Bandung, Indonesia 40132

If you want to learn more about the Global Partnership of Laboratories of Faecal Sludge Analysis, you know a laboratory which would like to join the Partnership, or you need assistance in setting up a faecal sludge lab, feel free to contact us.

Contact
Dr. Konstantina Velkushanova
Coordinator | Global Partnership of Laboratories of Faecal Sludge Analysis
k.velkushanova@un-ihe.org
+31 15 2151 715

https://sanitationeducation.org/laboratories

Annex 2

This appendix includes a link[1] to a virtual document titled 'Addendum of data related to drying of faecal sludge from on-site sanitation facilities and fresh faeces'. This document was compiled based on a project funded by the Bill & Melinda Gates Foundation (BMGF) - 'Characterisation of faecal material during drying' - after faecal sludge drying was recognised as a gap in the implementation of innovative sanitation technologies.

The addendum summarises the results from experiments that are directly or indirectly related to the drying process. The data was obtained from experimental work conducted from 2013 to date, involving several research institutions. As the addendum is an initiative led by the Pollution Research Group at the University of KwaZulu-Natal, most of the data comes from this organisation. Partner institutions joined in this initiative and shared their data, including: (*i*) Swansea University through the SPECIFIC research group; (*ii*) Cranfield University through their energy laboratory; (*iii*) Duke University through their WASH-AID centre; (*iv*) Laval University through their Department of Civil and Water Engineering; (*v*) Victoria University through their Public & Environmental Engineering laboratory; and (*vi*) the Swiss Federal Institute of Aquatic Science and Technology (Eawag) through their Sanitation, Water and Solid Waste for Development (Sandec) department.

The addendum includes a landscape and the fundamentals of faecal sludge drying, and a summary and discussion of the results presented in datasheets. These datasheets are categorised in eight sections according to the different drying processes or faecal sludge properties: thermodynamics, kinetics, physiochemical properties, morphology, mechanical properties, dewaterability, disinfection, and gas emission. Different types of faecal samples were used for the generation of the data: fresh faeces and faecal sludge from ventilated improved pit (VIP) latrines, urine diversion dry toilets (UDDT), and an anaerobic baffled reactor (ABR) from a decentralised wastewater treatment plant (DEWAT).

In each datasheet, the data is displayed as graphs and includes an interpretation. In addition to this, the datasheets contain basic information such as feedstock, laboratory equipment, experimental conditions and performed analysis to explain how the data was obtained. The datasheets also offer bibliographic references to refer the reader to the relevant literature publication, and hyperlinks to the raw data files.

[1] https://gatesopenresearch.org/documents/4-188